D1667265

Handbook of
Therapeutic Antibodies

Edited by
Stefan Dübel

1807–2007 Knowledge for Generations

Each generation has its unique needs and aspirations. When Charles Wiley first opened his small printing shop in lower Manhattan in 1807, it was a generation of boundless potential searching for an identity. And we were there, helping to define a new American literary tradition. Over half a century later, in the midst of the Second Industrial Revolution, it was a generation focused on building the future. Once again, we were there, supplying the critical scientific, technical, and engineering knowledge that helped frame the world. Throughout the 20th Century, and into the new millennium, nations began to reach out beyond their own borders and a new international community was born. Wiley was there, expanding its operations around the world to enable a global exchange of ideas, opinions, and know-how.

For 200 years, Wiley has been an integral part of each generation's journey, enabling the flow of information and understanding necessary to meet their needs and fulfill their aspirations. Today, bold new technologies are changing the way we live and learn. Wiley will be there, providing you the must-have knowledge you need to imagine new worlds, new possibilities, and new opportunities.

Generations come and go, but you can always count on Wiley to provide you the knowledge you need, when and where you need it!

William J. Pesce
President and Chief Executive Officer

Peter Booth Wiley
Chairman of the Board

Handbook of Therapeutic Antibodies

Volume III

Edited by
Stefan Dübel

WILEY-VCH Verlag GmbH & Co. KGaA

The Editor

Prof. Dr. Stefan Dübel
Technical University of Braunschweig
Institute of Biochemistry and Biotechnology
Spielmannstr. 7
38106 Braunschweig
Germany

All books published by Wiley-VCH are carefully produced. Nevertheless, authors, editors, and publisher do not warrant the information contained in these books, including this book, to be free of errors. Readers are advised to keep in mind that statements, data, drug dosages, illustrations, procedural details or other items may inadvertently be inaccurate.

Library of Congress Card No.:
applied for

British Library Cataloguing-in-Publication Data
A catalogue record for this book is available from the British Library.

Bibliographic information published by the Deutsche Nationalbibliothek
Die Deutsche Nationalbibliothek lists this publication in the Deutsche Nationalbibliografie; detailed bibliographic data are available in the Internet at <http://dnb.d-nb.de>.

© 2007 WILEY-VCH Verlag GmbH & Co. KGaA, Weinheim

All rights reserved (including those of translation into other languages). No part of this book may be reproduced in any form – by photoprinting, microfilm, or any other means – nor transmitted or translated into a machine language without written permission from the publishers. Registered names, trademarks, etc. used in this book, even when not specifically marked as such, are not to be considered unprotected by law.

Cover Design: Schulz Grafik-Design, Fussgönheim

Wiley Bicentennial Logo: Richard J. Pacifico

Composition SNP Best-set Typesetter Ltd., Hong Kong

Printing betz-druck GmbH, Darmstadt

Bookbinding Litges & Dopf GmbH, Heppenheim

Printed in the Federal Republic of Germany
Printed on acid-free paper

ISBN 978-3-527-31453-9

Dedication

To Inge, Hans, Ulrike and Tasso Dübel – the four best things in my life

Stefan Dübel

Contents

Volume III

Handbook of Therapeutic Antibodies. Edited by Stefan Dübel
Copyright © 2007 WILEY-VCH Verlag GmbH & Co. KGaA, Weinheim
ISBN 978-3-527-31453-9

Volume I

Volume II

Overview of Therapeutic Antibodies

Trade name	FDA name	Chapter	Page
Avastin	bevacizumab	3	781
Bexxar	tositumomab	15	1145
CEA-Scan	arcitumomab	15	1134
Cotara	–	15	1137
Erbitux	cetuximab	4	815
Herceptin	trastuzumab	14	1111
HumaSpect-Tc	votumumab	15	1147
Humira	adalimumab	1	699
Indimacis-125	igovomab	15	1142
LeukoScan	sulesomab	15	1144
Leukosite	alemtuzumab	2	735
MabCampath	alemtuzumab	2	735
Mylotarg	gemtuzumab	7	871
Neutrospec	fanolesomab	6	853
OncoScint	satumomab	15	1143
Oncorad	satumomab	15	1143
Orthoclone	muromonab	9	907
Panorex	edrecolomab	15	1139
ProstaScint	capromab	15	1135
Raptiva	efalizumab	5	829
Remicade	infliximab	8	887
ReoPro	abciximab	15	1133
Rituxan	rituximab	13	1035
Simulect	basiliximab	15	1137
Synagis	palivizumab	12	1001
Tysabri	natalizumab	10	943
Verluma	nofetumomab	15	1142
Xolair	omalizumab	11	953
Zenapax	daclizumab	15	1138
Zevalin	imbritumomab	15	1141

Handbook of Therapeutic Antibodies. Edited by Stefan Dübel
Copyright © 2007 WILEY-VCH Verlag GmbH & Co. KGaA, Weinheim
ISBN 978-3-527-31453-9

A Greeting by the Editor

Today, therapeutic antibodies are essential assets for physicians fighting cancer, inflammation, and infections. These new therapeutic tools are a result of an immense explosion of research sparked by novel methods in gene technology which became available between 1985 and 1995.

This handbook endeavors to present the fascinating story of the tremendous achievements that have been made in strengthening humanity's weapons arsenal against widespread diseases. This story not only includes the scientific and clinical basics, but covers the entire chain of therapeutic antibody production – from downstream processing to Food and Drug Administration approval, galenics – and even critical intellectual property issues.

A significant part is devoted to emerging developments of all aspects of this process, including an article showing that antibodies may only be the first generation of clinically used targeting molecules, making the IgG obsolete in future developments, and novel ideas for alternative therapeutic paradigms.

Finally, approved antibody therapeutics are presented in detail in separate chapters, allowing the clinicians to quickly gain a comprehensive understanding of individual therapeutics.

In such a fast-developing area, it is difficult to keep pace with the rapidly growing information. For example, a PubMed search with "Herceptin" yields more than 1500 citations. Consequently, we have tried to extract the essentials from this vast resource, offering a comprehensive basis of knowledge on all relevant aspects of antibody therapeutics for the researcher, the company expert, and the bedside clinician.

At this point, I express my deep gratitude to all the colleagues who wrote for these books. Without their enthusiasm this project would have never materialized. I would also like to thank Dr Pauly from the publisher's office, who paved the way for this three-volume endeavor, and the biologist Ulrike Dübel – my wife. Both played essential roles in keeping the project on track throughout the organizational labyrinth of its production. The hard work and continuous suggestions of all of these colleagues were crucial in allowing the idea of a comprehensive handbook on therapeutic antibodies to finally become a reality.

Braunschweig, December 2006 Stefan Dübel

Handbook of Therapeutic Antibodies. Edited by Stefan Dübel
Copyright © 2007 WILEY-VCH Verlag GmbH & Co. KGaA, Weinheim
ISBN 978-3-527-31453-9

Foreword

The most characterized class of proteins are the antibodies. After more than a century of intense analysis, antibodies continue to amaze and inspire. This *Handbook of Therapeutic Antibodies* is not just an assembly of articles but rather a state-of-the-art comprehensive compendium, which will appeal to all those interested in antibodies, whether from academia, industry, or the clinic. It is an unrivaled resource which shows how mature the antibody field has become and how precisely the antibody molecule can be manipulated and utilized.

From humble beginnings when the classic monoclonal antibody paper by Kohler and Milstein ended with the line, “such cultures could be valuable for medical and industrial use” to the current Handbook you hold in your hand, the field is still in its relative infancy. As information obtained from clinical studies becomes better understood then further applications will become more streamlined and predictable. This Handbook will go a long way to achieving that goal. With the application of reproducible recombinant DNA methods the antibody molecule has become as plastic and varied as provided by nature. This then takes the focus away from the antibody, which can be easily manipulated, to what the antibody recognizes. Since any type, style, shape, affinity, and form of antibody can be generated, then what the antibody recognizes now becomes important.

All antibodies have one focus, namely, its antigen or more precisely, its epitope. In the realm of antibody applications antigen means “target.” The generation of any sort of antibody and/or fragment is now a relatively simple procedure so the focus of this work has shifted to the target, and rightly so. Once a target has been identifed then any type of antibody can be generated to that molecule. Many of the currently US Food and Drug Administration (FDA) approved antibodies were obtained in this manner. If the target is unknown then the focus is on the specificity of the antibody and ultimately the antigen it recognizes.

As the field continues to mature the applications of antibodies will essentially mimic as much of the natural human immune response as possible. In this respect immunotherapy may become immunomanipulation, where the immune system is being manipulated by antibodies. With the success of antibody monotherapy the next phase of clinical applications is the use of antibodies with standard chemotherapy, and preliminary studies suggest the combination of

Handbook of Therapeutic Antibodies. Edited by Stefan Dübel
Copyright © 2007 WILEY-VCH Verlag GmbH & Co. KGaA, Weinheim
ISBN 978-3-527-31453-9

these two modalities is showing a benefit to the patient. When enough antibodies become available then cocktails of antibodies will be formulated for medical use. Since the natural antibody response is an oligoclonal response then cocktails of antibodies can be created by use of various *in vitro* methods to duplicate this in a therapeutic setting. In essence, this will be oligotherapy with a few antibodies. After all, this is what nature does and duplicating this natural immune response may be effective immunotherapy.

And all of this brings us back full circle to where it all starts and ends, the antibody molecule. No matter what version, isotype, form, or combination used the antibody molecule must first be made and shown to be biologically active. Currently, many of the steps and procedures to generate antibodies can be obtained in kit form and therefore are highly reproducible, making the creation of antibodies a straightfoward process. Once the antibody molecule has been generated it must be produced in large scale for clinical and industrial applications. More often than not this means inserting the antibody genes into an expression system compatible with the end use of the antibody (or fragment). Since many of the steps in generating clinically useful antibodies are labor intensive and costly, care must be used to select antibodies with the specificity and activity of interest before they are mass produced. For commercial applications the FDA will be involved so their guidelines must be followed.

Stating the obvious, it would have been nice to have this Handbook series in the late 1970s when I entered the antibody field. It certainly would have made the work a lot easier! And here it is, about 30 years later, and the generation of antibodies has become "handbook easy." In this respect I am envious of those starting out in this field. The recipies are now readily available so the real challenge now is not in making antibodies but rather in the applications of antibodies. It is hoped that this Handbook will provide a bright beacon where others may easily follow and generate antibodies which will improve our health. The immune system works and works well; those using this Handbook will continue to amaze and inspire.

Mark Glassy
Chairman & Professor, The Rajko Medenica Research Foundation, San Diego, CA, USA
Chief Executive Officer, Shantha West, Inc., San Diego, CA, USA
December 2006

List of Authors

James Allen
University College London
Department of Biochemistry and Molecular Biology
Gower Street
Darwin Building
London WC1E 6BT
UK

Christian Antoni
Schering-Plough
Clinical Research Allergy/Respiratory/Immunology
2015 Galloping Hill Rd
Kenilworth, NJ 07033
USA

Rosalin Arends
Pfizer Inc.
MS 8220-3323
Eastern Point Road
Groten, CT 06339
USA

Michaela A.E. Arndt
Department of Medical Oncology and Cancer Research
University of Essen
Hufelandstr. 55
45122 Essen
Germany

Kenneth D. Bagshawe
Department of Oncology
Charing Cross Campus
Imperial College London
Fulham Palace Road
London W6 8RF
UK

Harald Becker
Wetzbach 26 D
64673 Zwingenberg
Germany

Klaus Bergemann
Boehringer Ingelheim Pharma GmbH & Co. KG
BioPharmaceuticals
Birkendorfer Str. 65
88397 Biberach a.d. Riss
Germany

Thomas Beyer
University Schleswig-Holstein
Campus Kiel
Division of Nephrology
Schittenhelmstr. 12
24105 Kiel
Germany

Handbook of Therapeutic Antibodies. Edited by Stefan Dübel
Copyright © 2007 WILEY-VCH Verlag GmbH & Co. KGaA, Weinheim
ISBN 978-3-527-31453-9

Michael Braunagel
Affitech AS
Gaustadalléen 21
0349 Oslo
Norway

Marianne Brüggemann
The Babraham Institute
Protein Technologies Laboratory
Babraham
Cambridge CB22 3AT
UK

Kerry A. Chester
CR UK Targeting & Imaging Group
Department of Oncology
Hampstead Campus
UCL, Rowland Hill Street
London NW3 2PF
UK

Daan J.A. Crommelin
Utrecht University
Utrecht Institute for Pharmaceutical Sciences (UIPS)
Sorbonnelaan 16
3584 CA Utrecht
The Netherlands

Rathin C. Das
Affitech USA, Inc.
1945 Arsol Grande
Walnut Creek, CA 94595
USA

Jason L.J. Dearling
Royal Free and University College Medical School
University College London
Cancer Research UK Targeting & Imaging Group
Department of Oncology
Rowland Hill Street
Hampstead Campus
London NW3 2PF
UK

Michael Dechant
University Schleswig-Holstein
Campus Kiel
Division of Nephrology
Schittenhelmstr. 12
24105 Kiel
Germany

Peter Markus Deckert
Charité Universitätsmedizin Berlin
Medical Clinic III – Haematology, Oncology and Transfusion Medicine
Campus Benjamin Franklin
Hindenburgdamm 30
12200 Berlin
Germany

Eduardo Díaz-Rubio
Hospital Clínico San Carlos
Medical Oncology Department
28040 Madrid
Spain

Dimiter S. Dimitrov
Protein Interactions Group
Center for Cancer Research Nanobiology Program
CCR, NCI-Frederick, NIH
Frederick, MD 21702
USA

Stefan Dübel
Technical University of Braunschweig
Institute of Biochemistry and Biotechnology
Spielmannstr. 7
38106 Braunschweig
Germany

Christian Eckermann
Boehringer Ingelheim Pharma GmbH & Co. KG
BioPharmaceuticals
Birkendorfer Str. 65
88397 Biberach a.d. Riss
Germany

Paul Ellis
Department of Medical Oncology
Guy's Hospital
Thomas Guy House
St. Thomas Street
London SE1 9RT
UK

Thomas Elter
Department of Hematology and Oncology
University of Cologne
Kerpener Str. 62
50937 Köln
Germany

Andreas Engert
Department of Hematology and Oncology
University of Cologne
Kerpener Str. 62
50937 Köln
Germany

Yang Feng
Protein Interactions Group
Center for Cancer Research Nanobiology Program
CCR, NCI-Frederick, NIH
Frederick, MD 21702
USA

Markus Fiedler
Scil Proteins GmbH
Affilin Discovery
Heinrich-Damerow-Str. 1
06120 Halle an der Saale
Germany

Peter Fischer
Boehringer Ingelheim Pharma GmbH & Co. KG
Department of R&D Licensing & Information Management
Birkendorfer Str. 65, K41-00-01
88397 Biberach a.d. Riss
Germany

Martin Foerster
Friedrich-Schiller-University
Department of Pneumology and Allergy
Medical Clinics I
Erlanger Allee 101
07740 Jena
Germany

Patrick Garidel
Boehringer Ingelheim Pharma GmbH & Co. KG
BioPharmaceuticals
Birkendorfer Str. 65
88397 Biberach a.d. Riss
Germany

Guiseppe Giaccone
Vrije Universiteit Medical Center
Department of Medical Oncology
De Boelelaan 1117
1081 HV Amsterdam
The Netherlands

Ralf Gold
Department of Neurology
St. Josef Hospital
Ruhr University Bochum
Gudrunstr. 56
44791 Bochum
Germany

Martin Gramatzki
University of Schleswig-Holstein
Campus Kiel
Division of Stem Cell Transplantation and Immunotherapy
Schittenhelmstr. 12
24105 Kiel
Germany

Stefanos Grammatikos
Boehringer Ingelheim Pharma GmbH & Co. KG
BioPharmaceuticals
Birkendorfer Str. 65
88397 Biberach a.d. Riss
Germany

Larry Green
Abgenix, Inc.
6701 Kaiser Drive
Fremont, CA 94555
USA

Michael Hallek
Department of Hematology and Oncology
University of Cologne
Kerpener Str. 62
50937 Köln
Germany

Melanie Hartsough
Center for Drug Evaluation and Research
Food and Drug Administration
Division of Biological Oncology Products
10903 New Hampshire Ave.
Silver Spring, MD 20993
USA

Mingyue He
The Babraham Institute
Technology Research Group
Cambridge CB2 4AT
UK

Martin H. Holtmann
Johannes-Gutenberg-University
1st Department of Medicine
Rangenbeckstr. 1
55131 Mainz
Germany

Alexandra Huhalov
Royal Free and University College Medical School
University College London
Cancer Research UK Targeting & Imaging Group
Department of Oncology
Rowland Hill Street
Hampstead Campus
London NW3 2PF
UK

Michael Hust
Technical University of Braunschweig
Institute of Biochemistry and Biotechnology
Spielmannstr. 7
38106 Braunschweig
Germany

Alexander Jacobi
Boehringer Ingelheim Pharma GmbH & Co. KG
BioPharmaceuticals
Birkendorfer Str. 65
88397 Biberach a. d. Riss
Germany

Sigbert Jahn
Serono GmbH
Freisinger Str. 5
85716 Unterschleissheim
Germany

Wim Jiskoot
Gorlaeus Laboratories
Leiden/Amsterdam Center for Drug Research (LACDR)
Division of Drug Delivery Technology
P.O. Box 9502
2300 RA Leiden
The Netherlands

Thomas Jostock
Novartis Pharma AG
Biotechnology Development
Cell and Process R & D
CH-4002 Basel
Switzerland

Joachim R. Kalden
University of Erlangen-Nürnberg
Medical Clinic III
Rheumatology, Immunology & Oncology
Krankenhausstrasse 12
91052 Erlangen
Germany

Hitto Kaufmann
Boehringer Ingelheim Pharma GmbH & Co. KG
BioPharmaceuticals
Birkendorfer Str. 65
88397 Biberach a.d. Riss
Germany

Ralph Kempken
Boehringer Ingelheim Pharma GmbH & Co. KG
BioPharmaceuticals
Birkendorfer Str. 65
88397 Biberach a.d. Riss
Germany

Scott Klakamp
AstraZeneca Pharmaceuticals LP
24500 Clawiter Road
Hayward, CA 94545
USA

Roland E. Kontermann
University Stuttgart
Institute for Cell Biology and Immunology
Allmandring 31
70569 Stuttgart
Germany

Zoltán Konthur
Max Planck Institute for Molecular Genetics
Department of Vertebrate Genomics
Ihnestr. 63–73
14195 Berlin
Germany

Jürgen Krauss
Department of Medical Oncology and Cancer Research
University of Essen
Hufelandstr. 55
45122 Essen
Germany

Claus Kroegel
Friedrich-Schiller-University
Department of Pneumology and Allergy
Medical Clinics I
Erlanger Allee 101
07740 Jena
Germany

Hartmut Kupper
Abbott GmbH & Co. KG
Knollstr. 50
67061 Ludwigshafen
Germany

Meina Liang
AstraZeneca Pharmaceuticals LP
24500 Clawiter Road
Hayward, CA 94545
USA

Charito Love
Long Island Jewish Medical Center
Division of Nuclear Medicine
New Hyde Park, New York
USA

Andrew C.R. Martin
University College London
Department of Biochemistry and Molecular Biology
Darwin Building
Gower Street
London WC1E 6BT
UK

Christian Menzel
Technical University of Braunschweig
Institute of Biochemistry and Biotechnology
Spielmannstr. 7
38106 Braunschweig
Germany

Gerhard Moldenhauer
German Cancer Research Center
Department of Molecular Immunology
Tumor Immunology Program
Im Neuenheimer Feld 280
69120 Heidelberg
Germany

Dafne Müller
University Stuttgart
Institute for Cell Biology and Immunology
Allmandring 31
70569 Stuttgart
Germany

Markus F. Neurath
Johannes-Gutenberg-University
1st Department of Medicine
Langenbeckstr. 1
55131 Mainz
Germany

Dianne L. Newton
SAIC Frederick, Inc.
Developmental Therapeutics Program
National Cancer Institute at Frederick
Frederick, MD 21702
USA

Michael J. Osborn
The Babraham Institute
Protein Technologies Laboratory
Babraham
Cambridge CB22 3AT
UK

Christopher J. Palestro
Albert Einstein College of Medicine
Bronx, New York
USA
and:
Long Island Jewish Medical Center
New Hyde Park, New York
USA

Gabriele Pecher
Humboldt University Berlin
Medical Clinic for Oncology and Hematology
Charité Campus Mitte
Charitéplatz 1
10117 Berlin
Germany

Matthias Peipp
University Schleswig-Holstein
Campus Kiel
Division of Stem Cell Transplantation and Immunotherapy
Schittenhelmstr. 12
24105 Kiel
Germany

Edith A. Perez
Mayo Clinic Jacksonville
Division of Hematology/Oncology
4500 San Pablo Road
Jacksonville, FL 32224
USA

Sandra Pisch-Heberle
Boehringer Ingelheim Pharma GmbH & Co. KG
BioPharmaceuticals
Birkendorfer Str. 65
88397 Biberach a.d. Riss
Germany

Ponraj Prabakaran
Protein Interactions Group
Center for Cancer Research Nanobiology Program
CCR, NCI-Frederick
Frederick, MD 21702
USA

Gabriele Reich
Ruprecht-Karls-University
Institute of Pharmacy and Molecular Biotechnology (IPMB)
Department of Pharmaceutical Technology and Pharmacology
Im Neuenheimer Feld 366
69120 Heidelberg
Germany

Josephine N. Rini
Albert Einstein College of Medicine
Bronx, New York
USA
And:
Long Island Jewish Medical Center
Division of Nuclear Medicine
New Hyde Park, New York
USA

Lorin Roskos
AstraZeneca Pharmaceuticals LP
24500 Clawiter Road
Hayward, CA 94545
USA

Susanna M. Rybak
Bionamomics LLC
411 Walnut Street, #3036
Green Cove Springs, FL 32043
USA

José W. Saldanha
National Institute for Medical Research
Division of Mathematical Biology
The Ridgeway
Mill Hill
London NW7 1AA
UK

Jochen Salfeld
Abbott Bioresearch Center
100 Research Drive
Worcester, MA 01605
USA

Huub Schellekens
Utrecht University
Department of Pharmaceutics Sciences
Department of Innovation Studies
Sorbonnelaan 16
3584 CA Utrecht
The Netherlands

Sebastian Schimrigk
Ruhr University Bochum
St. Josef Hospital
Department of Neurology
Gudrunstr. 56
44791 Bochum
Germany

Thomas Schirrmann
Technical University Braunschweig
Institute of Biochemistry and Biotechnology
Department of Biotechnology
Spielmannstr. 7
38106 Braunschweig
Germany

Norbert Schleucher
Hematolgy and Medical Oncology
Marienkrankenhaus Hamburg
Alfredstr. 9
22087 Hamburg
Germany

Alexander C. Schmidt
National Institutes of Health
National Institute of Allergy and Infectious Diseases
Laboratory of Infectious Diseases
50 South Drive, Room 6130
Bethesda, MD 20892
USA
And:
Charité Medical Center at Free University and Humboldt University Berlin
Center for Perinatal Medicine and Pediatrics
Schumannstr 20/21
13353 Berlin
Germany

Karlheinz Schmitt-Rau
Serono GmbH
Freisinger Str. 5
85716 Unterschleissheim
Germany

Marjorie A. Shapiro
Center for Drugs Evaluation and Research
Food and Drug Administration
Division of Monoclonal Antibodies
HFD-123
5600 Fishers Lane
Rockville, MD 20872
USA

Surinder K Sharma
CR UK Targeting & Imaging Group
Department of Oncology
Hampstead Campus
UCL, Rowland Hill Street
London NW3 2PF
UK

Igor A. Sidorov
Center for Cancer Research Nanobiology Program
CCR, NCI-Frederick, NIH
Protein Interactions Group
P.O. Box B, Miller Drive
Frederick, MD 21702-1201
USA

Arne Skerra
Technical University Munich
Chair for Biological Chemistry
An der Saatzucht 5
85350 Freising-Weihenstephan
Germany

Jennifer A. Smith
The Babraham Institute
Protein Technologies Laboratory
Babraham
Cambridge CB22 3AT
UK

Patrick G. Swann
Center for Drug Evaluation and Research
Food and Drug Administration
Division of Monoclonal Antibodies
HFD-123
5600 Fishers Lane
Rockville, MD 20872
USA

Michael J. Taussig
The Babraham Institute
Technology Research Group
Cambridge CB2 4AT
UK

Lars Toleikis
RZPD, Deutsches Ressourcenzentrum für Genomforschung GmbH
Im Neuenheimer Feld 515
69120 Heidelberg
Germany

Daniel Tracey
Abbott Bioresearch Center
100 Research Drive
Worcester, MA 01605
USA

Martina M. Uttenreuther-Fischer
Boehringer Ingelheim Pharma GmbH & Co. KG
Department of Medicine
Clinical Research Oncology
Birkendorfer Str. 65
88397 Biberach a.d. Riss

Thomas Valerius
University Schleswig-Holstein
Campus Kiel
Division of Nephrology
Schittenhelmstr. 12
24105 Kiel
Germany

Udo Vanhoefer
Department of Medicine
Hematology and Medical Oncology, Gastroenterology and Infectious Diseases
Marienkrankenhaus Hamburg
Alfredstr. 9
22087 Hamburg
Germany

Michael Wenger
International Medical Leader
F. Hoffmann-La Roche Ltd.
Bldg. 74/4W
CH-4070 Basel
Switzerland

Maria Wiekowski
Schering-Plough
Clinical Research Allergy/Respiratory/Immunology
2015 Galloping Hill Road
Kenilworth, NJ 07033
USA

Xiangang Zou
The Babraham Institute
Protein Technologies Laboratory
Cambridge CB22 3AT
UK

Volume III
Approved Therapeutics

1 Adalimumab (Humira)

Hartmut Kupper, Jochen Salfeld, Daniel Tracey, and Joachim R. Kalden

1.1 Introduction

Adalimumab (Humira; Abbott Laboratories, Abbott Park, IL, USA) is the first fully human recombinant immunoglobulin G1 (IgG1) monoclonal antibody designed to inhibit tumor necrosis factor alpha (TNF-α or TNF). Previous anti-TNF monoclonal antibodies were composed of both murine and human components, which increased the potential for immune responses and limited the long-term use. In contrast, because adalimumab is fully human, it is able to elicit a long-term response with a low degree of immunogenicity, with or without concomitant administration of immunosuppressants such as methotrexate (MTX) (Rau 2002; van de Putte et al. 2003).

Adalimumab is composed of human-derived heavy and light chain variable regions and human IgG1:κ constant regions, engineered through phage display technology (US Humira PI 2006). Phage display technology is designed to recapitulate the physiologic antibody generation process in the laboratory. Comparable to the natural selection process for the B-cell displaying the appropriate antibody, phage display allows for the selection of a fully human antibody specific for an antigen, in this case TNF, from a large repertoire of antibodies. If the desired antibody is felt to be rare in the repertoire, a variant of the technology allows for more rapid "guided selection" in a two-stage process. Therefore, the first step in the generation of adalimumab was a guided selection approach using the murine anti-human TNF antibody MAK195 to isolate a human antibody that recognized the same neutralizing epitope as MAK195. MAK195 is a potent neutralizing monoclonal antibody, which has a high affinity and a low off-rate constant for human TNF. The MAK195 VH and VL (variable portion of heavy and light chains) were paired with human cognate repertoires, and these phage antibody libraries underwent antigen binding selection using recombinant human TNF as the antigen. The selected human VH and VL genes were then combined to generate a fully human anti-TNF antibody.

Handbook of Therapeutic Antibodies. Edited by Stefan Dübel
Copyright © 2007 WILEY-VCH Verlag GmbH & Co. KGaA, Weinheim
ISBN 978-3-527-31453-9

Early human anti-TNF antibodies were optimized in a process mirroring the natural process for antibody optimization. The final antibody, adalimumab, is a full-length IgG1:κ molecule with optimized heavy and light chains characterized by high specificity, affinity, and potency (van de Putte et al. 2003). Adalimumab contains no nonhuman components or artificially fused human peptide sequences. The resulting molecule is, therefore, indistinguishable in structure and function from a naturally occurring human IgG1, and has a comparable terminal half-life of approximately 2 weeks (Salfeld et al. 1998).

Adalimumab is produced in a Chinese hamster ovary (CHO) host cell that is transfected with a plasmid vector containing the expression cassettes for adalimumab heavy and light chains. Adalimumab is produced by a standard, well-controlled fermentation and purification process (US Humira PI 2006). Each batch of adalimumab is characterized rigorously in a series of biochemical and biophysical assays in order to meet prespecified release criteria.

1.2 Pharmacology

In the body, TNF is protective against infection and injury through multiple biologic mechanisms that also can play a role in inflammation. This cytokine interacts with endothelial cells, synovial fibrobasts, keratinocytes, dendritic cells, and other components of the immune system as part of a complex cascade of response-to-injury events (Mease and Goffe 2005). In this setting, TNF promotes the accumulation of inflammatory cells, activates endothelial adhesion molecules, and promotes the synthesis of other proinflammatory cytokines (e.g., interleukin [IL]-1, IL-6, granulocyte-macrophage colony-stimulating factor) and chemokines (Fig. 1.1) (Lee and Kavanaugh 2005; Maini and Taylor 2000; Mease and Goffe 2005). Despite a critical, protective role in the immune response, chronically elevated levels of TNF have been implicated as a pathogenic component of a number of disease states, including rheumatoid arthritis (RA), psoriasis, psoriatic arthritis (PsA), ankylosing spondylitis (AS), and Crohn's disease.

Three lines of evidence support a role for TNF in these conditions. First, high concentrations of TNF are found in the synovial fluid of patients with RA, psoriatic lesions of patients with psoriasis (Partsch et al. 1997; Ritchlin et al. 1998), sacroiliac joint biopsy specimens from patients with AS (Braun et al. 1995), and in stool, mucosa, and blood of patients with Crohn's disease (Braegger et al. 1992; Murch et al. 1991, 1993). Second, in animal models of RA, TNF has been shown to accelerate disease activity; and third, anti-TNF antibodies have been shown to decrease disease activity in animal models of RA (Cooper et al. 1992; Williams et al. 1992).

Adalimumab binds to soluble TNF with high specificity and high affinity ($K_d = 6 \times 10^{-10}$ M). Adalimumab binding to TNF prevents interaction with TNF-RI/II cell-surface receptors (Lee and Kavanaugh 2005; Salfeld et al. 1998; US Humira PI 2006). Adalimumab neutralizes TNF proinflammatory activity (Fig.

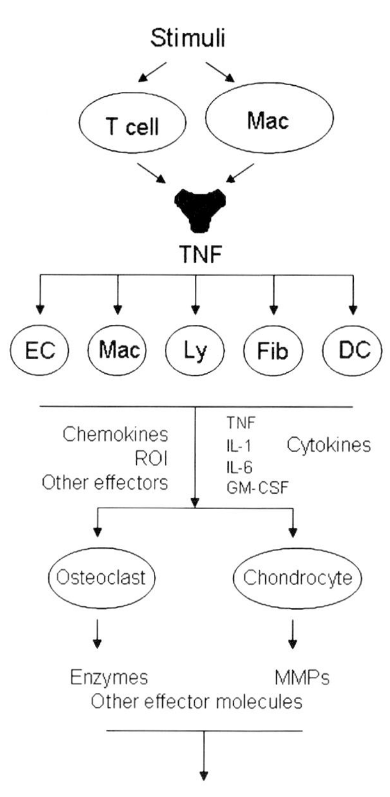

Fig. 1.1 The role of tumor necrosis factor (TNF) in the pathogenesis of rheumatoid arthritis. DC = dendritic cells; EC = endothelial cells; Fib = fibroblasts; GM-CSF = granulocyte-macrophage colony-stimulating factor; IL = interleukin; Ly = lymphocytes; Mac = macrophages; MMPs = matrix metalloproteases; ROI = reactive oxygen intermediates.

1.2) (Lee and Kavanaugh 2005; Salfeld et al. 1998), and also binds to and neutralizes the cell membrane-associated form of TNF, which may play a role in disease (Georgopoulos et al. 1996). Adalimumab does not bind to or inactivate lymphotoxin (LTα, formerly called TNF-β). The effects of adalimumab treatment on parameters of inflammatory disease are consistent with neutralization of TNF as the primary mechanism of action. The effects of adalimumab in patients with RA on various disease-related parameters were initially evaluated in a phase I, single-dose, placebo-controlled trial (den Broeder et al. 2002a). Patients were randomized to receive a single dose of 0.5, 1, 3, 5, or 10 $mg\,kg^{-1}$ adalimumab or placebo. At 1 week after injection of adalimumab, the acute-phase reactant C-reactive protein (CRP) and erythrocyte sedimentation rate (ESR) showed an impressive decrease from baseline at all doses (den Broeder et al. 2002a). Levels of IL-6, IL-1 receptor antagonist, and IL-1β messenger RNA also decreased rapidly (Barrera et al. 2001).

In patients receiving adalimumab, levels of matrix metalloproteases (MMP-1 and -3) decreased over 6 months (den Broeder et al. 2002a; Weinblatt et al. 2003). Long-term monotherapy with adalimumab modulated cartilage and synovium turnover, and was associated with improved radiologic outcomes (den Broeder

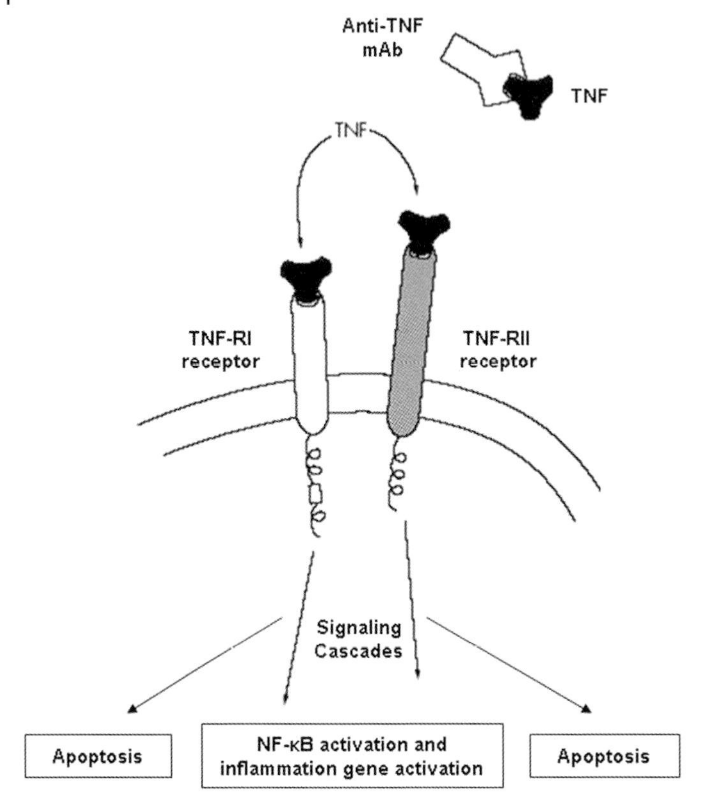

Fig. 1.2 Neutralization of tumor necrosis factor (TNF) by anti-TNF monoclonal antibody (mAb) prevents TNF binding to either the TNF-RI or the TNF-RII receptors. NF-κB = nuclear factor kappa B. (Adapted with permission from Mease 2002.)

et al. 2002b). Treatment with adalimumab also has been shown to alter expression of adhesion molecules responsible for leukocyte migration (endothelial leukocyte adhesion molecule-1, vascular cell adhesion molecule-1, and intercellular adhesion molecule-1) (US Humira PI 2006).

1.3 Pharmacokinetics

Adalimumab follows linear pharmacokinetic properties throughout the clinical dose range in healthy adults. The absolute bioavailability of adalimumab following a single 40-mg subcutaneous (SC) dose is 64%, and maximum plasma concentrations ($4.7 \pm 1.6\,\mu g\,mL^{-1}$) are achieved at 131 ± 56 h. Following the recommended dose of 40 mg SC every other week (EOW), steady-state serum trough adalimumab concentrations are three- to seven-fold higher than the expected effective dose in 50% of patients (Granneman et al. 2003). Concomitant MTX treatment does not significantly alter adalimumab pharmacokinetics. Neither does the presence of MTX not adversely affect adalimumab serum concentrations. At the recommended dose of 40 mg EOW, mean steady-state serum concentrations are 5 $\mu g\,mL^{-1}$ without MTX and 8–9 $\mu g\,mL^{-1}$ with MTX. After single and multiple dosing, MTX reduced adalimumab apparent clearance by 29% and 44%, respectively. Synovial fluid concentrations of adalimumab in patients with RA are 31% to 96% of those found in serum (US Humira PI 2006). Pharmacokinetic studies have not been conducted in children, or in patients with hepatic or renal impairment (US Humira PI 2006).

1.4 Adalimumab Comparisons with Infliximab and Etanercept

There are some structural and functional differences between adalimumab and the other TNF antagonists: infliximab, a chimeric murine/human monoclonal antibody; and etanercept, a TNF receptor fusion protein (Fig. 1.3). These include specificity, TNF binding affinity, immunogenicity, pharmacokinetics, and efficacy across different disease states. Adalimumab and infliximab are highly specific for TNF (in both soluble and membrane-bound forms), whereas etanercept binds and neutralizes LTα as well as TNF. Adalimumab binds to TNF with very high affinity (K_d ~85 pM) (Kaymakcalan et al. 2002), and potently neutralizes TNF in bioassays (IC_{50} ~130 pM) (Salfeld et al. 1998). The kinetic binding parameters of adalimumab and infliximab are similar; however, etanercept dissociates from TNF much more rapidly than adalimumab or infliximab (Kaymakcalan et al. 2002).

Immunogenicity is directly related to the structure of protein therapeutics, and may result in increased adverse effects and diminished efficacy. Antibodies to adalimumab, a fully human monoclonal IgG1 antibody indistinguishable from naturally occurring IgG1, have been observed in a small proportion of patients (US Humira PI 2006). This low level of immunogenicity is reflective of the natural process of anti-idiotypic antibody formation to endogenous antibodies characteristic of the natural immune network (Jerne 1974). Infliximab, a chimeric antibody developed by recombinant fusion of murine and human antibody components, can be immunogenic in a high proportion of patients, unless immunosuppressive drugs are administered concomitantly (Anderson 2005; Baert et al. 2003; Maini et al. 1998). Low levels of immunogenicity have been reported following administration of etanercept, which is composed of human components but in an artificial construct; however, considerable variability has been seen, depending on the assays used to detect anti-etanercept antibodies (Anderson 2005).

Pharmacokinetic differences between the three TNF antagonists are also apparent. Adalimumab has a longer half-life than etanercept (10–20 days versus 3.5–5 days, respectively), allowing for less-frequent dosing. In contrast, infliximab must

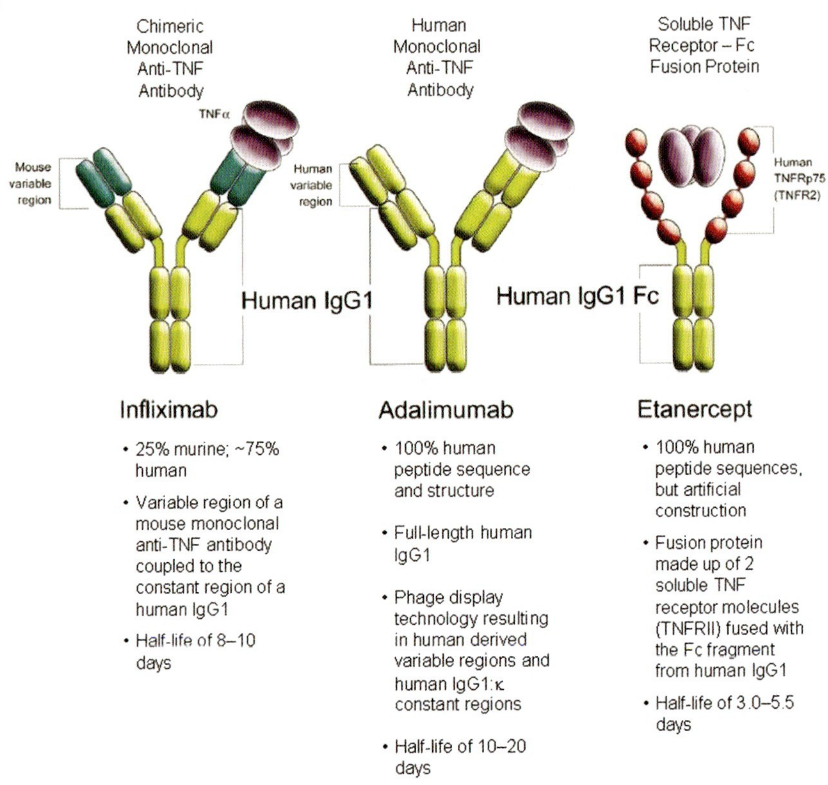

Fig. 1.3 Adalimumab structure in comparison to other tumor necrosis factor (TNF)-antagonists. IgG1 = immunoglobulin G1. (Adapted with permission from Anderson 2005.)

be administered by intravenous infusion, which produces peak concentrations significantly higher and trough concentrations significantly lower than adalimumab and etanercept (Granneman et al. 2003; Maini et al. 1998; Zhou et al. 2004). The clinical implication of these substantial fluctuations in concentrations is unknown.

Finally, there appear to be differences in efficacy among the three TNF antagonists, the efficacy of which has been evaluated in an animal model of RA driven by a human TNF transgene (Kaymakcalan et al. 2002). In this evaluation, adalimumab was more effective than infliximab and etanercept in preventing the development of arthritis, as demonstrated by suppressed histopathologic evidence of synovial inflammation, vascularity, and cartilage and bone erosion. Following administration of etanercept, human TNF was cleared more slowly from serum than after adalimumab or infliximab, suggesting that TNF–etanercept complexes persist longer in the circulation. The results of clinical studies evaluating each agent suggest that the efficacy of adalimumab is comparable to that of infliximab and etanercept in RA, but that infliximab and adalimumab may be more efficacious than etanercept in Crohn's disease and psoriasis, possibly a result of greater tissue penetration and/or effector function by the antibodies. Although it is accepted that all three agents bind to membrane TNF on transfected cells, *in-vitro* data on binding and initiation of Fc-mediated effector functions such as complement activation or antibody-induced cellular cytotoxicity (ADCC) by normal human cells are inconsistent (Scallon et al. 1995; van den Brande et al. 2003). There is also conflicting *in-vitro* data regarding the induction of apoptosis in T cells, monocytes, or other cells by TNF antagonists (Catrina et al. 2005; Mitoma et al. 2005; Shen et al. 2005; van den Brande et al. 2003); the relevance of these *in-vitro* studies to the efficacy or safety of these agents *in vivo* remains to be determined, however.

1.5 Indications

The currently approved indications for adalimumab as of September 2006 are summarized in Table 1.1. Note that there are subtle differences in the European and United States labeled indications.

1.5.1 Rheumatoid Arthritis

The first indication for adalimumab was for the treatment of RA. In the United States, adalimumab is indicated for reducing signs and symptoms, inducing major clinical response, inhibiting the progression of structural damage, and improving physical function in adult patients with moderately to severely active RA (US Humira PI 2006). The European labeling indicates that adalimumab should be used in the setting of disease-modifying antirheumatic drugs (DMARD) failure. However, in the case of severe, active and progressive RA, European labeling also includes a specific indication for the use of adalimumab in the treatment of adults not previously treated with MTX (EU Humira SPC 2006). US labeling indicates that adalimumab may be used alone or in combination with MTX and other DMARDs; European labeling includes administration in combination with MTX or as monotherapy (EU Humira SPC 2006; US Humira PI 2006).

1.5.2 Psoriatic Arthritis

Adalimumab is also indicated for use in the treatment of PsA. In the United States, the labeling states that adalimumab is indicated for reducing signs and symptoms of active arthritis in patients with PsA (US Humira PI 2006). Again, similar to the labeling in RA, the European labeling for this indication stipulates that adalimumab should be used when the response to previous DMARD therapy

Table 1.1 Approved indications (EU Humira SPC 2006; US Humira PI 2006).

EU Summary of Product Characteristics ***Therapeutic Indications***	***USA Package Insert*** ***Indications and Usage***
Rheumatoid arthritis (RA) Humira in combination with MTX, is indicated for • the treatment of moderate to severe, active RA in adult patients when the response to DMARDs including MTX has been inadequate. • the treatment of severe, active and progressive RA in adults not previously treated with MTX. Humira can be given as monotherapy in case of intolerance to MTX, or when continued treatment with MTX is inappropriate. Humira has been shown to reduce the rate of progression of joint damage as measured by radiography, and to improve physical function, when given in combination with MTX.	Humira is indicated for reducing signs and symptoms, inducing major clinical response, inhibiting the progression of structural damage and improving physical function in adult patients with moderately to severely active RA. Humira can be used alone or in combination with MTX or other DMARDs.
Psoriatic arthritis Humira is indicated for the treatment of active and progressive psoriatic arthritis in adults when the response to previous DMARD therapy has been inadequate.	Humira is indicated for reducing signs and symptoms of active arthritis in patients with psoriatic arthritis. Humira can be used alone or in combination with DMARDs.
Ankylosing spondylitis Humira is indicated for reducing signs and symptoms in patients with active ankylosing spondylitis.	Humira is indicated for the treatment of adults with severe active ankylosing spondylitis who have had an inadequate response to conventional therapy.

DMARDs = disease-modifying antirheumatic drugs; MTX = methotrexate.

has been inadequate (EU Humira SPC 2006), The US Food and Drug Administration labeled adalimumab for use alone or in combination with DMARDs; however, the labeling approved by the European Medicines Evaluation Agency for PsA does not mention the use of combination therapy at this time (EU Humira SPC 2006).

1.5.3 Ankylosing Spondylitis

Adalimumab is indicated for reducing signs and symptoms in patients with active ankylosing spondylitis (US Humira PI 2006). In the European Union, adalimumab is indicated for the treatment of adults with severe active ankylosing spondylitis who have had an inadequate response to conventional therapy (EU Humira SPC 2006).

1.5.4 Dosing and Administration

The recommended dose for RA, PSA, and AS is 40 mg SC, EOW (US Humira PI 2006; EU Humira SPC 2006). US labeling states that, in RA, some patients not taking concomitant MTX may derive additional benefit from increasing the dosing frequency to 40 mg weekly (US Humira PI 2006). Weekly dosing also is included in the European labeling for RA patients who are receiving monotherapy and experience a decrease in their response. Further, both labels state that glucocorticoids, salicylates, nonsteroidal anti-inflammatory drugs (NSAIDs), and analgesics may be continued during treatment with adalimumab.

Adalimumab is available as a preservative-free, sterile solution for SC injection, as 40 mg per 0.8 mL in 1-mL prefilled single-dose glass syringes. Under refrigeration (2–8°C), adalimumab has a shelf-life of 18 months. The drug formulation should not be frozen (US Humira PI 2006).

1.6 Clinical Experience

Adalimumab has been studied most extensively in patients with RA (Table 1.2), with more than 10000 patients having been enrolled in studies during the RA clinical development program. A total of 300 of these patients has been followed for at least 5 years, with some patients having been followed into their seventh year of treatment (Schiff et al. 2005b, 2006).

1.6.1 Studies in Rheumatoid Arthritis

The pivotal studies evaluated safety and efficacy of adalimumab treatment in a variety of settings: in combination with MTX in advanced disease (ARMADA [Anti-TNF Research Study Program of the Monoclonal Antibody Adalimumab, DE019), in combination with traditional DMARDs (STAR [The Safety Trial of Adalimumab in Rheumatoid Arthritis]), as monotherapy in patients with severe disease (DE011), and combined with MTX or as monotherapy in early disease (PREMIER). Following the double-blind phases of all of these studies, adalimumab treatment was continued during open-label extension (OLE) trials. In addition to these studies, more than 900 and more than 6600 patients were evaluated in the Act (Access to Therapy) and ReAct (Research in Active Rheumatoid Arthritis) trials, respectively, open-label trials designed to assess prospectively the safety and efficacy of adalimumab in real-life settings (Bombardieri et al. 2004; Burmester et al. 2004, 2005a; Schiff et al. 2006).

1.6.1.1 Adalimumab in Combination with MTX

Two large, randomized, double-blind, placebo-controlled, multicenter studies have evaluated the safety and efficacy of adalimumab in combination with MTX

Table 1.2 Summary of baseline data and 6-month efficacy endpoints for clinical trials evaluating adalimumab in patients with rheumatoid arthritis.* (Adapted with permission from Bansback et al. 2005.)

Trial name	*Treatment*	*Patients [n]*	*Mean age [years]*	*Disease duration [years]*	*Baseline HAQ*	*MTX dose [mg/week]*	*RF+ [%]*	*ACR20*	*ACR50*	*ACR70*
ARMADA (Weinblatt et al. 2003)	Placebo + MTX	62	56	11	1.6	17	79	15	8	10
	Adalimumab + MTX	67	57	12	1.6	17	80	67	55	27
DE019 (Keystone et al. 2004b,c)	Placebo + MTX	200	56	11	1.4	17	82	30	10	3
	Adalimumab + MTX	207	56	11	1.4	17	82	63	39	21
DE011 (van de Putte et al. 2004)	Placebo	110	54	12	1.9	–	90	19	8	2
	Adalimumab	113	53	11	1.8	–	90	46	22	12
STAR (Furst et al. 2003)	Placebo + DMARDs	318	56	12	1.4	NR†	62	35	11	4
	Adalimumab + DMARDs	318	55	9	1.4	NR†	63	53	29	15
PREMIER (Breedveld et al. 2005, 2006	MTX	257	52	1	1.5	20	81	63‡	46‡	28‡
	Adalimumab	274	52	1	1.5	–	83	54‡	42‡	26‡
	Adalimumab + MTX	268	52	1	1.5	20	87	73‡	62‡	46‡

* Note that only licensed indication dosages are shown (adalimumab 40 mg EOW).
† Methotrexate (MTX) used in only 56% of patients; dose not reported.
‡ Results from 12 months.
ACR = American College of Rheumatology; ARMADA = Anti-TNF Research Study Program of the Monoclonal Antibody Adalimumab; DMARDs = disease-modifying antirheumatic drugs; HAQ = Health Assessment Questionnaire; MTX = methotrexate; RF = rheumatoid factor; STAR = Safety Trial of Adalimumab in Rheumatoid Arthritis.

for treatment of long-standing RA in patients with insufficient response to MTX. In ARMADA, 271 patients received either adalimumab 20, 40, or 80 mg or placebo SC EOW in addition to MTX (Weinblatt et al. 2003). A total of 25% to 32% of patients receiving adalimumab achieved at least 20% improvement in American College of Rheumatology responses after 1 week (ACR20). At the end of 24 weeks, significantly more patients treated with adalimumab 20, 40, and 80 mg had achieved an ACR20 response (48%, 67%, and 66% respectively; $P < 0.001$ versus placebo for all comparisons) and an ACR50 response (32%, 55%, 43%, respectively; $P = 0.003$, $P < 0.001$, and $P < 0.001$, versus placebo, respectively) than those who received placebo (ACR20, 15%; ACR50, 8%). Some 27% of patients receiving adalimumab 40 mg ($P < 0.001$ versus placebo), and 19% of those receiving adalimumab 80 mg ($P = 0.02$ versus placebo) achieved an ACR70 response versus only 5% of patients receiving placebo. Scores on the Health Assessment Questionnaire Disability Index (HAQ-DI), fatigue scale, and the Short Form 36 (SF-36) also were improved over baseline, and these improvements were statistically greater than with placebo. Adalimumab-treated patients experienced up to a 40% reduction in the HAQ-DI scores. Clinically meaningful changes (≥10 points) in six domains of the SF-36 were attained by patients receiving adalimumab plus MTX as compared with two domains for those receiving placebo plus MTX. Adalimumab was well tolerated at all doses. Infections occurred at a similar rate in the adalimumab (1.55 per patient-year [PY]) and placebo groups (1.38/PY).

The OLE of this study is ongoing. Among patients completing 5 years of treatment, clinical efficacy was sustained, with 76%, 64%, and 39% of patients achieving ACR20, 50, 70 responses; 52% achieving clinical remission; and 28% having no physical limitations (Weinblatt et al. 2005, 2006). Furthermore, the majority of patients in the OLE were able to reduce corticosteroid and/or MTX dosages without adversely affecting long-term efficacy (Weinblatt et al. 2005). Serious adverse events in the OLE were similar to those observed during the controlled phase. The rate of serious infection was 2.03 events per 100 PY in the OLE phase compared with 2.30 during the blinded phase. The primary reasons for withdrawal from the study over time were lack of efficacy (8%), adverse events (11%), and other reasons (16%).

The long-term efficacy and safety of adalimumab in combination with DMARDs were evaluated in a rollover OLE study (DE020), enrolling 846 RA patients from ARMADA, STAR, and two phase I trials in which all patients received adalimumab and most received concomitant MTX. The significant clinical improvement seen after 6 months of treatment in controlled trials was sustained over 4 years, with patients maintaining significant reductions in Disease Activity Score in 28 joints (DAS28), swollen and tender joint counts, and HAQ-DI scores. ACR20, 50, and 70 response rates achieved at 6 months were also maintained through 4 years of therapy. At the last visit, 49% of patients achieved DAS28 <2.6, 24% had zero tender joints, 21% had zero swollen joints, and 44% achieved HAQ-DI ≤0.5 – parameters indicative of remission. A substantial percentage of patients treated with adalimumab plus MTX were able to reduce their use of steroids and decrease their MTX dose while maintaining control of their disease. Among those patients achieving remission, those with moderate disease and younger age tended to achieve remission more rapidly and sustain remission for longer periods than older patients and those with severe disease (Emery et al. 2004; Schiff et al. 2004).

DE019 was a randomized, double-blind, placebo-controlled, multicenter study assessing the ability of adalimumab to inhibit radiographic progression and reduce disease activity in 619 RA patients with active disease, despite therapy with MTX. Patients were randomized to receive adalimumab 40 mg EOW, adalimumab 20 mg weekly, or placebo, plus concomitant MTX (Keystone et al. 2004a). At the end of 52 weeks, as was seen in the ARMADA study, significantly more patients treated with adalimumab attained ACR20, 50, and 70 responses (40 mg EOW: 59%, 42%, 23% and 20 mg weekly: 55%, 38%, 21%, respectively) compared with patients receiving placebo (24%, 10%, 5%; $P \leq 0.001$).

Significantly less radiographic progression as measured by total Sharp scores (TSS) was seen after adalimumab treatment. At one year, the mean change in TSS of adalimumab-treated patients was 0.1 units compared with 2.7 units for placebo patients. Of the patients receiving adalimumab 40 mg EOW, 62% had no new erosions at Week 52, as did 58% of those receiving adalimumab 20 mg weekly. A significantly smaller percentage (46%; $P \leq 0.001$, $P \leq 0.05$, respectively) of patients in the placebo group had no new erosions at Week 52. Functional status and quality of life (HAQ-DI and SF-36) also improved significantly in the adalimumab groups, with approximately 40% decreases in HAQ-DI scores in the adalimumab groups versus 17% with placebo, and clinically meaningful improvement in the SF-36 for the patients receiving active treatment versus placebo (Keystone et al. 2004a).

In an OLE of this study, 40 mg adalimumab was administered EOW to 457 of the patients who completed DE019. Of these patients, 79% (n = 363) remained in the study at 3 years. The reasons for discontinuation were loss of efficacy (2%), adverse events (7%), and other reasons (11%). For those in the overall OLE population who remained in the study at 3 years, responses were maintained, with 58% achieving an ACR20 response, 42% achieving an ACR50 response, and 24% achieving an ACR70 response. The long-term effects of adalimumab treatment on disease progression were substantial. At the three-year follow-up, adalimumab plus MTX continued to inhibit structural damage. On radiographic examination of the 129 patients originally randomized to receive adalimumab 40 mg EOW, the mean TSS at the end of 3 years was 0.3. Some 62% of patients had no radiographic progression (defined as ≤0.5 unit increase in TSS from baseline), and 28% showed radiographic improvement (>0.5 unit decrease in TSS from baseline) at 3 years (Keystone et al. 2005). The sustained clinical benefit and inhibition of disease progression were accompanied by improvements in functional status as measured by HAQ (Keystone et al. 2004b,c).

1.6.1.2 Adalimumab with Traditional DMARDs

STAR was a randomized, double-blind, placebo-controlled, multicenter study that assessed the safety and efficacy of adalimumab in a heterogeneous group of RA patients with persistent disease activity despite concomitant DMARDs (Furst et al. 2003). The study was designed to reflect the general patient population and standard treatment regimens typically seen in clinical practice. This 24-week study enrolled 636 patients who had active disease while being treated with standard antirheumatic therapies. Participants received either placebo or adalimumab 40 mg EOW while continuing their standard therapy. MTX, antimalarials, and leflunomide were the most commonly used agents, alone or in combination.

The addition of adalimumab to standard antirheumatic therapy was well tolerated (Furst et al. 2003), with the majority of adverse events being mild or moderate in severity. Adverse event rates and withdrawal rates were similar in both treatment groups, with 91% of each group completing the study. Over the study period, no differences were seen in the incidence of any infection or serious infection between the treatment groups (Furst et al. 2003). Add-on treatment with adalimumab resulted in significantly higher percentages of patients achieving ACR20, 50, 70 responses (53%, 29%, 15%, respectively) compared with placebo plus standard therapy (35%, 11%, 3.5%, respectively; $P \leq 0.001$ for all comparisons) (Furst et al. 2003).

1.6.1.2.1 Additional Experience: Adalimumab with Traditional DMARDs

The Act and ReAct trials were 12-week, open-label studies designed to provide a prospective evaluation of efficacy and safety in real-life settings involving 936 and 6610 patients with active, insufficiently treated RA, respectively (Bombardieri et al. 2004; Burmester et al. 2004, 2005a; Schiff et al. 2006). Participants had various comorbidities, were treated with a broad range of antirheumatic therapies, and were enrolled in varied social care systems. Patients received adalimumab 40 mg EOW as classic add-on therapy (i.e., added to existing, but insufficient antirheumatic therapy). In ReAct, patients were given the option to continue adalimumab for an extension period.

At Week 12 in the ReAct study, 69%, 40%, and 18% of patients achieved ACR20, 50, and 70 responses, respectively. Moderate European League Against Rheumatism (EULAR) responses were achieved in 83% of patients, and 33% achieved a good EULAR response. The mean DAS28 was reduced by 2.1 units, with 20% of patients having a DAS28 <2.6 at Week 12. The mean HAQ scores decreased by 0.54 units. These data demonstrate that adalimumab was effective in more than 6000 difficult-to-treat patients with active RA (Burmester et al. 2005a; also data on file).

1.6.1.3 Adalimumab Monotherapy

The safety and efficacy of adalimumab as monotherapy in severe RA was assessed in 544 difficult-to-treat patients who had failed multiple DMARDs in a phase III randomized, double-blind, placebo-controlled, multicenter study (DE011) (van de Putte et al. 2004). Patients with active disease were randomized to receive adalimumab 20 mg EOW, 20 mg weekly, 40 mg EOW, 40 mg weekly, or placebo for 26 weeks. At 26 weeks, significantly more patients receiving adalimumab achieved the primary efficacy endpoint ACR20 compared with those in the placebo group ($P \leq 0.01$ for all treatment groups). An ACR20 response was achieved by 36%, 39%, 46%, 53%, and 19% of adalimumab 20 mg EOW, 20 mg weekly, 40 mg EOW, 40 mg weekly, and placebo groups, respectively. Response rates for ACR50 were 19%, 21%, 22%, 35% for adalimumab 20 mg EOW, 20 mg weekly, 40 mg EOW, 40 mg weekly, all of which were significantly higher than the 8% response seen with placebo. Treatment with adalimumab resulted in improved ACR70 rates compared with placebo (van de Putte et al. 2004). Adalimumab 40 mg weekly resulted in slightly higher ACR response than 40 mg EOW. Moderate EULAR response rates were significantly greater with adalimumab than with placebo at 26 weeks (42%, 48%, 56%, 61% versus 26%, $P \leq 0.05$). In addition, clinical improvement was achieved as early as 2 weeks after treatment initiation with adalimumab (van de Putte et al. 2004).

The long-term efficacy and safety of adalimumab as monotherapy were evaluated in a rollover OLE study (DE018), enrolling 794 RA patients from DE011 and three other adalimumab trials in which the majority of patients received monotherapy. The clinical responses achieved at Year 1 were sustained up to 5 years, with 67%, 40%, and 17% of patients maintaining ACR20, 50, and 70 responses, respectively. A total of 81% of these patients had a moderate EULAR response at 5 years, and initial reductions in tender and swollen joint counts were also maintained. High retention rates were achieved, which suggested high tolerability with adalimumab therapy (Burmester et al. 2003).

1.6.1.4 Pivotal Study in Early Rheumatoid Arthritis

Rheumatoid arthritis is characterized by rapid disease progression that is often insidious. By the time joint malalignment and functional disability are evident, substantial irreversible damage has occurred in many patients (Lee and Weinblatt 2001). In 70% of patients, radiographic evidence of joint destruction is present within the first 2 years (Lee and Weinblatt 2001; McQueen et al. 1998). As early as 4 months after disease onset, synovial hypertrophy, bone edema, and early erosion have been detected using magnetic resonance imaging (Lee and Weinblatt 2001; McGonagle et al. 1999). Moreover, active synovitis is detectable in biopsy specimens of symptomless knee joints in patients with early disease (Lee and Weinblatt 2001; Soden et al. 1989). These observations are consistent with data indicating that treatment with combination therapy with DMARDs before irreversible joint destruction has occurred may improve long-term clinical outcomes for patients with RA (Landewé et al. 2002; Möttönen et al. 2002; Tsakonas et al. 2000).

The outcomes of aggressive, early treatment of RA were tested in the PREMIER study, a randomized, double-blind, active comparator, multicenter trial of adalimumab in 799 MTX-naïve patients with very early RA (mean disease duration 0.7 years) (Breedveld et al. 2005, 2006). This trial compared treatment with adalimumab 40 mg EOW plus MTX, adalimumab 40 mg EOW alone, and MTX alone (rapidly escalated to 20 mg per week after study initiation) for 2 years. Combination therapy was consistently more effective than monotherapy with either agent for all outcomes measured. The ACR50 response rates at one year were 62% for the combination group versus 46% for MTX monotherapy ($P < 0.001$) and 42% for adalimumab monotherapy ($P < 0.001$). Similarly, a greater percentage of patients receiving the combination achieved an ACR70 response (46%) versus those receiving MTX monotherapy (28%; $P < 0.001$) or adalimumab monotherapy (26%; $P < 0.001$). These differences were maintained at 2 years, as shown in Fig. 1.4A.

In this population of MTX-naïve patients with recent-onset RA, inhibition of disease progression as measured radiographically was significantly greater with combination therapy or adalimumab alone than with MTX alone. The change

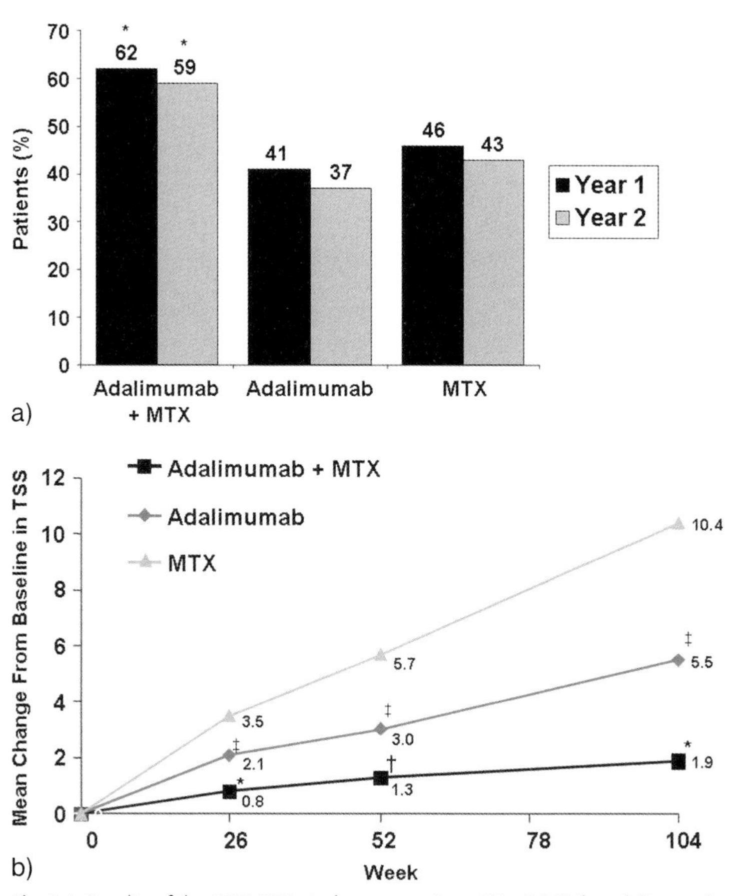

Fig. 1.4 Results of the PREMIER study. (a) American College of Rheumatology 50 response at Years 1 and 2. *$P < 0.001$ for adalimumab + methotrexate (MTX) versus MTX alone and adalimumab + MTX versus adalimumab alone. (b) Mean change from baseline in total Sharp scores (TSS) over time. *$P < 0.001$ for adalimumab + MTX versus MTX alone and adalimumab alone. †$P < 0.001$ for adalimumab + MTX versus MTX alone and $P = 0.002$ for adalimumab + MTX versus adalimumab alone. ‡$P < 0.001$ for adalimumab versus MTX alone. (Reproduced with permission from Breedveld et al. 2006.)

from baseline in mean TSS at one year was significantly lower for adalimumab plus MTX (1.3, $P < 0.001$) compared with MTX alone (5.7) and adalimumab alone (3.0). The differences in progression for patients treated with adalimumab versus MTX alone were even more marked at the end of Year 2, with an accumulated TSS change of 10.4 in the MTX monotherapy arm versus 5.5 in the adalimumab monotherapy arm and 1.9 in the combination arm. Although ACR responses were comparable between the two monotherapy arms, there was statistically less progression in the adalimumab monotherapy arm compared with the MTX monotherapy arm in both Year 1 (3.0 versus 5.7), and Year 2 (5.5 versus 10.4) ($P < 0.001$ for all comparisons; Fig. 1.4B). These findings emphasize that clinical assessment of signs and symptoms of RA may not fully reflect therapeutic benefit (Breedveld et al. 2006).

Interestingly, during Year 2, patients who were treated with MTX monotherapy continued to progress at approximately the same rate seen in Year 1 (5.7 Sharp unit progression in Year 1 and 4.7 Sharp unit progression in Year 2). In contrast, patients who received combination therapy had less than half the progression in Year 2 than they had experienced in Year 1 (1.3 Sharp unit progression in Year 1 and 0.6 Sharp unit progression in Year 2). Moreover, after 2 years, remission as measured by DAS28 <2.6 and major clinical response (continuous ACR70 response for ≥6 months) was achieved by almost half (49%) of the patients receiving combination therapy.

The results of PREMIER provide further evidence that early aggressive combination therapy of progressive RA with MTX and adalimumab can reduce disease progression to a greater extent than treatment of disease with MTX monotherapy. The benefits of early treatment with combination therapy are sustained for at least 2 years, and are associated with improved clinical outcomes.

1.6.1.5 **Adalimumab in Patients Previously Treated with other Anti-TNF Agents**

The ReAct data were analyzed based on use of prior anti-TNF therapy. Of the 6610 enrolled patients, 899 had received prior anti-TNF therapy with etanercept and/or infliximab (median prior treatment of 9.5 months). Despite higher disease activity and use of a greater number of prior DMARDs at baseline, patients with a history of prior anti-TNF therapy demonstrated robust responses with adalimumab. At Week 12, ACR20, 50, and 70 response rates were 60%, 33%, and 13%, respectively, among patients who had received prior anti-TNF therapy, as compared with 70%, 41%, and 19%, respectively, among patients who were naïve to anti-TNF therapy. Response to adalimumab among patients who had discontinued prior anti-TNF therapy due to loss of efficacy or intolerance was similar to that of patients who had never received anti-TNF agents, with 67% achieving ACR20 responses (Bombardieri et al. 2005, 2006; Burmester et al. 2005a).

The results from two additional studies of the use of adalimumab in RA patients who had failed infliximab therapy were consistent with those from ReAct, as described above (Nikas et al. 2006; van der Bijl et al. 2005). In the first study, adalimumab was administered at a dose of 40 mg EOW added to current therapy in 41 patients with long-standing moderate to severe RA who had discontinued infliximab therapy because of lack of efficacy, loss of efficacy after an initial response, and/or intolerance/adverse side effects (mean infliximab treatment duration, 17 months). After 16 weeks of adalimumab therapy, significant percentages of patients achieved ACR20, ACR50, and moderate EULAR responses (49%, 31%, 65%, respectively). Clinically meaningful decreases in tender and swollen joint counts were apparent by Week 2 of adalimumab therapy. Additionally, patients who switched to adalimumab demonstrated significant reductions in disease activity and disability, as assessed by changes in DAS28 and HAQ scores (Data on file; van der Bijl et al. 2005). In the second study, 24 patients with RA who discontinued therapy with infliximab (mean infliximab treatment duration, 18.5 months) were treated with adalimumab 40 mg EOW for 12 months. The results from these patients were compared with 25 patients receiving adalimumab who had not previously used anti-TNF therapy (controls). After 12 months of treatment, 75% of patients previously receiving infliximab and 76% of control patients achieved ACR20. Mean changes in DAS28 were similar for the two groups (−2.4 versus −2.7). A total of 71% of patients in the infliximab failure group and 72% of control patients achieved a EULAR response (Nikas et al. 2006). In both of these studies, adalimumab was well tolerated, even among patients with previous intolerance to infliximab.

Collectively these results suggest that, even after treatment failure with a TNF antagonist, a significant clinical response can be achieved with a different agent of the same class.

1.6.2
Pivotal Studies in Psoriatic Arthritis

ADEPT (Adalimumab Effectiveness in Psoriatic Arthritis Trial) evaluated adalimumab treatment in patients with moderate to severe PsA and a history of inadequate response to NSAIDs (Mease et al. 2005a–c). In this double-blind, randomized, parallel-group, multicenter, 24-week trial, 315 patients with PsA who had three or more swollen and three or more tender joints and either active psoriatic skin lesions or a documented history of psoriasis were randomized to receive adalimumab 40 mg SC EOW, or placebo (Mease et al. 2005a). Participants were stratified prior to randomization by MTX use (yes or no) and extent of psoriasis (<3% or ≥3%).

At Week 24, the ACR20 response rate was 57% with active treatment, which was significantly higher than with placebo (15%, $P < 0.001$) (Fig. 1.5A) (Mease et al. 2005a). Significantly more patients achieved ACR50 and 70 responses with adalimumab than with placebo. Among the 69 patients in each treatment group who were assessed using the Psoriasis Area and Severity Index (PASI), 59% of those in the adalimumab group achieved a 75% improvement in PASI compared with 1% of those in the placebo group (Fig. 1.5B). Some 42% of patients receiving adalimumab achieved the very marked response of 90% improvement in PASI;

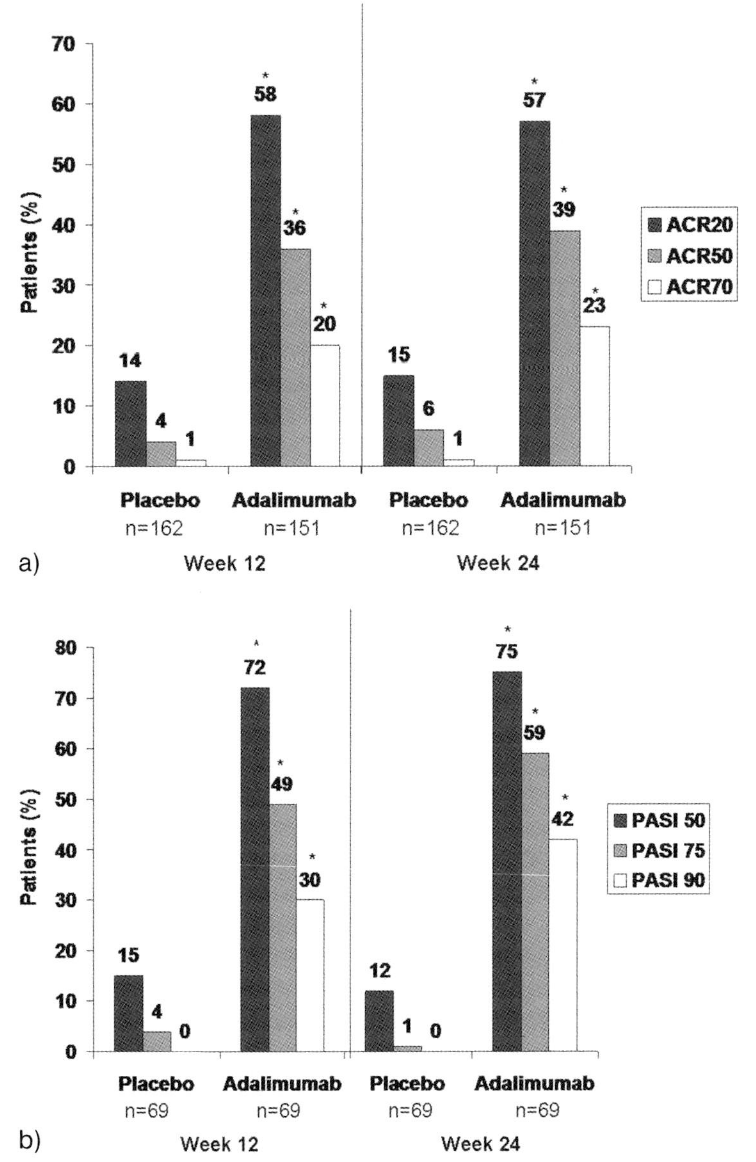

Fig. 1.5 Results from ADEPT (Adalimumab Effectiveness in Psoriatic Arthritis Trial). (a) Percentage of patients with psoriatic arthritis who met American College of Rheumatology (ACR) 20, 50, and 70 response criteria at Week 12 and Week 24. $^{*}P < 0.001$ placebo versus adalimumab, adjusted for baseline methotrexate use and extent of psoriasis at baseline. (b) Percentage of patients with psoriatic arthritis and ≥3% body surface area psoriasis involvement at baseline who met Psoriasis Area and Severity Index (PASI) 50, 75, and 90 response criteria at Week 12 and Week 24. $^{*}P < 0.001$ placebo versus adalimumab, adjusted for baseline methotrexate use. (Reproduced with permission from Mease et al. 2005.)

this level of response was reached by 0% of patients receiving placebo ($P < 0.001$). In addition, 67% of patients receiving active treatment were assessed as "clear" or "almost clear" using the Physician Global Assessment (PGA), compared with 10% of those in the placebo group.

Progression of radiographic damage in these PsA patients was inhibited with adalimumab treatment, yielding a mean change from baseline of the modified TSS equal to −0.2 at 24 weeks (Fig. 1.6) (Mease et al. 2005a). In the placebo group, the mean change in TSS was +1.0 at 24 weeks.

As would be expected, these clinical benefits were associated with improved levels of disability and quality-of-life responses (Mease et al. 2005a). The mean change from baseline of the HAQ-DI was −0.4 in the adalimumab group compared with −0.1 in the placebo group ($P < 0.001$). Physical function, as measured by the Functional Assessment of Chronic Illness Therapy, and quality of life, assessed with the Dermatology Life Quality Index, were similarly improved with active treatment (Mease et al. 2005a).

At the end of the 24-week blinded phase of ADEPT, patients were given the option of continuing in an OLE phase, during which all 285 patients who elected to continue received adalimumab 40 mg SC EOW (Mease et al. 2005b). Among the patients who had initially been randomized to the adalimumab group, the 24-week improvements in ACR, PASI, and HAQ scores were maintained at Week

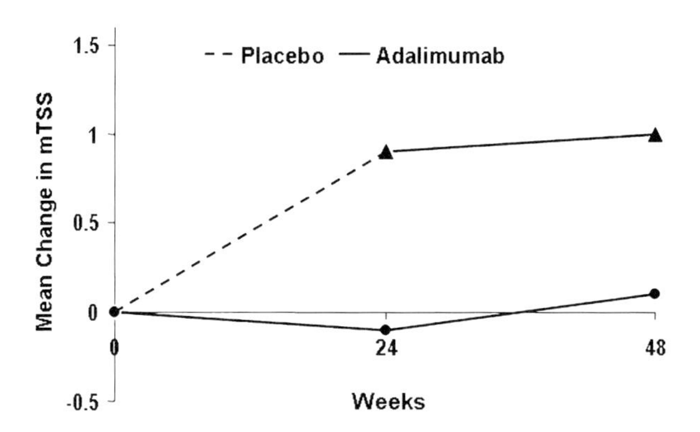

	N	Baseline	24 Wk Mean Change	48 Wk Mean Change
Placebo	152	21.8	0.9	1.0
Adalimumab	144	23.7	-0.1*	0.1

Fig. 1.6 Results from ADEPT (Adalimumab Effectiveness in Psoriatic Arthritis Trial). Changes in modified total Sharp score (TSS) at Weeks 24 and 48. *$P < 0.001$ placebo versus adalimumab using a ranked analysis of covariance. (From Mease et al. 2005a,c.)

48, as was inhibition of radiographic progression (mean change from baseline in TSS of 0.1 at Week 48) (Mease et al. 2005b,c). For those who had previously been randomized to the placebo group, ACR20, 50, and 70 were achieved at Week 48 of the OLE phase (24 weeks of active treatment) by 54%, 37%, and 21% of patients, respectively. In this group at 48 weeks, PASI50, 75, and 90 responses were attained by 76%, 63%, and 47% of patients respectively, similar to the responses after 24 weeks for patients originally randomized to adalimumab in the blinded phase of the study. Radiographic progression was decreased markedly compared with the 24 weeks on placebo, and the mean change in modified TSS was 0.1 during the first 24 weeks of active treatment with adalimumab (Fig. 1.6). Disability also improved among these patients, with a mean change in the HAQ-DI of −0.4 from baseline (Mease et al. 2005b,c). For patients who failed to demonstrate ≥20% decrease in swollen and total joint count response versus baseline after 12 weeks of treatment in the OLE, an option was given to increase the dose to 40 mg weekly. A total of 12 patients underwent a dose increase, with an ACR20 response resulting in three cases.

1.6.3 Ankylosing Spondylitis

ATLAS (Adalimumab Trial Evaluating Long-term Efficacy and Safety in Ankylosing Spondylitis) was a randomized, double-blind, placebo-controlled, multicenter study comparing adalimumab with placebo in patients with AS. A total of 315 patients was randomized in a 2:1 fashion to receive adalimumab 40 mg EOW versus placebo for 24 weeks. At Week 12, 58% of adalimumab patients versus 21% of placebo patients achieved a 20% improvement in the ASsessment in Ankylosing Spondylitis (ASAS20) International Working Group Improvement Criteria, a difference that was statistically significant (differences of 37.6% [95% CI, 27.4–47.8%], $P < 0.001$). A favorable clinical response was observed as early as 2 weeks after initiation of active treatment (van der Heijde et al. 2006).

Also at Week 12, 45.2% of adalimumab-treated patients achieved at least a 50% improvement in the Bath Ankylosing Spondylitis Disease Activity Index (BASDAI) compared with only 15.9% (17/107) of the placebo group (difference of 29.3% [95% CI, 19.6–39.0%], $P < 0.001$) This significant difference was sustained at Week 24 (van der Heijde et al. 2006).

The percentage of patients with at least a 20% improvement in five of the six ASAS assessment domains was significantly greater with adalimumab treatment (44.7%) at Week 24 compared with only 12.1% of the placebo patients (difference of 32.6% [95% CI, 23.4–41.7%], $P < 0.001$). Similar significantly better outcomes were observed with adalimumab for ASAS40 responders and percentage of patients achieving partial remission. These differences were seen at Week 12 and remained significant at Week 24 (Fig. 1.7) (van der Heijde et al. 2006).

The responses of patients with total spinal ankylosis (n = 11) were similar to those without total ankylosis (US Humira PI 2006).

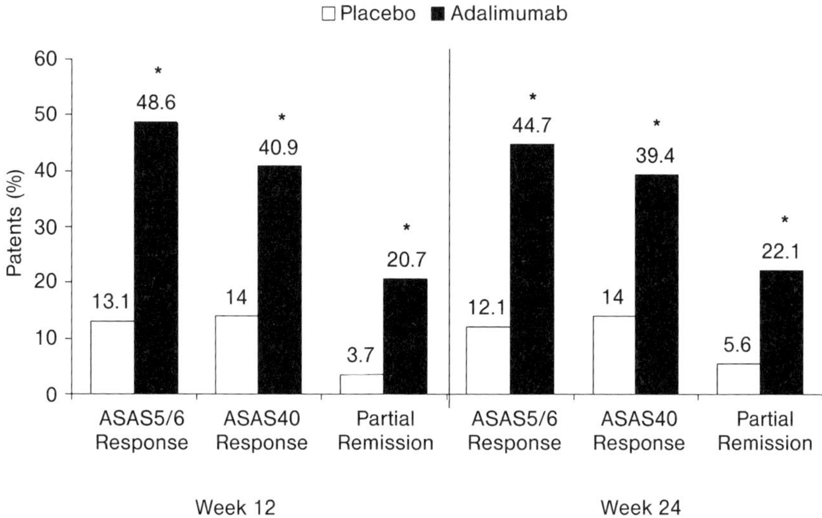

Fig. 1.7 Results from ATLAS (Adalimumab Trial Evaluating Long-term Efficacy and Safety in Ankylosing Spondylitis). ASsessment in Ankylosing Spondylitis (ASAS) 5/6 criteria response, ASAS40 response, and partial remissions. *$P < 0.001$ versus placebo. Values are imputed. (From van der Heijde et al. 2006.)

1.7 Review of Adalimumab Safety

In clinical trials, adalimumab has been generally well tolerated, with less than 10% of patients in clinical trials discontinuing treatment because of adverse events (Lee and Kavanaugh 2005; Furst et al. 2003; Keystone et al. 2004a). Moreover, data from a national registry demonstrate that adalimumab has a high adherence rate (82%) at one-year follow-up, comparable to that for other TNF inhibitors (Kristensen et al. 2004). Primary among the potential safety concerns for inhibitors of TNF are the risk of serious infections, including tuberculosis (TB) and opportunistic infections (Lee and Kavanaugh 2005; Schiff et al. 2006), autoimmune disease (Lee and Kavanaugh 2005), demyelination disorders (Magnano et al. 2004), and malignancies, particularly lymphoma (Lee and Kavanaugh 2005). Assessment of safety with long-term adalimumab suggests that the incidence of these serious adverse events with adalimumab treatment is low.

1.7.1 Safety in Specific Indications

1.7.1.1 Long-Term Safety in Rheumatoid Arthritis

Safety data are available for more than 10000 patients with moderate to severe RA who have been evaluated in clinical trials; 300 of these patients have had at least 5 years' exposure to adalimumab, with some followed for up to 7 years. These patients represent a total of 12 506 PY of adalimumab exposure (Schiff et al. 2006). In addition, adalimumab exposure during the US postmarketing period is estimated to be more than 78 500 PY (Schiff et al. 2006). The types of adverse event reported following adalimumab use in RA have been similar throughout the clinical development program and postmarketing surveillance (Schiff et al. 2005a).

Adverse events reported by ≥5% of patients treated with adalimumab during placebo-controlled phases of RA studies are listed in Table 1.3. Injection site reactions (ISR) are the most common adverse events associated with adalimumab, with an incidence of 20% versus 14% for placebo reported in four pivotal trials

Table 1.3 Adverse events reported by ≥5% of patients treated with adalimumab during placebo-controlled periods of rheumatoid arthritis studies (US Humira PI 2006).

Adverse event	*Adalimumab 40 mg EOW [n = 705] (%)*	*Placebo [n = 690] (%)*
Respiratory		
• Upper respiratory tract infection	17	13
• Sinusitis	11	9
• Flu syndrome	7	6
Gastrointestinal		
• Nausea	9	8
• Abdominal pain	7	4
Laboratory tests*		
• Laboratory test abnormal	8	7
• Hypercholesterolemia	6	4
• Hyperlipidemia	7	5
• Hematuria	5	4
• Increased alkaline phosphatase	5	3
Other		
• Injection site pain	12	12
• Headache	12	8
• Rash	12	6
• Accidental injury	10	8
• Injection site reaction†	8	1
• Back pain	6	4
• Urinary tract infection	8	5
• Hypertension	5	3

* Laboratory test abnormalities were reported as adverse events in European trials.
† Does not include erythema and/or itching, hemorrhage, pain, or swelling.
EOW = every other week.

for RA (ARMADA, DE011, DE019, and STAR) (US Humira PI 2006; EU Humira SPC 2006). Most ISRs were mild, were not recurrent, and rarely led to discontinuation from studies (US Humira PI 2006; Wells et al. 2003).

1.7.1.2 Safety Profile in Psoriatic Arthritis and Ankylosing Spondylitis

Although safety evaluations of adalimumab in patients with PsA and AS are less extensive than those for patients with RA, in clinical trials up to one year duration adalimumab has been generally well tolerated. Adverse event profiles are similar to those in RA, and no new safety concerns have arisen (Mease et al. 2005a,b; van der Heijde et al. 2006).

1.7.2 Immune System Function

An immunology substudy of DE019 determined that normal immune function is preserved during adalimumab therapy (Kavanaugh et al. 2002). Adalimumab did not significantly alter the total number of peripheral granulocytes, natural killer cells, monocytes/macrophages, B cells, T cells, or T-cell subsets. The ability of lymphocytes to proliferate, as evidenced by *in-vitro* proliferation to both mitogen and recall antigen, was not altered. There were no clinically relevant differences in delayed-type hypersensitivity reactivity, or in B-cell function, as assessed by total immunoglobulin levels, IgG levels, and antigen-specific antibody response to recall antigen. Phagocytic functions measured in oxidative burst assays and by phagocytosis of fluorescent particles were similar for neutrophils and macrophages from adalimumab- and placebo-treated patients (Kavanaugh et al. 2003).

1.7.3 Infections

In placebo-controlled RA trials, infections rates were approximately one per PY for both adalimumab- and placebo-treated groups. The most frequently reported infectious adverse events among RA trials were upper respiratory tract infections, rhinitis, bronchitis, and urinary tract infections (US Humira PI 2006; EU Humira SPC 2006). Rates of serious infections (5.1/100 PY) did not increase over the duration of treatment, were similar to those found with traditional DMARDs and other TNF antagonists, and were within the range reported among the general RA population (range ~3 to ~9/100 PY) (Schiff et al. 2003a, 2006). In controlled studies, the rate of serious infections among adalimumab-treated patients was not significantly different from placebo in either significance or type; the results of OLEs were similar to those of randomized trials (Lee and Kavanaugh 2005; Schiff et al. 2003a). In the ReAct study of 6610 patients with long-standing RA, the rate of serious infections was 5.5/100 PY, consistent with other data (G.R. Burmester et al., unpublished results).

1.7.3.1 Tuberculosis

Cases of TB have been seen with all three anti-TNF agents. During the initial phase of the development program, no TB screening was performed, but screening prior to treatment initiation was introduced during the phase II studies. Following the introduction of routine screening in European clinical trials, the rate of TB decreased by 75% (0.33/100 PY). In North American clinical RA trials, four cases had been reported after 4914 PY of exposures (0.08/100 PY) (Schiff et al. 2006). In the postmarketing period from December 21, 2002, through June 30, 2005, 17 cases of TB (0.02/100 PY) had been reported in the United States, of which five had extrapulmonary involvement. These data are consistent with that of other TNF inhibitors (Lee and Kavanaugh 2005). However, because of the small risk of reactivation of latent TB, a detailed medical history should be taken, and screening tests (i.e., tuberculin skin test, chest radiography) should be performed in all patients prior to initiating adalimumab therapy in order to rule out active and inactive (latent) TB infection. European labeling includes a contraindication for adalimumab in patients with active TB. Appropriate anti-TB prophylaxis should be initiated for patients with latent TB prior to beginning adalimumab therapy. Moreover, the risks and benefits of adalimumab in these patients should be considered carefully prior to initiating therapy (EU Humira SPC 2006).

1.7.3.2 Opportunistic Infections

Opportunistic infections have been seen with the use of all TNF antagonists. These infections are seen very rarely, and causative agents vary. There have been four cases of histoplasmosis reported in RA trials, all in areas where the condition is endemic (Schiff et al. 2006). Patients should be monitored closely for infections before, during, and after treatment with adalimumab (EU Humira SPC 2006).

1.7.4 Other Conditions

1.7.4.1 Lymphoma

The incidence of lymphoma in the almost 2500 adalimumab-treated patients in the clinical development program for RA was consistent with the rates of lymphoma among patients with moderate to severe RA (Schiff et al. 2003b). After 12 506 PY of adalimumab exposure, 15 lymphomas were seen (0.12/100 PY) The standardized incidence ratio (SIR) for lymphomas in adalimumab clinical trials in comparison with that of the normal RA population in the Surveillance, Epidemiology, and End Results database was 3.19, which is consistent with SIRs reported for RA populations naïve to anti-TNF therapy (Schiff et al. 2006).

1.7.4.2 Demyelinating Conditions

Six cases of multiple sclerosis (MS), two of nonspecific demyelination, and two of Guillain–Barré syndrome have been reported after 12 506 PY of exposure to adalimumab (Schiff et al. 2005a). Patients with MS have a statistically significantly higher coexistence of RA, psoriasis, and goiter than matched controls, suggesting that patients with these conditions may innately be at increased risk of MS as compared with the general population (Heinzlef et al. 2000; Magnano et al. 2004). The true impact of TNF inhibitors on the development of this disorder is unknown, however (Lee and Kavanaugh 2005; Magnano et al. 2004).

1.7.4.3 Autoantibodies and Autoimmune Diseases

Between 3% and 12% of patients treated with adalimumab develop autoantibodies to antinuclear antigen and double-stranded DNA (Lee and Kavanaugh 2005). The mechanism by which autoantibodies are generated – and the clinical implications of these antibodies – remain to be defined, because progression to lupus-like illness appears to be uncommon (Lee and Kavanaugh 2005). After 12 506 PY of adalimumab exposure, 13 cases of systemic lupus erythematosus (SLE) and lupus-like syndromes have been reported. However, none of these cases had significant internal organ involvement (Schiff et al. 2006).

1.8 Special Conditions and Populations

1.8.1 Pregnancy and Lactation

The use of adalimumab in pregnant women has not been evaluated. Although animal data have shown no indication of maternal toxicity, embryotoxicity, or teratogenicity, adalimumab should be used during pregnancy only if clearly needed. European labeling recommends that adequate contraception be used for at least 5 months after the last adalimumab treatment (EU Humira SPC 2006).

Whether or not adalimumab is secreted in breast milk or absorbed systemically after ingestion has not been determined. However, because human immunoglobulins are secreted in breast milk, women should not breastfeed for at least 5 months after the last adalimumab treatment (EU Humira SPC 2006).

1.8.2 Elderly

No dose adjustment is required for adalimumab treatment in individuals aged 65 years or more (EU Humira SPC 2006).

1.8.3 Children and Adolescents

Adalimumab is currently not indicated for use in children (US Humira PI 2006; EU Humira SPC 2006). Investigations in juvenile rheumatoid arthritis (JRA) are currently underway (Lovell et al. 2004); the preliminary results of a study evaluating the use of adalimumab in JRA are presented in Section 1.10.4.

1.8.4 Impaired Renal and/or Hepatic Function

At present, no dosing recommendations are available for patients with impaired renal or hepatic function (EU Humira SPC 2006).

1.8.5 Congestive Heart Failure

In several clinical trials of patients with congestive heart failure (CHF), TNF inhibitors (infliximab, etanercept) failed to benefit patents with this condition (Lee and Kavanaugh 2005). Clinical studies of adalimumab in treating CHF have not been conducted. In some studies in other indications, however, TNF inhibitors (including adalimumab) have been associated with worsening and new-onset CHF (EU Humira SPC 2006). During adalimumab trials, 44 patients reported a medical history of CHF, and three (7%) of these reported CHF events during the trials. Thirty-two cases of CHF were observed among 10006 (0.3%) patients who did not report any medical history of CHF. Over the 2.5-year period when the rate of CHF was evaluated, the rate at which events occurred appears to have remained stable (Schiff et al. 2006).

A causal association between TNF inhibitors and CHF has not been proven. Nonetheless, because of the potential risk found with other TNF inhibitors, adalimumab is contraindicated in patients with severe heart failure (New York Heart Association class III/IV). Adalimumab should be discontinued in patients who develop new or worsening symptoms of CHF (EU Humira SPC 2006).

1.9 Storage and Administration

Adalimumab must be refrigerated at between 2°C and 8°C (36–46°F), but the formulation should not be frozen. Prefilled syringes should be protected from light (US Humira PI 2006).

Adalimumab is administered via SC injection. Patients may self-inject adalimumab if their physician determines that it is appropriate and with medical follow-up, as necessary, after proper training in injection technique. In order to reduce the incidence of ISRs, the injection sites should be rotated. Injections should not be given in areas of the skin that are tender, bruised, red, or hard (US Humira PI 2006).

1.10 Outlook and New Indications

Adalimumab has been established as a safe and effective treatment for RA, PsA, and AS, and it is currently being evaluated for use in several related indications, including psoriasis (phase III), Crohn's disease (phase III), and JRA (phase III) (Hanauer et al. 2006; Langley et al. 2005; Lovell et al. 2004; Sandborn et al. 2005a).

1.10.1 Psoriasis

In a randomized, double-blind, placebo-controlled, multicenter study to evaluate the efficacy and safety of adalimumab in patients with moderate to severe plaque psoriasis, two adalimumab dosing regimens were evaluated (Langley et al. 2005). A total of 148 patients received either adalimumab 80 mg at Week 0 followed by 40 mg EOW beginning at Week 1 (Group A), adalimumab 80 mg at Weeks 0 and 1 followed by 40 mg weekly beginning at Week 2 (Group B), or placebo for 12 weeks. At Week 12, 53%, 80%, and 4% of patients in Groups A, B, and placebo, respectively, achieved PASI 75. The mean percentage improvements in PASI scores over time are illustrated in Fig. 1.8. The percentage of patients determined by the PGA to be clear or almost clear at Week 12 were 49% and 76% in the two adalimumab treatment groups, compared with 2% with placebo. Adalimumab also produced clinically important and statistically significant improvements in the quality of life of patients with psoriasis, based on Dermatology Life Quality Index, SF-36, and EuroQOL 5D scores (Wallace et al. 2005a–d). Adalimumab was safe and well tolerated, with no significant differences in the incidence of adverse events across the three groups (Langley et al. 2005).

Following the 12 weeks of blinded, placebo-controlled treatment, patients were allowed to continue in an OLE phase. Those who had received placebo during the blinded phase, received adalimumab 80 mg followed by 40 mg EOW beginning at Week 12. Treatment response was sustained throughout the OLE phase for those who had been treated with adalimumab during the blinded phase. After 24 weeks of continuous adalimumab treatment, 67% and 77% of patients in Groups A and B, respectively, achieved PASI 75. Among the patients who converted from placebo to adalimumab at the end of Week 12, 55% achieved PASI 75 at Week 24. Some 69% and 84% of patients in Groups A and B, respectively, and 46% of the placebo/adalimumab group, attained PGA ratings of clear or almost clear at Week 24 (Langley et al. 2005).

1.10.2 Crohn's Disease

The safety and efficacy of adalimumab in patients with Crohn's disease has been evaluated in two randomized, double-blind, placebo-controlled, multicenter trials.

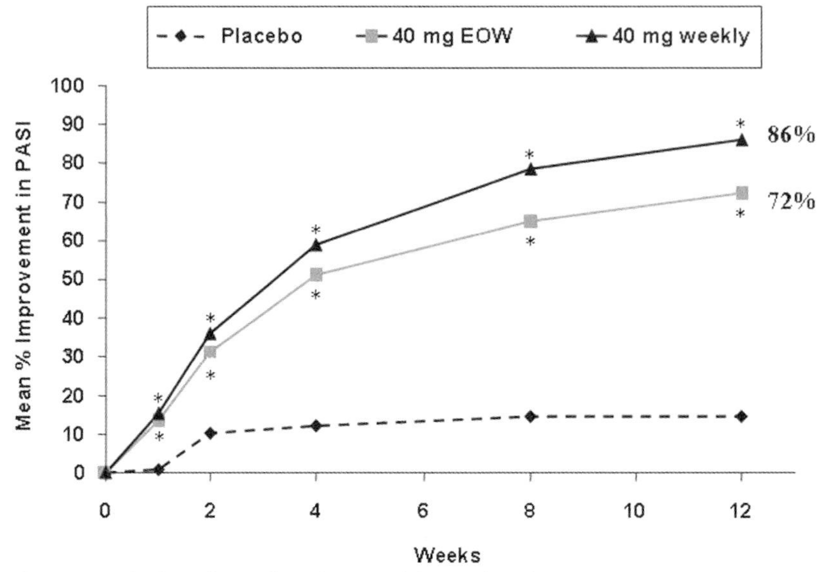

Fig. 1.8 Results from the study evaluating adalimumab in the treatment of moderate to severe chronic plaque psoriasis. Mean percentage improvement in Psoriasis Area and Severity Index (PASI) scores through 12 weeks of treatment using a modified intent-to-treat population. *$P < 0.001$ versus placebo. EOW = every other week. (From Langley et al. 2005.)

CLASSIC-I (Clinical assessment of Adalimumab Safety and efficacy Studied as Induction therapy in Crohn's disease) was a dose-ranging induction trial involving 299 patients with moderate to severe Crohn's disease naïve to anti-TNF therapy (Hanauer et al. 2006). Patients were randomized to one of four induction treatment groups: adalimumab 40 mg at Week 0/20 mg at Week 2 (Group A); 80 mg at Week 0/40 mg at Week 2 (Group B); 160 mg at Week 0/80 mg at Week 2 (Group C); or placebo at Weeks 0 and 2. Patients were followed through Week 4 (Hanauer et al. 2006).

Serum adalimumab concentrations increased proportionately with increasing dose. Adalimumab concentrations were similar at Weeks 2 and 4, indicating that the loading doses achieved stable serum levels by Week 2. Serum adalimumab concentrations achieved with the 160 mg/80 mg and 80 mg/40 mg doses were comparable to those achieved with 40 mg weekly and 40 mg EOW, respectively (Hanauer et al. 2006).

Rates of clinical remission (Crohn's disease activity index [CDAI] <150), a CDAI decrease from baseline of ≥70, and a CDAI decrease from baseline of ≥100 at Week 4 are shown in Fig. 1.9. The optimal induction dosing regimen in this study was adalimumab 160 mg at Week 0 followed by 80 mg at Week 2. The percentages of patients achieving remission at Week 4, the primary efficacy endpoint, in the adalimumab Groups A, B, C, and placebo were 18%, 24%, 36%, and 12%,

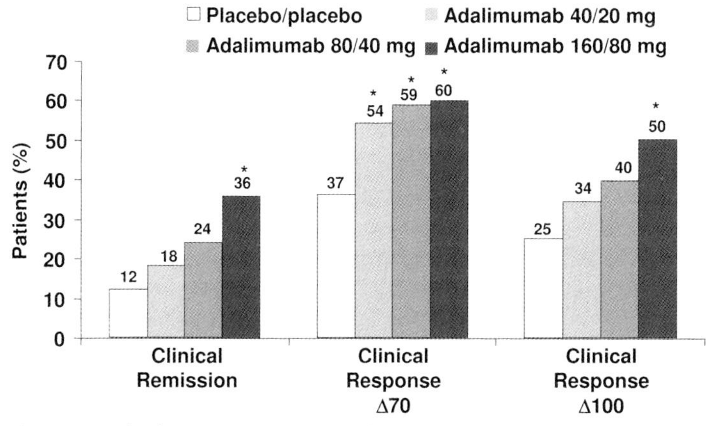

Fig. 1.9 Results from CLASSIC-I (CLinical assessment of Adalimumab Safety and efficacy Studied as Induction therapy in Crohn's disease). Clinical remission and response among adalimumab and placebo-treated patients at Week 4. $^{*}P < 0.05$ versus placebo. Clinical remission = Crohn's disease activity index (CDAI) <150. Clinical response Δ70/Δ100 = CDAI decrease from baseline ≥70 or ≥100. (From Hanauer et al. 2006.)

respectively. Percentages of patients achieving a CDAI decrease from baseline of ≥70 were 54% for Group A, 59% for Group B, 60% for Group C, and 37% for placebo. For a CDAI decrease from baseline of ≥100, the percentages were 34%, 40%, 50%, and 25%, for the respective treatment groups (Hanauer et al. 2006).

Adverse events occurred at similar frequencies in all four treatment groups, except that ISRs were more common among patients treated with adalimumab (26%, 24%, and 38% for Groups A, B, and C, respectively, versus 16% for the placebo group) (Hanauer et al. 2006).

Patients who completed the CLASSIC-I study were eligible to enroll in CLASSIC-II, a two-arm study (open-label and randomized arms) to evaluate adalimumab in the maintenance of clinical remission in subjects with Crohn's disease (Sandborn et al. 2005a). Fifty-five patients who were in remission at Weeks 0 and 4 were randomized to receive adalimumab 40 mg EOW, 40 mg weekly, or placebo for up to one year. The remaining subjects not in remission at both Weeks 0 and 4 received open-label adalimumab 40 mg EOW. The adalimumab dose could be increased to 40 mg weekly for flare or persistent nonresponse. In the open-label arm, the long-term efficacy and safety of adalimumab 40 mg EOW therapy was assessed in 220 patients with moderate to severe Crohn's disease. At 6 months, 71% remained on adalimumab 40 mg EOW (45%, n = 98) or weekly (26%, n = 58). Clinical remission was achieved in 33% of patients. The percentages of patients attaining CDAI decreases of 70 and 100 were 78% and 70%, respectively. Adverse events were mild to moderate in severity, and similar to those observed in previous studies of patients with RA (Sandborn et al. 2005a). Data available after one year supported the maintenance of remission, with clinical remission rates of 43% and clinical (CDAI) response changes of 70 and 100 in 69% and 61%

of patients, respectively (Sandborn et al. 2005b). Among the 55 patients in the randomized arm of the study, 83% of those receiving adalimumab 40 mg weekly, 68% receiving 40 mg EOW, and 39% receiving placebo maintained remission at one year (Sandborn et al. 2005c).

Similar to the data in RA regarding switching from a previous anti-TNF agent to adalimumab, preliminary evidence in patients with Crohn's disease also suggests that adalimumab is clinically beneficial in patients who have lost response or who are intolerant to infliximab (Barthel et al. 2005; Papadakis et al. 2005; Sandborn et al. 2004; Stallmach et al. 2004; Youdim et al. 2004). In one study of seven patients with Crohn's disease who had allergic reactions to infliximab, patients who had active disease and who had a previous response to infliximab (n = 6) also responded to adalimumab, as evidenced by improvement in the Harvey–Bradshaw index (HBI) and inflammatory markers (Youdim et al. 2004). In a second study, 15 patients with Crohn's disease who experienced an attenuated response to infliximab were treated with an adalimumab 80 mg loading dose following by 40 mg EOW for 6 months. Of the 13 patients who completed the study, seven had a complete response (HBI of ≤4 and withdrawal of corticosteroids), four had a partial response (decrease of ≥50% in HBI and reduction in corticosteroid dose), and two were nonresponders. In a third study of patients with Crohn's disease who had lost responsiveness or who had developed intolerance to infliximab and were switched to adalimumab, clinical remission (CDAI ≤150) was achieved by five of 17 patients, and a CDAI decrease of ≥70 was achieved by 10 of 17 patients at 12 weeks. In all of these studies, adalimumab was generally well tolerated, without producing signs or symptoms of allergic reactions (Papadakis et al. 2005; Sandborn et al. 2004; Youdim et al. 2004). These studies, although small in terms of patient numbers, suggest that subjects with Crohn's disease who have lost their response to, or are intolerant of, infliximab may benefit from switching to adalimumab.

1.10.3 Juvenile Rheumatoid Arthritis

Preliminary results from the initial open-label portion of a phase III, randomized, double-blind, placebo-controlled trial in children aged 4 to 17 years with JRA suggest that adalimumab provides rapid and substantial responses in these patients. Patients received 24 mg m^{-2} of adalimumab EOW, and concomitant MTX therapy was allowed. A total of 155 patients (81 receiving adalimumab plus MTX; 74 receiving adalimumab monotherapy) completed the 16-week open-label period. After 16 weeks of therapy, 95% of patients receiving adalimumab plus MTX achieved a pediatric ACR30, as did 88% of those receiving adalimumab monotherapy. ACR50 and ACR70 responses were achieved by 94% and 82%, respectively, of patients receiving the combination, and by 80% and 59%, respectively, of patients receiving adalimumab monotherapy. Adalimumab alone or in combination with MTX was generally safe and well tolerated during the 16-week interim study period (Lovell et al. 2004).

1.11 Conclusions

Despite its protective role in the immune response, TNF has been implicated in a number of chronic inflammatory disease states. Treatment with the fully human anti-TNF antibody, adalimumab, offers improvements in disease symptoms, as well as improvement in quality of life, functional status, and – most importantly – limits disease progression in a number of disease states. Although currently this agent is indicated for the treatment of RA, PsA, and AS, the safety and efficacy of adalimumab for other diseases, including plaque psoriasis, Crohn's disease and JRA, show great promise.

References

Adalimumab [PI; prescribing information] (2006) North Chicago, IL: Abbott Laboratories.

Adalimumab [SPC; summary of product characteristics] (2006) Queenborough, UK: Abbott Laboratories. Available at: http://www.emea.eu.int/humandocs/Humans/EPAR/humira/humira.htm. Accessed September 12, 2006.

Anderson, P.J. (2005) Tumor necrosis factor inhibitors: clinical implications of their different immunogenicity profiles. *Semin Arthritis Rheum* 34 (5 Suppl.1): 19–22.

Baert, F., Noman, M., Vermeire, S., et al. (2003) Influence of immunogenicity on the long-term efficacy of infliximab in Crohn's disease. *N Engl J Med* 348: 601–608.

Bansback, N., Brennan, A., and Anis, A.H. (2005) A pharmacoeconomic review of adalimumab in the treatment of rheumatoid arthritis. *Expert Rev Pharmacoeconomics Outcomes Res* 5: 519–529.

Barrera, P., Joosten, L.A., den Broeder, A.A., et al. (2001) Effects of treatment with a fully human anti-tumour necrosis factor α monoclonal antibody on the local and systemic homeostasis of interleukin 1 and TNFα in patients with rheumatoid arthritis. *Ann Rheum Dis* 60: 660–669.

Barthel, H.R., Gille, T., Halbsguth, A., Kramer, M. (2005) Successful treatment with adalimumab in infliximab-resistant Crohn's disease. *J Gastroenterol Hepatol* 20: 1464–1465.

Bombardieri, S., Moutsopoulos, H.M., McKenna, F., Michel, B., Webber, D.G., Kupper, H. (2004) Rapid response to adalimumab after first dose: the ReAct trial [abstract]. *Ann Rheum Dis* 63(Suppl.1): 261.

Bombardieri, S., Tzioufas, A.G., McKenna, F., et al. (2005) Adalimumab (HUMIRA) is effective in treating patients with rheumatoid arthritis who previously failed etanercept and/or infliximab in real-life clinical settings. *Arthritis Rheum* 52(Suppl.): S144.

Bombardieri, S., McKenna, F., Drosos, A., et al. (2006) Efficacy and safety of adalimumab (HUMIRA) in 899 patients with rheumatoid arthritis (RA) who previously failed etanercept and/or infliximab in clinical practice [abstract]. *Ann Rheum Dis* 65(Suppl. II): 178.

Braegger, C.P., Nicholls, S., Murch, S.H., Stephens, S., MacDonald, T.T. (1992) Tumour necrosis factor alpha in stool as a marker of intestinal inflammation. *Lancet* 339: 89–91.

Braun, J., Bollow, M., Neure, L., et al. (1995) Use of immunohistologic and in situ hybridization techniques in the examination of sacroiliac joint biopsy specimens from patients with ankylosing spondylitis. *Arthritis Rheum* 38: 499–505.

Breedveld, F.C., Weisman, M.H., Kavanaugh, A.F., et al. (2005) The efficacy and safety of adalimumab (Humira) plus methotrexate vs. adalimumab alone or methotrexate alone in the early treatment of rheumatoid arthritis: 1- and 2-year results of the PREMIER study [abstract]. *Ann Rheum Dis* 64(Suppl.III): 60.

Breedveld, F.C., Weisman, M.H., Kavanaugh, A.F., et al. for the PREMIER investigators (2006) The PREMIER study: a multicenter, randomized, double-blind clinical trial of combination therapy with adalimumab plus methotrexate versus methotrexate alone or adalimumab alone in patients with early, aggressive rheumatoid arthritis who had not had previous methotrexate treatment. *Arthritis Rheum* 54: 26–37.

Burmester, G.R., van de Putte, L.B.A., Rau, R., et al. (2003) Sustained efficacy of adalimumab monotherapy for more than four years in DMARD-refractory RA. *Ann Rheum Dis* 62(Suppl.1): 192–193.

Burmester, G.R., Monteagudo Sáez, I., Malaise, M.G., Canas de Silva, J., Webber, D.G., Kupper, H. (2004) Efficacy and safety of adalimumab (Humira) in European clinical practices: the ReAct trial [abstract]. *Ann Rheum Dis* 63(Suppl.1): 266.

Burmester, G.R., Monteagudo Sáez, I., Malaise, M.G., Kary, S., Kupper, H. (2005a) Adalimumab (Humira) is effective and safe in treating rheumatoid arthritis (RA) in real-life clinical practice: 1-year-results of the ReAct study [abstract]. *Arthritis Rheum* 52(Suppl.): S541–S542.

Burmester, G.R., Monteagudo Sáez, I., Malaise, M.G., et al (2005b) Adalimumab (Humira) is effective in patients who have previously been treated with TNF-antagonists (etanercept and/or infliximab) in widespread clinical practice: 12-week outcomes in the REACT Trial. *Ann Rheum Dis* 64(Suppl.III): 423–424.

Catrina, A.I., Trollmo, C., Klint, E., et al. (2005) Evidence that anti-tumor necrosis factor therapy with both etanercept and infliximab induces apoptosis in macrophages, but not lymphocytes, in rheumatoid arthritis joints: extended report. *Arthritis Rheum* 52: 61–72.

Cooper, W.O., Fava, R.A., Gates, C.A., Cremer, M.A., Townes, A.S. (1992) Acceleration of onset of collagen-induced arthritis by intra-articular injection of tumour necrosis factor or transforming growth factor-beta. *Clin Exp Immunol* 89: 244–250.

den Broeder, A.A., van de Putte, L.B.A., Rau, R., et al. (2002a) A single dose, placebo controlled study of the fully human anti-tumor necrosis factor-α antibody adalimumab (D2E7) in patients with rheumatoid arthritis. *J Rheumatol* 29: 2288–2298.

den Broeder, A.A., Joosten, L.A., Saxne, T., et al. (2002b) Long term anti-tumour necrosis factor alpha monotherapy in rheumatoid arthritis: effect on radiological course and prognostic value of markers of cartilage turnover and endothelial activation. *Ann Rheum Dis* 61: 311–318.

Emery, P., Schiff, M.H., Kalden, J.R., et al. (2004) Adalimumab (HUMIRA) plus methotrexate induces sustained remission in both early and long-standing rheumatoid arthritis. *Arthritis Rheum* 50(Suppl.): S183–S184.

Furst, D.E., Schiff, M.H., Fleischmann, R.M., et al. (2003) Efficacy and safety of the fully human anti-tumour necrosis factor-α monoclonal antibody, and concomitant standard antirheumatic therapy for the treatment of rheumatoid arthritis: results of STAR (Safety Trial of Adalimumab in Rheumatoid Arthritis). *J Rheumatol* 30: 2563–2571.

Georgopoulos, S., Plows, D., Kollias, G. (1996) Transmembrane TNF is sufficient to induce localized tissue toxicity and chronic inflammatory arthritis in transgenic mice. *J Inflamm* 46: 86–97.

Granneman, R.G., Zhang, Y., Noetersheuser, P.A., et al. (2003) Pharmacokinetic/pharmacodynamic (PK/PD) relationships of adalimumab (Humira, Abbott) in rheumatoid arthritis patients during Phase II/III clinical trials [abstract]. *Arthritis Rheum* 48 (Suppl.): S140.

Hanauer, S.B., Sandborn, W.J., Rutgeerts, P., et al. (2006) Human anti-tumor necrosis factor monoclonal antibody (adalimumab) in Crohn's disease: the CLASSIC-I Trial. *Gastroenterology* 130: 323–333.

Heinzlef, O., Alamowitch, S., Sazdovitch, V., et al. (2000) Autoimmune diseases in families of French patients with multiple sclerosis. *Acta Neurol Scand* 101: 36–40.

Jerne, N.K. (1974) Towards a network theory of the immune system. *Ann d'Immunologie* 125C: 373–389.

Kavanaugh, A., Grenwald, M., Zizic, T., et al. (2002) Treatment with adalimumab (D2E7) does not affect normal immune responsiveness [abstract]. *Arthritis Rheum* 45(Suppl.9): S132.

Kavanaugh, A., Cush, J.J., Matteson, E. ACR Hotline (2003) FDA meeting March 2003: update on the safety of new drugs for rheumatoid arthritis. Part II. CHF, infection, and other safety issues. Available at: http://www.rheumatology.org/publications/hotline/0803chf.asp?aud=mem. Accessed September 11, 2006.

Kaymakcalan, Z., Beam, C., Kamen, R., Salfeld, J. (2002) Comparison of adalimumab (D2E7), infliximab, and etanercept in the prevention of polyarthritis in a transgenic murine model of rheumatoid arthritis. *Arthritis Rheum* 46(Suppl.): S304.

Keystone, E.D., Kavanaugh, A.F., Sharp, J.T., et al. (2004a) Radiographic, clinical, and functional outcomes of treatment with adalimumab (a human anti-tumor necrosis factor monoclonal antibody) in patients with active rheumatoid arthritis receiving concomitant methotrexate therapy. *Arthritis Rheum* 50: 1400–1411.

Keystone, E.D., Kavanaugh, A.F., Perez, J.L., Spencer-Green, G.T. (2004b) Adalimumab (Humira) plus methotrexate provides sustained improvements in physical function over 2 years in treatment of patients with rheumatoid arthritis [abstract]. *Ann Rheum Dis* 64(Suppl.1): 278.

Keystone, E., Kavanaugh, A., Perez, J., Spencer-Green, G. (2004c) Adalimumab (Humira) plus methotrexate provides sustained improvements in physical function over 2 years of treatment in patients with rheumatoid arthritis [abstract]. *Ann Rheum Dis* 63(Suppl.1): 278.

Keystone, E.C., Kavanaugh, A.F., Sharp, J.T., et al. (2005) Inhibition of radiographic disease progression in patients with long-standing rheumatoid arthritis following 3 years of treatment with adalimumab (Humira) plus methotrexate [abstract]. *Ann Rheum Dis* 64(Suppl.III): 419.

Kristensen, L., Saxne, T., Geborek, P. (2004) Adherence to therapy of infliximab, etanercept, adalimumab, and anakinra in chronic arthritis patients treated in clinical practice in southern Sweden [abstract]. *Ann Rheum Dis* 64(Suppl.1): 263.

Landewé, R., Boers, M., Verhoeven, A., et al. (2002) COBRA combination therapy in patients with early rheumatoid arthritis: long-term structural benefits of a brief intervention. *Arthritis Rheum* 46: 347–356.

Langley, R., Leonardi, C., Toth, D., et al. (2005) Long-term safety and efficacy of adalimumab in the treatment of moderate to severe chronic plaque psoriasis [abstract]. *J Am Acad Dermatol* 52(3 Suppl.): 203.

Lee, D.M. and Weinblatt, M.E. (2001) Rheumatoid arthritis. *Lancet* 358: 903–911.

Lee, S.J. and Kavanaugh, A. (2005) Adalimumab for the treatment of rheumatoid arthritis. *Therapy* 2: 13–21.

Lovell, D.J., Ruperto, N., Goodman, S., et al. (2004) Preliminary data from the Study of Adalimumab in Children with Juvenile Idiopathic Arthritis (JIA). *Arthritis Rheum* 50(Suppl.): S436–S437.

Magnano, M., Robinson, W.H., Genovese, M.C. (2004) Demyelination and the use of TNF inhibition. *Clin Exp Rheumatol* 22(5 Suppl.35): S134–S140.

Maini, R.N. and Taylor, P.C. (2000) Anticytokine therapy for rheumatoid arthritis. *Annu Rev Med* 51: 207–229.

Maini, R.N., Breedveld, F.C., Kalden, J.R., et al. (1998) Therapeutic efficacy of multiple intravenous infusions of anti-tumor necrosis factor alpha monoclonal antibody combined with low-dose weekly methotrexate in rheumatoid arthritis. *Arthritis Rheum* 41: 1552–1563.

McGonagle, D., Conaghan, P.G., O'Connor, P., et al. (1999) The relationship between synovitis and bone changes in early untreated rheumatoid arthritis: a controlled magnetic resonance imaging study. *Ann Rheum Dis* 42: 1706–1711.

McQueen, F.M., Stewart, N., Crabbe, J., et al. (1998) Magnetic resonance imaging of the wrist in early rheumatoid arthritis reveals a high prevalence of erosions at four months after symptom onset. *Ann Rheum Dis* 57: 350–356.

Mease, P.J. (2002) Tumour necrosis factor (TNF) in psoriatic arthritis:

pathophysiology and treatment with TNF inhibitors. *Ann Rheum Dis* 61: 298–304.

Mease, P. and Goffe, B.S. (2005) Diagnosis and treatment of psoriatic arthritis. *J Am Acad Dermatol* 52: 1–19.

Mease, P.J., Gladman, D.D., Ritchlin, C.T., et al. for the ADEPT Study Group (2005a) Adalimumab for the treatment of patients with moderately to severely active psoriatic arthritis: results of a double-blind, randomized, placebo-controlled trial. ADEPT. *Arthritis Rheum* 52: 3279–3289.

Mease, P.J., Gladman, D.D., Ritchlin, C.T., et al. (2005b) Clinical efficacy and safety of adalimumab for psoriatic arthritis: 48-week results of ADEPT [abstract]. *Arthritis Rheum* 52(Suppl.): S215.

Mease, P., Sharp, J., Ory, P., et al. (2005c) Inhibition of joint destruction in PsA with adalimumab: 48-week results of ADEPT [abstract]. *Arthritis Rheum* 52(Suppl.): S631.

Mitoma, H., Horiuchi, T., Hatta, N., et al. (2005) Infliximab induces potent anti-inflammatory responses by outside-to-inside signals through transmembrane TNF-[945]. *Gastroenterology* 128: 376–392.

Möttönen, T., Hannonen, P., Korpela, M., et al. (2002) Delay to institution of therapy and induction of remission using single-drug or combination-disease-modifying antirheumatic drug therapy in early rheumatoid arthritis. *Arthritis Rheum* 46: 894–898.

Murch, S.H., Lamkin, V.A., Savage, M.O., Walker-Smith, J.A., MacDonald, T.T. (1991) Serum concentrations of tumour necrosis factor alpha in childhood chronic inflammatory bowel disease. *Gut* 32: 913–917.

Murch, S.H., Braegger, C.P., Walker-Smith, J.A., MacDonald, T.T. (1993) Location of tumour necrosis factor alpha by immunohistochemistry in chronic inflammatory bowel disease. *Gut* 34: 1705–1709.

Nikas, S.N., Voulgari, P.V., Alamanos, Y., et al. (2006) Efficacy and safety of switching from infliximab to adalimumab: a comparative controlled study. *Ann Rheum Dis* 65: 257–260.

Papadakis, K.A., Shaye, O.A., Vasiliauskas, E.A., et al. (2005) Safety and efficacy of adalimumab (D2E7) in Crohn's disease patients with an attenuated response to infliximab. *Am J Gastroenterol* 100: 75–79.

Partsch, G., Steiner, G., Leeb, B.F., Dunky, A., Broll, H., Smolen, J.S. (1997) Highly increased levels of tumor necrosis factor-α and other proinflammatory cytokines in psoriatic arthritis synovial fluid. *J Rheumatol* 24: 518–523.

Rau, R. (2002) Adalimumab (a fully human anti-tumour necrosis factor α monoclonal antibody) in the treatment of active rheumatoid arthritis: the initial results of five trials. *Ann Rheum Dis* 61(Suppl. III): ii70–ii73.

Ritchlin, C., Haas-Smith, S.A., Hicks, D., Cappuccio, J., Osterland, C.K., Looney, R.J. (1998) Patterns of cytokine production in psoriatic synovium. *J Rheumatol* 25: 1544–1552.

Salfeld, J., Kaymakcalan, Z., Tracey, D., Robert, A., Kamen, R. (1998) Generation of fully human anti-TNF antibody D2E7 [abstract]. *Arthritis Rheum* 41(Suppl.): S57.

Sandborn, W.J., Hanauer, S., Loftus, E.V. Jr., et al. (2004) An open-label study of the human anti TNF monoclonal antibody adalimumab in subjects with prior loss of response or intolerance to infliximab for Crohn's disease. *Am J Gastroenterol* 99: 1984–1989.

Sandborn, W., Hanauer, S., Lukas, M., et al. (2005a) Induction and maintenance of clinical remission and response in subjects with Crohn's disease treated during a 6-month open-label period with fully human anti-TNF-α monoclonal antibody adalimumab (Humira). *Gastroenterology* 128(4 Suppl.2): A-111.

Sandborn, W.J., Hanauer, S., Lukas, M., et al. (2005b) Clinical remission and response in patients with Crohn's disease treated with open-label for 1 year with fully human anti-TNF-α monoclonal antibody adalimumab. *Gut* 54(Suppl.VII): A18.

Sandborn, W.J., Hanauer, S., Enns, R., et al. (2005c) Maintenance of remission over 1 year in patients with active Crohn's disease treated with adalimumab: results of CLASSIC II, a blinded, placebo-controlled study. *Gut* 54(Suppl.VII): A81–A82.

Scallon, B.J., Moore, M.A., Trinh, H., Knigh, D.M., Ghrayeb, J. (1995) Chimeric anti-TNF-alpha monoclonal antibody cA2 binds recombinant transmembrane TNF-alpha and activates immune effector functions. *Cytokine* 7: 251–259.

Schiff, M., van de Putte, L.B., Breedveld, F.C., et al. (2003a) Rates of infection in adalimumab rheumatoid arthritis clinical trials [abstract]. *Ann Rheum Dis* 62(Suppl. I): 184.

Schiff, M.H., Chartash, E., Spencer-Green, G. (2003b) Malignancies in rheumatoid arthritis (RA) clinical trials with adalimumab (HUMIRA) [abstract]. *Arthritis Rheum* 48(9 Suppl.): S700.

Schiff, M.H., Weisman, M.H., Cohen, S.B., et al. (2004) Significant clinical improvement at 6 months are sustained over 4 years in patients with rheumatoid arthritis treated with adalimumab (Humira) plus methotrexate. *Arthritis Rheum* 50(Suppl.): S182–S183.

Schiff, M.H., Burmester, G.R., Kent, J., et al. (2005a) Safety of adalimumab in long-term treatment of patients with rheumatoid arthritis [abstract]. *J Am Acad Dermatol* 52(3 Suppl.): 203.

Schiff, M.H., Breedveld, F.C., Weisman, M. H., et al. (2005b) Adalimumab (Humira) plus methotrexate is safe and efficacious in patients with rheumatoid arthritis into the seventh year of therapy [abstract]. *Ann Rheum Dis* 64(Suppl. III): 438–439.

Schiff, M.H., Burmester, G.R., Kent, J.D., et al. (2006) Safety analyses of adalimumab (Humira) in global clinical trials with US postmarketing surveillance of patients with rheumatoid arthritis. *Ann Rheum Dis* 65: 889–894.

Shen, C., Assche, G.V., Colpaert, S., et al. (2005) Adalimumab induces apoptosis of human monocytes: a comparative study with infliximab and etanercept. *Aliment Pharmacol Ther* 21: 251–258.

Soden, M., Rooney, M., Cullen, A., Whelan, A., Feighery, C., Bresnihan, B. (1989) Immunohistological features in the synovium obtained from clinically uninvolved knee joints of patients with rheumatoid arthritis. *Br J Rheumatol* 28: 287–292.

Stallmach, A., Giese, T., Schmidt, C., Meuer, S.C., Zeuzem, S.S. (2004) Severe anaphylactic reaction to infliximab: successful treatment with adalimumab – report of a case. *Eur J Gastroenterol Hepatol* 16: 627–630.

Tsakonas, E., Fitzgerald, A.A., Fitzcharles, M.A., et al. (2000) Consequences of delayed therapy with second-line agents in rheumatoid arthritis: a 3 year follow up on the hydroxychloroquine in early rheumatoid arthritis HERA study. *J Rheumatol* 27: 623–629.

van der Bijl, A.E., Breedveld, F.C., Antoni, C.E., et al. (2005) Adalimumab (HUMIRA) is effective in treating patients with rheumatoid arthritis who previously failed infliximab treatment. *Ann Rheum Dis* 64(Suppl.III): 428.

van den Brande, J.M., Braat, H., van den Brink, G.R., et al. (2003) Infliximab but not etanercept induces apoptosis in lamina propria T-lymphocytes from patients with Crohn's disease. *Gastroenterology* 124: 1774–1785.

van der Heijde, D., Kivitz, A., Schiff, M.H., et al. for the Adalimumab Trial Evaluating Long-term Efficacy and Safety in Ankylosing Spondylitis Study Group (2006) Efficacy and safety of adalimumab in patients with ankylosing spondylitis: results of a randomized, placebo-controlled trial (ATLAS). *Arthritis Rheum* 54: 2136–2146.

van de Putte, L.B.A., Salfeld, J., Kaymakcalan, Z. (2003) Adalimumab. In: Moreland, L.W. Emery, P. (Eds.), *TNF-Inhibition in the Treatment of Rheumatoid Arthritis*. Martin Dunitz, London, pp. 71–93

van de Putte, L.B., Atkins, C., Malaise, M., et al. (2004) Efficacy and safety of adalimumab as monotherapy in patients with rheumatoid arthritis for whom previous disease modifying antirheumatic drug treatment has failed. *Ann Rheum Dis* 63: 508–516.

Wallace, K.L., Langley, R., Bissonnette, R., Melluli, L., Hoffman, R. (2005a) Quality of life in patients with psoriatic arthritis treated with adalimumab: subanalysis of studies in moderate to severe psoriasis [abstract]. *J Am Acad Dermatol* 52(3 Suppl.): 192.

Wallace, K.L., Gordon, K.B., Langley, M., et al. (2005b) Dermatologic quality of life in patients with moderate to severe plaque psoriasis receiving 48 weeks of adalimumab therapy [abstract]. *J Am Acad Dermatol* 52(3 Suppl.): 180.

Wallace, K.L., Bissonnette, R., Leonardi, C., et al. (2005c) Effects of adalimumab on health status as measured by EQ-5D in patients with moderate to severe plaque psoriasis [abstract]. *J Am Acad Dermatol* 52(3 Suppl.): 181.

Wallace, K.L., Leonardi, C., Kempers, S., et al. (2005d) General physical and mental health status in patients with moderate to severe plaque psoriasis receiving 48 weeks of adalimumab therapy [abstract]. *J Am Acad Dermatol* 52(3 Suppl.): 185.

Weinblatt, M.E., Keystone, E.C., Furst, D.E., et al. (2003) Adalimumab, a fully human anti-tumor necrosis factor α monoclonal antibody for the treatment of RA in patients taking concomitant methotrexate. The ARMADA trial. *Arthritis Rheum* 48: 35–45.

Weinblatt, M.E., Keystone, E.C., Furst, D.E., Kavanaugh, A.F., Cartash, E.K., Segurado, O.G. (2005) Long-term efficacy, remission, and safety of adalimumab (HUMIRA) plus methotrexate in patients with rheumatoid arthritis in the ARMADA trial [abstract]. *Arthritis Rheum* 52 (Suppl): 563.

Weinblatt, M.E., Keystone, E.C., Furst, D.E., Kavanaugh, A.F., Chartash, E.K., Segurado, O.G. (2006) Long-term efficacy and safety of adalimumab plus methotrexate in patients with rheumatoid arthritis: ARMADA 4-year extended study. *Ann Rheum Dis* 65: 753–759.

Wells, A.F., Kupper, H., Fischkoff, S., et al. (2003) Injection site reactions in adalimumab rheumatoid arthritis (RA) pivotal clinical trials [abstract]. *Ann Rheum Dis* 62(Suppl.I): 411.

Williams, R.O., Feldmann, M., Miani, R.N. (1992) Anti-tumour necrosis factor ameliorates joint disease in murine collagen-induced arthritis. *Proc Natl Acad Sci USA* 89: 974–978.

Youdim, A., Vasikiauskas, E.A., Targan, S.R., et al. (2004) A pilot study of adalimumab in infliximab-allergic patients. *Inflamm Bowel Dis* 10: 333–338.

Zhou, H., Buckwalter, M., Boni, J., et al. (2004) Population-based pharmacokinetics of the soluble TNFr etanercept: a clinical study in 43 patients with ankylosing spondylitis compared with post hoc data from patients with rheumatoid arthritis. *Int J Clin Pharmacol Ther* 42: 267–276.

2
Alemtuzumab (MabCampath)

Thomas Elter, Andreas Engert, and Michael Hallek

2.1
Introduction

Alemtuzumab (MabCampath) is a member of the Campath-1 family of monoclonal antibodies that recognize the CD52 glycoprotein on human lymphocytes. The development of alemtuzumab evolved from the detailed study of the immunologic properties of various rat monoclonal Campath-1 antibodies. Initially, these anti-CD52 antibodies were generated in the United Kingdom by investigators in the Cambridge University Department of Pathology (hence the name, Campath). At the time, the goal of these investigations was to target lymphocytes for removal from donor bone marrow prior to transplantation, in an effort to prevent graft-versus-host disease (GvHD) in transplant recipients.

The first Campath-1 antibody to be described was a rat monoclonal IgM antibody that could bind to T and B lymphocytes and some monocytes and fix human complement, while leaving stem cells intact (Hale et al. 1983a). In subsequent studies, Campath-1G antibodies of the IgG2b isotype, which also had properties for lymphocyte binding and complement fixation, were shown to have the added feature of inducing antibody-dependent cell-mediated cytotoxicity (ADCC) in human lymphocytes (Hale et al. 1985). This suggested that IgG2b might be an ideal candidate for depleting lymphocytes in patients with lymphoproliferative disorders. Data supporting this hypothesis were obtained from a preliminary study that compared IgM, IgG2a, and IgG2b to determine the variant that depleted lymphocytes most effectively. In patients with chronic lymphocytic leukemia (CLL), the IgM and IgG2a variants induced only a transient depletion of blood lymphocytes, whereas the IgG2b variant, termed Campath-1G, achieved a longlasting depletion of both peripheral blood and bone marrow lymphocytes (Dyer et al. 1989). Consequently, the rat monoclonal Campath-1G antibody was studied for further development.

An important limitation with rat monoclonal antibodies, however, is the development of human anti-murine antibody when administered repeatedly over longer periods of time. In order to minimize the antigenicity of Campath-1G and to optimize its potential as a therapeutic agent, Campath-1G was humanized by

Handbook of Therapeutic Antibodies. Edited by Stefan Dübel
Copyright © 2007 WILEY-VCH Verlag GmbH & Co. KGaA, Weinheim
ISBN 978-3-527-31453-9

inserting the hypervariable regions of the rat immunoglobulin into the human IgG1 framework – thus forming Campath-1H (alemtuzumab) (Riechmann et al. 1988). An initial pilot study in patients with non-Hodgkin's lymphoma showed significant responses to Campath-1H, with no detectable antiglobulin responses (Hale et al. 1988).

Alemtuzumab has since been studied in numerous clinical trials in a variety of diseases. It is currently indicated for the treatment of B-cell CLL in patients who have been treated with alkylating agents and who have failed fludarabine therapy (Campath-1H product information 2004).

2.2 Basic Principles

2.2.1 Antibody Features and Production

2.2.1.1 Features of Alemtuzumab (Campath-1H)

The humanization of Campath-1G was carried out by inserting the DNA sequences encoding the hypervariable regions (antigen-binding sites) of the rat IgG2 antibody to the human IgG1 isotype variable framework and constant domains. The human IgG1 isotype was selected based on its superiority to other isotypes in complement fixation, ADCC, and human IgG Fc receptor binding. The structure of the resultant humanized antibody, Campath-1H, is shown in Fig. 2.1 (Rai and Stephenson 2001). Campath-1H was shown to be as effective as Campath-1G at complement fixation, and to be more effective than Campath-1G at ADCC (Riechmann et al. 1988).

2.2.1.2 Antibody Production

The recombinant heavy-chain and light-chain DNAs of the antibody, containing both rat and human sequences as described above, were transfected into Chinese hamster ovary (CHO) cells to obtain clones to be used for the production of large quantities of Campath-1H via cell culture methods. Although the transfected cells are cultured in neomycin-containing medium, no detectable neomycin is present in the final solution of Campath-1H that is used clinically for injection (Campath-1H product information 2004).

2.3 Mechanism of Action

2.3.1 Molecular Target and Target Expression

The CD52 antigen, the target of the Campath-1 antibodies, contains 12 amino acids, with a large complex carbohydrate attached at residue 3 (asparagine) and a

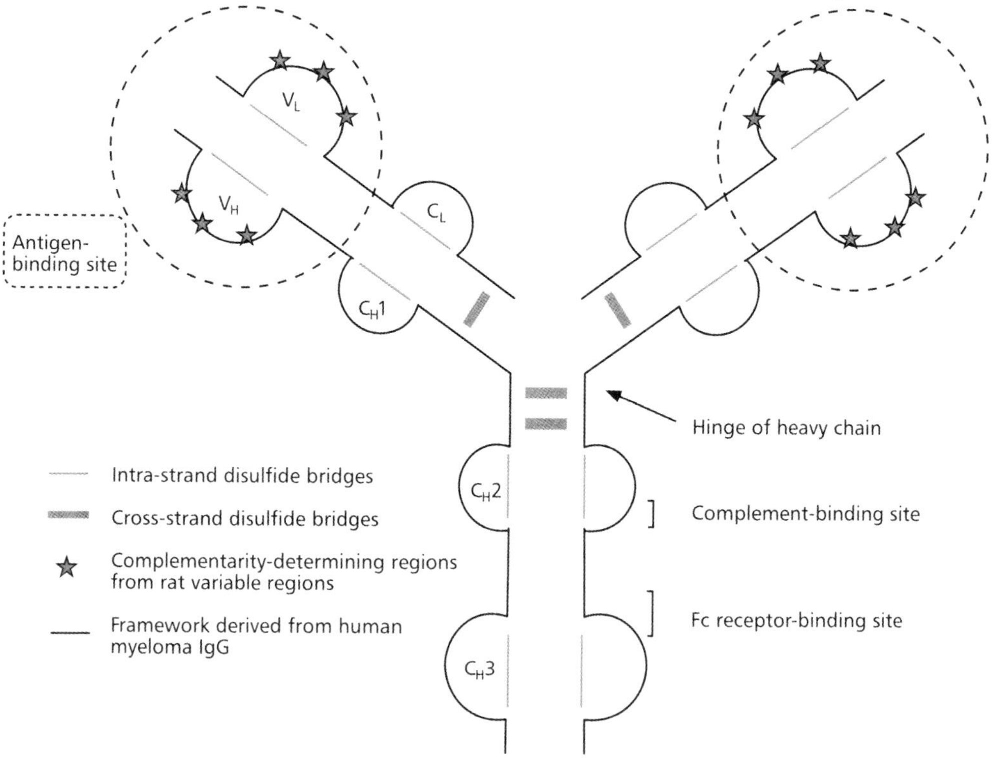

Fig. 2.1 Structure of the Campath-1H (alemtuzumab) monoclonal antibody. (Reproduced with permission from Rai and Stephenson 2001; © The Parthenon Publishing Group)

glycosylphosphatidylinositol (GPI) lipid anchor attached at the carboxy terminus. This anchor serves to maintain CD52 in the cell membrane, and portions of it – along with residues in the carboxy terminus – comprise the epitope recognized by Campath-1H (Fig. 2.2) (Hale 2001). Based upon structural analysis, it is thought that the proximity of this epitope to the cell surface contributes to the efficiency of complement-mediated cell lysis induced by Campath-1 antibodies (Xia et al. 1993).

The function of CD52 remains unknown. It is extensively expressed on lymphocytes, accounting for approximately 5% of the lymphocyte cell surface, and it is expressed at most stages of lymphocyte differentiation (Hale 2001); thus, CD52 may be an ideal target for the treatment of lymphoproliferative disorders. Its expression in the male reproductive tract occurs in the epithelial cells of the epididymis, vas deferens, and seminal vesicles; however, because it is shed from these cells, it is transferred to mature spermatozoa that pass through the genital tract. Nevertheless, no adverse effects of Campath-1H on reproductive function have been observed (Hale 2001; Kirchhoff 1996). Although CD52 is also expressed on monocytes, macrophages, and eosinophils, assessments of the effects of Campath-1H on bone marrow mononuclear cells and CD34(+) hematopoietic

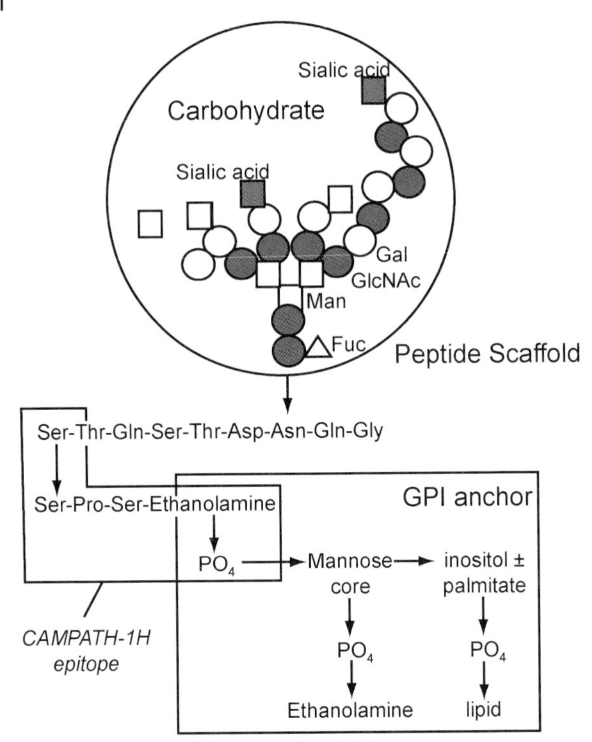

Fig. 2.2 Diagram of the CD52 antigen. CD52 is a glycoprotein comprising 12 amino acids with a complex carbohydrate attached to Asn-3. The protein is held in the outer layer of the cell membrane through a glycosylphosphatidylinositol (GPI)-lipid anchor attached to the C-terminus. (Reproduced with permission from Hale 2001.)

stem cells show that it does not reduce the number of progenitor cells, and importantly, does not affect their ability to proliferate, develop, or establish hematopoiesis (Gilleece and Dexter 1993).

In populations of normal lymphocytes, CD52 expression was shown to be higher on T than on B lymphocytes, and in patients with leukemia, expression was higher in those with T-prolymphocytic leukemia (T-PLL) than in those with B-cell CLL (Ginaldi et al. 1998). These differences in expression may potentially have a role in the variations observed in therapeutic response to Campath-1H among different patients and lymphoid malignancies; patients who responded to treatment with Campath-1H were found to have higher levels of CD52 expression than nonresponders (Ginaldi et al. 1998). The level of therapeutic response may also be influenced by the shedding of CD52 from lymphoid cells. In patients with CLL, high levels of soluble CD52 were found in plasma, and increased level of soluble CD52 correlated with disease severity and decreased survival outcomes. In the plasma of patients treated with alemtuzumab, soluble CD52 was found to form complexes with alemtuzumab, and results of *ex-vivo* experiments showed that patient plasma containing the soluble form of CD52 blocked the binding of alemtuzumab to CLL cells. Responders to therapy had lower plasma levels of soluble CD52 than nonresponders, prompting the investigators to suggest that levels of soluble CD52 be considered when planning treatment with alemtuzumab (Albitar et al. 2004).

Another consideration in the targeting of the CD52 antigen is the emergence of CD52-subclones. For example, although most cells of a high-CD52-expressing B-cell lymphoma line (Wien 133) incubated with alemtuzumab and crosslinked with anti-human IgG underwent growth inhibition and apoptosis, surviving cells expanded for 2 to 4 weeks were found to have low levels of cell surface CD52 expression. The growth of these cells was not inhibited by subsequent CD52 crosslinking, and analysis suggested that a defect in synthesis or attachment of the GPI anchor conferred resistance (Rowan et al. 1998). Clinically, three of 25 patients with rheumatoid arthritis (RA) treated with Campath-1H 25–80 mg intravenously daily for 5 to 10 days in a phase II clinical trial developed high levels of CD52(−) B and T cells, but the effect on B cells was transient (<3 months), and the persistence of elevated levels of CD52(−) T cells for at least 20 months did not interfere with the therapeutic effect of Campath-1H (Brett et al. 1996). A recent report described the emergence of CD52(−) T-cell subsets in patients with B-CLL who were receiving an 18-week course of alemtuzumab 3 to 30 mg subcutaneously as first-line therapy. These cells comprised 80% of all peripheral T cells at the end of treatment, declined gradually, but persisted 18 months after treatment cessation. Again, no relationship between the level of CD52(−) T cells and response to treatment was apparent (Lundin et al. 2004).

2.3.2 Mechanism of Cell Lysis

Alemtuzumab has been shown to induce cell death in CD52-positive cells by several different mechanisms. As mentioned above, the location of the CD52 alemtuzumab-binding epitope close to the cell surface enables efficient complement-dependent cellular toxicity. Complement consumption by Campath-1M (IgM) was first documented in early studies of monkeys and human patients with lymphoid malignancies (Hale et al. 1983b). More recently, *in-vitro* assays using alemtuzumab and freshly isolated human neoplastic cells demonstrated that, in the presence of human complement, alemtuzumab efficiently induced complement-dependent cytotoxicity in a variety of malignant cells including B-CLL, B-PLL, hairy cell leukemia, Burkitt's lymphoma, mantle cell lymphoma, and follicular lymphoma (Golay et al. 2004).

ADCC also has a role in cell lysis, as Campath-1G (Dyer et al. 1989) and the human IgG1 portion of alemtuzumab have been shown to mediate ADCC (Greenwood et al. 1993). In addition, in a murine model of adult T-cell leukemia (ATL) that did not contain human complement, it was established that the Fc receptor was required for tumor cell killing by alemtuzumab (Zhang et al. 2003).

Moreover, in cultures of CLL patient-derived peripheral blood mononuclear cells, cross-linking of the Fc region of alemtuzumab with an anti-Fc antibody under complement-free conditions enhanced the low levels of apoptosis induced by alemtuzumab alone, produced cell clustering, and stimulated the production of pro-apoptotic proteins (Nückel et al. 2005). These observations suggested that ADCC was involved in the cell death induced by alemtuzumab, in part, via a caspase-dependent apoptotic pathway.

Induction of apoptosis by alemtuzumab appears to be mediated via a caspase-independent pathway, which may provide an advantage in patients resistant to therapeutic agents that induce apoptosis via classic pathways (Nückel et al. 2005; Rowan et al. 1998; Stanglmaier et al. 2004). As described above, alemtuzumab-induced apoptosis *in vitro* is enhanced significantly by the addition of crosslinking antibodies (Nückel et al. 2005; Rowan et al. 1998; Stanglmaier et al. 2004).

2.3.3 Immunogenicity and Antiglobulin Response

Antiglobulin responses, which limit the repeated use of murine monoclonal antibodies in transplant applications, were primarily a concern with the rat-derived Campath-1 antibodies. Given that alemtuzumab is a humanized version of a rat monoclonal antibody, this concern has been greatly reduced. For example, one study in kidney transplant patients found that while 15 of 17 patients had antiglobulin responses to the rat monoclonal antibody Campath-1G, none of 12 patients had detectable antiglobulin responses to alemtuzumab (Rebello et al. 1999). An analysis of antiglobulin responses was performed in two separate clinical trials that evaluated intravenous (IV) or subcutaneous (SC) administration of alemtuzumab in patients with CLL (Hale et al. 2004). In 30 patients who had previously failed therapy with alkylating agents and fludarabine, IV administration of alemtuzumab thrice weekly for up to 12 weeks produced no detectable antigloblulin responses (Hale et al. 2004). In another trial in which an extended dosing regimen of SC alemtuzumab was administered (three times weekly for up to 18 weeks) as first-line treatment, only a limited occurrence of antiglobulin responses to alemtuzumab (in two of 32 patients) was documented (Lundin et al. 2002). Although these responses interfered with treatment efficacy, they were not associated with any serious adverse effects (Hale et al. 2004).

2.4 Clinical Studies with Alemtuzumab

2.4.1 Pharmacokinetic Studies

In the bone marrow transplant (BMT) setting, the evaluation of IV alemtuzumab 10 mg day^{-1} administered to patients with chronic myelogenous leukemia on either a 5-day schedule (from 10 to 5 days prior to BMT) or a 10-day schedule (from 5 days prior to 4 days after BMT) showed that mean peak serum concentrations were $2.5\,\mu g\,mL^{-1}$ after 5 days (5-day schedule) and $6.1\,\mu g\,mL^{-1}$ after 10 days (10-day schedule). The terminal half-lives were 21 and 15 days, respectively (Rebello et al. 2001). In patients with CLL the indicated dosing regimen of alemtuzumab is an initial 2-h IV infusion with 3 mg daily which, when tolerated, should be increased to 10 mg daily. When the 10-mg dose is tolerated, a final increase is made to a maintenance dose of $30\,mg\,day^{-1}$, three times per week on alternate days, for up to 12 weeks. The mean half-life of alemtuzumab obtained in patients with CLL following this schedule was 11 h after the first 30-mg dose, and 6 days after the last 30-mg dose; thus, alemtuzumab displayed nonlinear elimination kinetics (Campath-1H product information 2004). When this dosing schedule was used in CLL patients who had previously failed therapy with alkylating agents and fludarabine, the mean cumulative dose required to reach the therapeutic serum level of $1.0\,\mu g\,mL^{-1}$ was 90 mg (Fig. 2.3). In contrast, for patients in the study of SC alemtuzumab (see Section 2.3.3), the mean cumulative dose required to reach a serum level of $1.0\,\mu g\,mL^{-1}$ was 551 mg (Hale et al. 2004). In patients with relapsed/refractory CLL treated with IV alemtuzumab, the clinical response was positively correlated with the maximum trough concentration of alemtuzumab in patient serum samples (Fig. 2.4A) (Hale et al. 2004). Moreover, throughout 12 weeks of therapy, the mean trough concentration of alemtuzumab was higher among responding patients (defined as those achieving <0.4% CLL cells in the bone marrow) compared to that of nonresponding patients (Fig. 2.4B).

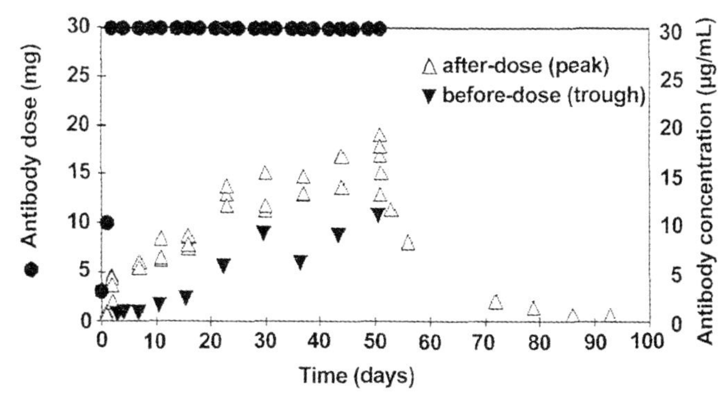

Fig. 2.3 Serum concentrations of alemtuzumab (intravenous administration) in a patient treated for relapsed/refractory chronic lymphocytic leukemia (CLL). Following dose escalation from 3 mg to 30 mg in Week 1, the patient received 30 mg alemtuzumab three times per week for 8 weeks. ● indicates the dose of alemtuzumab administered during the course of therapy. (Reproduced from Hale et al. 2004; © American Society of Hematology.)

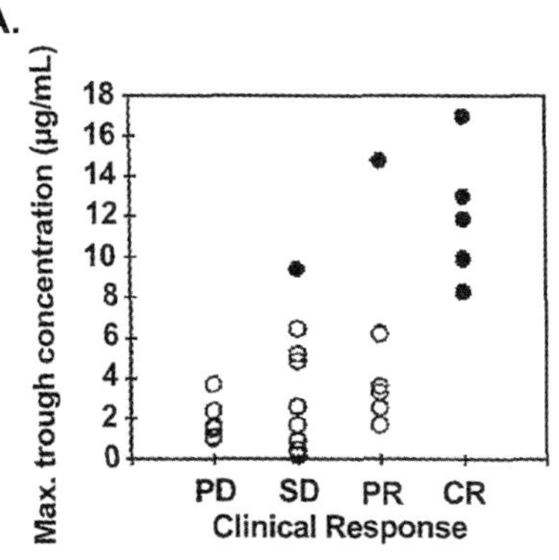

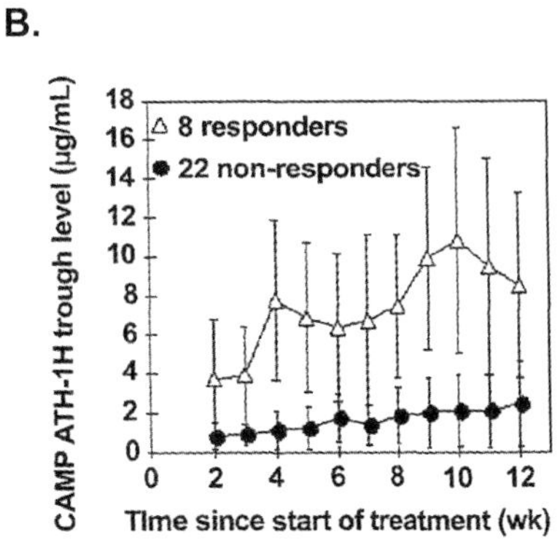

Fig. 2.4 Association between serum concentrations of alemtuzumab and response to therapy in patients with relapsed/refractory chronic lymphocytic leukemia (CLL) treated with intravenous alemtuzumab. (A) Maximum trough concentration of alemtuzumab and clinical response as assessed by NCI-WG criteria. Increase in response correlated with higher trough concentrations ($p = 0.006$; Kruskal-Wallis test). ● indicates patients with minimal residual disease (MRD)-negative response defined as <0.1% CLL cells in the bone marrow. Patients with MRD-negative response had significantly higher maximum trough concentrations compared to those with a MRD-positive response ($p < 0.0001$, Kruskal-Wallis test). (B) Mean trough concentrations of alemtuzumab during 12 weeks of intravenous administration. Responders were patients who achieved <0.4% CLL cells in the bone marrow. PD = progressive disease; SD = stable disease; PR = partial response; CR = complete response. (Reproduced from Hale et al. 2004; © American Society of Hematology.)

2.4.2 Chronic Lymphocytic Leukemia (CLL)

2.4.2.1 Relapsed/Refractory CLL

The safety and activity of alemtuzumab were initially evaluated in a phase II study of 29 patients with relapsed (n = 8) or refractory (n = 21) CLL. Most patients (76%) had advanced disease, and 72% had previously received two or more lines of therapy. Alemtuzumab was administered as a 2-h IV infusion, at an initial dose of 3 or 10 mg, based on patient characteristics. The target dose was 30 mg, given three times weekly for up to 12 weeks (Österborg et al. 1997). Responses to therapy were evaluated using the National Cancer Institute Working Group (NCIWG) criteria (Cheson et al. 1988). Alemtuzumab showed significant activity in this patient population, with an overall response rate (ORR) of 42% [4% complete response (CR), 38% partial response (PR)] and a median duration of response of 12 months. Rapid elimination of malignant cells was achieved in the blood (in 1–2 weeks) of 97% of patients, elimination of malignant cells was achieved in the bone marrow (in 6–12 weeks) of 36% of assessable patients, and splenomegaly resolved in 32%. In contrast, activity in the lymph nodes was limited, particularly for patients with bulky lymphadenopathy. Only 17% of patients had progressive disease (PD) after treatment. Of the 12 responders, nine were refractory to alkylating agents, and three had relapsed after an initial response to prior chemotherapy. All 12 weeks of the treatment period were completed by 12 patients; 10 had withdrawn from therapy due to lack of response, two after 6 weeks of therapy due to stable plateaus after achieving PR, four due to serious infections, and one withdrawal was of informed consent (Österborg et al. 1997). A brief summary of response rates in recent clinical trials of single-agent alemtuzumab in patients with CLL is provided in Table 2.1.

Safety evaluations showed moderate hematologic toxicity. Because some patients had low platelet and/or neutrophil counts at the start of therapy, the overall incidence of grade 4 thrombocytopenia and neutropenia was 24% and 20%, respectively. The incidence of grade 4 myelosuppression that developed during therapy included thrombocytopenia in 7% and neutropenia in 10% of patients. Grade 3/4 anemia developed in 38% of patients while receiving therapy. Nonhematologic toxicities included fever and rigors during the dose-escalation phase, rash, nausea, diarrhea, and hypotension. None of the toxicities was Grade 4, though one case of hypotension was Grade 3. Opportunistic infections were the main nonhematologic toxicity encountered, attributed largely to the longlasting lymphocytopenia induced by alemtuzumab. Infections included herpes simplex virus (HSV) reactivation, oral candidiasis, pneumonia, and septicemia; one case of pneumonia and two cases of septicemia were grade 3, and two cases of septicemia were grade 4. No deaths occurred either during treatment or within 6 months of follow-up (Österborg et al. 1997).

Further evaluation of the safety and efficacy of alemtuzumab was made in a pivotal, multicenter phase II trial of 93 patients with relapsed or refractory CLL. Heavily pretreated patients (median of three prior lines of therapy) who had failed prior therapy with alkylating agents and fludarabine received IV alemtuzumab 30 mg three times weekly for up to 12 weeks, using the same dose-escalation schedule described above, but with a starting dose of 3 mg for all patients (Keating et al. 2002b). Responses to therapy were evaluated using the revised guidelines of the 1996 NCIWG (Cheson et al. 1996). As in the above-described study, treatment with alemtuzumab resulted in a rapid elimination of malignant cells from the bone marrow (Fig. 2.5) (Keating et al. 2002b). The ORR was 33% (2% CR, 31% PR); in those patients who had never responded to fludarabine, the ORR was 29%, versus 38% in those who had previously responded. The median duration of response was 8.7 months. Median overall survival (OS) was 16 months for all patients, and 32 months for those who showed a response (Fig. 2.6) (Keating et al. 2002b). This result is particularly notable given the poor survival outcomes in patients who have become refractory to fludarabine. In a historical series of patients with CLL receiving first salvage therapy (with a variety of agents) for fludarabine-refractory disease, median OS was approximately 9–10 months (Keating et al. 2002c).

Alemtuzumab treatment resulted in the resolution of lymphocytosis (reduced to <30%) in 83% of all patients and in 93% of responders, while normal bone marrow biopsies were obtained in 26% of all patients and 48% of responders. In addition, 47% of patients with lymphadenopathy at baseline demonstrated at least

Table 2.1 Summary of response rates with single-agent alemtuzumab in chronic lymphocytic leukemia (CLL).

No. of patients	*Disease status*	*Route of administration*	*Median no. of prior therapies (range)*	*CR [%]*	*ORR [%]*	*Median response duration, months (range)*	*Reference*
29	Relapsed or refractory	IV	72% received ≥2 prior therapies	4	42	12(6–25+)	Österborg et al. (1997)
24	Primarily relapsed or refractory	IV	3 (1–8)	0	33	15.4(4.6–38+)	Rai et al. (2002)
93	Fludarabine-refractory	IV	3 (2–7)	2	33	8.7 (2.5–22.6+)	Keating et al. (2002b)
42	Advanced or refractory	IV	3 (1–9)	5	31	18 (range NR)* for CR patients	Ferrajoli et al. (2003)
36	Primarily advanced, fludarabine-refractory	IV	3 (1–12)	6	31	10 (3–36)	Lozanski et al. (2004)
91	Relapsed or refractory	IV/SC	3 (1–8)	36	54	TFS not reached for MRD (−) CR; 20 mo. for MRD (+) CR; 13 mo. for PR	Moreton et al. (2005)
41	Previously untreated	SC	None	19	87	TTF 18+ (7–44+)	Lundin et al. (2002)

* Among all responders in this study; the study included patients with various lymphoproliferative disorders (n = 78) including CLL (n = 42), T-cell prolymphocytic leukemia (n = 18), cutaneous T-cell lymphoma (n = 6), and others (n = 12).
CR = complete response; IV = intravenous; NR = not reported; ORR = overall response rate; SC = subcutaneous; TFS = treatment-free survival; TTF = time-to-treatment failure.

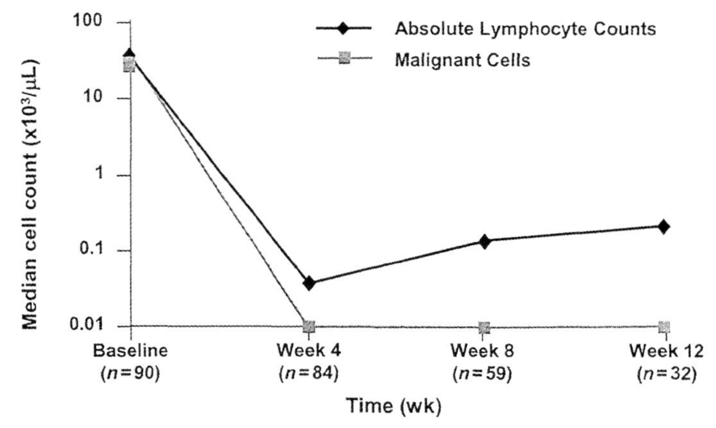

Fig. 2.5 Median number of malignant cells (CD19+/CD5+ cells) in peripheral blood during treatment with alemtuzumab. (Reproduced with permission from Keating et al. 2002b; © American Society of Hematology.)

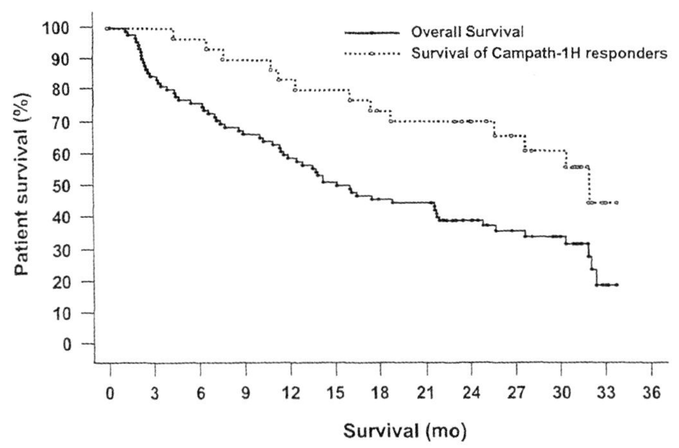

Fig. 2.6 Overall survival in patients with fludarabine-refractory (or fludarabine-failed) chronic lymphocytic leukemia (CLL) treated with alemtuzumab. The median survival was 16 months (95% CI: 11.8–21.9 months) for the intent-to-treat population (n = 93). The median survival for patients responding to treatment (n = 31) was 32 months. (Reproduced with permission from Keating et al. 2002b; © American Society of Hematology.)

a 50% reduction in enlarged nodes after treatment. However, as observed in previous studies, patients with bulky lymphadenopathy (lymph nodes >5 cm) were less likely to achieve resolution of disease in the lymph nodes. Among patients who had hepatomegaly at enrollment, resolution was achieved in 52%, and of those with splenomegaly, resolution was achieved in 54%. In terms of other measures of clinical benefit assessed at the end of treatment, 76% of all patients experienced resolution of B-CLL symptoms or fatigue, 55% achieved resolution of massive splenomegaly, 49% improved in performance score, and 76% experienced an improvement in anemia. The treatment course was completed according to the protocol by 65 patients, while 28 discontinued prematurely.

The toxicity of alemtuzumab in these severely myelosuppressed patients was acceptable, although some patients required dosing delays until the toxicities were managed. The discontinuation rate due to adverse events related to treatment was 24%. Hematologic toxicities while receiving alemtuzumab included neutropenia and thrombocytopenia. However, 82% of patients had hematologic abnormalities (neutropenia, thrombocytopenia, anemia) at enrollment, and by the 2-month follow-up visit 55% of these patients had improvements in at least one of those abnormalities. No significant shifts of IgG from baseline were noted. Nonhematologic toxicities during dose escalation included fever (17% Grade 3, 3% Grade 4), rigors (14% Grade 3), nausea (all Grade 1 or 2), vomiting (1% Grade 3), and rash (all Grade 1 or 2), which usually decreased in incidence by Week 2 of treatment. Other Grade 3/4 adverse events were dyspnea (28%), hypotension (17%), and hypoxia (3%).

Infections were the most common reason for discontinuation. Grade 3/4 infections, such as septicemia and cytomegalovirus (CMV) reactivation, occurred in 27% of all patients and were significantly more common in alemtuzumab nonresponders than in responders (36% versus 10%, respectively; $P < 0.01$). However, 53% of patients had a history of infection, and 33% had infections during the month prior to alemtuzumab treatment initiation. Nine deaths occurred during the study or within 30 days after treatment (five of these were due to infection and considered related to treatment), 19 occurred between days 30 and 180 after treatment, and 35 occurred more than 180 days after treatment (Keating et al. 2002b). The results of this trial led to the approval of alemtuzumab for the treatment of B-cell CLL in patients who have been treated with alkylating agents and in whom fludarabine therapy had failed.

A smaller phase II multicenter trial of alemtuzumab was conducted in 23 patients with advanced CLL and one patient with T-PLL, all of whom had previously received fludarabine and other regimens (Rai et al. 2002). Alemtuzumab was administered intravenously, starting with an initial dose of 10 mg daily and dose escalation to 30 mg, and a dose of 30 mg thrice weekly was then continued for a maximum of 16 weeks. Treatment response was evaluated using the 1996 NCIWG criteria. Investigators were encouraged to monitor patients for antiglobulin responses every 4 weeks during the treatment period and 28 days after treatment cessation. Patients were assessed monthly for 6 months post treatment and every 3 months thereafter. As in the pivotal phase II trial, the ORR was 33% (all PR). The median duration of response was 15.4 months. Median OS was 27.5 months for all patients, and 35.8 months for responders. Elimination of malignant cells was achieved in the blood of 75% of patients, and in the bone marrow of 37%. In patients with splenomegaly or hepatomegaly at baseline, splenomegaly was resolved in 38% and hepatomegaly in 50%. The platelet counts improved significantly from baseline to end of treatment, and continued to increase during follow-up (Rai et al. 2002).

The discontinuation rate due to adverse events was 37.5%. Hematologic toxicities were present in several patients at baseline; 20.8% had Grade 3/4 neutropenia and 41.7% had Grade 3 thrombocytopenia. During treatment, the rate of neutropenia increased to 59.1% but returned to near-baseline levels by the 2-month follow-up visit. Grade 3/4 nonhematologic toxicities included fever, rigors, and vomiting (16.7% for each), but these decreased in incidence after the first week of treatment. Grade 3 dyspnea was reported in one patient. Opportunistic infections (any grade) occurred in 41.7% of patients; similar to the pivotal trial, most of these incidences were observed in alemtuzumab nonresponders. Infections led to five treatment discontinuations and two deaths during treatment or within 35 days of treatment cessation. Of 10 patients evaluated for antiglobulin responses (total of 53 samples), one patient had a low-titer response after 7 weeks of treatment (one sample) (Rai et al. 2002).

In a phase II, single-site trial of patients with advanced CLL (n = 42), T-PLL (n = 18), or other chronic lymphoproliferative disorders (n = 18), treatment with alemtuzumab also demonstrated promising activity (Ferrajoli et al. 2003). Most patients were refractory to multiple therapies (median three prior lines of therapy); 55% of patients with CLL were refractory to alkylating agents and 55% to fludarabine, while 40% were refractory to both. The treatment protocol was similar to that used in the pivotal phase II trial, with an initial dose of 3 mg escalated to a final dose of 30 mg three times weekly for a maximum of 12 weeks of treatment. Responses to therapy were again evaluated using the 1996 NCIWG criteria. In those patients with NHL, treatment response was evaluated according to published NCIWG guidelines for NHL (Cheson et al. 1999). The ORR was 35% (13% CR, 22% PR), and the median duration of response was 18 months for patients who achieved a CR and 7 months for those who achieved a PR. The ORR in the subpopulation of patients with CLL (n = 42) was 31% (4% CR, 25% PR, 2% nodular PR). In those refractory to fludarabine, the ORR was 26%, and in those sensitive to fludarabine, the ORR was 37%. Patients with T-PLL showed a high ORR of 55% (44% CR, 11% PR) (Ferrajoli et al. 2003). Elimination of malignant cells was achieved in the blood of 84% of patients and in the bone marrow of 49%. Over half of all patients had a greater than 50% improvement with hepatomegaly and splenomegaly.

Grade 3/4 hematologic toxicities were transient neutropenia (34%) and thrombocytopenia (41%). As in the other trials, infusion-related adverse events (fever, rigors, skin rash, nausea, dyspnea, hypotension, and headache) were common, but decreased in incidence over time (Ferrajoli et al. 2003). Dyspnea of Grade 3/4 was observed in 11%; other toxicities were Grade 3/4 in 1% each (fever, rigors, hypotension, headache) or showed no Grade 3/4 toxicity (rash and nausea). Three patients with T-cell malignancies developed cardiovascular toxicities, and two of these patients discontinued alemtuzumab treatment. Infections and fever of unknown origin were common (experienced by 46% of all patients and 71% of those with CLL), with CMV reactivation being the most common infection reported (20% of all patients). Pneumonia occurred in 13% of patients, and two died of progressive pneumonia. Septicemia was also responsible for two deaths (Ferrajoli et al. 2003).

Because alemtuzumab demonstrated efficacy in several trials of patients with CLL refractory to conventional treatments, its activity was investigated further in other difficult-to-treat or high-risk populations of CLL patients. Mutations or deletions in the *p53* gene are predictive of a poor response to conventional therapy for CLL. Thus, it was important to determine whether alemtuzumab would provide benefits in patients with CLL who had *p53* abnormalities. A study using cryopreserved cells from previously treated CLL patients (n = 36), a majority (81%) of whom were refractory to fludarabine and who had received a median of three prior therapies, identified 15 patients (42%) who had *p53* mutations or deletions (Lozanski et al. 2004). Alemtuzumab was administered according to the indicated dosing schedule for up to 12 weeks, and responses were assessed using the 1996 NCIWG guidelines. While the ORR was 31% for all patients (6% CR, 25% PR), it was 40% (all PR) for patients with *p53* abnormalities. In all patients, the median duration of response was 10 months, and in those with *p53* abnormalities it was 8 months (Lozanski et al. 2004). Thus, alemtuzumab may be a promising therapeutic option for the subgroup of CLL patients with *p53* aberrations.

2.4.2.2 Minimal Residual Disease in CLL

The NCI-WG criteria (Cheson et al. 1996) is the current standard for assessing response to therapy in patients with CLL, and remains an important guideline for both the clinical trials setting and routine clinical practice. However, the NCI-WG definition for CR allows for <30% lymphocytes in the bone marrow, which may still harbor substantial levels of malignant cells, likely leading to disease relapse. Using a highly sensitive four-color flow cytometric assay, almost 25% of CLL patients achieving CR by NCI-WG criteria have been shown to have minimal residual disease (MRD), defined as >0.05% CLL cells among bone marrow leukocytes (Rawstron et al., 2001). MRD levels have been shown to be an important prognostic indicator; MRD-positive status (>0.05% CLL cells) is predictive of significantly decreased event-free survival and overall survival compared with MRD-negative response in patients with CLL (Rawstron et al. 2001).

In light of these findings, investigators in a recent clinical trial with alemtuzumab aimed to eradicate MRD in patients (n = 91) with relapsed/refractory CLL who had received a median of three prior therapies (Moreton et al. 2005). The majority of patients had received prior therapy with purine analogs, and 48% were refractory. After standard dose escalation, patients received alemtuzumab 30 mg (IV, n = 84; SC, n = 7) three times weekly until maximum response. Blood and bone marrow were examined by flow cytometry before, during, and after treatment for evidence of CLL and MRD negativity. Responses were assessed using the 1996 NCIWG criteria. Patients who had a CR by these criteria but were MRD-positive by four-color flow cytometry were designated as MRD-positive CR (Moreton et al. 2005). After a median of 9 weeks of treatment, the ORR was 54% (35% CR, 19% PR). MRD was eradicated in 20% of patients (MRD-negative CR);

15% had an MRD-positive CR. Patients achieving MRD-negative status had a significantly longer median OS compared with patients who achieved an MRD-positive CR, PR, or NR ($P = 0.0007$) (Fig. 2.7) (Moreton et al. 2005). Median treatment-free survival also was significantly longer among patients with MRD-negative CR ($P < 0.0001$), further illustrating the potential survival advantage of eradicating MRD. The results of this study also confirmed the effectiveness of alemtuzumab in patients with fludarabine-refractory disease. In this subgroup of difficult-to-treat patients, the ORR was 50% (27% CR, 23% PR), and MRD was eradicated in 18% (Moreton et al. 2005).

Common hematologic toxicities were Grade 3 and 4 neutropenia (18% and 30%, respectively) and Grade 3/4 thrombocytopenia (46%). Infusion-related adverse events were most frequently Grade 1 or 2 in severity, and they decreased in frequency after Week 3. The most common events were rigors and fever (76% overall, 13% Grade 3/4). Major infections (Grade 3/4), including pulmonary infection (seven patients), febrile neutropenia (four patients), herpesvirus infection (two patients), CMV reactivation (six patients), and fungal infection (three patients) occurred during therapy or within 1 month of completing therapy. CMV reactivation developed in eight patients and was fatal in one case (Moreton et al. 2005).

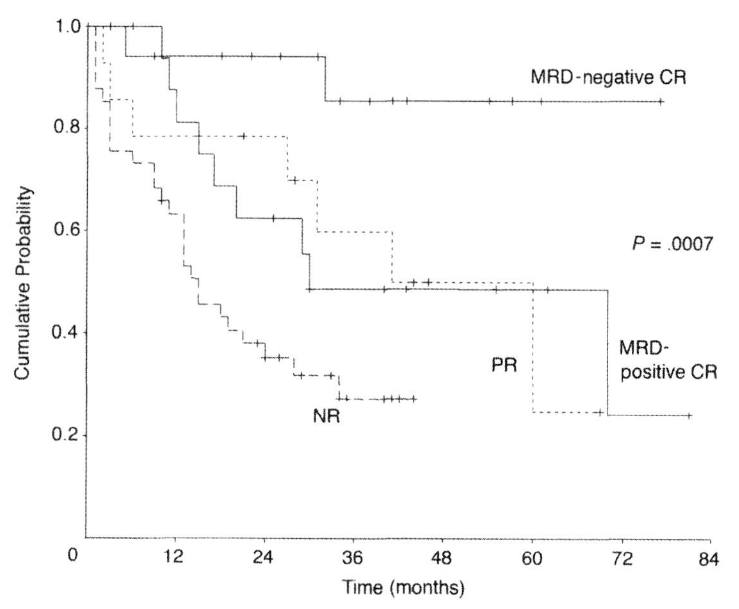

Fig. 2.7 Overall survival by response in patients with relapsed/refractory chronic lymphocytic leukemia (CLL) treated with alemtuzumab. Median survival was not reached for patients achieving MRD-negative complete response (CR), while median survival was 60 months for MRD-positive CR, 70 months for partial response (PR), and 15 months for nonresponders (NR) ($p = 0.0007$). (Reproduced with permission from Moreton et al. 2005.)

The ability of alemtuzumab to eliminate MRD also has been studied in CLL patients who received sequential therapy with alemtuzumab for consolidation following initial response to chemotherapy. These studies are discussed in Section 2.4.2.4.

2.4.2.3 Treatment-Naïve CLL

Alemtuzumab has also shown promising activity as first-line therapy for patients with untreated, progressive CLL. In an early pilot study, IV (n = 5) or SC (n = 4) administration of alemtuzumab (dose escalation to a target dose of 30 mg three times weekly for up to 18 weeks) was assessed in nine patients (Österborg et al. 1996). The ORR was 89% (33% CR, 56% PR). The median duration of response had not been reached by the time of publication, but ranged from 8+ to 24+ months. Complete elimination of CLL cells was obtained in 100% of patients, and bone marrow remission in 78% of patients. Hepatomegaly and/or splenomegaly was completely resolved in 67% of patients, and four patients experienced a >50% reduction in enlarged lymph nodes. One patient developed Grade 3 neutropenia, but all other hematologic toxicity was mild. Fever and rigor occurred with initial doses in 89% of patients. One patient developed CMV pneumonitis and oral candidosis (Österborg et al. 1996).

More recently, SC administration of alemtuzumab was evaluated in a phase II multicenter trial as first-line therapy in patients with CLL (n = 41) (Lundin et al. 2002). After dose escalation from 1 mg to 3 mg to the target dose of 30 mg, previously untreated patients received alemtuzumab 30 mg three times weekly for up to 18 weeks as a SC injection in the thigh. After 2 to 3 weeks of treatment, patients self-administered alemtuzumab at home. Responses were assessed using the 1996 NCIWG criteria. The ORR was 87% (19% CR, 68% PR). At the time of publication, the median time to treatment failure had not been reached (18+ months) (see Table 2.1). CLL cells were eliminated in 95% of patients at a median of 21 days, and CR or nodular PR was obtained in the bone marrow in 66% of patients by 18 weeks (Lundin et al. 2002).

Transient injection-site reactions (ISR) were noted in 90% of patients. Grade 1 reactions were erythema/edema, and Grade 2 reactions included pruritus and slight pain. Only 1 Grade 3 ISR was reported, with local pain that led to treatment discontinuation. Infusion-related toxicities frequently observed with the IV administration of alemtuzumab (i.e., rigors, nausea, hypotension) were absent or rare, with no episodes of rash/urticaria, bronchospasm, hypotension, or nausea, and a 17% incidence of transient rigor. Hematologic toxicities included Grade 4 neutropenia, which occurred in 21% of patients (Grade 2/3, 53%), and Grade 4 thrombocytopenia, which occurred in 5% (Grade 2/3, 11%). Although 39% of patients developed Grade 2/3 anemia, no Grade 4 anemia was reported. In addition, no major bacterial infections were observed. One patient who was allergic to prophylactic cotrimoxazole developed *Pneumocystis carinii* pneumonia after receiving treatment for 11 weeks. CMV reactivation occurred in 10% of patients, but did not lead to study withdrawal (Lundin et al. 2002). Importantly, the results of this study also established that single-agent alemtuzumab given subcutaneously could achieve response rates similar to those achieved with IV administration, but with an acceptable and more favorable safety profile.

A follow-up to this study was conducted to determine the effects of alemtuzumab on blood lymphocyte subsets at the end of treatment and during the posttreatment period (Lundin et al. 2004). By analyzing blood samples of 23 patients (all responders) with flow cytometry, the investigators found that treatment with alemtuzumab resulted in longlasting immune suppression (e.g., depression of CD4+ and CD8+ cells, as well as natural killer cells, normal B cells, granulocytes, and monocytes). The median end-of-treatment counts were low and remained less than 25% of baseline values for over 9 months after treatment cessation (Lundin et al. 2004). However, these low counts showed no significant association with late-occurring infections or autoimmune phenomena. In addition, no significant relationship between the cumulative dose of alemtuzumab and the severity or duration of lymphopenia was apparent (Lundin et al. 2004).

2.4.2.4 Chemoimmunotherapy Combinations

Because alemtuzumab has a mechanism of action which is distinct from those of chemotherapeutic agents used to treat B-CLL, combination therapy with chemotherapeutic agents may lead to synergistic activity and thereby result in improved response compared to either agent alone. Thus, protocols using concurrent and sequential administration of fludarabine and alemtuzumab have been investigated. This combination was evaluated initially in a study of six patients with previously treated CLL (Kennedy et al. 2002). Prior to receiving combination therapy, patients had become refractory to single-agent fludarabine (median eight courses of treatment) and single-agent alemtuzumab. Combination treatment with IV fludarabine (25 mg m^{-2} for 3 days every 28 days) and alemtuzumab (30 mg three times weekly) was given for a minimum of 8 weeks. Responses were assessed using the 1996 NCIWG guidelines. Although a range of responses was obtained, the ORR was 83% (16% CR; 67% PR), with only one patient showing PD. In addition, bone marrow normalized in three patients, with two showing MRD-negative response by flow cytometry, while lymphadenopathy was completely resolved in one patient (Kennedy et al. 2002). Responses in two patients enabled them to undergo subsequent successful autologous peripheral blood stem cell transplantation. Thus, the two agents appeared to act synergistically and were highly active as combination therapy. Acceptable toxicity was seen with this combination approach, with one case of *Pseudomonas* bronchopneumonia during neutropenia, while two patients experienced neutropenia (Kennedy et al. 2002).

In a recent clinical trial, the concurrent administration of fludarabine and alemtuzumab (the FluCam regimen) was evaluated in a phase II, single-center trial conducted in 36 CLL patients with relapsed or refractory disease (Elter et al. 2005). Patients had received a median of two prior therapies; four patients had previously received single-agent alemtuzumab, and 22 had previously received fludarabine as a single agent or in combination with another agent. Some 25%

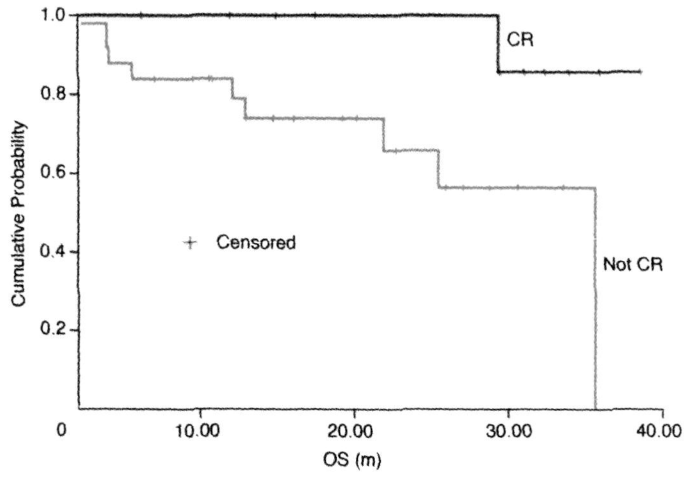

Fig. 2.8 Overall survival by response (CR versus non-CR) in patients with relapsed/refractory chronic lymphocytic leukemia (CLL) treated with combination therapy with alemtuzumab and fludarabine. Median survival was not reached for patients achieving CR, while median survival was approximately 35 months for those not achieving CR. (Reproduced with permission from Elter et al. 2005.)

of patients were refractory to prior fludarabine therapy. The dose of alemtuzumab was escalated on consecutive days from 3 mg to 10 mg to the target dose of 30 mg, after which FluCam (fludarabine 30 mg m^{-2} day for days 1–3 and alemtuzumab 30 mg for days 1–3) was administered every 4 weeks for a maximum of six treatment cycles. Responses were evaluated using the 1996 NCIWG criteria. The ORR was 83% (30% CR, 53% PR), with a median OS of 35.6 months for all patients in this study; the median OS for those who achieved a CR was not reached at the time of publication (Fig. 2.8) (Elter et al. 2005). The median TTP was 12.97 months for all patients, and calculated to be 21.9 months for those achieving a CR. When patients were analyzed by prior response to treatment, eight of the 12 refractory patients responded (4 CR, 4 PR), and 22 of the 24 relapsed patients responded (7 CR, 15 PR). Of those patients who had received fludarabine previously, 18 responded (6 CR, 12 PR), and six of the nine who previously were refractory to fludarabine responded to FluCam (Elter et al. 2005).

Analysis of the 140 treatment cycles assessable for toxicity showed that the major Grade 3/4 toxicities were leukopenia (44%), thrombocytopenia (30%), and neutropenia (26%). Grade 3/4 nonhematologic toxicities were rare (infectious complications in three patients), and most Grade 1/2 toxicities were infusion-related. Thus, the data from this larger trial further demonstrate the synergistic effect of alemtuzumab and fludarabine when given in combination, and indicate that this combination therapy is effective and well tolerated by patients with relapsed/refractory B-CLL (Elter et al. 2005).

Alemtuzumab has also been evaluated as sequential combination therapy (consolidation) in previously treated CLL patients with residual disease after initial chemotherapy. In a trial of 41 patients who had responded to prior chemotherapy with a PR, a nodular PR, or a CR with bone marrow evidence of residual disease, alemtuzumab 10 mg was administered three times weekly for 4 weeks, and if residual disease persisted, a dose of 30 mg three times weekly for 4 additional weeks was offered (after a 4-week rest period) (O'Brien et al. 2003). The rationale for the lower dose was that patients were likely to have less tumor burden compared with the typical population of relapsed/refractory patients for whom the full dose of alemtuzumab is indicated. However, after the first 24 patients were assessed, the remaining patients who enrolled were given the 30-mg dose to increase response rates. The response criteria in this trial differed from those in the trials described previously; patients were required to convert to a CR or a nodular PR by 1996 NCIWG criteria to be considered responders, and residual disease was allowed only in the bone marrow. For those patients who had a CR with bone marrow involvement at the start of treatment (n = 3), an MRD-negative status was required for them to be considered responders (O'Brien et al. 2003).

The ORR was 46%, comprising a 39% response with the 10-mg dose and a 56% response with the 30-mg dose. Failure to respond was due largely to residual lymphadenopathy. In patients with specific disease site involvement prior to treatment, alemtuzumab consolidation therapy achieved a 48% response in clearance of disease from bone marrow nodules, an 86% response in reducing bone marrow lymphocytes to <30%, and an 86% response by immunophenotyping. Achievement of a MRD-negative response, as assessed by polymerase chain reaction (PCR), was attained by 38% of 29 evaluable patients (O'Brien et al. 2003). At the time of publication, the median overall TTP had not been reached; between 24 and 38 months after therapy, six patients were still in remission. Patients who were PCR-negative had not reached their median TTP, whereas those who were PCR-positive had a median TTP of 15 months. Toxicities included Grade 3/4 neutropenia in 30% and Grade 3/4 thrombocytopenia in 14% of patients, infections (including CMV reactivation) in 15% of patients, and infusion-related events (all Grade 1/2) in most patients (O'Brien et al. 2003).

Alemtuzumab consolidation therapy was also evaluated in 34 patients who had previously responded to fludarabine-based therapy (Montillo et al. 2006). In this study, patients received consolidation therapy with alemtuzumab in an effort to eradicate MRD for peripheral stem cell collection in preparation for autologous transplantation. All patients had previously received first-line treatment with fludarabine-based therapy, including fludarabine monotherapy (n = 31) or fludarabine plus cyclophosphamide (n = 1). Two additional patients initially received fludarabine monotherapy, but were switched to fludarabine plus cyclophosphamide after suboptimal response in the lymph nodes. Responses were evaluated using the 1996 NCI-WG criteria. Patients had achieved a CR (35%), nodular PR (21%), or PR (44%) with their prior fludarabine-based therapy, but all had MRD positivity as measured by consensus primer PCR (Montillo et al. 2006). Alemtuzumab was administered SC with initial dose escalation up to 10 mg, three times weekly for 6 weeks. For stem cell mobilization, patients received granulocyte colony-stimulating factor (G-CSF) alone or with intermediate-dose ara-C ($800\,mg\,m^{-2}$ every 12h). The median time from the last dose of fludarabine to initiation of alemtuzumab was 16 weeks (range 12–76 weeks). Following consolidation with alemtuzumab, 53% of patients improved their responses to initial fludarabine-based therapy; overall, 27 patients (79%) achieved a CR with 19 (56%) having a MRD-negative CR. In addition, 24 of 26 patients were successfully able to mobilize peripheral stem cells with adequate stem cell collection in preparation for subsequent transplantation. At the time of publication, 18 patients had undergone stem cell transplantation; at a median follow-up of 28 months after alemtuzumab therapy (range, 11–42 months), 17 of these patients remain in CR (Montillo et al. 2006). The most common adverse events associated with SC alemtuzumab were first-dose reactions such as injection-site reactions (76%) and fever (41%), and were primarily Grade 1 or 2. No hematologic toxicities were associated with alemtuzumab therapy. Although CMV reactivation occurred in 53% of patients (as a result of routine monitoring for CMV antigenemia during the study), the development of active CMV infection was prevented in all cases by pre-emptive therapy with oral ganciclovir (or in three patients, by spontaneous resolution). No bacterial or fungal infections were reported (Montillo et al. 2006). This study demonstrates the activity of alemtuzumab consolidation therapy in eliminating MRD following fludarabine-based induction in patients with CLL, and the feasibility of stem cell mobilization and successful engraftment using this therapeutic approach.

A phase III randomized study compared outcomes with alemtuzumab consolidation therapy versus observation only in patients ($n = 21$) with residual disease following an initial response to fludarabine-based treatment (Wendtner et al. 2004). Consolidation with alemtuzumab 30mg three times weekly (for up to 12 weeks) ($n = 11$) resulted in superior responses compared to observation alone ($n = 10$) in terms of the proportion of patients achieving MRD negativity (five of six evaluable versus none of three evaluable, respectively; $P = 0.048$) and mean progression-free survival (no progression versus 24.7 months, respectively; $P = 0.036$). In both study arms, patients had previously received six cycles of fludarabine or fludarabine + cyclophosphamide as first-line therapy (Wendtner et al. 2004). Two patients who received alemtuzmab converted from a PR to a CR, and three showed bone marrow clearance. Grade 3/4 hematologic toxicities were common in the alemtuzumab arm, with seven patients with neutropenia, four with thrombocytopenia, and two with anemia. There were eight cases of Grade 3/4 infections (pulmonary aspergillosis, CMV reactivation, HSV, herpes zoster, and pulmonary tuberculosis reactivation). These infectious complications resulted in early termination of the study, such that only two patients completed the planned 12 weeks of therapy; however, all patients completed at least 3 weeks. No Grade 3/4 first-dose reactions were observed (Wendtner et al. 2004). The high incidence of infectious events observed in the alemtuzumab arm of this study was likely due to the relatively short recovery time between the initial therapy and alemtuzumab administration. Further studies are warranted to better establish the optimal rest period between induction and consolidation therapy that would allow for sufficient recovery of immune function. The benefits of alemtuzumab consolidation were demonstrated further in a recent update of the results of this trial. Thus, studies have shown the feasibility and promising activity of alemtuzumab in combination with chemotherapy, administered either concurrently or sequentially as consolidation therapy to eliminate residual disease following chemotherapy-based treatment.

2.4.2.5 **Immunotherapy Combination**

The feasibility of the combination of alemtuzumab with rituximab, a chimeric murine/human monoclonal antibody targeting the CD20 antigen, was investigated in 48 patients with relapsed/refractory lymphoid malignancies (Faderl et al. 2003). The CD20 antigen is expressed on normal and malignant B lymphocytes, and rituximab is approved for the treatment of patients with relapsed or refractory, low-grade or follicular, CD20-positive, B-cell NHL (Rituxan product information 2005). Patients in this study had poor prognoses and one of the following malignancies coexpressing CD52 and CD20: CLL, CLL/PLL, PLL, mantle cell leukemia/lymphoma, or Richter transformation. The treatment protocol was IV rituximab 375 $mg\,m^{-2}$ once weekly for 4 weeks, and IV alemtuzumab dose escalation from 3 mg to 10 mg to 30 mg on three consecutive days during Week 1, followed by 30 mg on days 3 and 5 of Weeks 2, 3, and 4. A second cycle was available to patients based on their status after the first cycle. Responses were assessed using the 1996 NCIWG criteria for CLL. The ORR was 52% (8% CR, 4% nodular PR, 40% PR), with 63% of CLL patients and 44% of patients with CLL/PLL responding to this combination regimen (Faderl et al. 2003). The one patient with PLL also responded; however, none of the patients with mantle cell leukemia/lymphoma (n = 4) or Richter transformation (n = 2) responded to therapy. Although the majority of patients had peripheral blood clearance of malignant cells, only 36% of those with CLL, 25% of those with CLL/PLL, and the one patient with PLL had bone marrow responses. At least a 50% decrease in lymphadenopathy and hepatosplenomegaly occurred in 59% of those with CLL and 67% of those with CLL/PLL. The median TTP was 6 months, and the median OS was 11 months (Faderl et al. 2003).

Hematologic toxicities and infusion-related toxicities were common, but most were Grade 2, or less. Over half (52%) of the patients had at least one infectious complication. CMV antigenemia was detected in 27% of patients; 15% of these were symptomatic. Other infections included pneumonia, sinusitis, and fever of unknown origin (Faderl et al. 2003). Overall, the combination was well tolerated, and results supported the feasibility of combining these two monoclonal antibodies.

2.4.2.6 **Safety with Alemtuzumab in CLL**

As can be seen from the toxicities detailed in the individual trials described in Sections 2.4.2.1 to 2.4.2.5, alemtuzumab administration is associated with infusion-related adverse events, hematologic toxicities, and infectious events. Although some of these events may be of Grade 3/4 severity, guidelines for managing these complications are available and enable patients to derive maximum benefit from alemtuzumab-based therapy.

2.4.2.6.1 Infusion-related Adverse Events

Intravenous administration of alemtuzmab is associated with fever, rigors, nausea, vomiting, skin rash, dyspnea, hypotension, headache, and hypoxia. Most of these adverse events tend to resolve after about 2 weeks from the time of initial dose administration (Keating et al. 2002b; Ferrajoli et al. 2003; Moreton et al. 2005). Recommendations for the prevention and management of infusion-related adverse events were recently published as part of the outcomes of an expert opinion roundtable that generated treatment guidelines for alemtuzumab in CLL (Table 2.2) (Keating et al. 2004). The authors stated that, in their experience,

Table 2.2 Guidelines for managing acute first-dose infusion reactions with intravenous alemtuzumab. (Modified from Keating et al. 2004, with permission.)

Reaction	*Recommendations*
Standard premedications	• Diphenhydramine (50 mg) or chlorpheniramine (≤35 mg) and acetaminophen (650 mg) 30 min before infusion. • Second dose if acetaminophen (4 h later) during the first week of treatment.
Severe infusion-related events	• Add premedication with hydrocortisone (intravenous dose 50–100 mg) to subsequent infusions. • Discontinue steroids as soon as possible, preferably after Week 1
Rigors	• Hold infusion and administer intravenous meperidine or pethidine (25 mg). • Premedicate with anti-emetics if patients experience meperidine/pethidine-induced nausea.
Rash	• Add 25–50 mg diphenhydramine/chlorpheniramine orally every 4 h as needed. • Hydrocortisone may be used if additional diphenhydramine/chlorpheniramine is not effective. • For severe rash, premedicate with H_2 receptor antagonists (e.g., cimetidine, ranitidine).
Hypotension	• Hydrate with normal saline solution unless contraindicated, based on underlying cardiac condition.
Dyspnea	• Hold infusion and treat with inhaled β_2-agonists. • For severe dyspnea, temporary use of steroids may be necessary.
Febrile neutropenia	• Antibiotics or G-CSF may be administered at the discretion of the clinician.

G-CSF = granulocyte colony-stimulating factor.

Table 2.3 Summary of common adverse events and first-dose reactions with intravenous and subcutaneous alemtuzumab. Data expressed as % of patients. (Modified from Keating et al. 2004, with permission.)

	IV administration (n = 93)		*SC administration (n = 41)*	
Events	Grade 1 or 2	Grade 3 or 4	Grade 1 or 2	Grade 3 or 4
Fever	69	13	15	2
Rigors	77	11	15	2
Rash/urticaria	44	4	0	0
Hypotension	14	1	0	0
Dyspnea	11	6	0	0
Nausea/vomiting	78	0	0	0
Diarrhea	12	1	0	0
Headache	13	0	0	0
Fatigue	14	3	5	2
Local injection-site reactions	0	0	88	0

these clinical guidelines have allowed most patients receiving alemtuzumab to continue with therapy. The administration of alemtuzumab via the subcutaneous route may avoid or minimize many of these infusion-related toxicities, as was demonstrated in the phase II trial of previously untreated patients with CLL (Lundin et al. 2002) described in Section 2.4.2.3. Although ISRs were observed in the majority of patients, the infusion-related toxicities described above were rare, or absent (Lundin et al. 2002). While direct comparisons cannot be made between trials due to differences in patient populations, differences in the incidence of infusion-related events in the SC alemtuzumab trial and that of the most common infusion-related events seen in the pivotal trial of IV alemtuzumab suggest that SC administration circumvents many of these toxicities (Table 2.3).

2.4.2.6.2 **Hematologic Toxicities**

Hematologic toxicities are associated with both the IV and SC administration of alemtuzumab, owing to its myelosuppressive properties. In previously treated CLL patients, the major hematologic toxicities reported were transient thrombocytopenia and neutropenia (Ferrajoli et al. 2003; Keating et al. 2002b; Moreton et al. 2005; Rai et al. 2002); these events, which also occur when alemtuzumab is administered to treatment-naïve CLL patients, are typically of higher incidence and greater severity in previously treated patients (Lundin et al. 2002). In addition, the disease itself is often associated with cytopenias, such that patients may frequently present with thrombocytopenia or neutropenia even prior to receiving alemtuzumab therapy.

Hematologic toxicities reported with concurrent chemoimmunotherapy with FluCam were leukopenia, thrombocytopenia, and neutropenia (Elter et al. 2005). With alemtuzumab consolidation therapy, Grade 3/4 hematologic toxicities did not occur in one small study (Montillo et al. 2002), but Grade 3/4 thrombocytopenia and neutropenia were common in a larger trial with a more heavily pretreated patient population (O'Brien et al. 2003), as well as in the randomized trial of alemtuzumab consolidation (Wendtner et al. 2004). The combination of alemtuzumab and rituximab, which was given for only one cycle in most patients in the study, produced myelosuppression in approximately 66% of patients, but events of Grade 3/4 severity were rare (Faderl et al. 2003).

Clinical recommendations for the management of hematologic toxicities are based on the fact that thrombocytopenia appears to be most common during Weeks 2 through 4 of therapy, and neutropenia appears to be most common during Weeks 4 through 8; thus, careful monitoring during these times may help clinicians to monitor and manage these events promptly. Severe events are managed with growth factor support during treatment, and experts do not recommend interrupting therapy (although this is suggested in the alemtuzumab product information), because patients will be vulnerable to infusion-related toxicities when re-starting therapy. The current guidelines for managing thrombocytopenia, neutropenia and anemia (which is not as common in incidence or as severe in presentation as thrombocytopenia and neutropenia) are listed in Table 2.4 (Keating et al. 2004). Experts have found that transient neutropenia at 4 weeks of therapy does not warrant discontinuation of therapy unless the patient has presented with major infections.

2.4.2.6.3 **Immunosuppression and Infectious Events**

In a retrospective review of the records of 27 fludarabine-refractory CLL/small-cell lymphocytic leukemia (SLL) or B-PLL patients, data analysis showed that, regardless of the salvage therapy administered, 89% of patients developed serious infections, 78.4% of which were bacterial (Perkins et al. 2002). Serious infections were defined as those that required hospital admission and IV antibiotic treatment. The median time to development of the first serious infection was 4 months from the time the patient became refractory to fludarabine, with a median time between hospital admissions of 42 days (Perkins et al. 2002). Thus, chemotherapy-refractory CLL patients are highly vulnerable to serious infections while receiving salvage therapy. Because these patients have poor immune function even before beginning alemtuzumab-based therapy, they may show a high incidence of infectious complications that may not be attributed entirely to the immunosuppressive nature of this agent.

Studies of alemtuzumab in patients with refractory/relapsed CLL have shown that serious infections have developed during therapy, including septicemia, CMV reactivation, pneumonia, and herpes virus infections, with infection-related deaths reported in all trials. There appeared to be a trend toward a lower incidence of infections in patients with the best responses to therapy (Ferrajoli et al. 2003; Keating et al. 2002b; Moreton et al. 2005; Rai et al. 2002). With the FluCam

Table 2.4 Guidelines for managing hematological toxicities. (Modified from Keating et al. 2004, with permission.)

Condition	*Recommendation(s)*
Anemia	
All patients with anemia	Alemtuzumab may be given with transfusion (with irradiated blood products) or erythropoietin support.
Neutropenia	Dose reduction or discontinuation of alemtuzumab is not recommended unless neutropenia is accompanied by a major infection.
ANC > 500 μL	Support may not be required as cytopenias are likely to resolve spontaneously.
ANC < 500 μL	G-CSF or GM-CSF support may be used at the discretion of the clinician.
ANC < 250 μL	Alemtuzumab treatment should be temporarily interrupted until resolution.
Febrile neutropenia	Alemtuzumab treatment should be temporarily interrupted until resolution.
Thrombocytopenia	
$<25\,000\,\mu L^{-1}$	If no hemorrhage is present, alemtuzumab should be continued at full dose.
$<10\,000\,\mu L^{-1}$	Transfusion support may be given at the discretion of the clinician.
Hemorrhagic event	Alemtuzumab treatment should be temporarily interrupted until resolution.

ANC = absolute neutrophil count; G-CSF = granulocyte colony-stimulating factor; GM-CSF = granulocyte macrophage colony-stimulating factor.

combination, three of 36 patients with relapsed/refractory disease developed infections (two with fungal pneumonia and one with sepsis, which was fatal) (Elter et al. 2005). Again, these patients either had poor performance status at study entry or did not respond to alemtuzumab therapy (i.e., they had progressive disease after treatment). Subclinical CMV reactivation was observed in three patients (Elter et al. 2005). When alemtuzumab was used as consolidation therapy, 37% of patients developed infections, including nine who experienced CMV reactivation. In addition, three patients developed Epstein–Barr virus (EBV) large cell lymphoma (O'Brien et al. 2003). The use of alemtuzumab as first-line therapy in patients with CLL was not associated with any major infections (no Grade >1 incidences) in patients receiving anti-infective prophylaxis, and CMV reactivations were manageable (Lundin et al. 2002). This finding further supports the notion that the immune function of CLL patients may be more intact in the frontline setting or in less-advanced stages of the disease. In the first-line consolidation setting, alemtuzumab was associated with CMV reactivation as well as pulmonary aspergillosis, herpes virus infections, and tuberculosis reactivation (Montillo et al. 2002; Wendtner et al. 2004). While T-cell counts may decrease to undetectable levels in some patients during treatment, observations in small numbers of patients suggested that no correlations existed between absolute decreases in T-cell counts and the risk of severe infection (Wendtner et al. 2004).

Clinical guidelines for the prevention and management of infectious events in patients receiving alemtuzumab include routine antimicrobial prophylaxis and weekly monitoring for CMV reactivation, which often occurs between Weeks 3 and 6 of therapy (Keating et al. 2004). If CMV reactivation is observed or suspected, then prompt treatment with ganciclovir (or equivalent) is recommended. More detailed recommendations are outlined in Table 2.5 (Keating et al. 2004).

2.4.3 T-Cell Leukemias and Lymphomas

2.4.3.1 T-Cell Lymphomas (Cutaneous/Peripheral T-Cell Lymphoma)

Patients with cutaneous T-cell lymphomas (CTCLs) have few treatment options and poor prognoses after their disease becomes refractory to topical therapies. Because malignant T cells typically have high levels of CD52 expression, studies were conducted to determine whether alemtuzumab had activity in patients with advanced CTCLs.

Promising results were observed when alemtuzumab was evaluated in a phase II, multicenter study of patients with advanced mycosis fungoides or Sézary syndrome (MF/SS), one of the most common types of CTCLs (Lundin et al. 2003). Patients with CD52-positive, advanced MF/SS (n = 22; median three prior therapies) received alemtuzumab 30 mg three times weekly for up to 12 weeks (after an initial dose escalation from 3 mg). The ORR was 55% (32% CR, 23% PR) (Table 2.6); in patients who had received one or two prior lines of therapy, the ORR was 80%, and in those who had received three or more prior lines of therapy it was 33%. Peripheral blood was cleared of tumor cells in 86% of patients, and tumor cells were cleared from lymph nodes and skin in 55% of patients for each disease site (Lundin et al. 2003). Erythroderma was ameliorated in 69% of patients, with 38% experiencing a CR. Patients also reported reductions in and/or disappearance of itching. Half of the patients completed all 12 weeks of treatment. The median TTF was 12 months in those who responded to therapy. Toxicities were characteristic of those seen previously with alemtuzumab in patients with CLL (i.e., infusion-related adverse events, hematologic toxicities, and infections), although the time frame for the appearance of neutropenia was somewhat delayed. The results of this trial demonstrated the activity of alemtuzumab in patients with advanced MF/SS (Lundin et al. 2003).

A subsequently reported case study of alemtuzumab treatment (10-week course) of a 32-year-old man with advanced-stage SS showed dramatic results supporting this activity. The patient had almost complete disappearance of itching after 1 week of treatment; disappearance of erythroderma and pruritis, almost complete disappearance of Sézary cells from peripheral blood and bone marrow, regression

Table 2.5 Guidelines for managing infections events. (Data from Keating et al. 2004.)

Infection	*Recommeddations*
Pneumocystis carinii pneumonia	• Prophylaxis with trimethoprim/sulfamethoxazole DS twice daily. • For patients not tolerating above regimen, other alternatives may include aerosolized pentamidine, oral dapsone, and oral atovaquone. • Prophylaxis should be continued for ≥2 months (preferably 4 months) after completion of alemtuzumab therapy, or until CD4 levels recover to 250000 μL^{-1}.
Viral infections	• Prophylaxis with agents including famciclovir, acyclovir, or valacyclovir. • Prophylaxis should be continued ≥2 months (preferably 4 months) after completion of alemtuzumab therapy, or until CD4 levels recover to 250000 μL^{-1}.
CMV reactivation	
Patients with fever of unknown origin	• Test for CMV antigen, preferably by PCR if available. • If PCR is not available, clinicians should treat all cases of fever of unknown origin with use of preemptive ganciclovir.
Symptomatic patients with positive PCR results	• Treat immediately with intravenous ganciclovir or foscarnet. • If unresponsive to ganciclovir alone, foscarnet may be added. • Hold alemtuzumab therapy until infection has cleared.
Asymptomatic patients with positive PCR results	• Test again by a second PCR test. • If patient tests positive in the second test, initiate treatment with ganciclovir and monitor with a quantitative PCR, if possible. • If unresponsive to ganciclovir alone, foscarnet may be added.
Symptomatic for pulmonary infection but negative for CMV on PCR test	• Test by bronchoscopy and lavage.

CMV = cytomegalovirus; DS = double strength; PCR = polymerase chain reaction.

of lymphadenopathy, and normal blood counts 4 months after starting treatment; there were no signs of disease at the 12-month follow-up assessment. No toxicities were noted (Gautschi et al. 2004).

In contrast, a trial of alemtuzumab in eight patients with advanced MF/SS showed less favorable results. Although the ORR was 38% (all PR), and patients reported improvements in pruritis, the response duration was less than 3 months, and a high incidence of Grade 4 hematologic toxicity and infectious complications

Table 2.6 Summary of response rates with alemtuzumab in T-cell malignancies.

No. of patients	*Disease type*	*Route of administration*	*Median no. of prior therapies (range)*	*CR [%]*	*ORR [%]*	*Median response duration, months (range)*	*Reference*
38	Primarily previously treated, refractory* T-PLL	IV	NR*	60	76	7 (4–45)	Dearden et al. (2001)
76	Primarily refractory† T-PLL	IV	2 (0–5)†	37.5	50	8.7 (0.1+ to 44.4) for patients with CR	Keating et al. (2002a)
22	Advanced mycosis fungoides/Sézary syndrome	IV	3 (1–5)	32	55	TTF 12 (5–32+)	Lundin et al. (2003)
8	Advanced, heavily pretreated mycosis fungoides/Sézary syndrome	IV	6.5 (1–17)	0	38	2.5 (2–3.5)	Kennedy et al. (2003)
14	Relapsed or refractory peripheral T-cell lymphoma	IV	2 (1–4)	21	36	6 (2–12) for patients with CR	Enblad et al. (2004)

* Two patients in this study were treatment-naïve.
† Four patients in this study were treatment-naïve.
CR = complete response; IV = intravenous; NR = not reported; ORR = overall response rate; T-PLL = T-cell prolymphocytic leukemia; TTF = time-to-treatment failure.

was observed (Kennedy et al. 2003). The authors postulated that the discrepancy between these results and those of others may be due to the more heavily pretreated patient population in this study (i.e., seven of eight patients had received at least four prior therapies).

Alemtuzumab has also been studied in patients with advanced peripheral T-cell lymphoma (PTCL), as these also have limited treatment options and poor prognoses. In a phase II study of 14 patients with relapsed/refractory PTCL who were heavily pretreated, alemtuzumab was dose escalated to 30 mg three times weekly for a maximum treatment period of 12 weeks (Enblad et al. 2004). The ORR was 36% (CR in three patients, PR in two), and CR durations were 2, 6, and 12 months in the three patients (Table 2.6). Only one patient completed all 12 weeks of treatment, and the median treatment duration was 6 weeks. Premature discontinuation occurred due to achievement of a CR (n = 3) and toxicity and/or progressive disease (n = 10). Hematologic toxicities (pancytopenia in four patients, hemophagocytosis syndrome in two) and infectious complications (CMV reactivation in six patients, pulmonary aspergillosis in two) were common, and five patients died due to serious adverse events related to alemtuzumab treatment in combination with advanced disease. Although the ORR was encouraging in this group of poor-prognosis patients, the high rates of toxicity prohibited the investigators from recommending alemtuzumab for the treatment of patients with PTCL unless they were part of clinical trials that are carefully designed and monitored (Enblad et al. 2004).

Another group of investigators used a reduced dose of alemtuzumab in patients with advanced PTCL. Ten patients with relapsed/refractory PTCL unspecified (n = 6) or MF (n = 4) were given alemtuzumab 10 mg three times weekly for 10 weeks (Zinzani et al. 2005). Patients had received a median of three prior treatments. The ORR was 60% (20% CR, 40% PR). Both CRs were in PTCL patients; three MF patients had a PR. The median duration of response was 7 months, and at this dose alemtuzumab was well tolerated (no Grade 3/4 hematologic toxicities, one CMV reactivation that was managed effectively).

A case report of alemtuzumab treatment of a heavily pretreated 74-year-old patient with CD52-positive Lennert's lymphoma, a PTCL/lymphoepithelioid cell variant, showed that after 5 weeks of treatment, there was a significant reduction in thoracic and abdominal lesions. Although the patient experienced CMV reactivation during treatment, it was managed successfully (Zeitlinger et al. 2005).

2.4.3.2 T-Cell Prolymphocytic Leukemia (T-PLL)

T-PLL cells also have high levels of CD52 expression. No approved treatment exists for T-PLL, and those that are used show high rates of relapse, with short median survival durations. Investigation of the activity of alemtuzumab in 39 patients with T-PLL (two naïve to therapy, 37 who had received prior treatment) showed that in 38 evaluable patients, alemtuzumab 30 mg three times weekly produced a high ORR of 76% (60% CR, 16% PR) (Dearden et al. 2001) (Table 2.6). Among patients previously resistant to chemotherapy, 34% achieved a CR. Responses occurred at most disease sites. The median OS was 10 months for all patients, and 16 months for those who achieved a CR. Nine patients were alive 29 months after completing therapy, and one patient survived 54 months. Toxicities were characteristic of those seen with alemtuzumab treatment of patients with CLL, and alemtuzumab was well tolerated in these patients (Dearden et al. 2001). Seven patients were able to undergo high-dose chemotherapy (HDT) and autologous stem cell transplantation after alemtuzumab treatment, as flow cytometry and PCR verified that their harvested stem cells were free of T-PLL cells. Owing to these encouraging results, the investigators concluded that alemtuzumab might be an effective first-line treatment option for T-PLL (Dearden et al. 2001).

In a retrospective study that included data from 18 patients in the above trial, data from a total of 76 patients with T-PLL who had been treated with alemtuzumab were analyzed to determine its activity and safety (Keating et al. 2002a). Patients were enrolled in a compassionate-use program, and all but four (who had not received prior therapy) had failed prior therapies. Alemtuzumab 30 mg was given three times weekly for 4–12 weeks after the initial dose escalation. In the 72 pretreated patients, the ORR was 50% (37.5% CR, 12.5% PR), and three of the four previously untreated patients achieved a CR, making the total ORR 50% (Keating et al. 2002a) (Table 2.6). In the bone marrow, 39% of patients achieved a CR; in the spleen, 33% of patients with splenomegaly achieved a CR; in the lymph nodes, the CR was 32%; and 30% of patients with hepatomegaly achieved a CR. The response in the skin was 43%. Of 27 pretreated patients who achieved a CR with alemtuzumab, only one had achieved a CR and 10 a PR when on prior therapy. The median OS was 7.5 months for all patients, and 14.8 months for those who had achieved a CR. Toxicities were not different from those seen patients with CLL. The response in untreated patients suggests that alemtuzumab should be studied further in chemotherapy-naïve patients (Keating et al. 2002a). A separate study of four patients with T-PLL showed that treatment with alemtuzumab produced an ORR of 75% (Fløisand et al. 2004). Overall, the outcomes obtained in these trials have demonstrated that alemtuzumab is efficacious in patients with T-PLL.

2.4.3.3 Adult T-Cell Leukemia

As alemtuzumab has demonstrated tumor cell killing and prolongation of survival demonstrated in a murine model of adult T-cell leukemia (Zhang et al. 2003), and because no effective therapy exists for this malignancy, investigations of the activity of alemtuzumab in patients are warranted. A recent case report of a 55-year-old woman with refractory adult T-cell leukemia who received treatment with alemtuzumab and the nucleoside analog pentostatin as part of an ongoing clinical trial described improvements in her condition. Alemtuzumab was given at a dose of 30 mg three times weekly, and pentostatin at a dose of 4 mg m^{-2} weekly for 4 weeks, followed by dosing every other week. The regimen was well tolerated, and resolution of palpable disease, normal blood counts, and improved computed tomography (CT) scans were observed after 2 months of therapy. In addition, at this time point and at several repeat examinations, the patient had no morphological or immunophenotypic evidence of disease in the bone marrow (Zhang et al. 2003). Hence, alemtuzumab might have promising activity in adult T-cell leukemia and should be investigated further in this patient population (Ravandi and Faderl 2005).

2.4.4 Non-Hodgkin's Lymphoma (NHL)

As described in Section 2.2, two patients with NHL were among the first to receive alemtuzumab therapy during its development process, and the compound has shown activity in blood, bone marrow, spleen, and lymph nodes in these patients (Hale et al. 1988). Relapsed or resistant low-grade NHL is difficult to treat. Rituximab is approved for this indication, and thus it is feasible that alemtuzumab might also have utility in these patients. In a phase II multicenter trial, 50 previously treated patients (25 relapsed, 25 resistant) were given alemtuzumab 30 mg three times weekly for up to 12 weeks (Lundin et al. 1998). The initial dose was either 3 or 10 mg, determined on the basis of each patient's baseline conditions. Patients completed a median of 8 weeks of treatment, with nine completing the full 12 weeks. Although 40% of patients did not respond, an ORR of 20% (4% CR, 16% PR) was obtained (Lundin et al. 1998) (Table 2.7). A CR was obtained in the blood of 94% of patients within a median of 7 days; 32% of patients had a CR in the bone marrow, splenomegaly resolved completely in 15% of patients, lymphadenopathy resolved completely in 5% of patients (and was reduced by over 50% in 11% of patients), and skin lesions completely regressed in 40% of patients. The median TTP was 4 months. Hematologic toxicities were moderate, with major Grade 4 toxicities being anemia (10%), neutropenia (28%), and thrombocytopenia (22%); however, 28% of patients had low neutrophil and/or platelet counts at baseline that improved with alemtuzumab therapy (Lundin et al. 1998). Infusion-related reactions were also common, but decreased in incidence over time. Infections included HSV reactivation, oral candidiasis, *Pneumocystis carinii* pneumonia, CMV pneumonitis, pulmonary aspergillosis, tuberculosis, and septicemia. Six patients died during treatment or follow-up for reasons including progressive disease, septicemia, pre-existing respiratory insufficiency, and pneumonia of unknown origin (Lundin et al. 1998). This study has demonstrated that, although a subset of patients with NHL responded to treatment with alemtuzumab, the response was not durable. The investigators hypothesized that the poor response in the lymph nodes and progression of lymphadenopathy in a high percentage of patients contributed to the shorter duration of response.

A study in 16 NHL patients with nonbulky disease and two with MRD showed slightly better responses (ORR = 22%, with two patients in CR alive over 4.5 years after treatment) (Table 2.7), but the trial was terminated early due to the high incidence and severity of infectious complications (Khorana et al. 2001). Thus, even in NHL patients with nonbulky disease or MRD, only a subset showed benefits with alemtuzumab treatment.

Table 2.7 Summary response rates with alemtuzumab in non-Hodgkin's lymphoma (NHL).

No. of patients	*Disease state*	*Route of administration*	*Median no. of prior therapies (range)*	*CR [%]*	*ORR [%]*	*Median response duration, months (range)*	*Reference*
50	Relapsed or refractory low-grade NHL	IV	70% received ≥2 prior therapies	4*	20	NR (time-to-progression 4 months for all responders; 10 months for patients with mycosis fungoides)	Lundin et al. (1998)
18	Previously treated nonbulky NHL	IV	2 (1–4)	17†	22	7 (5–54.+); 14+ months in 1 patient with PCR-negative CR	Khorana et al. (2001)
18	Relapsed or refractory low-grade and high-grade NHL	IV	3 (1–4)	0	44	NR	Uppenkamp et al. (2002)

* The NHL subtype for these patients was mycosis fungoides.

† Includes two patients with low-grade NHL who achieved CR, and one patient with minimal residual disease who achieved a PCR-negative CR in the bone marrow.

CR = complete response; IV = intravenous; NR = not reported; ORR = overall response rate; PCR = polymerase chain reaction.

Another study of alemtuzumab in 18 patients with relapsed/refractory NHL included two with high-grade NHL (Uppenkamp et al. 2002). Two patients were treated with a maximum of 75 mg and 240 mg alemtuzumab once weekly, and the remaining 16 received alemtuzumab 30 mg three times weekly for 6 weeks; a subset of the 16 were participants in the 50-patient, phase II, multicenter trial described above (Lundin et al. 1998). The response criteria were somewhat different, with responses classified as CR, major disease improvement (MDI), disease improvement (DI), or limited disease improvement (LDI). Responses obtained were DI (n = 2) and LDI (n = 6), for a response rate of 44%. Six of the eight responders showed a reduction of bone marrow infiltration, five a decrease in enlarged lymph nodes, and two a reduction in splenomegaly. However, the median duration of response was 3.5 months. Two patients died while on the study, and two were withdrawn due to bronchospasm. Hematologic toxicities were recorded, but few were Grade 3/4. The major treatment-limiting factor was infection, with 11 patients reporting a total of 12 different infections (Uppenkamp et al. 2002).

2.4.5 Stem Cell Transplantation (SCT)

2.4.5.1 Donor T-Cell Depletion (Prevention of GvHD) and Prevention of Graft Rejection

The Campath-1 family of antibodies was originally developed to provide a tool to deplete mature T lymphocytes from donor bone marrow allografts *in vitro* in preparation for transplantation, in the hope of reducing the risk of GvHD in transplant recipients. In an early study of 11 high-risk patients, HLA-matched sibling donor bone marrow was treated with rat monoclonal Campath-1 (autologous serum was used as the complement source) and transplanted, and no anti-GvHD prophylaxis was provided. Engraftment was rapid and successful in all patients, and over the 12-month follow-up period, none of the patients developed GvHD (Waldmann et al. 1984). These results indicated that Campath-1 and autologous human serum-provided complement could successfully remove mature T lymphocytes from allografts prior to transplantation. However, further study revealed that donor T-cell depletion increased the risk of graft rejection and disease relapse through minimization of the graft-versus-leukemia effect. Small numbers of donor T cells may be added to help restore the graft-versus-leukemia effect, although the risk of GvHD is then increased. This was illustrated in a study of 131 patients with acute leukemia (acute nonlymphocytic leukemia [ANLL] or acute lymphoblastic leukemia [ALL]) who received T-cell-depleted matched sibling allografts (121 allografts depleted using Campath-1), as 81 received donor T cells post transplant to restore graft-versus-leukemia activity (Naparstek et al. 1995). Of these patients, 39 developed acute GvHD and 21 developed chronic GvHD. The risk of developing GvHD was significantly higher in recipients of donor T cells than in nonrecipients ($P < 0.0001$). When the overall probability of relapse was examined, donor T cell administration provided no benefit over no administration (2-year probability of relapse, 25% versus 32%, respectively, $P = 0.64$). However, patients with ALL who received donor T cells and developed GvHD had a reduced 2-year probability of relapse (14%) compared with those who received donor T cells but had no GvHD (61%), and with those who did not receive donor T cells or develop GvHD (56%) (Naparstek et al. 1995). The results suggested that donor T cells mediated the graft-versus-leukemia effect in parallel with GvHD.

A further strategy to counter the increased risk of rejection of donor T-cell-depleted allografts is to also treat the recipient with Campath-1. Results were compared in 951 patients with leukemias and other malignancies, who received HLA-matched sibling transplants in which Campath-1 (IgM and IgG2b) was used in various protocols to deplete T cells in donor bone marrow *in vitro* and/or recipient tissues *in vivo* to control GvHD and graft rejection, respectively (Hale and Waldmann 1994). The analysis showed that the lowest rate of combined complications (defined as graft failure and/or GvHD) was achieved with a protocol of Campath-1G patient treatment 5–10 days prior to chemoradiotherapy and transplant of an allograft pretreated with Campath-1M (plus complement) *in vitro* (Hale and Waldmann 1994).

This protocol was investigated in a clinical trial of 70 patients with acute myelogenous leukemia (AML) receiving HLA-matched sibling transplants. There were two control groups – one group comprising 50 patients who received Campath-1M-depleted allografts but no Campath-1G; and a second group comprising the International Bone Marrow Registry data from 459 patients who had received nondepleted grafts and conventional GvHD prophylaxis (Hale et al. 1998). The Campath-1G *in vivo*/Campath-1M *in vitro* protocol provided superior results compared with those in the first control group in terms of decreased incidence of graft rejection (6% versus 31%), acute GvHD (4% versus 20%) and transplant-related mortality (15% versus 58% at 5 years), and compared with the second control group in terms of decreased incidence of acute GvHD (4% versus 35%, respectively) and transplant-related mortality (15% versus 26% at 5 years). The incidence of graft rejection was higher in the study group compared to the second control group (6% versus 2%). Patients in the study group achieved significantly longer 5-year survival (62% versus 35%; $P = 0.001$) and leukemia-free survival (60% versus 33%; $P = 0.002$) compared to the first control group; the risk of relapse did not differ between the treatment groups. Overall survival, leukemia-free survival, and risk of relapse did not significantly differ between the study group and the second control group (Hale et al., 1998). The results of this study confirmed that T-cell depletion is highly effective in preventing GvHD, and also showed that *in-vivo* Campath-1G administration provides additional depletion of host lymphocytes to help prevent graft rejection.

2.4.5.2 **Reduced-Intensity/Non-Myeloablative Conditioning**

Nonmyeloablative conditioning regimens have been developed to reduce transplant-related mortality and to facilitate engraftment of allogeneic stem cell transplants (SCT) in patients with hematologic malignancies. Existing regimens may be fludarabine-based, may also include melphalan or busulfan, or may use total body irradiation (TBI) only. Although these regimens allow for effective allogeneic engraftment with minimal nonhematologic toxicities, risks for severe acute GvHD remains high. The use of alemtuzumab as part of a novel conditioning regimen has been investigated to determine whether it could decrease the incidence of GvHD. In a study of 44 patients with hematological malignancies (14 with NHL, 10 with Hodgkin's disease, six AML, seven multiple myeloma, three hypoplastic myelodysplastic syndrome, and one each with ALL, CLL, chronic myeloid leukemia and plasma cell leukemia), alemtuzumab 20 mg was given on days −8 to −4, fludarabine $30\,mg\,m^{-2}$ on days −7 to −3, and melphalan $140\,mg\,m^{-2}$ on day −2 (Kottaridis et al. 2000). Patients received unmanipulated allogeneic peripheral blood stem cells from HLA-matched siblings ($n = 36$) or unrelated donors ($n = 8$), with conventional GvHD prophylaxis [cyclosporine (CsA) with or without methotrexate]. Of 43 evaluable patients, 42 achieved sustained engraftment. No instances of Grade 3/4 acute GvHD were reported, the incidence of Grade 2 acute GvHD was 5%, and only one patient developed chronic GvHD in liver and skin. The estimated OS at 12 months was 73.2% (Kottaridis et al. 2000). Thus, this novel nonmyeloablative regimen with alemtuzumab was feasible and facilitated allogeneic engraftment while minimizing the morbidity and mortality associated with GvHD.

This conditioning regimen was evaluated further in 47 patients with hematologic malignancies who received allogeneic BMT or SCT from unrelated donors, 20 of whom were mismatched for HLA class I and/or 2 alleles (Chakraverty et al. 2002). Patients were given CsA alone for GvHD prophylaxis. High rates of allogeneic engraftment were observed, with 95.7% of patients achieving sustained neutrophil recovery and engraftment (median 13 days) and 85.1% achieving platelet recovery (median 16.5 days) with independence from platelet transfusions. The results showed that a CR was achieved in 50% of the patients who had a partial response to chemotherapy at the time of transplantation. The incidence of Grade 2 acute GvHD was 14.9%, and that of Grade 3/4 GvHD 6.4%. Three of 38 evaluable patients developed limited chronic GvHD, no patient developed extensive chronic GvHD, and no patient died due to GvHD. Progression-free survival and OS at 1 year was 61.5% and 75.5%, respectively. No difference in the incidence of GvHD or overall 1-year survival was observed between patients with HLA-mismatched and HLA-matched transplants (Chakraverty et al. 2002).

Given that CMV reactivation is a frequent complication in the SCT setting, the incidence of CMV infection was investigated after the use of this alemtuzumab-containing, nonmyeloablative conditioning regimen. The pattern of CMV reactivation and outcome was investigated in 101 patients transplanted with either unmanipulated peripheral blood stem cells from a matched family donor or unmanipulated bone marrow from an unrelated donor (Chakrabarti et al. 2002). Those who were at risk for CMV infection were CMV-seropositive pretransplant, or received grafts from CMV-seropositive donors. Patients were then screened for CMV every week from transplantation to 100 days after transplantation. If two consecutive screenings revealed CMV positivity, then ganciclovir, foscarnet, or cidofovir were administered as preemptive therapy (Chakrabarti et al. 2002). The incidence of CMV infection was 51%, with a median time to first infection of 27 days. The majority of patients (90%) who developed an infection did so within 35 days of transplantation. Three patients who developed CMV disease died. Grade 3/4 acute GvHD developed in only 3.9% of patients. The OS probability at 12 months was 71.3% and at 18 months was 65%, and the presence of CMV infection did not influence survival rates (Chakrabarti et al. 2002).

A further study compared results from 129 patients treated in two prospective trials with a nonmyeloablative conditioning regimen with fludarabine and melphalan who received GvHD prophylaxis with either CsA and alemtuzumab or CsA and methotrexate (Perez-Simon et al. 2002). The alemtuzumab-containing prophylactic regimen was associated with a higher incidence of CMV reactivation (46.6% versus 22.7%; $P = 0.018$) despite the fact that, at the initiation of the study, a larger proportion of patients in the alemtuzumab study were CMV-negative donor/recipient cases compared to the methotrexate-based study (37% versus 9.1%; $P = 0.009$). The alemtuzumab-containing regimen was more effective than the methotrexate-containing regimen in reducing the incidence of both acute GvHD (21.7% versus 45.1%; $P = 0.006$) and chronic GvHD (5% versus 66.7%, respectively; $P < 0.001$). Both regimens were capable of inducing an early and sustained engraftment. No significant differences in OS were observed between the two regimens. Thus, both nonmyeloablative treatment regimens were effective for allogeneic engraftment, but were associated with different spectra of complications (Perez-Simon et al. 2002). Although the addition of alemtuzumab to a nonmyeloablative regimen effectively reduces the incidence of GvHD, careful monitoring of patients for CMV reactivation should be performed owing to the high rates of infection observed in the above studies.

2.4.5.3 Safety

It is important to bear in mind that these transplantation protocols are all investigative in nature, and that detailed safety studies on specific methods have not been carried out to date. For example, in donor T-cell-depletion protocols, the posttransplant administration of donor T cells to the patient carries an increased risk of induction of GvHD, and therefore an increased risk of mortality. In the 131-patient study described above, six patients treated with donor T cells posttransplant died of GvHD. In addition, 15 patients in that same trial died from organ failure induced by the conditioning regimen (Naparstek et al. 1995). Further, as detailed above, nonmyeloablative conditioning regimens incorporating alemtuzumab show a high incidence of CMV infection posttransplant, which may also be fatal when patients develop CMV disease (Chakrabarti et al. 2002). Further studies are clearly needed to optimize transplantation treatment protocols with alemtuzumab.

2.5 The Future Outlook

2.5.1 Other Hematologic Malignancies (Multiple Myeloma, Acute Leukemias)

Because CD52 is expressed on lymphocytes, it is feasible that alemtuzumab might have applications in many other hematologic malignancies. A preliminary study in 61 patients with the plasma cell proliferative disorders multiple myeloma (n = 23), primary systemic amyloidosis (n = 29), and monoclonal gammopathy of unspecified origin (MGUS, n = 9) showed that the CD52 antigen was highly expressed on these malignant cells (Kumar et al. 2003). The percentage of the CD38+/CD45+ subset of malignant plasma cells expressing CD52 was 73%, 88%, and 80% in the MGUS, multiple myeloma, and primary systemic amyloidosis groups, respectively, suggesting that alemtuzumab may have activity in these malignancies (Kumar et al. 2003).

A case study in a 58-year-old patient with relapsed ALL suggested that administration of alemtuzumab 30 mg three times weekly for 4 weeks, with subsequent follow-up of 30 mg three times weekly every 2 months, may induce long-term remission (Laporte et al. 2004). The patient remained free of disease 16 months after treatment initiation. Although he developed cytopenia, the condition was managed using supportive care with growth factors. Another study of three patients with ALL who had relapsed after SCT showed that alemtuzumab 30 mg every other day for five doses produced clinical responses in all patients (but no CRs) (Piccaluga et al. 2005). However, the responses were not durable, and all three patients died of progressive disease within a few months of treatment. Differences in treatment protocols and patient characteristics may have accounted for these differences in response. Nonetheless, both reports indicated that alemtuzumab has activity in patients with ALL, and this should prompt further investigation in this malignancy.

2.5.2 Prevention of GvHD in Solid Organ Transplantation

The immunosuppressive properties of alemtuzumab make it an attractive candidate for use in solid organ transplantation. Although calcineurin inhibitor-based therapies with steroids are part of the standard immunosuppressive regimens used in solid organ transplantation, long-term use of these regimens may be associated with nephrotoxicity, hyperlipidemia, and secondary diabetes. Several groups have investigated the activity and safety of alemtuzumab as induction therapy in kidney transplantation protocols. In 44 patients receiving cadaveric kidney transplants, alemtuzumab 0.3 $mg\,kg^{-1}$ was administered on the day of transplantation (day 0) and on postoperative day 4, with both doses preceded by IV prednisolone (Ciancio et al. 2004). The investigators assessed the feasibility of using lower (50%) doses of tacrolimus and mycophenolate mofetil (MMF) for maintenance immunosuppression and of eliminating corticosteroid therapy after the first week post transplantation. Patients also received CMV prophylaxis (Ciancio et al. 2004). Results showed that after 1 year, this protocol achieved 100% patient and graft survival, with no requirement for corticosteroids in 38 of 44 patients. Four patients developed biopsy-proven acute graft rejection during the first 12 months post transplantation, which was successfully treated with corticosteroid-based therapies including combination therapy with bolus corticosteroid and alemtuzumab 0.3 $mg\,kg^{-1}$ in two patients. No incidences of CMV or polyoma virus infections were noted. Adverse events requiring hospitalization occurred in four patients; these included atypical pneumonia, infected lymphocele, transient Epstein–Barr virus infection, and transient diarrhea (Ciancio et al. 2004).

A larger, retrospective study compared the effects of alemtuzumab induction therapy (30 $mg\,day^{-1}$) given on the day of, and the day following, kidney transplantation in 126 patients versus outcomes in 1115 historical controls who did not receive alemtuzumab. Patients in the comparison arms had received anti-CD25 monoclonal antibodies, basiliximab, or daclizumab ($n = 799$), polyclonal antithymocyte globulin ($n = 160$) or other therapies ($n = 156$) that included anti-CD3 monoclonal antibody, antithymocyte globulin, or non-antibody-based therapies (Knechtle et al. 2004). Alemtuzumab was given in combination with low-dose methylprednisolone (10 $mg\,day^{-1}$) and MMF (1000 mg twice daily) and either tacrolimus or CsA (initiated when serum creatinine was $<3.0\,mg\,dL^{-1}$). Overall, patients who received alemtuzumab induction therapy had significantly decreased incidence of graft rejection ($P = 0.037$) and improved graft survival ($P = 0.0159$). Similarly, among the subgroup of patients who experienced delayed graft function, alemtuzumab therapy resulted in significantly decreased incidence of graft rejection ($P = 0.0096$) and improvement in graft survival ($P = 0.0119$). Importantly, the use of alemtuzumab as induction therapy may allow for reduced reliance on steroids and nephrotoxic immunosuppressive agents, without associated increases in the incidence of infections or secondary malignancies (Knechtle et al. 2004).

Recent data have demonstrated the long-term efficacy and safety of alemtuzumab induction therapy. A five-year follow-up analysis of a cohort of 33 patients who had received alemtuzumab induction therapy for renal transplantation (20 $mg\,day^{-1}$ on the day of, and the day after, transplantation, followed by half-dose CsA from day 3 onwards, and no other immunosuppressants) showed that – compared with 66 controls who had not received alemtuzumab – no significant differences were seen in rates of patient survival, graft survival, graft function, acute rejection, infection, or serious adverse events (Watson et al. 2005).

A pilot study of patients receiving kidney and pancreas transplantation assessed alemtuzumab induction therapy (a single dose of 30 mg on the day of transplantation, given after a single dose of dexamethasone 100 mg) with rapid steroid elimination in comparison with rabbit antithymocyte globulin induction with a steroid-containing regimen (Sundberg et al. 2005). After transplantation, the alemtuzumab group received MMF 2 $mg\,day^{-1}$ followed by tacrolimus initiated after either a brisk diuresis or serum creatinine $<4.0\,mg\,dL^{-1}$; the control group received MMF $1\,mg\,day^{-1}$, delayed tacrolimus as above, and daily steroids tapered to $20\,mg\,day^{-1}$ at week 1 and to $5\,mg\,day^{-1}$ at month 2. Each group contained 16 patients of the following transplantation types: deceased donor kidney transplant (n = 9), living donor kidney transplant (n = 5), simultaneous kidney/pancreas transplant (n = 1), and sequential pancreas transplant after kidney transplant (n = 1). No significant differences were seen between groups in terms of delayed kidney graft function, incidence of thrombocytopenia, incidence of infection, creatinine clearance at 6 months, incidence of acute rejection, and patient and graft survival rates after 9 and 11 months of follow-up for the alemtuzumab and control groups, respectively (Sundberg et al. 2005). Although patients in the alemtuzumab group had a significantly lower white blood cell count and absolute lymphocyte count at the follow-up evaluations (all $P < 0.05$), there was no difference in the incidence of infections.

Another group sought to determine whether the use of alemtuzumab and MMF as induction therapy and as maintenance therapy posttransplant for pancreas/kidney or solitary pancreas transplantation would allow elimination of the use of calcineurin inhibitors and steroids. Patients (n = 75) in a pilot study were given alemtuzumab 30 mg on the day of transplantation and on postoperative days 2, 14, and 42 for induction therapy. In addition, one dose of rabbit antithymocyte globulin was given on postoperative day 4, and MMF $\geq 2\,g\,day^{-1}$ was given for at least 6 weeks (Gruessner et al. 2005). Maintenance therapy consisted of alemtuzumab 30 mg when the absolute lymphocyte count was $\geq 200\,mm^{-3}$ and MMF $\geq 2\,g\,day^{-1}$. Results were compared with those of 266 historical controls who received antithymocyte globulin and tacrolimus-based therapy. Patient survival at 6 months for each of the three types of recipient (pancreas transplant alone, pancreas after kidney transplant, simultaneous pancreas/kidney transplant) with alemtuzumab treatment was ≥90%, and did not differ significantly from that of historical controls (Gruessner et al. 2005). Pancreas graft survival did not differ between transplant recipient types or between the alemtuzumab group and controls. No pancreas graft rejection was seen at 6 months in the simultaneous pancreas/kidney and the pancreas after kidney transplant groups; the rate of pancreas graft rejection in the pancreas transplant alone group was 15%. At 6 months, the alemtuzumab and control groups did not differ in rates of abdominal and CMV infections, and no cases of secondary malignancies post transplant were noted with alemtuzumab therapy (Gruessner et al. 2005). Although a longer follow-up time is needed, the results of this study suggests that the combination of alemtuzumab and MMF is feasible and effective in the pancreas/kidney transplantation setting, and may eliminate the adverse effects associated with use of steroids and calcineurin inhibitors.

The above-described studies illustrate that, depending upon the dosing schedule used, alemtuzumab may enable elimination of the need for long-term steroids and reduction in or elimination of calcineurin inhibitor use posttransplant. Further studies are warranted to better establish the role of alemtuzumab in the solid organ transplant setting.

2.5.3
Applications in Autoimmune Disease

Because rheumatoid arthritis (RA) is thought to be a T-lymphocyte-mediated disorder, alemtuzumab is also being evaluated for the treatment of patients with RA. Forty-one patients with active RA refractory to treatment were given alemtuzumab 100, 250, or 400 mg over 5 or 10 days, and responses to therapy were monitored for 6 months (Isaacs et al. 1996). Treatment provided improvement in symptoms, as measured in 35 evaluable patients by 50% Paulus responses (achieved by 50% of patients at 31 days, by 43% at 59 days, by 29% at 87 days, and by 20% at 178 days) and reductions in median swollen and painful joint scores (from baseline up to day 59 at the 100-mg dose, up to day 178 at the 250-mg dose, and up to day 129 [swollen] and day 178 [painful] at the 400-mg dose). The higher doses of alemtuzumab (250 and 400 mg) appeared to be associated with improved response rates and more durable responses compared to the 100-mg dose. Infusion-related adverse events were experienced by most patients, and with the exception of a rare, fatal infection, infectious events were minor (Isaacs et al. 1996). Among 31 patients with sera available for antiglobulin testing, nine showed positive antiglobulin responses. However, the development of antiglobulin was not associated with therapeutic activity or adverse events (Isaacs et al. 1996). While there is evidence of activity of alemtuzumab in RA, its role in therapy remains to be defined with further investigation. Alemtuzumab is also being investigated in other autoimmune disease, including in patients with multiple sclerosis.

References

Albitar, M., Do, K.A., Johnson, M.M., Giles, F.J., Jilani, I., O'Brien, S., Cortes, J., Thomas, D., Rassenti, L.Z., Kipps, T.J., Kantarjian, H.M., Keating, M. (2004) Free circulating soluble CD52 as a tumor marker in chronic lymphocytic leukemia and its implication in therapy with anti-CD52 antibodies. *Cancer* 101: 999–1008.

Brett, S.J., Baxter, G., Cooper, H., Rowan, W., Regan, T., Tite, J., Rapson, N. (1996) Emergence of CD52-, glycosylphosphatidylinositol-anchor-deficient lymphocytes in rheumatoid arthritis patients following Campath-1H treatment. *Int Immunol* 8: 325–334.

CAMPATH (Alemtuzumab) product information, 2004.

Chakrabarti, S., Mackinnon, S., Chopra, R., Kottaridis, P.D., Peggs, K., O'Gorman, P., Chakraverty, R., Marshall, T., Osman, H., Mahendra, P., Craddock, C., Waldmann, H., Hale, G., Fegan, C.D., Yong, K., Goldstone, A.H., Linch, D.C., Milligan, D.W. (2002) High incidence of cytomegalovirus infection after nonmyeloablative stem cell transplantation: potential role of Campath-1H in delaying immune reconstitution. *Blood* 99: 4357–4363.

Chakraverty, R., Peggs, K., Chopra, R., Milligan, D.W., Kottaridis, P.D., Verfuerth, S., Geary, J., Thuraisundaram, D., Branson, K., Chakrabarti, S., Mahendra, P., Craddock, C., Parker, A., Hunter, A., Hale, G., Waldmann, H., Williams, C.D., Yong, K., Linch, D.C., Goldstone, A.H., Mackinnon, S. (2002) Limiting transplantation-related mortality following unrelated donor stem cell transplantation by using a nonmyeloablative conditioning regimen. *Blood*, 99: 1071–1078.

Cheson, B.D., Bennett, J.M., Grever, M., Kay, N., Keating, M.J., O'Brien, S., Rai, K.R. (1996) National Cancer Institute-sponsored Working Group guidelines for chronic lymphocytic leukemia: revised guidelines for diagnosis and treatment. *Blood* 87: 4990–4997.

Cheson, B.D., Bennett, J.M., Rai, K.R., Grever, M.R., Kay, N.E., Schiffer, C.A., Oken, M.M., Keating, M.J., Boldt, D.H., Kempin, S.J., et al. (1988) Guidelines for clinical protocols for chronic lymphocytic leukemia: recommendations of the National Cancer Institute-sponsored working group. *Am J Hematol* 29: 152–163.

Cheson, B.D., Horning, S.J., Coiffier, B., Shipp, M.A., Fisher, R.I., Connors, J.M., Lister, T.A., Vose, J., Grillo-Lopez, A., Hagenbeek, A., Cabanillas, F., Klippensten, D., Hiddemann, W., Castellino, R., Harris, N.L., Armitage, J.O., Carter, W., Hoppe, R., Canellos, G.P. (1999) Report of an international workshop to standardize response criteria for non-Hodgkin's lymphomas. NCI Sponsored International Working Group. *J Clin Oncol* 17: 1244.

Ciancio, G., Burke, G.W., Gaynor, J.J., Mattiazzi, A., Roohipour, R., Carreno, M. R., Roth, D., Ruiz, P., Kupin, W., Rosen, A., Esquenazi, V., Tzakis, A.G., Miller, J. (2004) The use of Campath-1H as induction therapy in renal transplantation: preliminary results. *Transplantation* 78: 426–433.

Dearden, C.E., Matutes, E., Cazin, B., Tjonnfjord, G.E., Parreira, A., Nomdedeu, B., Leoni, P., Clark, F.J., Radia, D., Rassam, S.M., Roques, T., Ketterer, N., Brito-Babapulle, V., Dyer, M.J., Catovsky, D. (2001) High remission rate in T-cell prolymphocytic leukemia with CAMPATH-1H. *Blood* 98: 1721–1726.

Dyer, M.J., Hale, G., Hayhoe, F.G., Waldmann, H. (1989) Effects of CAMPATH-1 antibodies in vivo in patients with lymphoid malignancies: influence of antibody isotype. *Blood* 73: 1431–1439.

Elter, T., Borchmann, P., Schulz, H., Reiser, M., Trelle, S., Schnell, R., Jensen, M., Staib, P., Schinkothe, T., Stutzer, H., Rech, J., Gramatzki, M., Aulitzky, W., Hasan, I., Josting, A., Hallek, M., Engert, A. (2005) Fludarabine in combination with alemtuzumab Is effective and feasible in patients with relapsed or refractory B-cell chronic lymphocytic leukemia: results of a phase II trial. *J Clin Oncol* 23: 7024–7031.

Enblad, G., Hagberg, H., Erlanson, M., Lundin, J., MacDonald, A.P., Repp, R., Schetelig, J., Seipelt, G., Osterborg, A. (2004) A pilot study of alemtuzumab (anti-CD52 monoclonal antibody) therapy for patients with relapsed or chemotherapy-refractory peripheral T-cell lymphomas. *Blood* 103: 2920–2924.

Faderl, S., Thomas, D.A., O'Brien, S., Garcia-Manero, G., Kantarjian, H.M., Giles, F.J., Koller, C., Ferrajoli, A., Verstovsek, S., Pro, B., Andreeff, M., Beran, M., Cortes, J., Wierda, W., Tran, N., Keating, M.J. (2003) Experience with alemtuzumab plus rituximab in patients with relapsed and refractory lymphoid malignancies. *Blood* 101: 3413–3415.

Ferrajoli, A., O'Brien, S.M., Cortes, J.E., Giles, F.J., Thomas, D.A., Faderl, S., Kurzrock, R., Lerner, S., Kontoyiannis, D. P., Keating, M.J. (2003) phase II study of alemtuzumab in chronic lymphoproliferative disorders. *Cancer* 98: 773–778.

Fløisand, Y., Brinch, L., Gedde-Dahl, T., Tjonnfjord, G.E. (2004) [Treatment of T-cell prolymphocytic leukemia with monoclonal anti-CD52 antibody (alemtuzumab]. *Tidsskr Nor Laegeforen* 124: 768–770.

Gautschi, O., Blumenthal, N., Streit, M., Solenthaler, M., Hunziker, T., Zenhausern, R. (2004) Successful treatment of chemotherapy-refractory Sezary syndrome with alemtuzumab (Campath-1H). *Eur J Haematol* 72: 61–63.

Gilleece, M.H. and Dexter, T.M. (1993) Effect of Campath-1H antibody on human hematopoietic progenitors in vitro. *Blood* 82: 807–812.

Ginaldi, L., De, M.M., Matutes, E., Farahat, N., Morilla, R., Dyer, M.J., Catovsky, D. (1998) Levels of expression of CD52 in normal and leukemic B and T cells: correlation with in vivo therapeutic responses to Campath-1H. *Leuk Res* 22: 185–191.

Golay, J., Manganini, M., Rambaldi, A., Introna, M. (2004) Effect of alemtuzumab on neoplastic B cells. *Haematologica* 89: 1476–1483.

Greenwood, J., Clark, M., Waldmann, H. (1993) Structural motifs involved in human IgG antibody effector functions. *Eur J Immunol* 23: 1098–1104.

Gruessner, R.W., Kandaswamy, R., Humar, A., Gruessner, A.C., Sutherland, D.E. (2005) Calcineurin inhibitor- and steroid-free immunosuppression in pancreas-kidney and solitary pancreas transplantation. *Transplantation* 79: 1184–1189.

Hale, G. (2001) The CD52 antigen and development of the CAMPATH antibodies. *Cytotherapy* 3: 137–143.

Hale, G., Waldmann, H. (1994) Control of graft-versus-host disease and graft rejection by T cell depletion of donor and recipient with Campath-1 antibodies. Results of matched sibling transplants for malignant diseases. *Bone Marrow Transplant* 13: 597–611.

Hale, G., Bright, S., Chumbley, G., Hoang, T., Metcalf, D., Munro, A.J., Waldmann, H. (1983a) Removal of T cells from bone marrow for transplantation: a monoclonal antilymphocyte antibody that fixes human complement. *Blood* 62: 873–882.

Hale, G., Swirsky, D.M., Hayhoe, F.G., Waldmann, H. (1983b) Effects of monoclonal anti-lymphocyte antibodies in vivo in monkeys and humans. *Mol Biol Med* 1: 321–334.

Hale, G., Clark, M., Waldmann, H. (1985) Therapeutic potential of rat monoclonal antibodies: isotype specificity of antibody-dependent cell-mediated cytotoxicity with human lymphocytes. *J Immunol* 134: 3056–3061.

Hale, G., Dyer, M.J., Clark, M.R., Phillips, J. M., Marcus, R., Riechmann, L., Winter, G., Waldmann, H. (1988) Remission induction in non-Hodgkin lymphoma with reshaped human monoclonal antibody CAMPATH-1H. *Lancet* 2: 1394–1399.

Hale, G., Zhang, M.J., Bunjes, D., Prentice, H.G., Spence, D., Horowitz, M.M., Barrett, A.J., Waldmann, H. (1998) Improving the outcome of bone marrow transplantation by using CD52 monoclonal antibodies to prevent graft-versus-host disease and graft rejection. *Blood* 92: 4581–4590.

Hale, G., Rebello, P., Brettman, L.R., Fegan, C., Kennedy, B., Kimby, E., Leach, M., Lundin, J., Mellstedt, H., Moreton, P., Rawstron, A.C., Waldmann, H., Österborg, A., Hillmen, P. (2004) Blood concentrations of alemtuzumab and antiglobulin responses in patients with chronic lymphocytic leukemia following intravenous or subcutaneous routes of administration. *Blood* 104: 948–955.

Hillmen, P., Skotnicki, A., Robak, T., Jaksic, B., Dmoszynska, A., Sirard, C., Mayer, J. (2006) Preliminary phase 3 efficacy and safety of alemtuzumab vs chlorambucil as front-lint therapy for patients with progressive B-cell chronic lymphocytic leukemia (BCLL). *J Clin Oncol* 24 (Suppl.1), 339s, Abstract 6511.

Isaacs, J.D., Manna, V.K., Rapson, N., Bulpitt, K.J., Hazleman, B.L., Matteson, E. L., St Clair, E.W., Schnitzer, T.J., Johnston, J.M. (1996) CAMPATH-1H in rheumatoid arthritis – an intravenous dose-ranging study. *Br J Rheumatol* 35: 231–240.

Keating, M.J., Cazin, B., Coutre, S., Birhiray, R., Kovacsovics, T., Langer, W., Leber, B., Maughan, T., Rai, K., Tjonnfjord, G., Bekradda, M., Itzhaki, M., Herait, P. (2002a) Campath-1H treatment of T-cell prolymphocytic leukemia in patients for whom at least one prior chemotherapy regimen has failed. *J Clin Oncol* 20: 205–213.

Keating, M.J., Flinn, I., Jain, V., Binet, J.L., Hillmen, P., Byrd, J., Albitar, M., Brettman, L., Santabarbara, P., Wacker, B., Rai, K.R. (2002b) Therapeutic role of alemtuzumab (Campath-1H) in patients who have failed fludarabine: results of a large international study. *Blood* 99: 3554–3561.

Keating, M., O'Brien, S., Kontoyiannis, D., Plunkett, W., Koller, C., Beran, M., Lerner, S., Kantarjian, H. (2002c) Results of first salvage therapy for patients refractory to a fludarabine regimen in chronic lymphocytic leukemia. *Leuk Lymphoma* 43: 1755–1762.

Keating, M., Coutre, S., Rai, K., Osterborg, A., Faderl, S., Kennedy, B., Kipps, T., Bodey, G., Byrd, J.C., Rosen, S., Dearden, C., Dyer, M.J., Hillmen, P. (2004)

Management guidelines for use of alemtuzumab in B-cell chronic lymphocytic leukemia. *Clin Lymphoma* 4: 220–227.

Kennedy, B., Rawstron, A., Carter, C., Ryan, M., Speed, K., Lucas, G., Hillmen, P. (2002) Campath-1H and fludarabine in combination are highly active in refractory chronic lymphocytic leukemia. *Blood* 99: 2245–2247.

Kennedy, G.A., Seymour, J.F., Wolf, M., Januszewicz, H., Davison, J., McCormack, C., Ryan, G., Prince, H.M. (2003) Treatment of patients with advanced mycosis fungoides and Sezary syndrome with alemtuzumab. *Eur J Haematol* 71: 250–256.

Khorana, A., Bunn, P., McLaughlin, P., Vose, J., Stewart, C., Czuczman, M.S. (2001) A phase II multicenter study of CAMPATH-1H antibody in previously treated patients with nonbulky non-Hodgkin's lymphoma. *Leuk Lymphoma* 41: 77–87.

Kirchhoff, C. (1996) CD52 is the 'major maturation-associated' sperm membrane antigen. *Mol Hum Reprod* 2: 9–17.

Knechtle, S.J., Fernandez, L.A., Pirsch, J.D., Becker, B.N., Chin, L.T., Becker, Y.T., Odorico, J.S., D'alessandro, A.M., Sollinger, H.W. (2004) Campath-1H in renal transplantation: The University of Wisconsin experience. *Surgery* 136: 754–760.

Kottaridis, P.D., Milligan, D.W., Chopra, R., Chakraverty, R.K., Chakrabarti, S., Robinson, S., Peggs, K., Verfuerth, S., Pettengell, R., Marsh, J.C., Schey, S., Mahendra, P., Morgan, G.J., Hale,G., Waldmann, H., de Elvira, M.C., Williams, C.D., Devereux, S., Linch, D.C., Goldstone, A.H., Mackinnon, S. (2000) In vivo CAMPATH-1H prevents graft-versus-host disease following nonmyeloablative stem cell transplantation. *Blood* 96: 2419–2425.

Kumar, S., Kimlinger, T.K., Lust, J.A., Donovan, K., Witzig, T.E. (2003) Expression of CD52 on plasma cells in plasma cell proliferative disorders. *Blood* 102: 1075–1077.

Laporte, J.P., Isnard, F., Garderet, L., Fouillard, L., Gorin, N.C. (2004) Remission of adult acute lymphocytic leukaemia with alemtuzumab. *Leukemia* 18: 1557–1558.

Lozanski, G., Heerema, N.A., Flinn, I.W., Smith, L., Harbison, J., Webb, J., Moran, M., Lucas, M., Lin, T., Hackbarth, M.L., Proffitt, J.H., Lucas, D., Grever, M.R., Byrd, J.C. (2004) Alemtuzumab is an effective therapy for chronic lymphocytic leukemia with p53 mutations and deletions. *Blood* 103: 3278–3281.

Lundin, J., Osterborg, A., Brittinger, G., Crowther, D., Dombret, H., Engert, A., Epenetos, A., Gisselbrecht, C., Huhn, D., Jaeger, U., Thomas, J., Marcus, R., Nissen, N., Poynton, C., Rankin, E., Stahel, R., Uppenkamp, M., Willemze, R., Mellstedt, H. (1998) CAMPATH-1H monoclonal antibody in therapy for previously treated low-grade non-Hodgkin's lymphomas: a phase II multicenter study. European Study Group of CAMPATH-1H Treatment in Low-Grade Non-Hodgkin's Lymphoma. *J Clin Oncol* 16: 3257–3263.

Lundin, J., Kimby, E., Bjorkholm, M., Broliden, P.A., Celsing, F., Hjalmar, V., Mollgard, L., Rebello, P., Hale, G., Waldmann, H., Mellstedt, H., Osterborg, A. (2002) phase II trial of subcutaneous anti-CD52 monoclonal antibody alemtuzumab (Campath-1H) as first-line treatment for patients with B-cell chronic lymphocytic leukemia (B-CLL). *Blood* 100: 768–773.

Lundin, J., Hagberg, H., Repp, R., Cavallin-Stahl, E., Freden, S., Juliusson, G., Rosenblad, E., Tjonnfjord, G., Wiklund, T., Osterborg, A. (2003) phase 2 study of alemtuzumab (anti-CD52 monoclonal antibody) in patients with advanced mycosis fungoides/Sezary syndrome. *Blood* 101: 4267–4272.

Lundin, J., Porwit-MacDonald, A., Rossmann, E.D., Karlsson, C., Edman, P., Rezvany, M.R., Kimby, E., Osterborg, A., Mellstedt, H. (2004) Cellular immune reconstitution after subcutaneous alemtuzumab (anti-CD52 monoclonal antibody, CAMPATH-1H) treatment as first-line therapy for B-cell chronic lymphocytic leukaemia. *Leukemia* 18: 484–490.

Montillo, M., Cafro, A.M., Tedeschi, A., Brando, B., Oreste, P., Veronese, S., Rossi, V., Cairoli, R., Pungolino, E., Morra, E. (2002) Safety and efficacy of subcutaneous Campath-1H for treating residual disease in patients with chronic lymphocytic leukemia responding to fludarabine. *Haematologica* 87: 695–700.

Montillo, M., Tedeschi, A., Miqeuleiz, S., Veronese, S., Rossi, V., Cairoli, R., Intropido L., Ricci, F., Colosimo, A., Scarpati, B., Montagna, M., Nichelatti, M., Regazzi, M., Morra, E. (2006) Alemtuzumab as consolidation after a response to fludarabine is effective in purging residual disease in patients with chronic lymphocytic leukemia. *J Clin Oncol* 24: 2337–2342.

Moreton, P., Kennedy, B., Lucas, G., Leach, M., Rassam, S.M., Haynes, A., Tighe, J., Oscier, D., Fegan, C., Rawstron, A., Hillmen, P. (2005) Eradication of minimal residual disease in B-cell chronic lymphocytic leukemia after alemtuzumab therapy is associated with prolonged survival. *J Clin Oncol* 23: 2971–2979.

Naparstek, E., Or, R., Nagler, A., Cividalli, G., Engelhard, D., Aker, M., Gimon, Z., Manny, N., Sacks, T., Tochner, Z., et al. (1995) T-cell-depleted allogeneic bone marrow transplantation for acute leukaemia using Campath-1 antibodies and post-transplant administration of donor's peripheral blood lymphocytes for prevention of relapse. *Br J Haematol* 89: 506–515.

Nückel, H., Frey, U.H., Roth, A., Duhrsen, U., Siffert, W. (2005) Alemtuzumab induces enhanced apoptosis in vitro in B-cells from patients with chronic lymphocytic leukemia by antibody-dependent cellular cytotoxicity. *Eur J Pharmacol* 514: 217–224.

O'Brien, S.M., Kantarjian, H.M., Thomas, D.A., Cortes, J., Giles, F.J., Wierda, W.G., Koller, C.A., Ferrajoli, A., Browning, M., Lerner, S., Albitar, M., Keating, M.J. (2003) Alemtuzumab as treatment for residual disease after chemotherapy in patients with chronic lymphocytic leukemia. *Cancer* 98: 2657–2663.

Österborg, A., Fassas, A.S., Anagnostopoulos, A., Dyer, M.J., Catovsky, D., Mellstedt, H. (1996) Humanized CD52 monoclonal antibody Campath-1H as first-line treatment in chronic lymphocytic leukaemia. *Br J Haematol* 93: 151–153.

Österborg, A., Dyer, M.J.S., Bunjes, D., Pangalis, G.A., Bastion, Y., Catovsky, D., Mellstedt, H. (1997) phase II multicenter study of human CD52 antibody in previously treated chronic lymphocytic leukemia. *J Clin Oncol* 15: 1567–1574.

Perez-Simon, J.A., Kottaridis, P.D., Martino, R., Craddock, C., Caballero, D., Chopra, R., Garcia-Conde, J., Milligan, D.W., Schey, S., Urbano-Ispizua, A., Parker, A., Leon, A., Yong, K., Sureda, A., Hunter, A., Sierra, J., Goldstone, A.H., Linch, D.C., San Miguel, J.F., Mackinnon, S. (2002) Nonmyeloablative transplantation with or without alemtuzumab: comparison between 2 prospective studies in patients with lymphoproliferative disorders. *Blood* 100: 3121–3127.

Perkins, J.G., Flynn, J.M., Howard, R.S., Byrd, J.C. (2002) Frequency and type of serious infections in fludarabine-refractory B-cell chronic lymphocytic leukemia and small lymphocytic lymphoma: implications for clinical trials in this patient population. *Cancer* 94: 2033–2039.

Piccaluga, P.P., Martinelli, G., Malagola, M., Rondoni, M., Bonifazi, F., Bandini, G., Visani, G., Baccarani, M. (2005) Alemtuzumab in the treatment of relapsed acute lymphoid leukaemia. *Leukemia* 19: 135.

Rai, K.R., Stephenson, J. (2001) *Campath-1H: emerging frontline therapy in chronic lymphocytic leukemia.* Parthenon Publishing, Pearl River, NY, p. 62.

Rai, K.R., Freter, C.E., Mercier, R.J., Cooper, M.R., Mitchell, B.S., Stadtmauer, E.A., Santabarbara, P., Wacker, B., Brettman, L. (2002) Alemtuzumab in previously treated chronic lymphocytic leukemia patients who also had received fludarabine. *J Clin Oncol* 20: 3891–3897.

Ravandi, F. and Faderl, S. (2006) Complete response in a patient with adult T-cell leukemia (ATL) treated with combination of alemtuzumab and pentostatin. *Leuk Res* 30: 103–105.

Rawstron, A.C., Kennedy, B., Evans, P.A., Davies, F.E., Richards, S.J., Haynes, A.P., Russel, N.H., Hale, G., Morgan, G.J., Jack, A.S., Hillmen, P. (2001) Quantitation of minimal disease levels in chronic lymphocytic leukemia using a sensitive flow cytometric assay improves the prediction of outcome and can be used to optimize therapy. *Blood* 98: 29–35.

Rebello, P.R., Hale, G., Friend, P.J., Cobbold, S.P., Waldmann, H. (1999) Anti-globulin responses to rat and humanized CAMPATH-1 monoclonal antibody used to treat transplant rejection. *Transplantation* 68: 1417–1420.

Rebello, P., Cwynarski, K., Varughese, M., Eades, A., Apperley, J.F., Hale, G. (2001) Pharmacokinetics of CAMPATH-1H in BMT patients. *Cytotherapy* 3: 261–267.

Riechmann, L., Clark, M., Waldmann, H., Winter, G. (1988) Reshaping human antibodies for therapy. *Nature* 332: 323–327.

Rituxan (Rituximab) product information, 2005. Genentech, South San Francisco, CA.

Rowan, W., Tite, J., Topley, P., Brett, S.J. (1998) Cross-linking of the CAMPATH-1 antigen (CD52) mediates growth inhibition in human B- and T-lymphoma cell lines, and subsequent emergence of CD52-deficient cells. *Immunology* 95: 427–436.

Stanglmaier, M., Reis, S., Hallek, M. (2004) Rituximab and alemtuzumab induce a nonclassic, caspase-independent apoptotic pathway in B-lymphoid cell lines and in chronic lymphocytic leukemia cells. *Ann Hematol* 83: 634–645.

Sundberg, A.K., Roskopf, J.A., Hartmann, E. L., Farney, A.C., Rohr, M.S., Stratta, R.J. (2005) Pilot study of rapid steroid elimination with alemtuzumab induction therapy in kidney and pancreas transplantation. *Transplant Proc* 37: 1294–1296.

Uppenkamp, M., Engert, A., Diehl, V., Bunjes, D., Huhn, D., Brittinger, G. (2002) Monoclonal antibody therapy with CAMPATH-1H in patients with relapsed high- and low-grade non-Hodgkin's lymphomas: a multicenter phase I/II study. *Ann Hematol* 81: 26–32.

Waldmann, H., Polliak, A., Hale, G., Or, R., Cividalli, G., Weiss, L., Weshler, Z., Samuel, S., Manor, D., Brautbar, C., et al. (1984) Elimination of graft-versus-host disease by in-vitro depletion of alloreactive lymphocytes with a monoclonal rat anti-human lymphocyte antibody (CAMPATH-1). *Lancet* 2: 483–486.

Watson, C.J., Bradley, J.A., Friend, P.J., Firth, J., Taylor, C.J., Bradley, J.R., Smith, K.G., Thiru, S., Jamieson, N.V., Hale, G., Waldmann, H., Calne, R. (2005) Alemtuzumab (CAMPATH 1H) induction therapy in cadaveric kidney transplantation – efficacy and safety at five years. *Am J Transplant* 5: 1347–1353.

Wendtner, C.M., Ritgen, M., Schweighofer, C.D., Fingerle-Rowson, G., Campe, H., Jager, G., Eichhorst, B., Busch, R., Diem, H., Engert, A., Stilgenbauer, S., Dohner, H., Kneba, M., Emmerich, B., Hallek, M. (2004) Consolidation with alemtuzumab in patients with chronic lymphocytic leukemia (CLL) in first remission – experience on safety and efficacy within a randomized multicenter phase III trial of the German CLL Study Group (GCLLSG). *Leukemia* 18: 1093–1101.

Xia, M.Q., Hale, G., Lifely, M.R., Ferguson, M.A., Campbell, D., Packman, L., Waldmann, H. (1993) Structure of the CAMPATH-1 antigen, a glycosylphosphatidylinositol-anchored glycoprotein which is an exceptionally good target for complement lysis. *Biochem J* 293 (Pt 3): 633–640.

Zeitlinger, M.A., Schmidinger, M., Zielinski, C.C., Chott, A., Raderer, M. (2005) Effective treatment of a peripheral T-cell lymphoma/lymphoepitheloid cell variant (Lennert's lymphoma) refractory to chemotherapy with the CD-52 antibody alemtuzumab. *Leuk Lymphoma* 46: 771–774.

Zhang, Z., Zhang, M., Goldman, C.K., Ravetch, J.V., Waldmann, T.A. (2003) Effective therapy for a murine model of adult T-cell leukemia with the humanized anti-CD52 monoclonal antibody, Campath-1H. *Cancer Res* 63: 6453–6457.

Zinzani, P.L., Alinari, L., Tani, M., Fina, M., Pileri, S., Baccarani, M. (2005) Preliminary observations of a phase II study of reduced-dose alemtuzumab treatment in patients with pretreated T-cell lymphoma. *Haematologica* 90: 702–703.

3
Bevacizumab (Avastin)

Eduardo Díaz-Rubio, Edith A. Perez, and Guiseppe Giaccone

3.1
Introduction

3.1.1
Angiogenesis is Vital for Tumor Development

Angiogenesis, the process whereby new blood vessels are formed from pre-existing vessels, is essential for the growth and development of both normal tissues and tumors (Carmeliet 2003; Carmeliet and Jain 2000). Physiologically, angiogenesis plays a critical role in embryonic and infant growth and development, but in healthy adults this role is limited to wound healing and the menstrual cycle (Ferrara et al. 2003). The malignant progression of solid tumors is dependent on pathological angiogenesis: tumors with a volume of less than $2\,mm^3$ receive their nutrients by diffusion, but in order to grow any larger, tumors must establish a vascular supply. Tumor cells not only rely on angiogenesis to ensure a steady supply of nutrients and growth factors, but also use the newly formed blood vessels as a route for metastatic spread.

Angiogenesis is a complex process that is tightly controlled by factors that stimulate or inhibit the formation of new blood vessels (Carmeliet 2003; Carmeliet and Jain 2000; Jain 2003). In order to grow, tumors must trigger the development of their own blood supply, which they do by disrupting the delicate balance of pro- and anti-angiogenic factors (Ferrara et al. 2003). Tumor mutations can tip the balance in favor of pro-angiogenic factors, stimulating the development of new blood vessels in and around the lesion – this is termed the "angiogenic switch" (Bergers and Benjamin 2003). Because tumor angiogenesis results from abnormalities in the tightly controlled expression of pro- and anti-angiogenic factors, the resulting tumor vasculature has abnormal structure and function. Compared with vasculature in healthy tissues, tumor blood vessels are immature, abnormally branched, irregularly shaped, and prone to hemorrhage. Tumor vessels are also leaky, because they lack the tight endothelial layer seen in normal vasculature. This leads to a rise in interstitial fluid pressure in tumor tissue,

Handbook of Therapeutic Antibodies. Edited by Stefan Dübel
Copyright © 2007 WILEY-VCH Verlag GmbH & Co. KGaA, Weinheim
ISBN 978-3-527-31453-9

which compromises the penetration of oxygen and macromolecules, such as chemotoxic agents, into the lesion.

3.1.2
Vascular Endothelial Growth Factor: A Key Angiogenic Factor

Of the different pro-angiogenic factors, vascular endothelial growth factor (VEGF) is the key mediator of both normal and tumor angiogenesis (Fig. 3.1) (Ferrara 2001; Jain 2003; Yancopoulos et al. 2000). VEGF is a homodimeric heparin-binding glycoprotein with a molecular weight of approximately 45 kDa. VEGF, also known as VEGF-A, belongs to a subfamily that includes the growth factors VEGF-B, VEGF-C, VEGF-D and platelet growth factor (Shibuya 2001). At least four main VEGF isoforms exist, the most common being $VEGF_{165}$, which consists of 165 amino acids (Ferrara 2001; Ferrara et al. 2003). VEGF binds primarily to two receptors situated on vascular endothelial cells: VEGF receptor-1, also known as Flt-1, and VEGF receptor-2, also known as Flk-1 or KDR (Ferrara et al. 2004).

Activation of the VEGF pathway through binding of VEGF to VEGF receptor-2 triggers a series of signaling events that promote vascular endothelial cell growth, migration and survival, as well as the mobilization of endothelial progenitor cells from the bone marrow into the circulation (Ferrara et al. 2003, 2004). Under normal conditions, the balanced production of VEGF and other pro-angiogenic factors and anti-angiogenic factors ensures that blood vessels are formed when

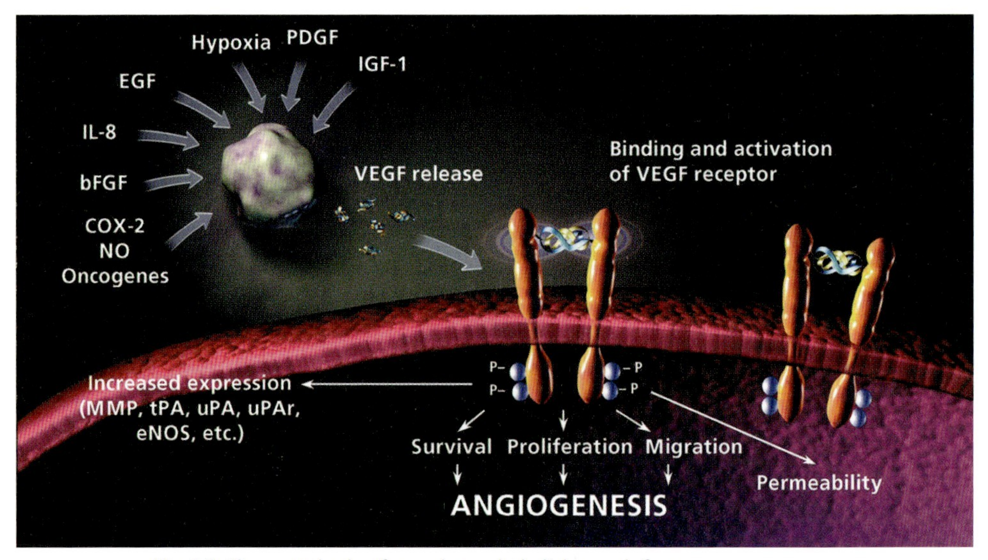

Fig. 3.1 The central role of vascular endothelial growth factor (VEGF) in angiogenesis.

and where they are physiologically required. However, in tumors the uncontrolled release of VEGF drives many of the processes central to tumor angiogenesis: it stimulates vascular endothelial cells to grow and migrate to form blood vessels; and it enables the resulting morphologically and functionally abnormal vessels to survive inside tumors. VEGF expression has also been shown to correlate with microvessel density and metastatic spread in some tumor types (Bergers and Benjamin 2003; Ferrara et al. 2004). The induction of VEGF expression appears to be characteristic of many tumor types, and studies have demonstrated that increased VEGF levels are a predictor of poor prognosis for cancer patients (Poon et al. 2001).

VEGF is also known as vascular permeability factor because initially it was shown to induce vascular permeability; in tumors, it drives the formation of holes in the walls of the vasculature and increases the permeability of blood vessels to circulating macromolecules (Ferrara et al. 2003). This allows plasma proteins to leak out of tumor blood vessels to form an extravascular matrix, which enhances the environment for subsequent endothelial cell growth. Increasing the permeability of the tumor vasculature also results in higher interstitial pressure, which reverses the normal pressure gradients within tissues and impairs the delivery of oxygen and chemotherapeutic drugs to the tumor (Jain 2002).

A variety of factors up-regulate VEGF expression (Bergers and Benjamin 2003; Ferrara 2002; Ferrara et al. 2004). Hypoxia is the primary stimulus, triggering vessel growth by signaling through hypoxia-inducible factors (HIFs), especially HIF-1α. As a tumor grows, it outstrips its blood supply and becomes hypoxic, which prompts an increase in HIF-1α levels. HIF-1α induces VEGF expression and the stimulation of further angiogenesis. In this way, VEGF production by tumors creates a positive feedback loop that supports the development of further vasculature and thus tumor growth. Other stimuli that up-regulate VEGF production include major growth factors that are frequently expressed by tumors; these include epidermal growth factor (EGF), keratinocyte growth factor 1, insulin-like growth factor 1 (IGF-1), fibroblast growth factor (FGF) and platelet-derived growth factor (PDGF) (Frank et al. 1995). Hormones such as thyroid-stimulating hormone (TSH), angiotensin II and adrenocorticotropic hormone (ACTH) also induce VEGF expression (Shifren et al. 1998; Soh et al. 1996), as do inflammatory cytokines such as interleukin (IL)-1α, IL-6 and prostaglandin E_2 (Ben-Av et al. 1995; Cohen et al. 1996). In addition, mutations in tumor suppressor genes including *p53* and oncogenes such as *ras* up-regulate VEGF expression.

3.1.3
Targeting Angiogenesis and VEGF: A Rational Treatment Option

The central role of VEGF in tumor angiogenesis makes the former an attractive target for anticancer therapy. VEGF inhibition has the potential to cause regression of immature tumor blood vessels, to inhibit the development of new tumor vasculature, and to reverse the structural and functional abnormalities in existing tumor vasculature.

Importantly, inhibiting VEGF is likely to block tumor angiogenesis without compromising healthy vasculature. As angiogenesis plays a limited physiological role in healthy adults, the inhibition of VEGF should have minimal undesired effects on normal physiological processes and thus have relatively few adverse side effects.

Two additional factors make VEGF an attractive therapeutic target. First, because VEGF circulates in the blood and acts directly on vascular endothelial cells, drugs that target VEGF do not need to penetrate tumor tissue in order to inhibit tumor angiogenesis. Second, VEGF acts on endothelial cells, which are relatively stable and therefore less likely to mutate to a treatment-resistant phenotype than genetically unstable tumor cells. This makes endothelial cells a more attractive target than tumor cells for long-term therapy (Kerbel and Folkman 2002).

Two different approaches to targeting VEGF have been investigated: one approach involves targeting the VEGF molecule itself, while the other focuses on the cell-surface receptors for VEGF (Gabrilovich et al. 1999; Vitaliti et al. 2000). One disadvantage of receptor-targeted approaches is that the VEGF receptors may also bind members of the VEGF superfamily other than VEGF, and their inhibition could influence physiological systems other than angiogenesis. Targeting the VEGF molecule itself enables selective action on the best-known angiogenic factor, and is also likely to be the safer option, with minimal adverse side effects on normal physiology.

The most advanced approach to VEGF inhibition is the humanized monoclonal antibody bevacizumab (Avastin), which inhibits the VEGF ligand and is currently the only anti-angiogenic antibody approved for the treatment of cancer. Other anti-angiogenic agents in relatively advanced stages of development include sunitinib (Sutent), sorafenib (Nexavar) and PTK/ZK (vatalanib), all of which are small-molecule tyrosine kinase inhibitors targeting one or more of the VEGF receptors as well as other tyrosine kinase receptors. In order that the information regarding bevacizumab can be considered in context, the agents are briefly described at this point:

- Sunitinib is an orally active compound that inhibits VEGF receptor-2 as well as the PDGF, Flt3 and kit receptors; as such, it has the potential to inhibit both angiogenesis and tumor cell growth. Based on positive interim data from phase III trials, sunitinib has recently been submitted for regulatory approval in the USA as monotherapy for patients with refractory gastrointestinal (GI) stromal tumors and renal cell carcinoma (RCC) (Demetri et al. 2005; Motzer et al. 2005).
- Sorafenib is an oral cytostatic pan-kinase inhibitor targeting the VEGF, PDGF and Raf kinase receptors. It has recently been launched as monotherapy for the treatment of patients with advanced refractory RCC in the USA following a phase III trial demonstrating improved disease

control rates and progression-free survival over placebo (Escudier et al. 2005). Further trials of both sunitinib and sorafenib are ongoing.
- Vatalanib is an oral VEGF receptor tyrosine kinase inhibitor. Data from two phase III trials of vatalanib in combination with FOLFOX4 (5-fluorouracil [5-FU]/leucovorin [LV] plus oxaliplatin [Eloxatin]) as either first- or second-line therapy have been reported; neither of these trials has demonstrated any benefit of the addition of vatalanib to FOLFOX4 in terms of response or progression-free survival, and it seems unlikely that there will be an overall survival benefit (Hecht et al. 2005; Schering AG 2005). Vatalanib continues to be assessed in clinical trials for colorectal cancer (CRC), RCC and non-small cell lung cancer (NSCLC).

Based on these data, it is clear that the availability of agents with anti-angiogenic activity is expanding rapidly. However, the data raise questions regarding the utility of targeting only the VEGF receptors, and whether inhibition of receptors involved in other tumor growth and survival mechanisms is required to have anticancer activity. For a number of reasons, the data mentioned do not provide answers to these issues. However, as described below, inhibition of the VEGF ligand using the humanized anti-VEGF monoclonal antibody has proved to be clinically effective in combination with chemotherapy in metastatic CRC, NSCLC and breast cancer.

3.2 The Development of Bevacizumab

3.2.1 Origin and Genetic Engineering

Bevacizumab is a humanized monoclonal antibody developed from A4.6.1, a murine antibody that targets human VEGF. A4.6.1 recognizes a particular sequence of amino acids within the human VEGF molecule (Wiesmann et al. 1997) and binds at this specific site with high affinity [dissociation constant (K_d) = 8×10^{-10} M]. Binding of A4.6.1 to VEGF prevents VEGF from binding to and activating its receptors (Ferrara et al. 2003).

As a murine protein, A4.6.1 is likely to provoke an immune response in humans and is therefore unsuitable for use in patients. To resolve this problem, the A4.6.1 antibody was humanized to form bevacizumab, which is 93% human and 7% mouse in origin. It was engineered by site-directed mutagenesis of a human DNA framework; in this process, the six regions that determine the binding specificity of A4.6.1 were transferred to a human DNA framework (Presta et al. 1997). In order to provide equivalent binding affinity, seven amino acid residues within the framework were changed to their corresponding murine residues.

3.2.2
Mode of Action

Bevacizumab blocks the binding of all VEGF isoforms to their receptors by binding to free VEGF, thus removing VEGF from the circulation (Presta et al. 1997). By targeting circulating VEGF rather than a specific receptor, bevacizumab inhibits VEGF activity at all the receptors to which VEGF binds.

Anti-VEGF therapy inhibits the development of new tumor vasculature, which is essential for further tumor growth and metastasis. VEGF inhibition in human tumor cell lines has been shown to inhibit vascular endothelial cell proliferation and migration, and to suppress new vascular sprouting within 24 h of administration (Baluk et al. 2005). In a mouse model, anti-VEGF therapy blocked new vessel formation and caused the collapse of pre-existing vessels within 9 days (Osusky et al. 2004). Preclinical studies have demonstrated that targeting VEGF inhibits tumor angiogenesis, blocks tumor growth (Kim et al. 1993), and reduces the number and size of liver metastases in nude mice (Warren et al. 1995). The removal of anti-VEGF therapy leads to rapid regrowth of capillaries and the resumption of tumor growth (Baluk et al. 2005), suggesting that continuous VEGF inhibition is required to gain the greatest antitumor benefit.

As well as inhibiting new tumor blood vessel formation, anti-VEGF therapy causes regression of existing tumor vasculature. Throughout the early stages of tumor development, the formation of new blood vessels remains dependent upon a constant supply of VEGF (Bergers and Benjamin 2003). VEGF is a survival factor for vascular tumor cells, protecting them from apoptosis and promoting tumor growth (Harmey and Bouchier-Hayes 2002). Without a steady supply of VEGF, vascular endothelial cells undergo apoptosis, and any recently formed tumor microvasculature disintegrates. A single infusion of anti-VEGF therapy has been shown to reduce tumor microvascular density in humans (Willett et al. 2004).

VEGF inhibition may also result in the remodeling of tumor vasculature: immature vessels are pruned away, leaving only mature, normally functioning blood vessels (Jain 2001, 2005). This process leads to a more ordered tumor vasculature, a more efficient blood supply, and decreased interstitial fluid pressure inside the tumor (Jain 2001), such that chemotoxic agents and oxygen can penetrate the tumor more effectively. Thus, anti-VEGF therapy may improve the efficacy of chemotherapy and radiotherapy.

3.2.3
Preclinical Activity of Bevacizumab

Bevacizumab and A4.6.1 have demonstrated almost identical activity in preclinical tumor growth models; bevacizumab inhibited VEGF-induced proliferation of endothelial cells *in vitro* with an ED_{50} (effective dose) of $50 \pm 5\,ng\,mL^{-1}$, while the ED_{50} for A4.6.1 is $48 \pm 8\,ng\,mL^{-1}$ (Presta et al. 1997). Both antibodies also block angiogenesis and suppress the growth of human tumor xenografts in mice to similar extents (Presta et al. 1997). Thus, in human cells and tissues, these two antibodies are pharmacologically equivalent.

Bevacizumab and A4.6.1 have also demonstrated synergistic activity in combination with chemotherapy in preclinical models. A synergistic antitumor effect was observed when bevacizumab ($2.5\,mg\,kg^{-1}$) was combined with capecitabine (Xeloda) ($360\,mg\,kg^{-1}$); the combination inhibited the growth of colorectal tumor xenografts in nude mice more effectively and for longer than either agent alone (Shen et al. 2004). Bevacizumab has also shown synergy with paclitaxel (Taxol) and trastuzumab (Herceptin), an inhibitor of the human epidermal growth factor receptor (HER)2. Several other studies have demonstrated the potential of bevacizumab in combination with standard chemotherapies. *In vitro*, bevacizumab overcame VEGF-induced protection of endothelial cells against docetaxel (Taxotere) treatment (Sweeney et al. 2001). *In vivo*, bevacizumab enhanced tumor suppression in animals when added to cisplatin (Platinol) (Kabbinavar et al. 1995), topotecan (Soffer et al. 2001), or capecitabine (Shen et al. 2004). Overall, the preclinical toxicity data indicate that both single and repeated doses of bevacizumab are likely to be well tolerated (Ryan et al. 1999). These encouraging preclinical data prompted an extensive clinical trial program, the details of which are outlined in the following section.

3.3 Clinical Trials of Bevacizumab in Key Tumor Types

The safety of bevacizumab as monotherapy (Gordon et al. 2001) and in combination with chemotherapy (Margolin et al. 2001) was assessed in two phase I trials. In the monotherapy trial, 25 patients with various types of solid tumor received 0.1, 0.3, 1, 3, or $10\,mg\,kg^{-1}$ bevacizumab on days 0, 28, 35, and 42 (Gordon et al. 2001). There were no grade 3 or 4 adverse effects, but a small number of patients reported grade 1 or 2 asthenia, headache, fever, rash and nausea. In the phase I trial of bevacizumab combined with chemotherapy, 12 patients with solid tumors received $3\,mg\,kg^{-1}$ bevacizumab weekly for 8 weeks with carboplatin (Paraplatin)/paclitaxel, doxorubicin (Adriamycin) or 5-fluorouracil (5-FU)/leucovorin (LV) (Margolin et al. 2001). Bevacizumab did not increase the frequency or severity of the anticipated adverse side effects of the concomitant chemotherapy regimens. In addition, no significant bevacizumab-related toxicity was observed. Of the 12 patients, three responded to treatment and remained on bevacizumab therapy for up to 40 weeks. No additional toxicity was observed in these patients, suggesting that long-term bevacizumab use may be well tolerated (Margolin et al. 2001).

The promising clinical activity and favorable toxicity profile demonstrated in these trials led to phase II and III trials of bevacizumab combined with standard chemotherapy regimens in a range of indications including CRC, breast cancer, NSCLC, RCC, pancreatic cancer, and ovarian cancer. Given that anti-VEGF therapy is most likely to have an effect in earlier disease, when the tumor is most dependent on VEGF for further growth and development, bevacizumab is currently being investigated in the adjuvant setting and as first-line therapy for metastatic disease in combination with standard regimens. To date, more than 7000 patients have been treated with bevacizumab in clinical trials.

3.3.1 Colorectal Cancer

Until relatively recently, 5-FU with or without LV defined the standard of care for the first-line treatment of patients with metastatic CRC. Irinotecan (Camptosar) combined with bolus 5-FU/LV (IFL) became a new standard for chemotherapy-naïve patients with metastatic CRC based on the findings of two randomized trials. These trials demonstrated improved response rates and survival for patients treated with irinotecan combined with 5-FU/LV compared with bolus (Saltz et al. 2000) or infusional (Douillard et al. 2000) 5-FU treatment alone. Oxaliplatin was established as another agent for first-line treatment of metastatic CRC following a randomized trial (de Gramont et al. 2000) that demonstrated an improved response rate and progression-free survival in patients treated with oxaliplatin plus infusional 5-FU/LV (FOLFOX4) compared with 5-FU/LV alone. A subsequent study suggested that FOLFOX4 might improve median survival compared with IFL in patients with metastatic CRC (Goldberg et al. 2004).

While these new drugs have improved clinical outcomes for patients with CRC, the poor prognosis of this disease means that new therapies are still needed. Novel targeted therapies such as bevacizumab have been developed to improve survival without significantly increasing the toxicity of current therapies. On the basis of the evidence provided by the clinical trials of bevacizumab outlined in the following section, the US Food and Drug Administration (FDA) approved bevacizumab for the first-line treatment of metastatic CRC in combination with 5-FU-based chemotherapy in February 2004. The recommended dose is $5\,mg\,kg^{-1}$ bevacizumab given once every 14 days as an intravenous (IV) infusion until disease progression is detected. A European license was granted in January 2005 for the first-line use of bevacizumab ($5\,mg\,kg^{-1}$ once every 14 days until disease progression) in combination with 5-FU/LV, with or without irinotecan.

3.3.1.1 Efficacy of Bevacizumab in CRC

An initial phase II randomized, controlled, open-label clinical trial (AVF0780) compared bevacizumab plus 5-FU/LV with chemotherapy alone (Kabbinavar et al. 2003). In this trial, 104 patients with untreated metastatic CRC were randomized to one of three arms: control (5-FU/LV alone, n = 36); bevacizumab $5\,mg\,kg^{-1}$ every 2 weeks plus 5-FU/LV (n = 35); and bevacizumab $10\,mg\,kg^{-1}$ every 2 weeks plus 5-FU/LV (n = 33). Patients received 5-FU/LV according to the Roswell Park regimen: 5-FU $500\,mg\,m^{-2}$ and LV $500\,mg\,m^{-2}$ weekly for the first 6 weeks of each 8-week cycle. Patients in the control arm were allowed to cross over to receive bevacizumab on disease progression.

The addition of bevacizumab to 5-FU/LV improved outcomes compared with 5-FU/LV alone, with the $5\,mg\,kg^{-1}$ bevacizumab dose appearing to be most effective. Bevacizumab $5\,mg\,kg^{-1}$ significantly increased time to disease progression, the primary endpoint, by 73% from 5.2 months in the control arm to 9.0 months for patients treated with bevacizumab plus 5-FU/LV ($p = 0.005$). The median time to progression for patients receiving bevacizumab $10\,mg\,kg^{-1}$ was 7.2 months. The risk of disease progression was 61% lower in the $5\,mg\,kg^{-1}$ arm ($p = 0.002$ versus control) and 46% lower in the $10\,mg\,kg^{-1}$ arm ($p = 0.052$ versus control). The tumor response rate increased from 17% in the 5-FU/LV arm to 24% in the bevacizumab $10\,mg\,kg^{-1}$ arm ($p = 0.434$ versus control) and 40% in the bevacizumab $5\,mg\,kg^{-1}$ arm ($p = 0.029$ versus control). Bevacizumab led to a trend towards increased median overall survival (13.8 months for the control arm and 16.1 and 21.5 months for the bevacizumab $10\,mg\,kg^{-1}$ and $5\,mg\,kg^{-1}$ arms, respectively), but this increase failed to reach statistical significance. This may have been due to the small sample size and the fact that patients in the 5-FU/LV arm were allowed to cross over to bevacizumab on disease progression. Based on these data, a bevacizumab dose of $5\,mg\,kg^{-1}$ every 2 weeks was selected for subsequent trials.

A second phase II randomized, double-blind, controlled clinical trial (AVF2192) evaluated the efficacy of bevacizumab in combination with 5-FU/LV in patients with untreated metastatic CRC who were not optimal candidates for first-line irinotecan treatment (Kabbinavar et al. 2005a). The toxicity of irinotecan makes this agent unsuitable for some patients, in particular those with poor performance status or prior radiation therapy to the abdomen or pelvis. It should be noted that these patients are generally less healthy and tend to have a poorer prognosis than those eligible for irinotecan therapy. A total of 209 patients was randomized to receive 5-FU/LV (Roswell Park regimen) combined with either placebo ($n = 105$) or bevacizumab $5\,mg\,kg^{-1}$ ($n = 104$) every 2 weeks.

Bevacizumab led to a trend towards improved median overall survival, the primary endpoint, by 29% from 12.9 months for patients receiving placebo to 16.6 months for patients in the bevacizumab arm ($p = 0.16$). Median progression-free survival increased from 5.5 months for the placebo arm to 9.2 months for the bevacizumab arm ($p = 0.0002$), while tumor response rates increased from 15.2% to 26% ($p = 0.055$). It should be noted that this trial was originally powered to identify an improvement in survival from 8.5 months in the 5-FU/LV arm to 14 months in the bevacizumab arm. Survival was considerably longer in both arms, making the trial underpowered for the overall survival endpoint. Nevertheless, the addition of bevacizumab to 5-FU/LV provides clinical benefit, as demonstrated by the observation that bevacizumab prolonged median and progression-free survival by 3.7 months when compared with 5-FU/LV alone.

A phase III randomized, double-blind, controlled clinical trial in patients with metastatic CRC (AVF2107) assessed whether combining bevacizumab with first-line irinotecan-containing chemotherapy improves overall survival compared with chemotherapy alone (Hurwitz et al. 2004). The study was designed to compare the efficacy of bevacizumab (5 $mg\,kg^{-1}$ IV every 2 weeks) combined with irinotecan (125 $mg\,m^{-2}$ by IV infusion), 5-FU (500 $mg\,m^{-2}$ by IV bolus) and LV (20 $mg\,m^{-2}$ by IV bolus) (IFL) with that of IFL alone. A total of 923 patients was randomly assigned to one of three arms: IFL/placebo (n = 411); IFL/bevacizumab (n = 402); or 5-FU (500 $mg\,m^{-2}$ by IV bolus)/LV (500 $mg\,m^{-2}$ by IV infusion)/bevacizumab (n = 110). IFL was administered once weekly for the first four weeks of a 6-week cycle, while 5-FU/LV was administered once weekly for the first six weeks of an 8-week cycle. The 5-FU/LV/bevacizumab arm was a control to ensure that safety in the IFL/bevacizumab arm was acceptable. This control arm was closed after approximately 100 patients had been enrolled to each arm, and an independent review of safety data had concluded that the safety profiles of the IFL/bevacizumab and 5-FU/LV/bevacizumab arms were comparable.

Adding bevacizumab to IFL significantly increased overall survival by 30%; patients receiving IFL plus placebo had a median survival of 15.6 months, which increased to 20.3 months in patients receiving IFL plus bevacizumab ($p < 0.001$) (Fig. 3.2). Furthermore, bevacizumab increased progression-free survival by 71%, from 6.2 to 10.6 months ($p < 0.001$). The overall response rate was 44.8% for the IFL/bevacizumab arm and 34.8% for the IFL/placebo arm ($p = 0.004$) (Hurwitz et al. 2004). Furthermore, improvements in survival were seen in all prospectively defined patient subgroups based on baseline patient and tumor characteristics and irrespective of VEGF expression status (Fyfe et al. 2004; Holden et al. 2005).

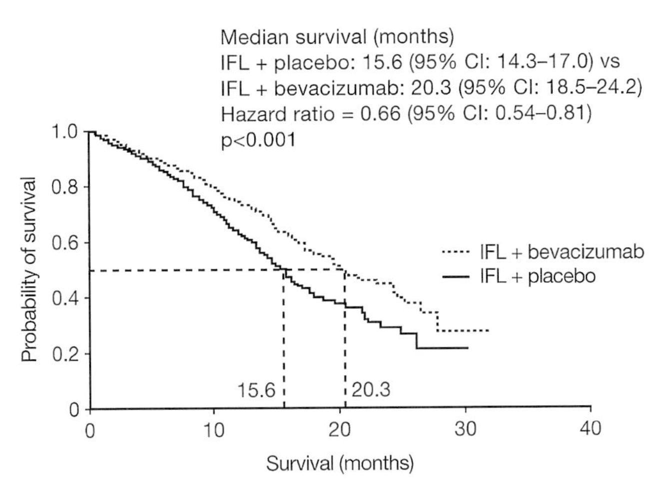

CI = confidence interval

Fig. 3.2 Overall survival in patients with metastatic colorectal cancer (CRC) treated with irinotecan + 5-fluorouracil [5-FU]/leucovorin [LV] (IFL), with or without bevacizumab. (Reproduced with permission; © 2004 Massachusetts Medical Society. All rights reserved.)

Hurwitz et al. (2004) also reported that in the 25% of patients in this trial who were treated second line with oxaliplatin, overall survival was 25.1 months in the IFL plus bevacizumab arm and 22.2 months in the IFL plus placebo arm. This finding supports the conclusion of others (Grothey et al. 2004), that patients who receive several active anticancer agents survive longer than those who receive only one or two agents.

Efficacy data for the third arm of this trial, in which patients were treated with bevacizumab plus 5-FU/LV, have been reported (Hurwitz et al. 2005). Interim efficacy and safety analyses were conducted after approximately 100 patients had been randomly assigned to each treatment arm (100 to the IFL/placebo arm, 103 to the IFL/bevacizumab arm and 110 to the 5-FU/LV/bevacizumab arm). In this analysis, median overall survival was 18.3 and 15.1 months with 5-FU/LV/bevacizumab and IFL/placebo, respectively ($p = 0.25$), while progression-free survival was 8.8 and 6.8 months ($p = 0.42$) and the overall response rate was 40% and 37% ($p = 0.66$). These data confirm the activity of bevacizumab in CRC beyond combination with IFL and suggest that the 5-FU/LV/bevacizumab regimen is an active alternative treatment regimen for CRC patients.

Each of the 5-FU/LV/bevacizumab arms of these three clinical trials (Hurwitz et al. 2004; Kabbinavar et al. 2003, 2005a) recruited a relatively small number of patients. This may explain why none of these trials showed a statistically significant increase in overall survival for patients treated with bevacizumab plus 5-FU/LV, even though the increases were clinically significant. To evaluate the efficacy of the 5-FU/LV/bevacizumab regimen in a larger patient population, a combined analysis of pooled raw data from the two phase II trials and the 5-FU/LV/bevacizumab arm of the pivotal phase III trial was conducted. The combined control group comprised patients randomized to 5-FU/LV or IFL, while the bevacizumab arm included patients randomized to 5-FU/LV plus bevacizumab $5\,mg\,kg^{-1}$ every 2 weeks. The inclusion of IFL patients in the control arm biased the analysis against seeing a benefit with bevacizumab because IFL is a more effective regimen than 5-FU/LV (Kabbinavar et al. 2005a).

The combined analysis (Kabbinavar et al. 2005b) revealed that adding bevacizumab ($5\,mg\,kg^{-1}$ every 2 weeks) to 5-FU/LV significantly increased overall survival; median overall survival in the pooled 5-FU/LV/bevacizumab group was 17.9 months, compared with 14.6 months in the combined control group ($p = 0.008$). Progression-free survival also increased from 5.6 to 8.8 months ($p = 0.0001$), as did the response rate (from 24.5% to 34.1%, $p = 0.019$). The improvements in overall and progression-free survival seen with bevacizumab in this combined analysis were as good as or better than those observed when irinotecan or oxaliplatin is added to 5-FU/LV (de Gramont et al. 2000; Saltz et al. 2000).

A recent trial (Eastern Cooperative Oncology Group [ECOG] 3200) examined bevacizumab plus FOLFOX4 as second-line therapy for metastatic CRC (Giantonio et al. 2005). In this trial, 829 CRC patients previously treated with an irinotecan-based regimen received bevacizumab ($10\,mg\,kg^{-1}$ every 2 weeks), FOLFOX4 or FOLFOX4 plus bevacizumab. Adding bevacizumab to FOLFOX4 significantly improved median overall survival, which increased from 10.8 months with

FOLFOX4 alone to 12.9 months for patients who received FOLFOX4 plus bevacizumab (hazard ratio = 0.76; p = 0.0018). Progression-free survival also increased (from 4.8 months for FOLFOX4 alone to 7.2 months for FOLFOX4 plus bevacizumab, $p < 0.0001$), as did the response rate (from 9.2% to 21.8%, $p < 0.0001$). In this study, the safety profile of bevacizumab plus oxaliplatin appeared similar to that observed in other trials of bevacizumab in CRC.

3.3.1.2 Safety and Management of Bevacizumab in CRC

Bevacizumab appears to be generally well tolerated by patients with metastatic CRC (Table 3.1). The most common adverse effects observed in clinical trials of CRC that are attributable to bevacizumab therapy include hypertension, proteinuria, arterial thrombosis, wound-healing complications, and bleeding. While higher incidences of these events were reported in patients receiving

Table 3.1 Incidence of selected adverse events in clinical trials of bevazumab in combination with chemotherapy for metastatic CRC (NB: data are not adjusted for different periods of therapy).

	Kabbinavar et al. (2005a) AVF2192		*Hurwitz et al. (2004) AVF2107*		*Hurwitz et al. (2005) Third arm of AVF2107*	
Adverse events (% of patients)	*5-FU/LV + placebo (n = 104)*	*5-FU/LV + bevacizumab 5 mg kg⁻¹ (n = 100)*	*IFL + placebo (n = 397)*	*IFL + bevacizumab 5 mg kg⁻¹ (n = 393)*	*IFL + placebo (n = 98)*	*5-FU/LV + bevacizumab 5 mg kg⁻¹ (n = 109)*
Hypertension						
Any	5	32	8.3	22.4	14.3	33.9
Grade 3	3	16	2.3	11.0	3.1	18.3
Proteinuria						
Any	19	38	21.7	26.5	25.1	34.9
Grade 3	0	1	0.8	0.8	0	1.8
Any thrombotic event	18	18	16.2	19.4	19.4	13.8
Bleeding event						
Grade 3/4	3	5	2.5	3.1	1.0	6.4
Diarrhea						
Grade 3/4	40	39	24.7	32.4	25.5	37.6
Leukopenia						
Grade 3/4	7	5	31.1	37.0	37.7	5.5
GI perforation	0	2	0.0	1.5	0	0

IFL = 5-fluorouracil [5-FU]/leucovorin [LV].

bevacizumab than in control patients, most of these adverse effects were manageable.

Hypertension is the most common adverse side effect reported in trials of bevacizumab in CRC, with 22.4 to 33.9% of all bevacizumab-treated patients reporting hypertension of any grade. However, very few grade 4 events were reported, and only 0.7% of all patients in clinical trials discontinued therapy due to hypertension. Grade 3 hypertension was reported by 11 to 18.3% of bevacizumab-treated patients. Although hypertension can occur at any time during the course of treatment, in most cases it can be effectively managed with oral anti-hypertensive agents. However, if these are not effective in lowering blood pressure, then bevacizumab treatment should be discontinued. Treatment should also be permanently discontinued in patients who develop hypertensive crises. Caution should be exercised before administering bevacizumab to patients with uncontrolled hypertension, and the patients' blood pressure should be monitored during treatment.

In CRC trials, 26.5 to 38% of bevacizumab-treated patients reported proteinuria of any grade, compared with 19 to 25.1% of control patients; proteinuria is usually asymptomatic, and grade 4 proteinuria was rarely reported. Proteinuria should be monitored with dipstick urinalysis before and during bevacizumab treatment. Proteinuria improves when bevacizumab therapy is stopped, and it is currently recommended that bevacizumab therapy be interrupted if patients develop proteinuria ≥2 g per 24 h, and discontinued if patients develop nephrotic syndrome.

In clinical trials, the overall incidence of any thromboembolism in patients with metastatic CRC was similar in bevacizumab-treated (13.8–19.4%) and control patients (16.2–19.4%). The incidence of grade 3/4 thromboembolic events was also similar in the two groups (16.0–17.6% with bevacizumab versus 14.4–18.3% with control). However, a higher incidence of arterial thromboembolic events was reported for bevacizumab-treated patients (3.3–10%) than for control patients (1.3–4.8%). In light of these findings, a retrospective analysis examined five controlled, completed clinical trials of bevacizumab to identify whether bevacizumab treatment carries an increased risk of arterial thromboembolism: three CRC trials (Hurwitz et al. 2004; Kabbinavar et al. 2003, 2005a), one NSCLC trial (Johnson et al. 2004), and one metastatic breast cancer (MBC) trial (Miller et al. 2005a) were used for the analysis.

The results revealed an increased risk of developing arterial thromboembolic events in bevacizumab-treated patients, 3.8% of whom reported an arterial thromboembolic event compared with 1.7% of patients receiving chemotherapy alone. In the phase III trial of IFL with or without bevacizumab, some patients who developed thromboembolic events were treated using anticoagulation therapy (full-dose warfarin); an analysis of data from these patients suggested that co-administration of warfarin and bevacizumab did not increase the incidence of bleeding events (Hambleton et al. 2004). The risk of developing an arterial thromboembolic event was higher for patients aged over 65 years or who had a history of such events. Patients who experience an arterial thromboembolic event should discontinue bevacizumab therapy.

One of the known roles of VEGF in adults is its involvement in wound healing. However, bevacizumab-treated patients who underwent surgery between 28 and 60 days before starting therapy did not report any wound-healing complications during bevacizumab treatment. In contrast, 10 to 20% of bevacizumab-treated patients who underwent major surgery (emergency and elective) while receiving bevacizumab treatment reported adverse events indicative of wound-healing complications. For this reason, it is recommended that bevacizumab therapy be discontinued at least 6 weeks prior to elective surgery. Therapy should also be discontinued in any patients who develop wound-healing complications that require medical intervention.

Minor mucocutaneous hemorrhage was reported relatively frequently in CRC clinical trials, with 22 to 31.2% of bevacizumab-treated patients experiencing epistaxis. This was most commonly grade 1 in severity, lasting less than 5 min, and was easily managed using standard first-aid techniques. Bevacizumab did not increase the risk of severe hemorrhagic events; the incidence of grade 3 and 4 bleeding events in bevacizumab-treated patients was similar to that in control patients. It is important to note that the risk of central nervous system (CNS) hemorrhage in patients with CNS metastases has not been fully evaluated because these patients were excluded from bevacizumab clinical trials. Bevacizumab is therefore contraindicated in patients with untreated CNS metastases. Furthermore, no information is available on the safety profile of bevacizumab in patients with congenital bleeding diathesis or acquired coagulopathy, and those receiving full-dose anticoagulation for the treatment of thromboembolism prior to starting bevacizumab treatment. These patients should be treated with bevacizumab with care, although patients who developed venous thrombosis while receiving bevacizumab therapy did not appear to be at increased risk of serious bleeding when treated with full-dose warfarin and bevacizumab concomitantly.

Finally, GI perforation is a rare but life-threatening side effect reported in patients with metastatic CRC treated with bevacizumab. Across clinical trials in CRC, GI perforation was observed in 1.5 to 2% of bevacizumab-treated patients; of these events, <0.5% were fatal. In some cases predisposing factors were present, such as intra-abdominal inflammation due to gastric ulcer disease, tumor necrosis or chemotherapy-associated colitis (Hurwitz et al. 2004; Johnson et al. 2004). Thus, bevacizumab-treated patients with metastatic CRC and an intra-abdominal inflammation may be at increased risk of developing a GI perforation. If a patient develops a GI perforation, bevacizumab should be permanently discontinued.

In summary, bevacizumab is generally well tolerated and does not increase the frequency or severity of chemotherapy-related toxicities. While some specific safety concerns have been identified, most adverse side effects are either easily manageable, or are not clinically significant.

3.3.1.3 **Future Use of Bevacizumab in CRC**

Clinical data indicate that bevacizumab has considerable potential in combination with existing therapeutic options; indeed, it is the only agent to date that has demonstrated a survival benefit in metastatic CRC when combined with first-line chemotherapy. Following these encouraging data, several ongoing trials are examining bevacizumab combined with oxaliplatin-based chemotherapy as first-line therapy for metastatic CRC (e.g. studies NO16966, and DREAM). Bevacizumab may also confer clinical benefit in the adjuvant setting in view of the fact that VEGF inhibition should prevent the angiogenic switch, which is a key factor in malignant growth, but further clinical studies must be conducted are required before reaching any conclusions. Several trials are ongoing (BO17920, NSABP C-08 and QUASAR2) to examine the potential role of this agent as part of adjuvant treatment in CRC.

The phase III trial NO16966 enrolled 2,035 patients, and is designed to compare the safety and efficacy of XELOX (oxaliplatin 130 mg m^{-2} day 1, capecitabine 1000 mg m^{-2} twice daily for 14 days, every 3 weeks) with or without bevacizumab (7.5 mg kg^{-1} every 3 weeks) with that of FOLFOX4 with or without bevacizumab (5 mg kg^{-1} every 2 weeks) as first-line treatment for patients with metastatic CRC. The primary objectives of this trial are to show that the XELOX and FOLFOX4 regimens provide equivalent progression-free survival, and that adding bevacizumab to XELOX/FOLFOX4 improves progression-free survival compared with XELOX/FOLFOX4 alone. Both endpoints were met; XELOX is as effective as FOLFOX4 in terms of progression-free survival (8.0 versus 8.5 months, respectively), and the addition of bevacizumab to either chemotherapy (XELOX or FOLFOX4) significantly improved progression-free survival compared with chemotherapy alone (9.4 versus 8.0 months) (Cassidy et al., 2006).

TREE-2 is a randomized study comparing three oxaliplatin-based regimens plus bevacizumab in 223 CRC patients (Hochster et al. 2006). Patients received bolus 5 FU/LV combined with oxaliplatin (bFOL) and bevacizumab (5 mg kg^{-1} every 2 weeks), XELOX plus bevacizumab (7.5 mg kg^{-1} every 3 weeks) or a modified FOLFOX6 regimen plus bevacizumab (5 mg kg^{-1} every 2 weeks). Recently presented data indicate that combining these regimens with bevacizumab improved response rates compared with the chemotherapy regimens on their own; 53% for modified FOLFOX6 plus bevacizumab, 48% for the XELOX combination, and 41% for the bFOL combination. The addition of bevacizumab also improved time to progression (9.9, 10.3 and 8.3 months) and overall survival (26.0, 27.0 and 20.7 months) compared with modified FOLFOX6, XELOX and bFOL alone.

The toxicity of chemotherapy regimens remains an issue, and the OPTIMOX or "stop-and-go" approach has been developed to reduce the cumulative sensory neurotoxicity observed with oxaliplatin (de Gramont et al. 2004). Given that bevacizumab increases progression-free survival and treatment duration, cumulative neurotoxicity is likely to become an issue when bevacizumab is combined with oxaliplatin-based regimens. To examine the OPTIMOX approach as a way to limit toxicity and maintain efficacy, the phase III DREAM trial will recruit 640 patients with metastatic CRC. Patients will be randomized to one of three arms in which chemotherapy is paused and reintroduced: modified FOLFOX7 or XELOX4 plus bevacizumab (with bevacizumab plus erlotinib (Tarceva) during chemotherapy pause) or modified FOLFOX7 plus bevacizumab (with bevacizumab alone during chemotherapy pause).

BRiTE and First BEAT are two large studies assessing bevacizumab safety and efficacy in large patient populations; 1,987 and 1,927 patients with metastatic CRC, receiving bevacizumab with first-line chemotherapy (regimen choice at physician's discretion) were recruited, respectively. The safety profile of bevacizumab in these studies appears to be consistent with the prospective randomised clinical trials (Cunningham et al. 2006; Hedrick et al. 2006); bevacizumab-related serious adverse effects were reported in 9–12% of the study populations, GI perforations were uncommon, and no new safety signals were identified. The estimated median progression-free survival of 10.2 months observed in BRiTE compares favourably with prospective randomised clinical trials (Kozloff et al., 2006).

In the adjuvant setting, Roche is currently conducting an open-label phase III trial BO17920 of bevacizumab treatment of colon cancer (Fig. 3.3). This trial will recruit 3450 patients with high-risk stage II or III colon cancer (more than half have been enrolled to date) randomized to three treatment arms: FOLFOX4 for 24 weeks; FOLFOX4 plus bevacizumab ($5\,\mathrm{mg\,kg^{-1}}$ every 2 weeks) for 24 weeks followed by 24 weeks of bevacizumab monotherapy ($7.5\,\mathrm{mg\,kg^{-1}}$ every 3 weeks); and XELOX plus bevacizumab ($7.5\,\mathrm{mg\,kg^{-1}}$ every 3 weeks) for 24 weeks followed by 24 weeks of bevacizumab monotherapy ($7.5\,\mathrm{mg\,kg^{-1}}$ every 3 weeks). The primary endpoint of the study is disease-free survival.

In the USA, the National Surgical Adjuvant Breast and Bowel Project (NSABP) is conducting a phase III trial to assess the efficacy of bevacizumab added to FOLFOX6 (NSABP C-08). The trial aims to recruit 2700 patients who will receive modified FOLFOX6 (every 2 weeks for 12 cycles) either alone or combined with bevacizumab ($5\,\mathrm{mg\,kg^{-1}}$ every 2 weeks for a year). The primary endpoint of the study is disease-free survival.

Initial data from clinical trials to date suggest that bevacizumab is effective when combined with any standard chemotherapy regimen for the first-line treatment of metastatic CRC, including regimens containing 5-FU, capecitabine, irinotecan, or oxaliplatin. Data from a number of ongoing trials of first-line bevacizumab with regimens such as FOLFOX, XELOX, and FOLFIRI (5-FU/LV plus irinotecan) will be reported during 2006, and will confirm the safety and efficacy of bevacizumab as part of these first-line treatment regimens. Numerous other

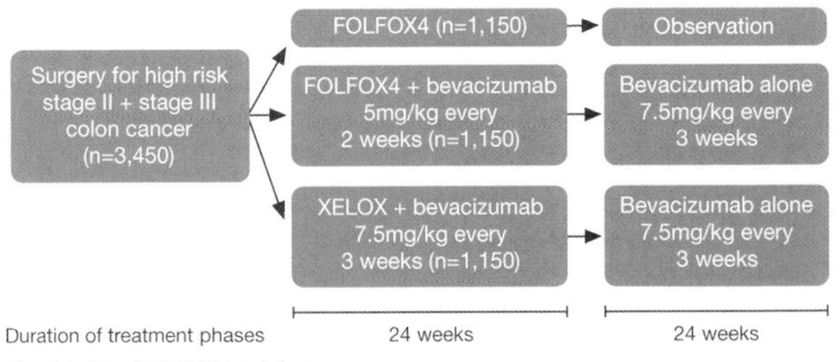

Fig. 3.3 The BO17920 trial design.

ongoing trials will determine the efficacy and safety of bevacizumab in combination with other chemotherapy regimens and biological therapies in this setting. Based on its mechanism of action, bevacizumab may be most effective as adjuvant therapy, and a number of ongoing phase III trials involving >6000 patients are designed to produce evidence to demonstrate this.

3.3.2 Non-Small Cell Lung Cancer

Patients with NSCLC have a poor prognosis and high mortality. The standard treatment for advanced NSCLC is platinum-based chemotherapy, which to date has demonstrated only modest survival benefit. The median overall survival with first-line platinum-based chemotherapy is only 8 to 10 months for advanced NSCLC patients, and novel treatment strategies are certainly urgently needed in this condition. Targeted therapies such as bevacizumab have the potential to improve patient survival without significantly adding to the toxicity of conventional chemotherapy. Indeed, bevacizumab plus carboplatin/paclitaxel is now the ECOG standard of care for patients with nonsquamous NSCLC (Sandler et al. 2005).

3.3.2.1 Efficacy of Bevacizumab in NSCLC

In the first randomized phase II trial of bevacizumab in NSCLC, patients with previously untreated advanced or recurrent NSCLC received chemotherapy (carboplatin/paclitaxel) alone or with bevacizumab at a dose of either 7.5 or 15 $mg\,kg^{-1}$ every 3 weeks (Johnson et al. 2004). The primary efficacy endpoints were time to disease progression and tumor response rate, while the secondary efficacy endpoints were overall survival and duration of response. In order to be eligible, patients had to be chemotherapy naïve, have histologically confirmed stage IIIB (with pleural effusion), stage IV or recurrent NSCLC, and have an ECOG performance status of 0, 1, or 2. In this trial, 99 patients were randomly assigned to one of three treatment arms: carboplatin/paclitaxel alone (n = 32); carboplatin/paclitaxel plus 7.5 $mg\,kg^{-1}$ bevacizumab every 3 weeks (n = 32); or carboplatin/paclitaxel plus 15 $mg\,kg^{-1}$ bevacizumab every 3 weeks (n = 35). Paclitaxel (200 $mg\,m^{-2}$) was administered over 3 h every 3 weeks, and carboplatin was given at an area under the concentration–time curve (AUC) of 6. Patients received up to six cycles of carboplatin/paclitaxel. On completion of the planned chemotherapy, patients without progressive disease were allowed to continue on bevacizumab. Patients in the chemotherapy-alone arm were permitted to receive bevacizumab (15 $mg\,kg^{-1}$ every 3 weeks) on disease progression.

In the 15 $mg\,kg^{-1}$ bevacizumab arm, time to disease progression was significantly increased to 7.4 months from 4.2 months in controls, which represents a 46% reduction in the risk of progression ($p = 0.023$). An apparent, but nonsignificant, improvement of overall survival was also observed: patients receiving chemotherapy plus bevacizumab 15 $mg\,kg^{-1}$ every 3 weeks survived for 17.7 months, compared with 14.9 months for control patients ($p = 0.63$). The lack of significance was not surprising, given the small number of patients enrolled in this study. Furthermore, 19 of the 32 control patients crossed over to single-agent bevacizumab on disease progression, which may have positively influenced survival in the control arm. Indeed, mean survival in the control arm of this study (14.9 months) was considerably higher than that in several other comparable trials of carboplatin/paclitaxel (8–10 months) (Kelly et al. 2001; Schiller et al. 2002).

Bevacizumab at a dose of $7.5\,mg\,kg^{-1}$ every 3 weeks did not improve time to progression or survival compared with the control arm (4.3 and 4.2 months, respectively). In this arm, five severe bleeding events associated with squamous cell histology occurred, which may have decreased median survival (10/32 patients in this arm had squamous cell histology, compared with 3/34 patients in the $15\,mg\,kg^{-1}$ arm). A more detailed discussion of the bleeding episodes is presented below.

These data provided the basis for a phase III trial (E4599) to further investigate bevacizumab at a dose of $15\,mg\,kg^{-1}$ every 3 weeks. In this trial, patients with previously untreated advanced nonsquamous NSCLC were randomized to receive carboplatin/paclitaxel alone or with bevacizumab at a dose of $15\,mg\,kg^{-1}$ every 3 weeks (Sandler et al. 2005). The primary endpoint of this study was overall survival, while the secondary endpoints included response rate, time to disease progression, and tolerability. In order to be eligible, patients had to be chemotherapy-naïve, have histologically confirmed stage IIIB (with pleural effusion), stage IV or recurrent nonsquamous NSCLC, and have an ECOG performance status of 0 or 1. They also needed to have adequate hematologic, renal and hepatic function. Patients with squamous cell histology, brain metastases and hemoptysis were excluded from this trial because the phase II trial had identified squamous histology as a risk factor for severe hemoptysis.

A total of 878 patients has been recruited and randomly assigned to two treatment arms: 444 patients received carboplatin/paclitaxel alone, and 434 received carboplatin/paclitaxel plus bevacizumab $15\,mg\,kg^{-1}$ every 3 weeks. Patients in the control arm are not permitted to cross over to bevacizumab on disease progression. The results presented in Table 3.2 are from the second planned interim analysis after 469 deaths; the final analysis is planned at 650 deaths.

Table 3.2 Interim results of the E4599 phase III trial of carboplatin/paclitaxel with or without bevacizumab.

	Carboplatin/ paclitaxel	*Carboplatin/ paclitaxel + bevacizumab*	*p-value*
No. of patients	350	357	
Complete response, n (%)	0 (0)	5 (1.4)	
Partial response, n (%)	35 (10.0)	92 (25.8)	
Overall response rate, n (%)	35 (10.0)	97 (27.2)	<0.0001
Median survival (months)	10.2	12.5	0.0075
Progression-free survival (months)	4.5	6.4	<0.0001

The interim analysis conducted demonstrated that adding bevacizumab to chemotherapy significantly increased overall survival compared with chemotherapy alone: patients receiving bevacizumab had a median survival of 12.5 months, compared with 10.2 months for control patients (hazard ratio = 0.77; $p = 0.0075$). In addition, bevacizumab significantly increased progression-free survival (6.4 months for patients receiving bevacizumab versus 4.5 months for control patients; $p < 0.0001$) as well as response rate (27.2% versus 10%; $p < 0.0001$).

In summary, the preliminary results of this phase III trial demonstrate that adding bevacizumab to carboplatin/paclitaxel in patients with nonsquamous NSCLC provides a statistically and clinically significant survival advantage. The final data, which confirm the preliminary efficacy results, will be published soon. Following these positive results, the regimen of bevacizumab plus carboplatin/paclitaxel was adopted as the ECOG standard of care for first-line therapy in patients with nonsquamous NSCLC.

Recent clinical trials have also assessed the efficacy and safety of bevacizumab in combination with other biological agents in the absence of chemotherapy. A phase I/II trial evaluated the dose-limiting toxicity, efficacy and tolerability of bevacizumab in combination with erlotinib, an epidermal growth factor receptor (EGFR) tyrosine kinase inhibitor (Herbst et al. 2005; Sandler et al. 2004). The primary endpoint of phase I of this study was to establish dose-limiting toxicity, while that of phase II was to assess the efficacy and tolerability of the bevacizumab and erlotinib combination. The secondary endpoints included overall response rate and pharmacokinetics. To be eligible, patients had to have histologically confirmed stage IIIB (with pleural effusion), stage IV or recurrent nonsquamous NSCLC, and have a Karnofsky performance status of at least 70%. They also needed to have adequate hematologic, renal and hepatic function, and have relapsed after at least one platinum-based chemotherapy regimen for recurrent or metastatic disease. Patients with squamous cell histology, CNS metastases and hemoptysis were excluded from the trial, as were those who had prior anti-VEGF and/or anti-HER1/EGFR therapy, recent major surgery or radiation therapy.

A total of 40 patients was recruited; in a two-stage design, 12 patients were assigned to the dose-finding study in phase I, and a further 28 patients were recruited to assess the efficacy and tolerability of bevacizumab and erlotinib at the established dose in the second-line setting. Patients in phase I were assigned to one of three dose regimens: erlotinib (100 mg day^{-1} for 3 weeks) plus bevacizumab (7.5 mg kg^{-1} once every 3 weeks), erlotinib (100 mg day^{-1} for 3 weeks) plus bevacizumab (15 mg kg^{-1} once every 3 weeks), or erlotinib (150 mg day^{-1} for 3 weeks) plus bevacizumab (15 mg kg^{-1} once every 3 weeks). The dose chosen for further investigation in phase II was erlotinib 150 mg day^{-1} plus bevacizumab 15 mg kg^{-1} every 3 weeks.

The results of this trial indicate that combined bevacizumab and erlotinib therapy shows some early evidence of antitumor activity in patients with advanced or recurrent nonsquamous NSCLC. All 40 patients were available for efficacy assessment as determined by RECIST (Response Evaluation Criteria in Solid Tumors). No patient had a complete response, but eight patients had a partial response (20%; 95% CI: 7.6–32.4%) and 26 had stable disease (65%;

95% CI: 50.2–79.8%). The median overall survival for all patients was 12.6 months, and progression-free survival was 7.0 months.

A phase II trial evaluated bevacizumab (15 mg/kg every 3 weeks) combined with chemotherapy (docetaxel 75 mg/m^2 or pemetrexed 500 mg/m^2) or with erlotinib (150 mg daily) in patients with recurrent or refractory NSCLC (Herbst et al. 2006). Median progression-free survival was 3 months for chemotherapy alone, 4.8 months for bevacizumab plus chemotherapy (HR versus chemotherapy alone = 0.66), and 4.4 months for bevacizumab plus erlotinib (HR versus chemotherapy alone = 0.72). Median overall survival was 8.6 months for chemotherapy alone, 12.6 months for bevacizumab plus chemotherapy (HR versus chemotherapy alone = 0.74), and 13.7 months for bevacizumab plus erlotinib (HR versus chemotherapy alone = 0.76). Further investigation of the combination of bevacizumab and erlotinib as treatment for NSCLC is warranted.

3.3.2.2 Safety and Management of Bevacizumab in NSCLC

Most bevacizumab-related adverse side effects observed in NSCLC trials are similar in incidence and severity to those reported for CRC, and should be managed in the same manner. Severe and sometimes fatal hemoptysis has been observed in patients with advanced NSCLC, precluding the use of bevacizumab in patients with squamous cell NSCLC.

Patients with NSCLC treated with bevacizumab are at increased risk of severe bleeding events. In the phase II trial of carboplatin/paclitaxel with or without one of two doses of bevacizumab (Johnson et al. 2004), six of the 66 bevacizumab-treated patients (9%) experienced a major, life-threatening bleeding event (pulmonary hemorrhage and/or hemoptysis), and four of these died. A retrospective multivariate analysis revealed that patients with squamous cell histology and/or tumors located in the center of the chest close to major blood vessels were at greatest risk of developing serious bleeding events. In some cases, these major hemorrhagic events were preceded by cavitation or necrosis of the tumor.

As a result of these findings, the entry criteria for future NSCLC trials were adjusted to exclude patients with squamous cell histology (ca. 30% of NSCLC patients). By excluding patients with squamous cell histology, the incidence of severe bleeding events was reduced in the pivotal phase III trial (Sandler et al. 2005): eight of the 420 bevacizumab-treated patients (1.9%) in this trial experienced grade ≥3 hemoptysis, compared with 9% of patients in the phase II trial. Combining bevacizumab and erlotinib also appeared to reduce the incidence of these events (Herbst et al. 2005): no serious dose-limiting toxicities were reported in this trial, and no hemoptysis was observed.

In summary, a specific safety recommendation for clinical trials of NSCLC is the exclusion of patients with squamous cell histology, as this is associated with a greater risk of severe bleeding events. Patients requiring medical intervention for the management of a bleeding event should discontinue bevacizumab. Otherwise, the side-effect profile of bevacizumab is similar to that observed in trials in other indications. It is important to note that GI perforations are uncommon (<1%) in patients with NSCLC treated with bevacizumab. Furthermore, trials of bevacizumab in NSCLC add data demonstrating that bevacizumab can be combined with many chemotherapy regimens as well as with other targeted therapies.

3.3.2.3 Future Use of Bevacizumab in NSCLC

Conventional chemotherapy regimens have reached a plateau in the treatment of NSCLC, and it is unlikely that new combinations of chemotoxic agents will offer significant improvements in patient survival. Thus, there is a clinical need for more effective therapy to treat this disease. Both VEGF and EGFR inhibition have been demonstrated to extend survival in NSCLC, leading to the expectation that dual inhibition will have greater efficacy than inhibition of single agents. Therefore, combinations of bevacizumab with other biologic agents may be effective in the treatment of NSCLC. To examine the therapeutic benefit of bevacizumab in combination with chemotherapy and/or biologic agents, three phase III trials of first-line bevacizumab in NSCLC are currently under way.

A phase III trial (BO17704) will evaluate the efficacy and safety of gemcitabine (Gemzar)/cisplatin with or without bevacizumab as first-line therapy in advanced nonsquamous NSCLC. In this trial, 1,150 patients will be randomised (1:1:1) to receive gemcitabine/cisplatin with or without bevacizumab (either 7.5 or 15 $mg\,kg^{-1}$ every 3 weeks). Recruitment is now complete and results are expected in 2007. The primary endpoint of the study is progression-free survival, and secondary endpoints include overall survival and response rate. No cross-over is allowed.

Two phase III trials will assess the combination of bevacizumab and erlotinib. The first will investigate erlotinib with or without bevacizumab in advanced nonsquamous NSCLC. In this trial, 650 patients will be enrolled to assess the efficacy of bevacizumab combined with erlotinib versus erlotinib alone after the failure of standard first-line therapy. The primary endpoint is overall survival, and secondary endpoints will include progression-free survival, response rate, safety, and pharmacokinetics. The second trial will evaluate the efficacy of first-line bevacizumab with or without erlotinib in 1,150 patients with stage IIIB/IV non-squamous NSCLC. The primary endpoint is progression-free survival, while the secondary endpoint is the safety of bevacizumab plus erlotinib.

3.3.3 Breast Cancer

VEGF expression is inversely correlated with overall survival in patients with breast cancer (Gasparini et al. 1997, 1999; Linderholm et al. 2001), providing a rationale for inhibiting VEGF in MBC. Several phase II and III trials have investigated the efficacy and safety of bevacizumab therapy in breast cancer.

3.3.3.1 Efficacy of Bevacizumab in Breast Cancer

In a phase II dose-escalation study of bevacizumab in patients with MBC (AVF0776g), 75 patients were treated with one of three doses of bevacizumab (3,

10, or 20 mg kg^{-1}) every 2 weeks (Cobleigh et al. 2003). After 22 weeks of bevacizumab therapy, 16% of patients had stable disease or an ongoing objective response. The objective response rate was 5.6%, 12.2% and 6.3% for bevacizumab doses of 3, 10, and 20 mg kg^{-1}, respectively. The highest numerical response rate was seen at a dose of bevacizumab 10 mg kg^{-1} every 2 weeks. This observation, combined with a higher incidence of severe headaches and nausea at the 20 mg kg^{-1} dose level, led the authors to suggest bevacizumab 10 mg kg^{-1} every 2 weeks as an appropriate dose for future trials. Whilst no formal phase III studies comparing the benefit of different doses of this agent have been conducted, Roche is planning a multicenter phase III trial (AVADO) that will examine the efficacy of different bevacizumab doses in patients with MBC (see Section 3.3.3.3).

In the AVF2119 phase III trial, 462 patients with MBC were randomly assigned to receive capecitabine alone (2500 mg m^{-2} day^{-1} twice daily on Days 1 through 14, every 3 weeks) or the same dose of capecitabine plus bevacizumab (15 mg kg^{-1} every 3 weeks) (Miller et al. 2005a). The patients enrolled into this trial had been pretreated with anthracyclines and taxanes in the adjuvant and/or metastatic setting. The primary endpoint was progression-free survival as determined by an independent review facility (IRF), while secondary endpoints included progression-free survival based on investigator assessment, objective response rate as determined by both IRF and investigators, and overall survival.

Combining bevacizumab with capecitabine did not increase progression-free survival, which was 4.86 months for the combination arm and 4.17 months for the capecitabine arm ($p = 0.857$). Additionally, bevacizumab did not improve overall survival when added to capecitabine in this pretreated patient population (15.1 versus 14.5 months). There are a number of potential explanations for these observations, including the concept that other angiogenic growth factors come into play in later stage disease, making VEGF inhibition a less effective strategy. However, it was interesting that the objective response rate observed in patients receiving bevacizumab with capecitabine was significantly higher than that reported for patients receiving capecitabine alone, as assessed by both the IRF (19.8% versus 9.1%, $p = 0.001$) and the investigators (30.2% versus 19.1%, $p = 0.006$).

Given the mechanism of action of bevacizumab and the decreased relative influence of VEGF on tumor growth as breast cancer progresses, the authors suggest that anti-angiogenic agents may be more effective as first-line treatment for breast cancer. To test this theory, ECOG has conducted a randomized, open-label phase III intergroup trial (E2100) in collaboration with members of the North American Breast Intergroup to study the efficacy of bevacizumab combined with weekly paclitaxel as first-line treatment for MBC (Miller et al. 2005b). The results of this trial were presented at the 13th European Cancer Conference (ECCO) 2005 and the 28th San Antonio Breast Cancer Symposium (SABCS) 2005 following a data cut-off date of 27 September 2005. A total of 715 patients was randomized to one of two arms: paclitaxel alone (90 mg m^{-2} weekly for 3 weeks followed by one week without treatment); or the same dose of paclitaxel plus bevacizumab 10 mg kg^{-1} every 2 weeks. Interim analysis demonstrated that progression-free survival (the primary endpoint) improved from 6.11 months for patients treated with paclitaxel alone to 11.4 months for those receiving paclitaxel plus bevacizumab ($p < 0.0001$). The addition of bevacizumab to paclitaxel also led to a twofold increase in response rate for all patients (13.8% to 29.9%, $p < 0.0001$) and for patients with measurable disease only (16% to 37.7%, $p < 0.0001$) (Table 3.3). Assessment of overall survival requires a longer follow-up, as the differences are not yet statistically different between the two arms.

Additional phase II data have been reported for bevacizumab in combination with vinorelbine (Navelbine) (Burstein et al. 2002), docetaxel (Ramaswamy and Shapiro 2003), erlotinib (Dickler et al. 2004; Rugo et al. 2005), and letrozole (Femara) (Traina et al. 2005) (Table 3.4). In summary, bevacizumab has been shown to have clinical activity in breast cancer, and may soon play a role as part of standard treatment regimens.

Table 3.3 Interim results of the E2100 trial (paclitaxel with and without bevacizumab in metastatic breast cancer).

	Paclitaxel	*Paclitaxel + bevacizumab*	*p-value*
No. of patients	350	365	
Overall response rate (%)			
• All patients	13.8	29.9	<0.0001
• Patients with measurable disease	16.0	37.7	<0.0001
Progression-free survival (months)	6.11	11.4	<0.0001
Overall survival (months)	25.2	28.4	0.12

Table 3.4 Interim response rates of patients treated with bevacizumab in combination with other agents in phase II clinical trials of metastatic breast cancer.

	Burstein et al. (2002)	*Ramaswamy and Shapiro (2003)*	*Rugo et al. (2005)*	*Traina et al. (2005)*
Bevacizumab plus	Vinorelbine	Docetaxel	Erlotinib	Letrozole
Total no. of patients enrolled	56	18	37	23
No. of patients with				
• Complete response	1	0	0	0
• Partial response	16	10	1	2
• Stable disease	25	6	11	13
• Progressive disease	12	0	25	3
• Too early for assessment	1	0	0	5
• Withdrew due to toxicity	1	2	0	0

3.3.3.2 Safety and Management of Bevacizumab in Breast Cancer

Bevacizumab was generally well tolerated in clinical trials of MBC. Some adverse side effects occurred more frequently in bevacizumab-treated patients than control patients, but these did not limit therapy as they consisted mainly of manageable hypertension and proteinuria. More bleeding events were reported in bevacizumab-treated patients than control patients, but these were mainly grade 1/2 epistaxis. Serious hemorrhagic events, such as those observed in the NSCLC trials, were rare and did not occur more frequently in bevacizumab-treated patients than in control patients in MBC trials.

Thromboembolic events were infrequent in breast cancer trials, and occurred at similar rates in bevacizumab-treated and control patients. In the phase II dose-escalation trial, two patients developed axillary/subclavian vein thrombosis, but no cases of thrombosis in the lower extremities or pulmonary emboli were seen. Future phase III studies will further examine whether bevacizumab increases the risk of thromboembolic events.

In the phase III trial of bevacizumab combined with capecitabine (Miller et al. 2005a), 3% of bevacizumab-treated patients reported congestive heart failure (CHF) or cardiomyopathy, compared with 1% of patients in the control group (capecitabine alone). The severity of these incidents ranged from asymptomatic declines in left ventricular ejection fraction (LVEF) to symptomatic CHF requiring hospitalization. All of the bevacizumab-treated patients had previously been treated with anthracyclines, and some had received left chest wall radiation. In addition, three of the seven affected patients had a baseline LVEF <50%. In the phase II dose-escalation trial, two of 75 patients developed CHF; both of these had also received prior anthracyclines and chest wall radiation, and one patient had metastatic involvement of the pericardium. Due to the small number of events, the association between cardiac dysfunction, prior anthracycline exposure and bevacizumab treatment remains unclear. Cardiac dysfunction will continue to be monitored in clinical trials in patients with breast cancer.

In summary, bevacizumab is generally well tolerated in breast cancer, producing similar adverse events to those reported for other indications. Ongoing trials will improve our understanding of the potential role of this agent in breast cancer.

3.3.3.3 Future Use of Bevacizumab in Breast Cancer

The results of the phase III trial of bevacizumab combined with capecitabine indicate that patients with tumors refractory to chemotherapy may not represent the optimal setting for anti-angiogenic agents. Therefore, future trials of MBC will focus on patients with less-advanced disease.

Ongoing and planned clinical trials of bevacizumab in breast cancer include the AVADO trial (BO17708). This randomized, double-blind, placebo-controlled phase III trial conducted by Roche will compare the efficacy of docetaxel in combination with bevacizumab or placebo. A total of 705 patients who have received no prior chemotherapy for MBC will be recruited and randomized to one of three arms: docetaxel ($100\,mg\,m^{-2}$ every 3 weeks) plus placebo; docetaxel plus bevacizumab (7.5 mg kg^{-1} every 3 weeks); or docetaxel plus bevacizumab (15 mg kg^{-1} every 3 weeks). The primary endpoint of this trial is progression-free survival.

Perez and colleagues in the North Central Cancer Treatment Group (NCCTG) have recently completed accrual to a phase II study of docetaxel 75 mg m^{-2} plus capecitabine 1650 mg m^{-2} per day for 14 days plus bevacizumab 15 mg kg^{-1}, with cycles repeated every 3 weeks. Data are expected during 2006.

In the USA, a pilot safety phase II trial of bevacizumab plus chemotherapy in lymph node-positive breast cancer (E2104) will evaluate the incidence of cardiotoxicity when bevacizumab is added to an anthracycline-based regimen in the adjuvant setting. Patients must have undergone surgery to remove the tumor 4 to 12 weeks before joining the trial, and have received no prior chemotherapy or radiotherapy. The investigators will randomize 204 patients to one of two treatment arms: doxorubicin (60 mg m^{-2})/cyclophosphamide (Cytoxan) (600 mg m^{-2}) with or without bevacizumab 10 mg kg^{-1} every 2 weeks (four cycles). Patients in both arms will then receive paclitaxel (175 mg m^{-2}) plus bevacizumab 10 mg kg^{-1} every 2 weeks (four cycles) followed by 18 or 22 cycles of bevacizumab alone. The primary endpoint is to determine the incidence of cardiac dysfunction, while the secondary endpoints are to determine changes in LVEF in these patients and the noncardiac toxicity of this regimen.

The benefit of bevacizumab plus chemotherapy in the neoadjuvant setting will be assessed by a second double-blind, placebo-controlled phase II US trial (TORI B-02). Patients will receive a single dose of either bevacizumab or placebo (7.5 or 15 mg kg^{-1} every 3 weeks), followed by a six-cycle treatment phase of docetaxel, doxorubicin and cyclophosphamide plus either bevacizumab or placebo. Patients who received bevacizumab in the first phase of the study will then receive bevacizumab alone until disease progression, while patients who received placebo will receive no further therapy.

3.4 The Potential of Bevacizumab

As tumors cannot progress without a blood supply, angiogenesis plays a central role in the growth and development of cancer. In addition, in patients with solid tumors and hematological diseases, VEGF overexpression correlates with poor prognosis and survival rates. Consequently, bevacizumab offers the potential to increase survival without adding to the toxicity of conventional therapy across all tumor types.

A number of recently completed and ongoing clinical trials indicate that bevacizumab has the potential to be used across multiple tumor types. This agent significantly improves the survival of patients receiving first-line therapy for metastatic CRC, with proven efficacy when combined with 5-FU/LV with or without irinotecan. Ongoing clinical trials in metastatic CRC are assessing the efficacy of bevacizumab in the adjuvant setting and when combined first line with other chemotherapy regimens, including oxaliplatin- and capecitabine-based regimens. Clinical benefit has also been demonstrated for bevacizumab when combined with standard first-line chemotherapy regimens in patients with metastatic NSCLC (Sandler et al. 2005) and breast cancer (Miller et al. 2005b). It is important to note that the role of bevacizumab in patients pretreated with chemotherapy has not been clearly established.

The future potential of bevacizumab therapy is illustrated by the number of tumor types for which data are already available and in which phase III trials are planned or ongoing. These include RCC, pancreatic cancer, and ovarian cancer.

Anti-VEGF therapy is a rational approach in RCC because many patients with RCC have a mutation of the von Hippel-Lindau (VHL) gene that causes VEGF overexpression, resulting in hypervascularization. Bevacizumab monotherapy has been shown to increase time to disease progression in patients with pretreated metastatic RCC (Yang et al. 2003). In addition, a phase II trial in RCC combining bevacizumab and erlotinib achieved a 25% response rate (Hainsworth et al. 2005a), which is impressive in a disease where responses are uncommon even with standard therapy (interferon-α2a). Two ongoing phase III trials are evaluating the impact of bevacizumab plus interferon or interferon alone as first-line therapy on overall survival.

Bevacizumab has also demonstrated activity in a number of trials in pancreatic cancer. A phase I trial combining bevacizumab with capecitabine and radiotherapy in pancreatic cancer patients reported partial or minor responses in 40% of patients (Crane et al. 2005). A phase II trial of bevacizumab combined with gemcitabine as first-line treatment for metastatic pancreatic cancer achieved a response rate of 19% (Kindler et al. 2004). Two phase III trials of bevacizumab in pancreatic cancer are ongoing. The first trial, BO17706, will examine overall survival in 600 previously untreated patients treated with gemcitabine plus erlotinib with or without bevacizumab $5\,mg\,kg^{-1}$ every 2 weeks. The second trial, CALGB 80303, will compare overall survival in 530 previously untreated patients receiving either gemcitabine ($1000\,mg\,m^{-2}$ weekly for 3 weeks of a 4-week cycle) plus bevacizumab ($10\,mg\,kg^{-1}$ every 2 weeks) or gemcitabine alone.

In a phase II trial of ovarian cancer, bevacizumab monotherapy ($15\,mg\,kg^{-1}$ every 3 weeks) produced a response rate of 17.7% and a 6-month progression-free survival rate of 38.7%, which compares favorably with historical control data (Burger et al. 2005). However, another phase II trial of single-agent bevacizumab was halted after five GI perforations were reported in the first 11 patients; these may have occurred because the study population consisted of patients with more advanced disease.

Two phase III trials are planned in ovarian cancer. In the first trial, ICON 7, 500 patients with International Federation of Gynecology and Obstetrics (FIGO) stage IIb–IV epithelial ovarian carcinoma will receive carboplatin and paclitaxel with or without bevacizumab $7.5\,mg\,kg^{-1}$ every 3 weeks. In the second trial, GOG 0218, patients with FIGO stage III/IV ovarian or peritoneal carcinoma will be randomized to one of three arms. All patients will receive six cycles of chemotherapy (paclitaxel 175 mg m^{-2}, carboplatin AUC 6 every 3 weeks). Arm 1 patients will receive placebo with the chemotherapy, followed by placebo alone. Patients in arms 2 and 3 will receive bevacizumab (15 mg kg^{-1} every 3 weeks) with the chemotherapy, followed by placebo alone for arm 2 and bevacizumab alone for arm 3 for the duration of the 15 months of treatment.

As well as these tumor types in which large phase III trials are already ongoing or planned, encouraging data are emerging from clinical studies of bevacizumab alone or in combination with other agents for the treatment of hematological malignancies, including non-Hodgkin's lymphoma and acute myelogenous leukemia (Karp et al. 2004; Stopeck et al. 2005), prostate cancer (Rini et al. 2005), melanoma (Carson et al. 2003), hepatocellular carcinoma (Schwartz et al. 2004; Zhu et al. 2005) and gastric cancer (Shah et al. 2005).

3.5 Conclusions

Anti-angiogenic agents, of which bevacizumab is the most well-developed and well-understood, have the potential to improve patient survival with fewer adverse side effects than conventional chemotherapy regimens. Bevacizumab has been demonstrated to significantly improve survival in first-line therapy for metastatic CRC and NSCLC. Progression-free survival or time to progression for patients with MBC or RCC is improved by the addition of bevacizumab to first-line therapy. Ongoing trials are examining the efficacy and tolerability of bevacizumab with the objective of optimizing its use to maximize the benefit for patients. It appears likely that bevacizumab will become a key element of first-line regimens for a variety of tumor types, whether in combination with chemotherapy, immunotherapy, radiotherapy, or other novel agents. Future and ongoing clinical trials will assess the efficacy of bevacizumab in combination with a range of chemotherapy regimens, as well as in combination with other biological therapies (Chen 2004). Several phase I/II clinical studies of combined inhibition of the EGFR and the VEGF signaling pathways have already reported encouraging findings (Hainsworth et al. 2005b; Herbst et al. 2005). These trials will elucidate the efficacy and safety of bevacizumab as both first-line therapy for metastatic disease and in the adjuvant setting.

Bevacizumab is generally well tolerated, and serious adverse events are rare. Across all clinical trials to date – irrespective of tumor type – the most frequently observed side effects related to bevacizumab are hypertension, proteinuria and minor bleeding events; these are usually mild to moderate in severity, and clinically manageable using standard therapies or first-aid techniques. The most serious bevacizumab-related adverse events reported in clinical trials are GI perforations (for GI and ovarian cancers), severe hemorrhagic events (for NSCLC), and arterial thromboembolic events. These events occur in a small minority of patients, and research to further identify risk factors for these events is ongoing.

In summary, bevacizumab is an anti-angiogenic agent which has been proven to improve clinical outcomes, including survival and progression-free survival, in patients with some malignancies. It is already approved for clinical use first line in patients with metastatic CRC, and data are currently under evaluation to determine its availability as part of first-line use in NSCLC and MBC.

Acknowledgments

The authors thank F. Hoffmann-La Roche for providing funding for the preparation of this work and Yfke van Bergen of Gardiner-Caldwell Communications for providing editorial assistance.

References

Baluk, P., Hashizume, H., McDonald, D.M. (2005) Cellular abnormalities of blood vessels as targets in cancer. *Curr Opin Genet Dev* 15: 102–111.

Ben Av, P., Crofford, L.J., Wilder, R.L., Hla, T. (1995) Induction of vascular endothelial growth factor expression in synovial fibroblasts by prostaglandin E and interleukin-1: a potential mechanism for inflammatory angiogenesis. *FEBS Lett* 372: 83–87.

Bergers, G. and Benjamin, L.E. (2003) Tumorigenesis and the angiogenic switch. *Nat Rev Cancer* 3: 401–410.

Burger, R.A., Sill, M., Monk, B.J., Greer, B., Sorosky, J. (2005) phase II trial of bevacizumab in persistent or recurrent epithelial ovarian cancer (EOC) or primary peritoneal cancer (PPC): a gynecologic oncology group (GOC) study. *J Clin Oncol* 23: 457S–457S.

Burstein, H.J., Parker, L.M., Savoie, J., Younger, J., Kuter, I., Ryan, P.D., Garber, J.E., Campos, S.M., Shulman, L.N., Harris, L.N., Gelman, R., Winer, E. (2002) phase II trial of the anti-VEGF antibody bevacizumab in combination with vinorelbine for refractory advanced breast cancer. *Breast Cancer Res Treat* 79: Abstract 446.

Carmeliet, P. (2003) Angiogenesis in health and disease. *Nat Med* 9: 653–660.

Carmeliet, P. and Jain, R.K. (2000) Angiogenesis in cancer and other diseases. *Nature* 407: 249–257.

Carson, W.E., Biber, J., Shah, N., Reddy, K., Kefauver, C., Leming, P.D., Kendra, K., Walker, M. (2003) A phase 2 trial of a recombinant humanized monoclonal anti-vascular endothelial growth factor (VEGF) antibody in patients with malignant melanoma. *Proc Am Soc Clin Oncol* 22: 715 (Abstract 2873).

Cassidy, J., Clarke, S., Díaz-Rubio, E., Scheithauer, W., Figer, A., Wong, R., Koski, S., Lichinitser, M., Yang, T.-S., Saltz, L. (2006) First efficacy and safety results from XELOX-1/NO16966, a randomised 2 × 2 factorial phase III trial of XELOX vs. FOLFOX4 + bevacizumab or placebo in first-line metastatic colorectal cancer (MCRC). *Ann Oncol* 17(Suppl. 9): abstract LBA3.

Chen, H.X. (2004) Expanding the clinical development of bevacizumab. *Oncologist* 9(Suppl. 1): 27–35.

Cobleigh, M.A., Langmuir, V.K., Sledge, G.W., Miller, K.D., Haney, L., Novotny, W.F., Reimann, J.D., Vassel, A. (2003) A phase I/II dose-escalation trial of bevacizumab in previously treated metastatic breast cancer. *Semin Oncol* 30(5 Suppl. 16): 117–124.

Cohen, T., Nahari, D., Cerem, L.W., Neufeld, G., Levi, B.Z. (1996) Interleukin 6 induces the expression of vascular endothelial growth factor. *J Biol Chem* 271: 736–741.

Crane, C., Ellis, L., James, A., Henry, X., Douglas, E., Linus, H., Eric, T., Chaan, N., Jeffrey, L., Robert, W. (2005) phase I trial of bevacizumab (BV) with concurrent radiotherapy (RT) and capecitabine (CAP) in locally advanced pancreatic adenocarcinoma (PA). *Proceedings WCGC* (Abstract O-009).

de Gramont, A., Figer, A., Seymour, M., Homerin, M., Hmissi, A., Cassidy, J., Boni, C., Cortes-Funes, H., Cervantes, A., Freyer, G., Papamichael, D., Le Bail, N., Louvet, C., Hendler, D., de Braud, F., Wilson, C., Morvan, F., Bonetti, A. (2000) Leucovorin and fluorouracil with or without oxaliplatin as first-line treatment in advanced colorectal cancer. *J Clin Oncol* 18: 2938–2947.

de Gramont, A., Cervantes, A., Andre, T., Figer, A., Lledo, G., Flesch, M., Mineur, L., Russ, G., Quinaux, E., Etienne, P.-L. (2004) OPTIMOX study: FOLFOX 7/LV5FU2 compared to FOLFOX 4 in patients with advanced colorectal cancer. *J Clin Oncol* 22(Suppl.): 251s (Abstract 3525).

Demetri, G.D., van Oosterom, A.T., Blackstein, M., Garrett, C., Shah, M., Heinrich, M., McArthur, G., Judson, I., Baum, C.M., Casali, P.G. (2005) phase III, multicenter, randomized, double-blind, placebo-controlled trial of SU11248 in patients (pts) following failure of imatinib for metastatic GIST. *J Clin Oncol* 23(Suppl.): 308s (Abstract 4000).

Dickler, M., Rugo, H., Caravelli, J., Brogi, E., Sachs, D., Panageas, K., Flores, S., Moasser, M., Norton, L., Hudis, C. (2004) phase II trial of erlotinib (OSI-774), an epidermal growth factor receptor (EGFR)-tyrosine kinase inhibitor, and bevacizumab, a recombinant humanized monoclonal antibody to vascular endothelial growth factor (VEGF), in patients (pts) with metastatic breast cancer (MBC). *J Clin Oncol* 22(Suppl.): 127s (Abstract 2001).

Douillard, J.Y., Cunningham, D., Roth, A.D., Navarro, M., James, R.D., Karasek, P., Jandik, P., Iveson, T., Carmichael, J., Alakl, M., Gruia, G., Awad, L., Rougier, P. (2000) Irinotecan combined with fluorouracil compared with fluorouracil alone as first-line treatment for metastatic colorectal cancer: a multicentre randomised trial. *Lancet* 355: 1041–1047.

Escudier, B., Szczylik, C., Eisen, T., Stadler, W.M., Schwartz, B., Shan, M., Bukowski, R.M. (2005) Randomized phase III trial of the Raf kinase and VEGFR inhibitor sorafenib (BAY 43-9006) in patients with advanced renal cell carcinoma (RCC). *J Clin Oncol* 23(Suppl.): 380s (Abstract LBA4510).

Ferrara, N. (2001) Role of vascular endothelial growth factor in regulation of physiological angiogenesis. *Am J Physiol Cell Physiol* 280: C1358–C1366.

Ferrara, N. (2002) VEGF and the quest for tumour angiogenesis factors. *Nat Rev Cancer* 2: 795–803.

Ferrara, N., Gerber, H.P., LeCouter, J. (2003) The biology of VEGF and its receptors. *Nat Med* 9: 669–676.

Ferrara, N., Hillan, K.J., Gerber, H.P., Novotny, W. (2004) Discovery and development of bevacizumab, an anti-VEGF antibody for treating cancer. *Nat Rev Drug Discov* 3: 391–400.

Frank, S., Hubner, G., Breier, G., Longaker, M.T., Greenhalgh, D.G., Werner, S. (1995) Regulation of vascular endothelial growth factor expression in cultured keratinocytes. Implications for normal and impaired wound healing. *J Biol Chem* 270: 12607–12613.

Fyfe, G.A., Hurwitz, H., Fehrenbacher, L., Cartwright, T., Cartwright, J., Heim, W., Berlin, J., Kabbinavar, F., Holmgren, E., Novotny, W. (2004) Bevacizumab plus irinotecan/5-FU/leucovorin for treatment of metastatic colorectal cancer results in survival benefit in all pre-specified patient subgroups. *J Clin Oncol* 22(Suppl.): 274s (Abstract 3617).

Gabrilovich, D.I., Ishida, T., Nadaf, S., Ohm, J.E., Carbone, D.P. (1999) Antibodies to vascular endothelial growth factor enhance the efficacy to cancer immunotherapy by improving endogenous dendritic cell function. *Clin Cancer Res* 5: 2963–2970.

Gasparini, G., Toi, M., Gion, M., Verderio, P., Dittadi, R., Hanatani, M., Matsubara, I., Vinante, O., Bonoldi, E., Boracchi, P., Gatti, C., Suzuki, H., Tominaga, T. (1997) Prognostic significance of vascular endothelial growth factor protein in node-negative breast carcinoma. *J Natl Cancer Inst* 89: 139–147.

Gasparini, G., Toi, M., Miceli, R., Vermeulen, P.B., Dittadi, R., Biganzoli, E., Morabito, A., Fanelli, M., Gatti, C., Suzuki, H., Tominaga, T., Dirix, L.Y., Gion, M. (1999) Clinical relevance of vascular endothelial growth factor and thymidine phosphorylase in patients with node-positive breast cancer treated with either adjuvant chemotherapy or hormone therapy. *Cancer J Sci Am* 5: 101–111.

Giantonio, B.J., Catalano, P.J., Meropol, N.J., O'Dwyer, P.J., Mitchell, E.P., Alberts, S.R., Schwartz, M.A., Benson, A.B. (2005) High-dose bevacizumab improves survival when combined with FOLFOX4 in previously treated advanced colorectal cancer: results from the Eastern Cooperative Oncology Group (ECOG) study E3200. *J Clin Oncol* 23(Suppl.): 1s (Abstract 2).

Goldberg, R.M., Sargent, D.J., Morton, R.F., Fuchs, C.S., Ramanathan, R.K., Williamson, S.K., Findlay, B.P., Pitot, H.C., Alberts, S.R. (2004) A randomized controlled trial of fluorouracil plus leucovorin, irinotecan, and oxaliplatin combinations in patients with previously untreated metastatic colorectal cancer. *J Clin Oncol* 22: 23–30.

Gordon, M.S., Margolin, K., Talpaz, M., Sledge, G.W.J., Holmgren, E., Benjamin, R., Stalter, S., Shak, S., Adelman, D. (2001) phase I safety and pharmacokinetic study of recombinant human anti-vascular endothelial growth factor in patients with advanced cancer. *J Clin Oncol* 19: 843–850.

Grothey, A., Sargent, D., Goldberg, R.M., Schmoll, H.J. (2004) Survival of patients with advanced colorectal cancer improves with the availability of fluorouracil-leucovorin, irinotecan, and oxaliplatin in the course of treatment. *J Clin Oncol* 22: 1209–1214.

Hainsworth, J.D., Sosman, J.A., Spigel, D.R., Edwards, D.L., Baughman, C., Greco, A. (2005a) Treatment of metastatic renal cell carcinoma with a combination of bevacizumab and erlotinib. *J Clin Oncol* 23: 7889–7896.

Hainsworth, J.D., Sosman, J.A., Spigel, D.R., Patton, J.F., Thompson, D.S., Sutton, V., Hart, L.L., Yost, K., Greco, F.A. (2005b) Bevacizumab, erlotinib, and imatinib in the treatment of patients (pts) with advanced renal cell carcinoma (RCC): a Minnie Pearl Cancer Research Network phase I/II trial. *J Clin Oncol* 23(Suppl.): 388s (Abstract 4542).

Hambleton, J., Novotny, W.F., Hurwitz, H., Fehrenbacher, L., Cartwright, T., Hainsworth, J., Heim, W., Berlin, J., Kabbinavar, F., Holmgren, E. (2004) Bevacizumab does not increase bleeding in patients with metastatic colorectal cancer receiving concurrent anticoagulation. *J Clin Oncol* 22(Suppl.): 252s (Abstract 3528).

Harmey, J.H. and Bouchier-Hayes, D. (2002) Vascular endothelial growth factor (VEGF), a survival factor for tumour cells: implications for anti-angiogenic therapy. *BioEssays* 24: 280–283.

Hecht, J.R., Trarbach, T., Jaeger, E., Hainsworth, J., Wolff, R., Lloyd, K., Bodoky, G., Borner, M., Laurent, D., Jacques, C. (2005) A randomized, double-blind, placebo-controlled, phase III study in patients (Pts) with metastatic adenocarcinoma of the colon or rectum receiving first-line chemotherapy with oxaliplatin/5-fluorouracil/leucovorin and PTK787/ZK 222584 or placebo (CONFIRM-1). *J Clin Oncol* 23(Suppl.): 2s (Abstract LBA3).

Hedrick, E., Kozloff, M., Hainsworth, J., Badarinath, S., Cohn, A., Flynn, P., Dong, W., Suzuki, S., Sugrue, M., Grothey, A. (2006) Safety of bevacizumab plus chemotherapy as first-line treatment of patients with metastatic colorectal cancer: updated results from a large observational registry in the us (BRiTE). *J Clin Oncol* 24(Suppl.): 155s (Abstract 3536).

Herbst, R.S., Fehrenbacher, L., Belani, C.P., Bonomi, P.D., Hart, L., Melnyk, O., Sandler, A., Lin, M., O'Neill, V.J. (2006) A phase II, multicenter, randomized clinical trial to evaluate the efficacy and safety of bevacizumab (Avastin) in combination with either chemotherapy (docetaxel or pemetrexed) or erlotinib hydrochloride (Tarceva) compared with chemotherapy alone for treatment of recurrent or refractory non-small-cell lung cancer. *Eur J Cancer Suppl* 4: 20 (Abstract 53).

Herbst, R.S., Johnson, D.H., Mininberg, E., Carbone, D.P., Henderson, T., Kim, E.S., Blumenschein, G., Jr., Lee, J.J., Liu, D.D.,

Truong, M.T., Hong, W.K., Tran, H., Tsao, A., Xie, D., Ramies, D.A., Mass, R., Seshagiri, S., Eberhard, D.A., Kelley, S.K., Sandler, A. (2005) Phase I/II trial evaluating the anti-vascular endothelial growth factor monoclonal antibody bevacizumab in combination with the HER-1/epidermal growth factor receptor tyrosine kinase inhibitor erlotinib for patients with recurrent non-small-cell lung cancer. *J Clin Oncol* 23: 2544–2555.

Hochster, H.S., Hart, L.L., Ramanathan, R.K., Hainsworth, J.D., Hedrick, E.E., Childs, B.H. (2006) Safety and efficacy of oxaliplatin/fluoropyrimidine regimens with or without bevacizumab as first-line treatment of metastatic colorectal cancer (mCRC): Final analysis of the TREE-Study. *J Clin Oncol* 24(Suppl.): 148s (Abstract 3510).

Holden, S., Ryan, E., Kearns, A., Holmgren, E., Hurwitz, H. (2005) Benefit from bevacizumab (BV) is independent of pretreatment plasma vascular endothelial growth factor-A (pl-VEGF) in patients (pts) with metastatic colorectal cancer (mCRC). *J Clin Oncol* 23(Suppl.): 259s (Abstract 3555).

Hurwitz, H., Fehrenbacher, L., Novotny, W., Cartwright, T., Hainsworth, J., Heim, W., Berlin, J., Baron, A., Griffing, S., Holmgren, E., Ferrara, N., Fyfe, G., Rogers, B., Ross, R., Kabbinavar, F. (2004) Bevacizumab plus irinotecan, fluorouracil, and leucovorin for metastatic colorectal cancer. *N Engl J Med* 350: 2335–2342.

Hurwitz, H.I., Fehrenbacher, L., Hainsworth, J.D., Heim, W., Berlin, J., Holmgren, E., Hambleton, J., Novotny, W. F., Kabbinavar, F. (2005) Bevacizumab in combination with fluorouracil and leucovorin: an active regimen for first-line metastatic colorectal cancer. *J Clin Oncol* 23: 3502–3508.

Jain, R.K. (2001) Normalizing tumor vasculature with anti-angiogenic therapy: a new paradigm for combination therapy. *Nat Med* 7: 987–989.

Jain, R.K. (2002) Tumor angiogenesis and accessibility: role of vascular endothelial growth factor. *Semin Oncol* 29: 3–9.

Jain, R.K. (2003) Molecular regulation of vessel maturation. *Nat Med* 9: 685–693.

Jain, R.K. (2005) Normalization of tumor vasculature: an emerging concept in antiangiogenic therapy. *Science* 307: 58–62.

Johnson, D.H., Fehrenbacher, L., Novotny, W.F., Herbst, R.S., Nemunaitis, J.J., Jablons, D.M., Langer, C.J., DeVore, R.F., III, Gaudreault, J., Damico, L.A., Holmgren, E., Kabbinavar, F. (2004) Randomized phase II trial comparing bevacizumab plus carboplatin and paclitaxel with carboplatin and paclitaxel alone in previously untreated locally advanced or metastatic non-small-cell lung cancer. *J Clin Oncol* 22: 2184–2191.

Kabbinavar, F.F., Wong, J.T., Ayala, R.E., Wintroub, A.B., Kim, K.J., Ferrara, N. (1995) The effect of antibody to vascular endothelial growth factor and cisplatin on the growth of lung tumors in nude mice. *Proc Am Assoc Cancer Res* 36: 488 (Abstract 2906).

Kabbinavar, F., Hurwitz, H.I., Fehrenbacher, L., Meropol, N.J., Novotny, W.F., Lieberman, G., Griffing, S., Bergsland, E. (2003) phase II, randomized trial comparing bevacizumab plus fluorouracil (FU)/leucovorin (LV) with FU/LV alone in patients with metastatic colorectal cancer. *J Clin Oncol* 21: 60–65.

Kabbinavar, F.F., Schulz, J., McCleod, M., Patel, T., Hamm, J.T., Hecht, J.R., Mass, R., Perrou, B., Nelson, B., Novotny, W.F. (2005a) Addition of bevacizumab to bolus fluorouracil and leucovorin in first-line metastatic colorectal cancer: results of a randomized phase II trial. *J Clin Oncol* 23: 3697–3705.

Kabbinavar, F.F., Hambleton, J., Mass, R.D., Hurwitz, H.I., Bergsland, E., Sarkar, S. (2005b) Combined analysis of efficacy: the addition of bevacizumab to fluorouracil/leucovorin improves survival for patients with metastatic colorectal cancer. *J Clin Oncol* 23: 3706–3712.

Karp, J.E., Gojo, I., Pili, R., Gocke, C.D., Greer, J., Guo, C., Qian, D., Morris, L., Tidwell, M., Chen, H., Zwiebel, J. (2004) Targeting vascular endothelial growth factor for relapsed and refractory adult acute myelogenous leukemias: therapy with sequential 1-beta-d-arabinofuranosylcytosine, mitoxantrone, and bevacizumab. *Clin Cancer Res* 10: 3577–3585.

Kelly, K., Crowley, J., Bunn, P.A., Jr., Presant, C.A., Grevstad, P.K., Moinpour, C.M., Ramsey, S.D., Wozniak, A.J., Weiss, G.R., Moore, D.F., Israel, V.K., Livingston, R.B., Gandara, D.R. (2001) Randomized phase III trial of paclitaxel plus carboplatin versus vinorelbine plus cisplatin in the treatment of patients with advanced non-small-cell lung cancer: a Southwest Oncology Group trial. *J Clin Oncol* 19: 3210–3218.

Kerbel, R. and Folkman, J. (2002) Clinical translation of angiogenesis inhibitors. *Nat Rev Cancer* 2: 727–739.

Kim, K.J., Li, B., Winer, J., Armanini, M., Gillett, N., Phillips, H.S., Ferrara, N. (1993) Inhibition of vascular endothelial growth factor-induced angiogenesis suppresses tumour growth in vivo. *Nature* 362: 841–844.

Kindler, H.L., Friberg, G., Stadler, W.M., Singh, D.A., Locker, G., Nattam, S., Kozloff, M., Kasza, K., Vokes, E.E. (2004) Bevacizumab (B) plus gemcitabine (G) in patient (pts) with advanced pancreatic cancer (PC): updated results of a multi-center phase II trial. *J Clin Oncol* 22(Suppl.): 315s (Abstract 4009).

Kozloff, M., Hainsworth, J., Badarinath, S., Cohn, A., Flynn, P., Steis, R., Dong, W., Suzuki, S., Sugrue, M., Grothey, A. (2006) Efficacy of bevacizumab plus chemotherapy as first-line treatment of patients with metastatic colorectal cancer: updated results from a large observational registry in the us (BRiTE). *J Clin Oncol* 24(Suppl.): 155s (Abstract 3537).

Linderholm, B.K., Lindahl, T., Holmberg, L., Klaar, S., Lennerstrand, J., Henriksson, R., Bergh, J. (2001) The expression of vascular endothelial growth factor correlates with mutant p53 and poor prognosis in human breast cancer. *Cancer Res* 61: 2256–2260.

Margolin, K., Gordon, M.S., Holmgren, E., Gaudreault, J., Novotny, W., Fyfe, G., Adelman, D., Stalter, S., Breed, J. (2001) phase Ib trial of intravenous recombinant humanized monoclonal antibody to vascular endothelial growth factor in combination with chemotherapy in patients with advanced cancer: pharmacologic and long-term safety data. *J Clin Oncol* 19: 851–856.

Miller, K.D., Chap, L.I., Holmes, F.A., Cobleigh, M.A., Marcom, P.K., Fehrenbacher, L., Dickler, M., Overmoyer, B.A., Reimann, J.D., Sing, A.P., Langmuir, V., Rugo, H.S. (2005a) Randomized phase III trial of capecitabine compared with bevacizumab plus capecitabine in patients with previously treated metastatic breast cancer. *J Clin Oncol* 23: 792–799.

Miller, K.D., Wang, M., Gralow, J., Dickler, M., Cobleigh, M.A., Perez, E.A.S., Davidson, N.E. (2005b) Eastern Cooperative Oncology Group. A randomised phase III trial of paclitaxel versus paclitaxel plus bevacizumab as first-line therapy for locally recurrent or metastatic breast cancer. Presented at the 41st Annual Meeting of the American Society of Clinical Oncology, Orlando, Fl, USA, May 13–17.

Motzer, R.J., Rini, B.I., Michaelson, M.D., Redman, B.G., Hudes, G.R., Wilding, G., Bukowski, R.M., George, D.J., Kim, S.T., Baum, C.M. (2005) phase 2 trials of SU11248 show antitumor activity in second-line therapy for patients with metastatic renal cell carcinoma (RCC). *J Clin Oncol* 24(Suppl.): 380s (Abstract 4508).

Osusky, K.L., Hallahan, D.E., Fu, A., Ye, F., Shyr, Y., Geng, L. (2004) The receptor tyrosine kinase inhibitor SU11248 impedes endothelial cell migration, tubule formation, and blood vessel formation in vivo, but has little effect on existing tumor vessels. *Angiogenesis* 7: 225–233.

Poon, R.T.-P., Fan, S.-T., Wong, J. (2001) Clinical implications of circulating angiogenic factors in cancer patients. *J Clin Oncol* 19: 1207–1225.

Presta, L.G., Chen, H., O'Connor, S.J., Chisholm, V., Meng, Y.G., Krummen, L., Winkler, M., Ferrara, N. (1997) Humanization of an anti-vascular endothelial growth factor monoclonal antibody for the therapy of solid tumors and other disorders. *Cancer Res* 57: 4593–4599.

Ramaswamy, B. and Shapiro, C.L. (2003) phase II trial of bevacizumab in combination with docetaxel in women with advanced breast cancer. *Clin Breast Cancer* 4: 292–294.

Rini, B., Weinberg, V., Fong, L., Small, E. (2005) A phase 2 study of prostatic acid phosphatase-pulsed dendritic cells

(APC8015; Provenge) in combination with bevacizumab in patients with serologic progression of prostate cancer after local therapy. Presented at the Prostate Cancer Symposium, Orlando, Fl, USA, February 17–19 (Abstract 251).

Rivera, F., Cunningham, D., Michael, M., DiBartolomeo, M., Kretzschmar, A., Berry, S., Mazier, M., Lutiger, B., Van Cutsem, E. (2006) Preliminary safety of bevacizumab with first-line Folfox, CapOx, Folfiri and Capecitabine for mCRC-First BEAT. Presented at 8th World Congress on Gastrointestinal Cancer 2006 (Abstract O-017).

Rugo, H.S., Dickler, M.N., Scott, J.H., Moore, D.H., Melisko, M., Yeh, B.M., Caravelli, J., Brogi, E., Hudis, C., Park, J.W. (2005) Change in circulating endothelial cells (CEC) and tumor cells (CTC) in patients (pts) receiving bevacizumab and erlotinib for metastatic breast cancer (MBC) predicts stable disease at first evaluation. *J Clin Oncol* 23(Suppl.): 10s (Abstract 525).

Ryan, A.M., Eppler, D.B., Hagler, K.E., Bruner, R.H., Thomford, P.J., Hall, R.L., Shopp, G.M., O'Neill, C.A. (1999) Preclinical safety evaluation of rhuMABVEGF, an antiangiogenic humanized monoclonal antibody. *Toxicol Pathol* 27: 78–86.

Saltz, L.B., Cox, J.V., Blanke, C., Rosen, L.S., Fehrenbacher, L., Moore, M.J., Maroun, J. A., Ackland, S.P., Locker, P.K., Pirotta, N., Elfring, G.L., Miller, L.L. (2000) Irinotecan plus fluorouracil and leucovorin for metastatic colorectal cancer. Irinotecan Study Group. *N Engl J Med* 343: 905–914.

Sandler, A.B., Blumenschein, G.R., Henderson, T., Lee, J., Truong, M., Kim, E., Mass, B., Garcia, B., Johnson, D.H., Herbst, R.S. (2004) phase I/II trial evaluating the anti-VEGF MAb bevacizumab in combination with erlotinib, a HER1/EGFR-TK inhibitor, for patients with recurrent non-small cell lung cancer. *J Clin Oncol* 22(Suppl.): 127s (Abstract 2000).

Sandler, A.B., Gray, R., Brahmer, J., Dowlati, A., Schiller, J.H., Perry, M.C., Johnson, D.H. (2005) Randomized phase II/III trial of paclitaxel (P) plus carboplatin I with or without bevacizumab (NSC # 704865) in patients with advanced non-squamous non-small cell lung cancer (NSCLC): an Eastern Cooperative Oncology Group (ECOG) trial – E4599. *J Clin Oncol* 23(Suppl.): 2s (Abstract LBA4).

Schering AG. (2005) Schering and Novartis's vatalanib unlikely to meet phase II cancer trial endpoint. Press Release Posted on 28 July 2005.

Schiller, J.H., Harrington, D., Belani, C.P., Langer, C., Sandler, A., Krook, J., Zhu, J., Johnson, D.H. (2002) Comparison of four chemotherapy regimens for advanced non-small-cell lung cancer. *N Engl J Med* 346: 92–98.

Schwartz, J.D., Schwartz, M., Goldman, J., Lehrer, D., Coll, D., Kinkabwala, M., Wadler, S. (2004) Bevacizumab in hepatocellular carcinoma in patients without metastasis and without invasion of the portal vein. *J Clin Oncol* 22(Suppl.): 355s (Abstract 4088).

Shah, M.A., Ilson, D., Ramanathan, R.K., Levner, A., D'Adamo, D., Schwartz, L., Casper, E., Schwartz, G.K., Kelsen, D.P. (2005) A multicenter phase II study of irinotecan (CPT), cisplatin (CIS), and bevacizumab (BEV) in patients with unresectable or metastatic gastric or gastroesophageal junction (GEJ) adenocarcinoma. *J Clin Oncol* 23(Suppl.): 314s (Abstract 4025).

Shen, B.Q., Stainton, S., Li, D., Pelletier, N., Zioncheck, T.F. (2004) Combination of Avastin and Xeloda synergistically inhibits colorectal tumor growth in a COLO205 tumor xenograft model. *Proc Am Assoc Cancer Res* 45: 508 (Abstract 2203).

Shibuya, M. (2001) Structure and function of VEGF/VEGF-receptor system involved in angiogenesis. *Cell Struct Funct* 26: 25–35.

Shifren, J.L., Mesiano, S., Taylor, R.N., Ferrara, N., Jaffe, R.B. (1998) Corticotropin regulates vascular endothelial growth factor expression in human fetal adrenal cortical cells. *J Clin Endocrinol Metab* 83: 1342–1347.

Soffer, S.Z., Moore, J.T., Kim, E., Huang, J., Yokoi, A., Manley, C., O'Toole, K., Stolar, C., Middlesworth, W., Yamashiro, D.J., Kandel, J.J. (2001) Combination antiangiogenic therapy: increased efficacy in a murine model of Wilms tumor. *J Pediatr Surg* 36: 1177–1181.

Soh, E.Y., Sobhi, S.A., Wong, M.G., Meng, Y.G., Siperstein, A.E., Clark, O.H., Duh, Q. Y. (1996) Thyroid-stimulating hormone promotes the secretion of vascular endothelial growth factor in thyroid cancer cell lines. *Surgery* 120: 944–947.

Stopeck, A.T., Bellamy, W., Unger, J., Rimsza, L., Iannone, M., Fisher, R.I., Miller, T.P. (2005) phase II trial of single agent bevacizumab (Avastin) in patients with relapsed, aggressive non-Hodgkin's lymphoma (NHL): Southwest Oncology Group Study S0108. *J Clin Oncol* 23: 583s.

Sweeney, C.J., Miller, K.D., Sissons, S.E., Nozaki, S., Heilman, D.K., Shen, J., Sledge, G.W.J. (2001) The antiangiogenic property of docetaxel is synergistic with a recombinant humanized monoclonal antibody against vascular endothelial growth factor or 2-methoxyestradiol but antagonized by endothelial growth factors. *Cancer Res* 61: 3369–3372.

Traina, T.A., Dickler, M.N., Caravelli, J.F., Yeh, B.M., Brogi, E., Scott, J., Moore, D., Panageas, K., Flores, S.A., Norton, L., Park, J., Hudis, C., Rugo, H. (2005) A phase II trial of letrozole in combination with bevacizumab, an anti-VEGF antibody, in patients with hormone receptor-positive metastatic breast cancer. *Breast Cancer Res Treat* 94: Abstract 2030.

Vitaliti, A., Wittmer, M., Steiner, R., Wyder, L., Neri, D., Klemenz, R. (2000) Inhibition of tumor angiogenesis by a single-chain antibody directed against vascular endothelial growth factor. *Cancer Res* 60: 4311–4314.

Warren, R.S., Yuan, H., Matli, M.R., Gillett, N.A., Ferrara, N. (1995) Regulation by vascular endothelial growth factor of human colon cancer tumorigenesis in a mouse model of experimental liver metastasis. *J Clin Invest* 95: 1789–1797.

Wiesmann, C., Fuh, G., Christinger, H.W., Eigenbrot, C., Wells, J.A., de Vos, A.M. (1997) Crystal structure at 1.7 A resolution of VEGF in complex with domain 2 of the Flt-1 receptor. *Cell* 91: 695–704.

Willett, C.G., Boucher, Y., di Tomaso, E., Duda, D.G., Munn, L.L., Tong, R.T., Chung, D.C., Sahani, D.V., Kalva, S.P., Kozin, S.V., Mino, M., Cohen, K.S., Scadden, D.T., Hartford, A.C., Fischman, A.J., Clark, J.W., Ryan, D.P., Zhu, A.X., Blaszkowsky, L.S., Chen, H.X., Shellito, P.C., Lauwers, G.Y., Jain, R.K. (2004) Direct evidence that the VEGF-specific antibody bevacizumab has antivascular effects in human rectal cancer. *Nat Med* 10: 145–147.

Yancopoulos, G.D., Davis, S., Gale, N.W., Rudge, J.S., Wiegand, S.J., Holash, J. (2000) Vascular-specific growth factors and blood vessel formation. *Nature* 407: 242–248.

Yang, J.C., Haworth, L., Sherry, R.M., Hwu, P., Schwartzentruber, D.J., Topalian, S.L., Steinberg, S.M., Chen, H.X., Rosenberg, S.A. (2003) A randomized trial of bevacizumab, an anti-vascular endothelial growth factor antibody, for metastatic renal cancer. *N Engl J Med* 349: 427–434.

Zhu, A.X., Sahani, D., Norden-Zfoni, A., Holalkere, N.S., Blaszkowsky, L., Ryan, D. P., Clark, J., Taylor, K., Heymach, J.V., Stuart, K. (2005) A phase II study of gemcitabine, oxaliplatin in combination with bevacizumab (GEMOX-B) in patients with hepatocellular carcinoma. Presented at: 2005 Gastrointestinal Cancers Symposium; 27–29 January; Hollywood, Fl. Abstract 131. Available at: http://www.asco.org. Accessed 15 February.

4
Cetuximab (Erbitux, C-225)

Norbert Schleucher and Udo Vanhoefer

4.1
Introduction

Cetuximab is a chimeric monoclonal IgG1 antibody (Fig. 4.1) targeting the epidermal growth factor receptor (EGF-R).

The EGF-R (HER1, ErbB1) is one member of the erbB family consisting of four members, HER1 to HER4. HER2 is a truncated receptor, while HER3 has a point mutation in the ATP binding site resulting in a nonfunctional tyrosine kinase. The EGF-R is a transmembrane protein activated by various ligands such as transforming growth factor alpha (TGF-α), amphiregulin, betacellulin, or epidermal growth factor (EGF). Activation causes dimerization, homo-dimerization with another EGR-R, or hetero-dimerization with another member of the erbB family, and phosphorylization on the internal tyrosine-kinase domain. Activation of the internal tyrosine kinase domain induces a signal transduction network leading to cellular proliferation, dedifferentiation and protection from apoptosis, and further to angiogenesis, migration, invasion, and metastases. One important pathway here is the activation of phosphoinosin-3-kinase (PI3K) and the serine-threonine kinase AKT, causing a suppression of the PTEN tumor suppressor gene. One the other hand, protein kinase C (PKC) and mitogen-activated protein (MAP) kinase are activated by G-protein-dependent mechanisms. Thus, a variety of transcriptional factors are induced, leading to neoplastic proliferation as a post-translational effect.

The expression of EGF-R is usually determined by immunohistochemical staining. Thus, EGF-R is seen to be expressed in gastrointestinal malignancies, especially colorectal cancer (CRC), gynecologic tumors (breast cancer and ovarian cancer), genitourinary tumors (prostate-, bladder- and renal cell cancer; RCC), non-small cell lung cancer (NSCLC), and in squamous cell cancer of the head and neck (see Table 4.1). In general, EGF-R is expressed in approximately 60 to 80% of these malignancies.

Immunohistochemical staining of EGF-R expression on the cellular surface is limited by the receptor recycling. EGF-R is internalized and re-expressed by either

Handbook of Therapeutic Antibodies. Edited by Stefan Dübel
Copyright © 2007 WILEY-VCH Verlag GmbH & Co. KGaA, Weinheim
ISBN 978-3-527-31453-9

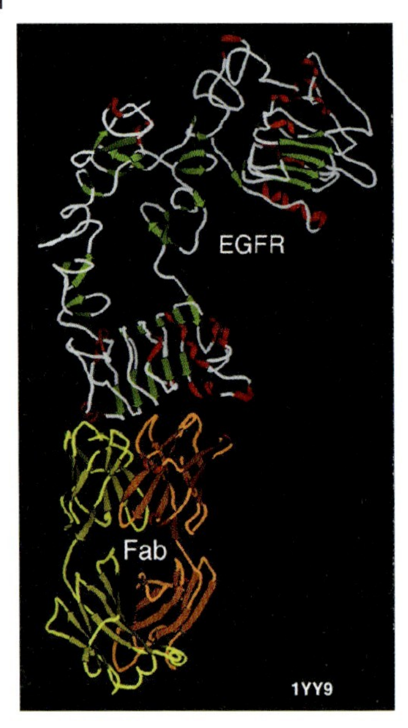

Fig. 4.1 Three-dimensional structure of the Fab fragment of cetuximab with its antigen, according to X-ray crystallographic data [49].

Table 4.1 EGFR expression in human malignancies.

Tumor	*EGF-R expression [%]*	*Reference(s)*
Colorectal cancer	25–82	Salomon (1995); Messa (1998) Goldstein (2001), Cunningham (2003)
Head and neck	80–100	Salomon (1995); Grandis (1996)
Pancreatic cancer	30–95	Salomon (1995); Uegaki (1997) Abbruzzese (2001)
NSCLC	40–81	Fujino (1996); Rusch (1997); Fontanini (1998); Gatzemeier (2003)
Renal cell cancer	50–90	Salomon (1995); Yoshida (1997)
Breast cancer	14–91	Klijn (1992); Beckman (1996); Bucci (1997); Walker (1999)
Ovarian cancer	35–70	Bartlett (1996); Fischer-Colbrie (1997)
Malignant glioma	40–63	Salomon (1995); Watanabe (1996); Rieske (1998)
Bladder cancer	31–48	Salomon (1995); Chow (1997)

NSCLC = non-small cell lung cancer.

smooth pits or coated pits. After internalization, the receptor may be degraded, with the consequent formation of multivesicular bodies, while the EGF-R is degraded in the lysosomes. In the case of recycling, a Golgi vesicle transports a recycled receptor to the cellular surface [1,2].

The expression of EGF-R has been inversely correlated with the prognosis of cancer patients; hence, patients with EGF-R-positive tumors have in general shorter survival times compared to patients with EGF-R-negative tumors [3–5]. In 1999, Inada evaluated 40 patients with esophageal cancer who underwent surgery; the 5-year survival among the EGF-R positive group was approximately 40%, compared to 70% in the EGF-R-negative group [6]. In general, patients with EGF-R-positive tumors are diagnosed at an advanced stage of the disease.

4.2 Preclinical Activity

In preclinical settings, cetuximab shows anti-tumor activity in a variety of cancer cell lines, the activity being comparable to that of conventional cytotoxic drugs. Furthermore, cetuximab can enhance the antitumor activity of irinotecan, platinum-derivatives, and fluoropyrimidines.

In human HT29 colon carcinoma nude mice xenografts, cetuximab induced a threefold reduction in tumor volume compared to untreated mice, and showed an almost equivalent antitumor activity as irinotecan. In this cell line the combination of cetuximab and irinotecan was synergistic such that no tumor growth was measured. Moreover, the combination of both drugs was even synergistic in irinotecan-refractory HT29 colon carcinoma xenografts, inducing a fourfold reduction in tumor growth compared to the irinotecan-treated controls [7].

Cisplatin resistance was reverted by the murine precursor drug M-225 in an A431 squamous cell cancer cell line in nude mice [8]. In this cisplatin-resistant cell line, a more than fourfold reduction in tumor volume resulted following addition of the noncytotoxic antibody M-225 to cisplatin treatment.

Cetuximab showed activity against RCC both *in-vitro* and *in-vivo*. In the RCC cell lines A498, Caki-1, SK-RC-4, SK-RC-29 and SW839, DNA synthesis was inhibited in a dose-dependent manner by treatment with cetuximab, while in Caki-1 ascites xenografts the survival of mice treated with cetuximab was prolonged compared to that of untreated mice [9].

In prostate cancer, cetuximab showed antitumor activity either alone or in combination with doxorubicin, with a significant reduction of tumor growth (DU145 and PC-3 xenografts) [10].

Cetuximab showed preclinical activity in p53 wild-type HepG2 hepatocellular carcinoma, and also induced cell cycle arrest in the G_0/G_1 phase, an increase of expression of the cyclin-dependent kinase inhibitors p21 (Waf1/CIP1) and p27 (Kip1), and a decrease in cyclin D1 expression. The combination with erlotinib, fluvastatin or doxorubicin resulted in synergistic antiproliferative effects [37].

Similar cell-cycle effects were seen in various cetuximab-sensitive cell lines of NSCLC, whereby synergistic effects were described for the combination of cetuximab with cisplatin, paclitaxel, or radiation [38].

Differential results were seen in pancreatic cancer cell lines. For example, after exposure to cetuximab the sensitivity of the MiaPaCa-2 pancreatic cancer cell line to gemcitabine and radiation was increased, but in BxPC-3 the antiproliferative effects of gemcitabine and radiation were independent of cetuximab exposure. In MiaPaCa-2 the expression of the pro-apoptotic BAX was up-regulated, while resistance in BxPC-3 cells was mediated by an alternate pathway of Ras-MAPK activation [39]. Likewise, additive cytotoxic effects of cetuximab plus gemcitabine were observed in the human pancreatic carcinoma L3.6pl nude mouse xenografts [40].

The combination of cetuximab with EGF-R tyrosine-kinase inhibitors (gefitinib, erlotinib) is of clinical and preclinical interest. For example, in head and neck-, vulvar-, prostate- and NSCLC -cancer cell lines the activity of the combination was higher than the single-agent activity of these drugs [47].

The contribution by radiosentization was demonstrated by Milas in the A431 cell line in nude mice, when the addition of three doses of the C-225 antibody to 18 Gy radiation resulted in an impressive delay in tumor growth [11].

4.3
Clinical Data I: Outcome of Monotherapy

The single-agent activity of cetuximab in patients with refractory CRC was demonstrated in two phase II studies by Saltz [12] (in 57 patients) and Lenz (in 346 patients), and also in a randomized phase II study by Cunningham [14] (in 111 patients). In each of these studies, cetuximab induced 10% of the objective remissions according to RECIST criteria, and approximately 35% of stable diseases. The median overall survival ranged from 6.9 to 8.7 months. In these studies the patients were heavily pretreated and refractory against fluoropyrimidines, irinotecan, and oxaliplatin, while EGF-R positivity was mandatory. In another study, Lenz et al. [13] included nine patients without EGF-R expression, and two of these responded to cetuximab monotherapy. The resultant toxicity was usually mild and consisted of skin rash (6% grade 3/4) and asthenia (11% grade 3/4) [12–14].

Single-agent activity in patients with cisplatin-refractory head and neck cancer was evaluated by Trigo (n = 103) [15], wherein objective remissions and disease stabilizations were observed in 12.6% and 45.2% of patients, respectively. The median overall survival was 5.9 months, and toxicity due to cetuximab was generally mild, with grade 3 toxicities according to NCI-CTC criteria generally below 5% (grade 3 skin rash 1%, fever 2%, fatigue 4%). Similar data were presented by Vermorken at al. (16).

In recurrent NSCLC, single-agent therapy with cetuximab is effective in salvage treatment. For example, when 66 patients were entered into a phase II study (71% pretreated with a platinum derivative, 25% with docetaxel), partial remissions were obtained in 4.5% of patients, while 30.3% achieved disease stabilization. The median survival was 8.1 months; toxicity was generally mild, with 6.1% of patients reporting skin rash, 13.6% fatigue, and 15.2% dyspnea grade 3/4 [17].

Table 4.2 Cetuximab Grade 3 and 4 toxicities (%) in various studies.

	Cetuximab combination with					
	Mono [14]	*Irinotecan [14]*	*Irinotecan [18]*	*IFL [19]*	*FOLFIRI [21]*	*FOLFOX [22]*
Diarrhea	1.7	21.2	22	28	14	26
Asthenia	10.4	13.7	14	10	n.s.	9
Skin rash	5.2	9.4	8	19	7	21
Neutropenia	0	9.5	17	28	17	14
Nausea/vomiting	4.3	7.1	6	n.s.	11	5
Anemia	2.7	4.8	n.s.	n.s.	n.s.	n.s.
Abdominal pain	5.2	3.3	n.s.	n.s.	n.s.	n.s.
Thrombopenia	0.9	0.5	n.s.	n.s.	n.s.	n.s.
Hypersensitivity	3.5	0	3	0	4.3	n.s.

n.s. = not stated, due to abstract publication. For other abbreviations, see text.

In metastatic RCC, when 54 patients were treated with cetuximab in the first-line setting, 7% tumor responses (partial and minor responses) and 25% disease stabilizations for a minimum of 6 months were observed, in association with a favorable toxicity profile [36].

4.4 Clinical Data II: Outcome of Combination Therapy

4.4.1 Colorectal Cancer

In metastatic CRC the efficacy of the combination cetuximab + irinotecan was evaluated in 138 patients by Saltz et al. [12], and in 218 patients by Cunningham et al. [14] (the BOND [Bowel Oncology with Cetuximab Antibody] study). Both studies included patients with EGF-R positive tumors that were refractory to chemotherapy with 5-fluorouracil (5-FU), irinotecan, and oxaliplatin. Notably, in the BOND study more than 65% of patients were pretreated with all three drugs. The objective response rate was 23% in both studies, and the disease control rate 30 to 56%. In the BOND study the median time to disease progression was 4.1 months compared to 1.5 months for monotherapy ($p < 0.0001$). The median overall survival was 8.6 months compared to 6.9 months for single-agent treatment ($p = 0.48$). The grade 3 and 4 toxic effects encountered were manageable, and included diarrhea (22%), asthenia/fatigue (14%), neutropenia (10–17%), and skin rash (10–11%) [14,18]. Cetuximab was approved for the treatment of metastatic EGF-R positive colorectal cancer in combination with irinotecan (not for monotherapy in Europe) following the failure of irinotecan and fluoropyrimide-based therapies.

Cetuximab was introduced into the first line treatment of metastatic CRC in a variety of phase II studies, and combined with all relevant combination regimes. Rosenberg et al. [19] evaluated the combination of cetuximab with weekly irinotecan, bolus 5-FU and leucovorin (LV) (the IFL regimen). A total of 29 patients was included, and the partial response rate was 48%, with an additional 41% disease stabilization. The toxicity was related to the IFL bolus regimen and consisted of grade 3 and 4 diarrhea and neutropenia in 28% of cases. A skin rash was observed in 21% of the patients [19]. Folprecht et al. [20] reported the results of a phase I/II study with a combination of cetuximab and weekly irinotecan/infusional 5-FU/LV (the AIO-irinotecan regimen) in 21 patients. The response was 67% and the stabilization rate of diseases was 29%. The median time to disease progression and median overall survival were impressive at 9.9 months and 33 months, respectively [20]. The combination of cetuximab with biweekly irinotecan and 5-FU/LV (FOLFIRI, CRYSTAL study) was described by Rougier et al. in 42 patients. Here, partial remissions were achieved in 46% of patients, and disease stabilization in 41%. The rates of high grade skin rash, diarrhea and neutropenia were 12%, 14%, and 12%, respectively [21]. Of major clinical interest was the combination of cetuximab with oxaliplatin and 5-FU/LV (FOLFOX-4 regimen). In this study (the ACROBAT study), the rate of complete response was 10%, but partial remissions were seen in 69% and 12% of patients were stable; this resulted in a clinical benefit rate of 91% (verified by an independent reviewer). Ten of 42 patients underwent metastatic surgery due to major response, and nine of these were completely (R0) resected. Grade III/IV toxicity was tolerable, with 33% of patients reporting skin-rash, 26% diarrhea, 26% neurotoxicity due to oxaliplatin, 21% neutropenia, and 16% mucositis [22].

The efficacy of cetuximab in combination with FOLFOX-4 in the second-line treatment of metastatic CRC after failure of an irinotecan-based first-line chemotherapy was analyzed in a randomized phase III trial (the EXPLORE study). The preliminary data after treatment of 102 patients showed a partial response rate of 20% in the experimental arm compared to 15.4% for FOLFOX-4 alone. Stable diseases were seen in 44% of patients versus 50% for patients treated with FOLFOX alone. Median progression-free survival and toxicity were similar in both arms [23]. The combination of cetuximab and irinotecan in second-line chemotherapy of CRC is currently under investigation (EPIC study). Following the analysis of data acquired from 400 patients, the combination appears to be feasible, though data relating to efficacy are not yet available. The study is ongoing until 1300 patients have been enrolled [24].

The efficacy of the antibody-combination of cetuximab and bevacizumab, with or without irinotecan, was analyzed by Saltz and colleagues in a randomized phase II trial (BOND-2 study) which included 81 patients with metastatic and refractory CRC [43]. The objective response rate for cetuximab and bevacizumab was 20%, with a progression-free survival of 5.6 months. The addition of irinotecan resulted in a response rate of 37% and a progression-free survival of 7.9 months. The expression of EGF-R was not mandatory in this study [43].

Recently, an adjuvant study with FOLFOX-4 plus cetuximab has been initiated in CRC, as has an investigation into the first-line treatment of a metastatic setting in a randomized phase III study (FOLFOX-4 + cetuximab versus FOLFOX-4 alone; the OPUS study). An additional study evaluating the efficacy of cetuximab in patients with EGF-R-negative tumors (the OPERA study) has also been initiated; the rationale here is the observation of Chung et al., who described a 25% response rate in patients with EGF-R-negative tumors [44].

4.4.2 Head and Neck Cancer

The combination of cetuximab and cisplatin/carboplatin was evaluated in recurrent squamous cell cancer of head and neck refractory to platinum-based therapy. Baselga and Kies analyzed 96 and 79 patients, respectively, in two nonrandomized phase II trials [25,26]. The results were similar in both studies, with 10% partial remissions and 55% stable disease. The median survival ranged from 5.2 to 6.0 months, and grade 3/4 toxicities were less than 10% [25,26]. In another trial, 56 patients with nasopharyngeal cancer refractory to radiation and progressive within 12 months after a platinum-based chemotherapy were treated with cetuximab and carboplatin [27]. A partial response of 17% was observed, and 47% were stable. The main toxicities observed (grade 3/4) were skin rash (13%), anemia (18%), thrombocytopenia (11%), and asthenia (9%) [27].

4.4.3 Non-Small Cell Lung Cancer (NSCLC)

In first-line chemotherapy of NSCLC (adenocarcinoma and squamous cell carcinoma) the addition of cetuximab to cisplatin and vinorelbine probably enhances the antitumor activity (LUCAS study). A total of 86 patients was randomized, and the confirmed response rate was 31% for the cetuximab/cisplatin/vinorelbin combination compared to 20% in the conventional arm of the study. The stable disease rate and time to progression were similar in both arms. Leukopenia (64% versus 51%), asthenia (17% versus 5%), infections (12% versus 5%), and skin rash (5% versus 0%) were enhanced in the cetuximab arm [28]. The combination of cetuximab with carboplatin/paclitaxel or carboplatin/gemcitabine is also feasible [29,30].

In second-line therapy of NSCLC, cetuximab was combined with docetaxel in 47 patients progressive within 3 months after first-line chemotherapy and with EGF-R-positive tumors. The partial response rate was 28% and the rate of stable diseases 18%. The most common grade 3 toxicities were skin rash (19%), infection (21%), and fatigue (21%). Interestingly, four patients showed allergic reactions that led to a discontinuation of treatment [31].

4.4.4
Other Tumors

A phase II trial of cetuximab plus gemcitabine in advanced pancreatic cancer showed 12% partial responses and 63% stable disease, with a one-year overall survival rate of 32% and a median overall survival of 7.1 months. Grade III/IV toxicities were neutropenia 39%, asthenia 22%, abdominal pain 22%, and thrombocytopenia 17% [41].

In advanced ovarian cancer, primary peritoneal carcinoma and cancer of the Fallopian tube, cetuximab was combined with carboplatin and paclitaxel (six cycles) followed by a cetuximab maintenance (6 months). This schedule was safely applied to 25 patients, with clinical responses being shown in 68% (17/25) [32].

4.5
Clinical Data III: Cetuximab and Radiotherapy/Chemo-Radiotherapy

In advanced squamous cell carcinoma of the head and neck, the combination of cetuximab and high-dose radiation prolongs survival compared to radiotherapy alone. In a phase III study, 424 patients were randomized to standard treatment, which consisted of 70–76 Gy radiation, or to the experimental arm to which cetuximab was added to the radiation for 8 weeks [33]. The median overall survival was 54 months in the radiotherapy/cetuximab group compared with 28 months in patients receiving radiotherapy alone ($p = 0.02$). The 2- and 3-year survival rates were 62% and 57% compared to 55% and 44%, respectively. Cetuximab did not enhance the toxicity of radiotherapy. Based on these data, cetuximab in combination with radiotherapy was approved for head and neck cancer in 2006.

The combination of cetuximab, cisplatin and 70-Gy radiation was also evaluated [35]. Here, 21 patients received eight doses of standard-dose cetuximab, cisplatin $100\,\text{mg}\,\text{m}^{-2}$ in Weeks 1 and 4, and concomitant radiotherapy. The 2-year overall survival rate was 76%, but 46% of patients reported grade 4 toxicities (anaphylaxis, anorexia, arrhythmia, bacteremia, hypokalemia, hyponatremia, myocardial infarction and mucositis), including two toxic deaths. Thus, this treatment schedule was not recommended due to problems of toxicity.

In NSCLC the combination of cetuximab with carboplatin/paclitaxel and radiation appears to be feasible with dominant hematologic toxicity (50% grade 3/4 blood/bone marrow toxicity). The efficacy data are not yet available, however [34].

4.5.1
The PARC Study

A current randomized phase II trial in advanced pancreatic cancer is comparing simultaneous to sequential cetuximab in combination with gemcitabine-based chemoradiation therapy (the PARC study) [42].

4.5.1.1 Dosing Schedule

Cetuximab treatment is started with a loading dose of 400 mg m^{-2} as a 2-h infusion, followed by a maintenance dose of 250 mg m^{-2} as a 1-h infusion once weekly until disease progression. This dosing schedule does not require dose reduction in combination with radiotherapy [48] or chemotherapy for CRC [14,18–24], head and neck cancer [26,27], and NSCLC [28–31]. The dosing schedule for cetuximab in combination with chemoradiotherapy is currently under investigation.

Due to the possibility of allergic reactions, premedication with an antihistaminic drug (e.g., 2 mg clemastine or 50 mg diphenhydramine) is recommended. The use of steroids is possible without inhibiting the efficacy of cetuximab, but in general this is not necessary.

Cetuximab is a low-emetogenic drug; therefore, the anti-emetic treatment depends on the emetogenicity of the concomitant chemotherapy, and usually consists of 5-HT_3 antagonists.

4.5.1.2 Determinants of Cetuximab Efficacy

The efficacy of cetuximab treatment in CRC is independent of the EGF-R staining activity [14,18]. In the BOND study, the response rates for cetuximab/irinotecan were 21 to 25% for patients with weak, moderate or strong EGF-R staining. For the monotherapy with cetuximab the response rates were more heterogeneous, but not different.

One important clinical predictor of cetuximab activity is the skin rash. In the BOND study, patients without skin reactions did not respond to monotherapy, and 6% responded to the combination of cetuximab and irinotecan. Patients with skin rash responded to the combination in 26% of cases, and in 13% of cases to the monotherapy. In the combination arm the median survival was 9.1 months for patients with skin rash compared to 3.0 months for those without skin toxicity [14].

Currently, few data are available regarding the molecular determinants of cetuximab efficacy. High gene expression levels of vascular endothelial growth factor (VEGF) have been associated with cetuximab resistance. The combination of low gene expression levels of cyclooxygenase-2, EGF-R and IL-8 is a significant predictor for overall survival in refractory CRC. Patients with the low gene expression combination showed a 13.5 months survival after cetuximab treatment compared to 2.3 months for patients with high gene expression levels. These finding were independent of skin toxicity. Patients with lower levels of EGF-R messenger-RNA had a longer overall survival than patients who showed a higher EGF-R messenger-RNA levels (7.3 months versus 2.2 months). Interestingly, patients with lower expression of cyclooxygenase-2 had a significantly higher rate of skin toxicity due to cetuximab treatment [45]. Another potential prognostic molecular marker may be the polymorphism of the cyclin D1 gene A870G [46].

The impact of EGF-R mutations is currently under investigation.

4.5.1.3 Treatment of Skin Toxicity

The treatment of skin rash is not evidence-based and consists mainly of treatment elements from juvenile acne. In the acute exudative phase, desinfective or antibiotic ointments can be used (e.g., benzoyl peroxide 5%, metronidazole gel, erythromycin 2% gel). Grade II or III skin toxicity mostly requires systemic antibiotic treatment, for example doxycycline (1–2 × 100 mg daily), tetracycline (2 × 250 mg), minocycline (2 × 50 mg), or the use of gyrase-inhibitors. Fatty ointments should not be used during the exudative phase, but in the xerotic phase skin care with fatty lotions is necessary; these include dexpanthenol-containing ointments, urea lotions, or bath oils. Antibiotic therapy is only required in case of superinfections, while antihistaminic drugs can be used for pruritus. It is also important to protect the skin from sunshine.

The treatment of nail toxicity (paronychia) consists of ichthyol ointments, local antibiotic treatment and puncture of abscesses. In the worst case the nail must be extracted.

References

1 Resat, H., Ewald, J.A., Dixan, D.A., et al. (2003) An integrated model of epidermal growth factor receptor trafficking and signal transduction. Biophys J 85: 730–745.

2 Wiley, H.S. (2003) Trafficking of the ErbB receptors and its influence on signalling. Exp Cell Res 284: 78–88.

3 Goldstein, N.S., Armin, M. (2001) Epidermal growth factor receptor immunohistochemical reactivity in patients with American Joint Committee on Cancer stage IV colon adenocarcinoma: implication for a standardized scoring system. Cancer 92: 1331–1346.

4 De Jong, K.P., Stellema, R., Karrenheld, A., et al. (1998) Clinical relevance of transforming growth factor alpha, epidermal growth factor receptor, p53 and Ki67 in colorectal liver metastases and corresponding primary tumours. Hepatology 28: 971–979.

5 Mayer, A., Takimoto, M., Fritz, E., et al. (1993) The prognostic significance of proliferating cell nuclear antigen, epidermal growth factor receptor, and mdr gene expression in colorectal cancer. Cancer 71: 2454–2460.

6 Inada, S., Koto, T., Futami, K., et al. (1999) Evaluation of malignancy and the prognosis of esophageal cancer based on an immunohistochemical study (p53, E-cadherin, epidermal growth factor receptor). Surg Today 29: 493–503.

7 Prewett, M.C., Hooper, A.T., Bassi, R., et al. (2002) Enhanced antitumour activity of anti-epidermal growth factor receptor monoclonal antibody IMC-C225 in combination with irinotecan (cpt-11) against human colorectal tumor xenografts. Clin Cancer Res 8: 994–1003.

8 Fan, Z., Baselga, J., Masui, H., et al. (1993) Antitumour effect of anti-epidermal growth factor receptor monoclonal antibodies plus cis-diamminedichloroplatinum on well established a431 cell xenografts. Cancer Res 53: 4637–4642.

9 Prewett, M.C., Rothman, M., Waksal, H., et al. (1998) Mouse-human chimeric anti-epidermal growth factor receptor antibody C225 inhibits the growth of human renal cell carcinoma xenografts in nude mice. Clin Cancer Res 4: 2957–2966.

10 Prewett, M.C., Rockwell, P., Rockwell, R.F., et al. (1996) The biological effects of C225, a chimeric monoclonal antibody to the EGF-R, on human prostate carcinoma. J Immunother Emphasis Tumor Immunol 19: 419–427.

11 Milas, L., Mason, K., Hunter, N., et al. (2000) In vivo enhancement of tumour radioresponse by C225 antiepidermal

growth factor receptor antibody. Clin Cancer Res 6: 701–708.

12 Saltz, L., Meropol, N.J., Needle, M.N., et al. (2004) phase II trial of cetuximab in patients with refractory colorectal cancer that expresses the epidermal growth factor receptor. J Clin Oncol 22: 1201–1208.

13 Lenz, H.J., Mayer, R.J., Gold, P.J., et al. (2004) Activity of cetuximab in patients with colorectal cancer refractory to both irinotecan and oxaliplatin. Proc Am Soc Clin Oncol 23: Abstract 3510.

14 Cunningham, D., Humblet, Y., Siena, S., et al. (2004) Cetuximab monotherapy and cetuximab plus irinotecan in irinotecan refractory metastatic colorectal cancer. N Engl J Med 35: 337–345.

15 Trigo, J., Hitt, R., Koralewski, P., et al. (2004) Cetuximab monotherapy is active in patients with platinum refractory recurrent/metastatic squamous cell carcinoma of the head and neck: results of a phase II study. Proc Am Soc Clin Oncol 23: Abstract 5502.

16 Vermorken, J.B., Bourhis, J., Trigo, I., et al. (2005) Cetuximab in recurrent/ metastatic squamous cell carcinoma of head and neck refractory to first line platinum based therapies. Proc Am Soc Clin Oncol 24: Abstract 5505.

17 Lilenbaum, R., Bonomi, P., Ansari, R., et al. (2005) A phase II trial of cetuximab as therapy for recurrent non-small cell lung cancer. Proc Am Soc Clin Oncol 24: Abstract 7036.

18 Saltz, L., Rubin, M., Hochster, H., et al. (2001) Cetuximab (IMC-C225) plus irinotecan (CPT-11) is active in CPT-11 refractory colorectal cancer that expresses epidermal growth factor receptor. Proc Am Soc Clin Oncol 20: Abstract 7.

19 Rosenberg, A.H., Loehrer, P.J., Needle, M.N., et al. (2002) Erbitux (IMC-C225) plus weekly irinotecan, fluorouracil and leucovorin in colorectal cancer that expresses the epidermal growth factor receptor. Proc Am Soc Clin Oncol 21: Abstract 536.

20 Folprecht, G., Lutz, M., Schöffski, P., et al. (2005) Cetuximab and irinotecan/ 5-FU/FA (AIO) is active and safe in the first line treatment of epidermal growth factor receptor expressing metastatic colorectal cancer. World Conference on Gastrointestinal Cancer, Abstract 053.

21 Rougier, P., Raoul, J.L., Van Laethem, J.L., et al. (2004) Cetuximab + FOLFIRI as first line treatment for metastatic colorectal cancer. Proc Am Soc Clin Oncol 23: Abstract 3513.

22 Cervantes, A., Casado, E., van Cutsem, E., et al. (2005) Phase II study of cetuximab plus FOLFOX-4 in first line setting for epidermal growth factor receptor expressing metastatic colorectal cancer. Eur J Cancer Suppl. Vol. 3(2), p 181.

23 Polikoff, J., Mitchell, E.P., Badarinath, S., et al. (2005) Cetuximab plus FOLFOX for colorectal cancer (EXPLORE): preliminary efficacy analysis of a randomized phase III trial. Proc Am Soc Clin Oncol 24: Abstract 3574.

24 Sobrero, A., Scheithauer, W., Laurel, J., et al. (2005) Cetuximab plus irinotecan for metastatic colorectal cancer: safety analysis of the first 400 patients in a randomized phase III trial (EPIC). Proc Am Soc Clin Oncol 24: Abstract 3580.

25 Kies, M.S., Arquette, M.A., Nabell, L., et al. (2002) Final report of the efficacy and safety of the anti-epidermal growth factor antibody Erbitx (IMC-C225), in combination with cisplatin in patients with recurrent sqaumous cell carcinoma of the head and neck refractory to cisplatin containing chemotherapy. Proc Am Soc Clin Oncol 21: Abstract 925 update.

26 Baselga, J., Trigo, J.M., Bourhis, J., et al. (2004) phase II multicenter study of the antiepidermal growth factor receptor monoclonal antibody cetuximab in combination with platinum-based chemotherapy in patients with platinum-refractory metastatic and/ or recurrent squamous cell carcinoma of the head and neck. J Clin Oncol 23: 5568–5577.

27 Chan, A.T.C., Hsu, M.M., Goh, B.C., et al. (2003) A phase II study of cetuximab in combination with carboplatin in patients with recurrent or metastatic nasopharyngeal carcinoma

who failed to a platinum based chemotherapy. Proc Am Soc Clin Oncol 22: Abstract 2000.

28 Rosell, R., Daniel, C., Ramlau, R., et al. (2004) Randomized phase II study of cetuximab in combination with cisplatin © and vinorelbine (V) vs. CV alone in the first line treatment of patients with epidermal growth factor expressing advanced non-small cell lung cancer. Proc Am Soc Clin Oncol 23: Abstract 7012.

29 Kelly, K., Hanna, N., Rosenberg, A., et al. (2003) A multi-centered phase I/II study of cetuximab in combination with paclitaxel and carboplatin in untreated patients with stage IV non-small cell lung cancer. Proc Am Soc Clin Oncol 22: Abstract 2592.

30 Robert, F., Blumenschein, G., Dicke, K., et al. (2003) phase Ib/IIa study of anti-epidermal growth factor receptor antibody, cetuximab in combination with gemcitabine/ carboplatin in patients with advanced non-small cell lung cancer. Proc Am Soc Clin Oncol 22: Abstract 2587.

31 Kim, E.S., Mauer, A.M., Tran, H.T., et al. (2003) A phase II stud of cetuximab, an epidermal growth factor receptor blocking antibody, in combination with docetaxel in chemotherapy refractory/ resistant patients with advanced non-small cell lung cancer: final report. Proc Am Soc Clin Oncol 22: Abstract 2581.

32 Aghajanian, C., Sabbatini, P., De Rosa, F., et al. (2005) A phase II study of cetuximab/ paclitaxel/ carboplatin for the initial treatment of advanced stage ovarian, primary peritoneal and fallopian tube cancer. Proc Am Soc Clin Oncol 24: Abstract 5047.

33 Bonner, J.A., Giralt, J., Harari, P.M., et al. (2004) Cetuximab prolongs survival in patients with locoregional advanced squamous cell carcinoma of head and neck: a phase III study of high-dose radiation therapy with or without cetuximab. Proc Am Soc Clin Oncol 23: Abstract 5507.

34 Werner-Wasik, M., Swann, S., Curran, W., et al. (2005) A phase II study of cetuximab in combination with chemoradiation in patients with stage IIIA/ B non-small cell lung cancer: An interim overall toxicity report of the RTOG 0324 trial. Proc Am Soc Clin Oncol 24: Abstract 7135.

35 Pfister, D.G., Aliff, T.B., Kraus, D.H., et al. (2003) Concurrent cetuximab, cisplatin, and concomitant boost radiation therapy for locoregionally advanced, squamous cell head and neck cancer: preliminary evaluation of a new combined-modality paradigm. Proc Am Soc Clin Oncol 22: Abstract 1993.

36 Gunnett, K., Motzer, R., Amato, R., et al. (1999) phase II study of anti-epidermal growth factor receptor antibody C225 alone in patients with metastatic renal cell carcinoma. Proc Am Soc Clin Oncol 18: Abstract 1309.

37 Huether, A., Hopfner, M., Baradari, V., et al. (2005) EGF-R blockade by cetuximab alone or as combination therapy for growth control of hepatocellular cancer. Biochem Pharmacol 70: 1568–1578.

38 Raben, D., Helfrich, B., Chan, D.C., et al. (2005) The effects of cetuximab alone and in combination with radiation and/ or chemotherapy in lung cancer. Clin Cancer Res 11(2 Pt.1): 795–805.

39 Huang, Z.Q., Buchsbaum, D.J., Raisch, K.P., et al. (2003) Differential responses by pancreatic carcinoma cell lines to prolonged exposure to Erbitux (IMC-C225) anti-EGF-R antibody. J Surg Res 111: 274–283.

40 Bruns, C.J., Harbison, M.T., Davis, D.W., et al. (2000) Epidermal growth factor receptor blockade with C225 plus gemcitabine results in regression of human pancreatic carcinoma growing orthotopically in nude mice by antiangiogenic mechanisms. Clin Cancer Res 6: 1936–1948.

41 Xiong, H.S., Rosenberg, A., Lo Buglio, A., et al. (2004) Cetuximab, a monoclonal antibody targeting the epidermal growth factor receptor, in combination with gemcitabine for advanced pancreatic cancer: a multicenter phase II trial. J Clin Oncol 22: 2610–2616.

42 Krempien, R., Muenter, M.W., Huber, P.E., et al. (2005) Randomized phase II – study evaluating EGF-R targeting therapy

with cetuximab in combination with radiotherapy and chemotherapy for patients with locally advanced pancreatic cancer – PARC study protocol. BMC Cancer 5: 131.

43 Saltz, L., Lenz, H., Hochster, H., et al. (2005) Randomized phase II trial of cetuximab, bevacizumab/ irinotecan versus cetuximab/ bevacizumab in irinotecan refractory colorectal cancer. Proc Am Soc Clin Oncol 24: Abstract 3508.

44 Chung, K.Y., Shia, J., Kemeny, N.E., et al. (2005) Cetuximab shows activity in colorectal cancer patients with tumors that do not express the epidermal growth factor receptor by immunohistochemistry. J Clin Oncol 23: 1–8.

45 Vallböhmer, D., Zhang, W., Gordon, M., et al. (2005) Molecular determinants of cetuximab efficacy. J Clin Oncol 23: 3536–3544.

46 Zhang, W., Yun, J., Press, A., et al. (2004) Association of cyclin D1 gene A870G polymorphism and clinical outcome of EGF-R positive metastatic colorectal cancer patients treated with epidermal growth factor receptor inhibitor cetuximab. Proc Am Soc Clin Oncol 23: Abstract 3518.

47 Huang, S., Armstrong, E.A., Benavente, S., et al. (2004) Dual-agent molecular targeting of the epidermal growth factor receptor. Cancer Res 64: 5355–5362.

48 Robert, F., Ezekiel, M.P., Spencer, S.A., et al. (2001) phase I study of anti-epidermal growth factor receptor antibody cetuximab in combination with radiation therapy in patients with advanced head and neck cancer. J Clin Oncol 19: 3234–3243.

49 Li, S., Schmitz, K.R., Jeffrey, P.D., Wiltzius, J.J.W., Kussie, P., Ferguson, K.M. (2005) Structural basis for inhibition of the epidermal growth factor receptor by cetuximab. Cancer Cell 7: 301–311.

5
Efalizumab (Raptiva)

Karlheinz Schmitt-Rau and Sigbert Jahn

5.1
Introduction

Efalizumab is a humanized monoclonal antibody to lymphocyte function-associated antigen (LFA-1). LFA-1 belongs to the family of the β_2 integrins and is expressed on the surface of T cells (CD4+ cells, T-helper cells). It is involved in several T-cell activities, such as T-cell activation and migration (Janeway et al. 2001; Kuypers and Roos 1989), as well as T-cell adhesion during cellular interactions that are important to the inflammatory processes (Lo et al. 1989).

LFA-1, Mac-1 and p159,95 – all of which are members of the β_2 integrin family – are heterodimeric molecules consisting of a β-subunit (CD18), common to all three molecules, which is linked noncovalently to the respective a-chain CD11a (LFA-1), CD11b (Mac-1), and CD11c (p159,95). T-cells mainly express LFA-1 (CD11a/CD18).

The ligands for LFA-1 are intercellular adhesion molecules (ICAM), and include:

- ICAM-1, which is expressed on leucocytes, vascular endothelium cells and epithelial cells, including keratinocytes (Dustin et al. 1986);
- ICAM-2, which is expressed on resting endothelium and lymphocytes (de Fougerolles et al. 1991); and
- ICAM-3, which is expressed on monocytes and resting lymphocytes (de Fougerolles et al. 1994).

ICAM expression can be triggered by proinflammatory mediators, such as tumor necrosis factor-α (TNF-α) and interferon-γ (IFN-γ) (Janeway et al. 2001). Monoclonal antibodies (mAbs) to LFA-1 inhibit LFA-1/ICAM interaction, leading to the inhibition of several T-cell-dependent immune functions. Results derived from animal models imply that inhibiting LFA-1/ICAM interaction by anti-CD11a antibodies might have a clinical benefit in certain T-cell-dependent diseases.

A humanized monoclonal anti-CD11a immunoglobulin (Ig)G1 antibody, efalizumab, could therefore show clinical efficacy in psoriasis.

Handbook of Therapeutic Antibodies. Edited by Stefan Dübel
Copyright © 2007 WILEY-VCH Verlag GmbH & Co. KGaA, Weinheim
ISBN 978-3-527-31453-9

5.2 Development and Characterization of the Antibody

Efalizumab (rhuMAb CD11a, hu1124) is a full-length, IgG1 kappa isotype antibody composed of two identical light chains each consisting of 214 amino acid residues, and two heavy chains each consisting of 451 residues. Each light chain is covalently coupled through a disulfide link to a heavy chain. The two heavy chains are covalently coupled to each other via inter-chain disulfide bonds consistent with the structure of human IgG1. The molecular weight of intact efalizumab is 148 841 Da.

Originally developed as a murine anti-CD11a monoclonal antibody (MHM24; Hildreth et al. 1983), efalizumab has been prepared by substituting human DNA sequences using genetic engineering methods to reduce immunogenicity. This results in a "humanized" mAb (HuIgG1) in which the complementarity-determining regions (CDRs) of the murine antibody – which are important for antigen recognition – are preserved.

Previous studies on murine MHM24 have shown that, similar to other anti-CD11a antibodies, it is able to inhibit T-cell function (Hildreth and August 1985; Dougherty and Hogg 1987).

The consensus sequences for the human heavy chain subgroup III (V_H CH1) and the light chain subgroup k 1 were used as the framework for the humanization; subsequently, several humanized variants were made and screened for binding as Fabs. To construct the first Fab variant of humanized MHM24, all six CDR residues were transferred from the murine antibody to the human framework.

Further variants were constructed by targeted exchange of either framework residues, or residues within CDRs using the first variant (Fab-1), as a template. For that purpose, both light and heavy chains were completely sequenced for each variant. Plasmids containing the sequences were then transformed to *Escherichia coli* for protein expression (Werther et al. 1996).

All variants were tested for CD11a binding in the Jurkat cell assay. V_L and V_H domains of the variant with optimal binding characteristics were then transferred to human IgG1 constant domains, producing the full-length intact humanized antibody (Werther et al. 1996).

Several *in-vitro* assays were performed to compare efalizumab with its parent murine antibody MHM24, including the keratinocyte cell-adhesion assay and the mixed lymphocyte response assay (MLR). The results showed that, in these assays, efalizumab was equally effective as MHM24. In addition, the apparent K_d values, as determined by saturation binding using peripheral blood mononuclear cells (PBMCs) of two human donors, were similar for both MHM24 and efalizumab (0.16 ± 0.01 and 0.13 ± 0.02 versus 0.11 ± 0.08 and 0.18 ± 0.03, respectively) (Werther et al. 1996).

5.3
Efalizumab in the Treatment of Psoriasis

5.3.1
Psoriasis: Prevalence, Characteristics, and Therapeutic Options

Psoriasis is one of the most common dermatological diseases. Although there is a great variation in the prevalence of psoriasis in different countries, due mainly to environmental and genetic factors, it can be said that the condition affects approximately 2–3% of the world's population (Jung and Moll 2003). Moreover, between 20 to 25% of these people suffer from moderate-to-severe forms of the disease (Weinstein and Menter 2003). The majority of patients (75%) show the first signs of disease manifestation before the age of 40 years, with a peak in the second decade of life (Gollnick and Bonnekoh 2001). These patients usually have a family history of psoriasis, their disease is more severe, and it is characterized by frequent relapses.

The most common form of psoriasis, with a prevalence of about 70%, is plaque psoriasis or "psoriasis vulgaris" (Jung and Moll 2003). Plaque psoriasis is characterized by hyperkeratosis, parakeratosis, and the presence of inflammatory lesions in the skin. There is a predilection for symmetrical involvement of the scalp, elbows, knees, and lower back, although it can occur anywhere on the body. Other, less frequent, morphologies of psoriasis include guttate, inverse, pustular, and erythrodermic forms. These may occur individually, concomitantly or sequentially.

There is agreement today that immunological mechanisms play an important role in the pathogenesis of many chronic relapsing inflammatory skin diseases such as psoriasis (Weinstein and Menter 2003; Schön and Boehncke 2005). Evidence of the pivotal role played by T cells in the pathology of psoriasis is accumulating, based on the following points:

- Activated T cells are found in psoriatic lesions (Bos et al. 1983; Ferenczi et al. 2000).
- T cells have the ability to induce the altered keratinocyte growth and differentiation pattern typical of psoriasis. This has been demonstrated in a SCID (severely compromised immunodeficient) mouse model by injecting autologous immunoctyes into the dermis of mice that have received grafts of human skin. Plaques typical of those seen in psoriasis are observed when immunocytes from a patient with psoriasis are injected into a mouse possessing a graft of symptom-free skin from the same patient (Gilhar et al. 1997; Wrone-Smith and Nickoloff 1996).
- T-cell-targeted immune suppressive drugs, such as cyclosporine (Lebwohl et al. 1998), and antibodies against the CD25 receptor (Gottlieb et al. 1995) and CD4 (Prinz et al. 1991), have been shown to improve psoriasis.

- In bone marrow transplantation, psoriasis can be transferred from a donor suffering from the disease to a healthy recipient. Also, psoriasis can be "cured" when bone marrow is transplanted from a healthy donor to a person with psoriasis (Gollnick and Bonnekoh 2001).
- When symptomless prepsoriatic human skin was engrafted onto AGR129 mice, deficient in type I and type II interferon receptors and for the recombination activating gene 2, resident human T cells in the skin grafts underwent local proliferation, demonstrating the importance of resident immune cells in the development of psoriasis (Boyman et al. 2004).

Today, a whole battery of treatment options is available for the treatment of psoriasis (Fig. 5.1). The mild forms of psoriasis are usually treated with topical preparations, including vitamin D_3 analogs, corticosteroids, and retinoids. When the disease becomes severe, phototherapeutic regimens are applied. Finally, there is the option to use an oral immune suppressive drug, such as methotrexate, cyclosporine, oral retinoids, or fumaric acid esters, either as monotherapy or in combination. As the majority of patients develop their disease before the age of 40 years, many of those with moderate-to-severe psoriasis will require decades of continuous systemic treatment or phototherapy. Unfortunately, none of the available therapies can be used chronically, because of long-term safety and toxicity problems.

Recent advances in the understanding of T-cell interactions in the pathogenesis of psoriasis have led to the development of several biological substances, such as the targeted T-cell modulator efalizumab, for continuous immune therapy of this disease, without the safety problems of the traditional systemic preparations.

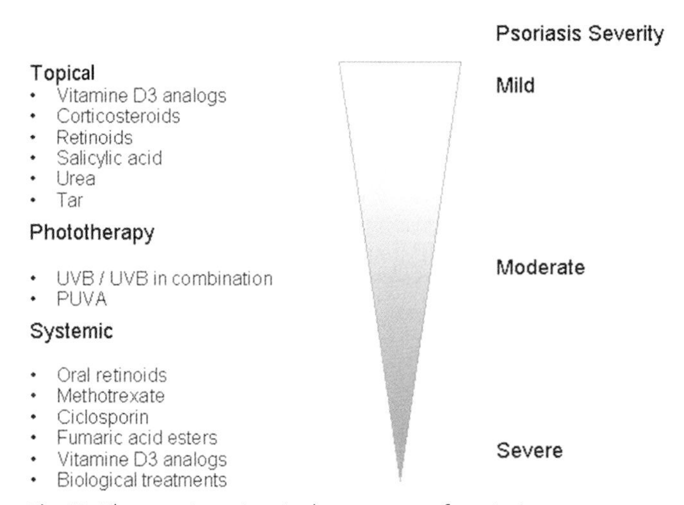

Fig. 5.1 Therapeutic options in the treatment of psoriasis.

5.3.2
Pathogenesis of Psoriasis

5.3.2.1 T-Cell Activation

The first step in the immune pathogenic cascade of psoriasis is the capture and processing of auto-antigen by Langerhans cells or dendritic cells ("antigen-presenting cells"; APCs) in the epidermis or dermis, respectively. Although the nature of the antigen is still unknown, there is some evidence that in genetically predisposed subjects keratinocyte-derived peptides may be involved (Bos and De Rie 1999; Valdimarsson et al. 1995).

The APC–antigen complex migrates to a skin-draining lymph node, where the antigen is presented via major histocompatibility complex II (MHC II) on the surface of the APC to the T-cell receptor (TCR) of the specific naïve CD4+ cell. Interaction of the APC-MHC-II/antigen–TCR complex is, however, not sufficient for T-cell activation.

The initial binding of T cells and APCs, as well as stabilization of the cell pair, is mediated by LFA-1 on the T cell and ICAM-1 on the APC. This stabilized structure has been referred to as the "immunological synapse" (Grakoui et al. 1999; Fig. 5.2), the formation of which is followed by delivery of the antigen-specific signal (signal 1) and a co-stimulatory signal (signal 2), which are also mediated by LFA-1 and ICAM-1. Delivery of signals 1 and 2 is followed by binding of cytokines that induce T cells to proliferate (i.e., to undergo clonal expansion) and to differentiate (Krueger 2002). Proliferation is thought to be largely mediated by the cytokine interleukin-2 (IL-2) (Janeway et al. 2001).

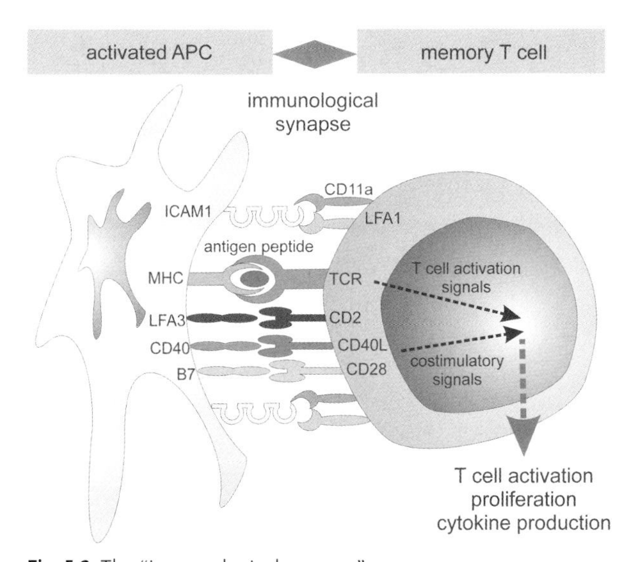

Fig. 5.2 The "immunological synapse".

5.3.2.2 T-Cell Migration and Extravasation

Following activation, T cells leave the lymph node and start trafficking via the bloodstream to the dermal vessels (Fig. 5.3). Stimulation by keratinocyte-derived cytokines (IL-8, CCL27) leads to increased expression of adhesion molecules, including E-selectin and ICAM-1, in the post-capillary venules of inflamed skin. E-selectin is the target for cutaneous lymphocyte antigen (CLA) on the T-cell surface. Binding of CLA to E-selectin, as well as interaction of the lymphocyte chemokine receptor (CCR10) with its ligand CCL27, slows circulating lymphocytes and causes them to "roll" along the endothelial wall (Fuhlbrigge et al. 2002; Homey et al. 2002).

As a result of increased exposure to chemokines, the affinity of LFA-1 for ICAM-1 is increased, probably mediated by a conformational change in the LFA-1 molecule. Bound T cells flatten and pass through the epithelium into the surrounding tissue – a process known as diapedesis (Bradley and Watson 1996). Once they have left the venule, T cells respond to chemokines, drawing them towards the site of inflammation in the dermis, and from there into the epidermis.

5.3.2.3 T-Cell Reactivation

Following transmigration from the circulation into dermal and epidermal tissue, memory T cells are brought into contact with antigen-presenting dendritic and Langerhans cells, as well as with keratinocytes, which – probably due to genetic

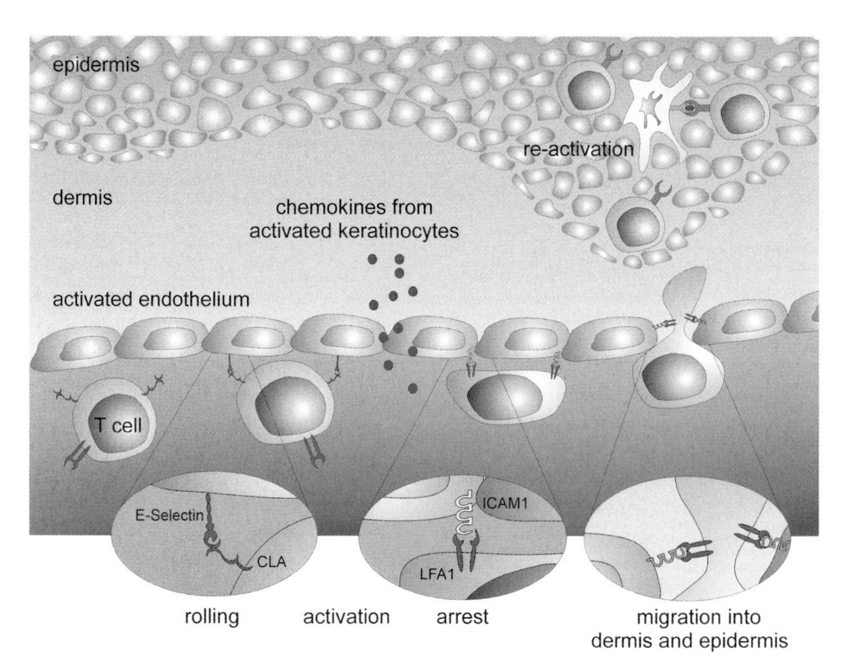

Fig. 5.3 T-cell migration, extravasation, and reactivation.

alterations – are also able to act as APCs (see Fig. 5.3). This interaction leads to the reactivation of T cells and is followed by increased production and secretion of cytokines, including IL-2, IFN-γ and TNF-α, which show the TH-1 cytokine pattern typically observed in psoriasis:

- IFN-γ leads to the induction of ICAM-1 in keratinocytes and endothelium, and consequently supports binding of lymphocytes to keratinocytes.
- TNF-α induces keratinocyte proliferation and stimulates keratinocyte-induced IL-8 production. This leads to chemotaxis of lymphocytes and neutrophils into the epidermis. TNF-α is also able to induce activation and proliferation of endothelial cells.

However, the exact mechanism by which (re-)activated T cells cause psoriatic lesions is not known. Direct lymphocyte/keratinocyte interactions might be involved (Bos and De Rie 1999), or cytokines produced during T-cell reactivation may be responsible for the most prominent alteration in psoriatic skin, namely hyperproliferation (Prinz et al. 1994).

In summary, a combination of different mechanisms may be responsible for the changes observed histologically and clinically in psoriatic skin. These include:

- increased proliferation of keratinocytes (hyperkeratosis);
- neutrophil and mast-cell migration into epidermis and dermis, respectively;
- increased production of defensins by activated keratinocytes; and
- increased angiogenesis.

5.3.3 Efalizumab: The Mechanism of Action

As described above, efalizumab is a humanized monoclonal antibody against CD11a, the a-subunit of LFA-1. Binding of efalizumab to CD11a leads to blockade of the interaction between LFA-1 and ICAM-1, the blockade of this interaction having several consequences (Krueger 2002; Jullien et al. 2004), including:

- the activation of T lymphocytes in lymph nodes;
- inhibition of extravasation of circulating lymphocytes in inflammatory skin;
- blockade of T-lymphocyte reactivation in skin by APCs; and
- reduction of the keratinocyte interaction with activated T-lymphocytes.

By inhibiting these processes, efalizumab may be able to prevent abnormal cytokine production, keratinocyte hyperproliferation and abnormal keratinocyte differentiation, which are characteristic of the psoriatic phenotype.

5.4 Pharmacology and Toxicology of Efalizumab

5.4.1 Preclinical Studies

A number of preclinical studies have been conducted with efalizumab in order to evaluate its pharmacodynamic and pharmacokinetic properties, as well as its toxicity (Menter and Griffiths 2004).

In-vitro studies performed with efalizumab have shown that it is able to bind to human and chimpanzee leucocytes, to inhibit lymphocyte binding to ICAM-1 on keratinocytes, and to inhibit T-cell proliferation. In murine and chimpanzee *in-vivo* models, efalizumab was able to down-regulate LFA-1 expression on lymphocytes, while efalizumab has been shown to have two-compartment kinetics, with nonlinear elimination in monkeys.

Anti-mouse CD11a antibodies have been used to study developmental toxicity in mice. In these studies, doses were administered at up to 30 times the equivalent recommended clinical dose of $1\,mg\,kg^{-1}$. However, no adverse effects were observed on mating, fertility, or reproductive parameters; neither was there any evidence of maternal toxicity, embryotoxicity, or teratogenicity.

Similarly to other immunoglobulins, anti-mouse CD11a antibody is secreted in the milk of lactating mice that are exposed to the antibody during gestation and lactation. Furthermore, a significant reduction was observed in the ability of their offspring to generate an antibody response at 11 weeks of age, which was – at least partially – reversible at Week 25. There were, however, no adverse effects on behavior, growth and reproductive function of the offspring.

Immunization of chimpanzees exposed to high doses of efalizumab (>10 times the clinical dose) with tetanus toxoid produced an impaired antibody response compared with control animals. However, to date no animal reproduction studies and long-term carcinogenicity studies have been conducted with efalizumab.

5.4.2 Pharmacodynamics

The pharmacodynamic properties of efalizumab were investigated in several phase I and phase II studies following intravenous and subcutaneous administration, either as a single dose or repeated weekly administration.

In the single-dose intravenous study, doses of 0.03 to $10\,mg\,kg^{-1}$ were administered (Gottlieb et al. 2000). Within 24 h, treatment with efalizumab reduced the level of CD11a expression on T cells to 25% of pretreatment levels, and this suppression persisted as long as efalizumab was present in the circulation. CD11a expression returned to baseline within 7 to 10 days following clearance of efalizumab, without showing any signs of lymphocyte depletion. The total white blood cell (WBC) count was slightly increased within about 8 h of efalizumab administration; circulating lymphocyte counts were increased (to about twofold the Upper

Limit of Normal) by day 7. Following multiple weekly dosing, lymphocyte levels remained elevated but returned to baseline after efalizumab clearance. This elevation of the lymphocyte count is probably due to demargination – the blocked entry of efalizumab-bound cells to tissues.

In order to achieve the full pharmacodynamic effect, intravenous doses exceeding 0.3 $mg\,kg^{-1}$ were necessary (Gottlieb et al. 2000). Complete saturation and maintenance of CD11a binding site down-regulation on lymphocytes required weekly intravenous doses of 0.6 $mg\,kg^{-1}$, which corresponds to an efalizumab plasma concentration of 5 $\mu g\,mL^{-1}$.

Several histologic changes were observed in psoriatic plaques following efalizumab administration. A marked reduction of keratin-16, corresponding to decreased disease activity, was noted. Keratinocyte ICAM-1 levels were also reduced, indicating reduced cytokine-mediated inflammation. Furthermore, a significant thinning of the epidermis and restoration of normal skin was observed after 28 days of treatment, in concordance with reductions of over 50% in cutaneous T-cell infiltration and reduced CD11a availability (Gottlieb et al. 2002). These data demonstrate that, by reducing CD11a on the surface of circulating and cutaneous T cells, efalizumab is able to reverse both the histological signs of inflammation and the pathological hyperplasia characteristic of plaque psoriasis.

In general, the effects of subcutaneous efalizumab on lymphocytes were comparable to those observed after intravenous dosing. Subcutaneous doses of 1 $mg\,kg^{-1}$ per week (or above) produced the required efalizumab plasma concentrations of 5 $\mu g\,mL^{-1}$ for binding site down-regulation and saturation. In addition to reduced CD11a expression on the surface of CD3+ T cells, binding of efalizumab also causes a reduced expression of other adhesion molecules, such as CD11b, L-selectin or β_7-integrin (Vugmeyster et al. 2004). The down-modulation of these adhesion molecules likely contributes to the anti-adhesive effects of efalizumab. There is also a decrease of $\alpha\beta$+ T-cell receptors and of TCR-associated co-receptors, such as CD4, CD8 or CD2. The inhibition of TCR-mediated activation therefore seems also to play a role in efalizumab's mode of action.

5.4.3 Pharmacokinetics

The pharmacokinetic properties of subcutaneous efalizumab were determined in an open, multicenter, phase I study of 70 patients suffering from moderate-to-severe plaque psoriasis. Patients received weekly doses of either 1 $mg\,kg^{-1}$ (n = 33) or 2 $mg\,kg^{-1}$ (n = 37) efalizumab for 12 weeks (Mortensen et al. 2005).

5.4.3.1 Absorption

After subcutaneous administration of efalizumab, peak plasma concentrations are reached after 2 to 3 days. The average estimated bioavailability was about 50% at the recommended dose level of subcutaneous efalizumab, 1.0 $mg\,kg^{-1}$ per week.

5.4.3.2 Distribution

Steady-state serum concentrations of efalizumab were achieved after four weekly doses of efalizumab at $1\,mg\,kg^{-1}$, and after eight weeks in patients receiving $2\,mg\,kg^{-1}$. At this dose level (with an initial dose of $0.7\,mg\,kg^{-1}$ in the first week), the mean efalizumab plasma trough values were $9.1 \pm 6.7\,\mu g\,mL^{-1}$ in the $1\,mg\,kg^{-1}$ group and $23.5 \pm 12.2\,\mu g\,mL^{-1}$ in the $2\,mg\,kg^{-1}$ group. The volumes of distribution of the central compartment after single intravenous doses were $110\,mL\,kg^{-1}$ at a dose level of $0.03\,mg\,kg^{-1}$, and $58\,mL\,kg^{-1}$ at a dose of $10\,mg\,kg^{-1}$ (Bauer et al. 1999).

5.4.3.3 Biotransformation

The metabolism of efalizumab is through internalization followed by intracellular degradation as a consequence of either binding to cell surface CD11a, or through endocytosis. The expected degradation products are small peptides and individual amino acids which are eliminated by glomerular filtration. Cytochrome P450 enzymes, as well as conjugation reactions, are not involved in the metabolism of efalizumab (Coffey et al. 2004).

5.4.3.4 Elimination

Efalizumab is cleared by dose-dependent nonlinear saturable elimination (Gottlieb et al. 2000). The mean steady-state clearance is 24.3 ± 18.5 and $15.7 \pm 12.6\,mL\,kg^{-1}$ per day for the 1 and $2\,mg\,kg^{-1}$ per week groups, respectively. The elimination half-life was about 6.21 ± 3.11 days for the $1\,mg\,kg^{-1}$ per week group, and 7.4 ± 2.5 days for the $2\,mg\,kg^{-1}$ per week group. T_{end} at steady-state is 25.5 ± 1.6 days and 44 ± 10 days at dose levels of 1 and $2\,mg\,kg^{-1}$ per week, respectively.

Efalizumab shows dose-dependent nonlinear pharmacokinetics, which can be explained by its saturable specific binding to cell-surface receptors CD11a. Clearance was more rapid at lower doses, suggesting a receptor-mediated mechanism at drug levels below $10\,\mu g\,mL^{-1}$ (Gottlieb et al. 2000).

In a population pharmacokinetic analysis of 1088 patients, body weight was found to be the most significant covariate affecting efalizumab clearance. Other covariates such as baseline Psoriasis Area and Severity Index (PASI), baseline lymphocyte count and age had modest effects on clearance; however, gender and ethnic origin had no effect (Sun et al. 2005).

Additional pharmacokinetic data are available from an open-label extended-treatment trial in which patients who responded to an initial treatment of efalizumab ($2\,mg\,kg^{-1}$ for 12 weeks) received the drug in a maintenance phase for up to 33 months at a dose of $1\,mg\,kg^{-1}$ (Gottlieb et al. 2003). Pharmacokinetic analysis of each 12-week treatment period for up to 15 months showed that steady-state trough levels remained constant during continuous efalizumab dosing. There was no evidence of efalizumab accumulation or alteration of the pharmacokinetic profile of efalizumab during long-term continuous dosing (Gottlieb et al. 2003).

5.5 Clinical Development of Efalizumab

The therapeutic efficacy, safety and tolerability, and the changes in health-related quality of life (HRQoL) of subcutaneous efalizumab have been evaluated in more than 3000 adult patients with moderate-to-severe plaque psoriasis. Five large, randomized, placebo-controlled, double-blind phase III studies have been conducted in which all patients included had a minimum affected body surface area (BSA) of 10%, and a PASI of ≥12. The presence of guttate, pustular or erythrodermic psoriasis as the sole or predominant form of psoriasis was an exclusion criterion. Patients were randomized to receive placebo or efalizumab, administered subcutaneously at doses of 1 or 2 mg kg^{-1} once weekly for 12 weeks, with an initial conditioning dose of 0.7 mg kg^{-1}. Patients could receive low-dose topical corticosteroids concomitantly with efalizumab; however, no other concomitant systemic therapies were allowed.

The primary endpoint was the proportion of patients with a ≥75% improvement in the PASI score (PASI 75) relative to baseline when assessed after a 12-week treatment course. Secondary endpoints included the Physician's Global Assessment of change (PGAc), the static PGA (sPGA), the proportion of patients with a ≥50% improvement in PASI score (PASI 50) relative to baseline after 12 weeks of treatment, the proportion of subjects who achieved a rating of "minimal" or "clear" on a static global assessment by the physician, and the overall lesion severity (OLS). Additional secondary endpoints were the improvement in the Dermatology Life Quality Index (DLQI) and the Psoriasis Symptom Assessment (PSA).

The physician-assessed PASI is a complex scoring system to quantify the efficacy of psoriasis treatments. The system takes into account the affected BSA, as well as the severity of erythema, scaling and infiltration in each of the four body areas; head, trunk, upper and lower extremities. The combined score has a numerical range of 0 to 72, with higher values indicating more serious disease. Although PASI is used by regulatory authorities (e.g., the US FDA and the European EMEA) as a precondition for the approval of new anti-psoriasis drugs, its relevance in clinical practice has been questioned by many experts (Carlin et al. 2004; Langley et al. 2004). Due to its complexity, the PASI measurement is therefore not frequently used in clinical practice.

In addition to the first 12 weeks of treatment, extended treatment and re-treatment periods have been studied in addition to an open-label long-term study in which patients received continuous efalizumab for up to 3 years.

5.5.1 Clinical Efficacy

5.5.1.1 Randomized, Placebo-Controlled, Double-Blind Studies

In one clinical study, 498 patients received subcutaneous efalizumab (1 or 2 mg kg^{-1} per week) or placebo for 12 weeks (Leonardi et al. 2005a). After 12 weeks of treatment, significantly more patients who received efalizumab showed at least 75% PASI improvement in both dose groups compared with patients receiving placebo (39% and 27% in the 1 and 2 $mg\,kg^{-1}$ groups respectively, versus 2% after placebo). Some 61% and 51% of efalizumab-treated patients achieved at least a 50% PASI improvement in the 1 and 2 $mg\,kg^{-1}$ groups compared with 15% in the placebo group.

In another study, 556 patients were randomized to receive placebo (n = 187) or efalizumab 1 $mg\,kg^{-1}$ per week (n = 369) (Gordon et al. 2003). The mean PASI improvement relative to baseline at Week 12 in this study was 52% in the efalizumab group versus 19% in the placebo group (Gordon et al. 2003). A significant improvement in PASI was seen at Week 4, and was further improved through to Week 12. Comparable improvements were observed in sPGA ("minimal" or "clear"; 26% versus 3%, respectively) and PGA ("excellent" or "cleared"; 33% versus 5%, respectively).

The efficacy of efalizumab after 12 weeks was confirmed in the other phase III trials conducted in North America (Lebwohl et al. 2003; Papp et al. 2006) and in the European CLEAR (Clinical Experience acquired with Raptiva) study (Sterry et al. 2004). The most important results of these studies are summarized in Table 5.1. As can been seen, increasing the dose to 2 $mg\,kg^{-1}$ per week (Lebwohl et al. 2003; Leonardi et al. 2005a) did not result in any additional benefit for the patients.

The CLEAR study, conducted in 103 centers in 19 countries, evaluated the efficacy and safety of efalizumab in patients with chronic plaque psoriasis, and was similar to the other phase III studies. However, this trial included a prospectively defined patient cohort for which at least two existing systemic therapies were unsuitable or inadequate due to contraindications, lack of efficacy, or intolerance. The results of this study demonstrate that even more severely affected patients, for whom treatment options are limited, showed treatment benefits with efalizumab, comparable to those seen in the broader population (Sterry et al. 2004).

5.5.1.2 Extended Treatment and Re-Treatment

In order to investigate the effect of extended treatment, efalizumab patients who did not achieve PASI 75 at week 12 were re-randomized to receive efalizumab 1 $mg\,kg^{-1}$ per week or placebo for another 12 weeks (Leonardi et al. 2005b). At Week 24, patients who received extended treatment showed improved responses; 20.2% achieved PASI 75 compared with 6.7% who received placebo during Weeks 13 to 24.

In another randomized, double-blind, placebo-controlled 12-week study (Menter et al. 2005), patients had the option to enter a second 12-week, open-label, extended efalizumab treatment period. At Week 24, 66.6% (n = 368) and 43.8% (n = 368) of the efalizumab patients achieved PASI 50 and PASI 75 responses, respectively.

In the CLEAR study (Ring et al. 2005), extended treatment of patients without PASI 75 response at first treatment (FT) in Week 12 maintained and improved

Table 5.1 The 12-weeK efficacy of efalizumab in randomized, placebo-controlled, double-blind studies.

Study protocol/ treatment	*Gordon et al. (2003) (n = 187/369) [%]*	*Lebwohl et al. (2003) (n = 122/232/243) [%]*	*Leonardi et al. (2005a) (n = 170/162/166) [%]*	*Sterry et al. (2004) (n = 264/529) [%]*	*Papp et al. (2006) (n = 236/450) [%]*
PASI 75 Placebo	4	5	2	4	3
PASI 75 Efalizumab 1 mg kg^{-1} per week	27	22	39	31	24
PASI 75 Efalizumab 2 mg kg^{-1} per week		28	27		
PASI 50 Placebo	14	16	15	14	14
PASI 50 Efalizumab 1 mg kg^{-1} per week	59	52	61	54	52
PASI 50 Efalizumab 2 mg kg^{-1} per week		57	51		
sPGA minimal or clear Placebo	3	3	3	3	4
sPGA minimal or clear 1 mg kg^{-1} per week	26	19	32	26	20
sPGA minimal or clear 2 mg kg^{-1} per week		23	22		

PASI = Psoriasis Area and Severity Index; sPGA = static Physician's Global Assessment of change.
n = number of patients in each treatment arm.

response. Among FT efalizumab patients with >50% but <75% improvement at FT Week 12, almost half (47.5%) showed PASI 75 response after a second 12-week period. Among FT efalizumab patients with PGA ratings of "good" at FT Week 12, 43.9% had ratings of "excellent" or "cleared" by FT Week 12.

When patients who received placebo during the first 12 weeks of treatment were switched to efalizumab in Weeks 13 to 24, a rapid improvement in their PASI was observed; 60.3% achieved PASI 50 and 24.1% PASI 75 at the end of the 12-week open-label efalizumab treatment phase (Menter et al. 2005).

The clinical benefit of efalizumab treatment is gradually lost after discontinuation of treatment. The median time to relapse (loss of ≥50% of PASI improvement) has been shown to be between 58 days (Ring et al. 2005) and 84 days (Lebwohl et al. 2003). When patients are re-treated at a dose of 1 mg kg^{-1} after relapse, efalizumab re-establishes disease control: at the end of the re-treatment period after 12 weeks, PASI scores in these patients showed mean improvements of 62% from baseline (Ring et al. 2005).

5.5.1.3 **Long-Term Efficacy**

An open-label trial was conducted to evaluate the safety, tolerability and efficacy of continuous subcutaneous efalizumab therapy for 3 years in patients who achieved a PASI 50 response or an sPGA grade of "mild", "minimal" or "clear" after an initial 12-week treatment period (Gottlieb et al. 2005). During the first 12-week treatment period, patients received 2.0 mg kg^{-1} efalizumab each week. During the third month, patients were also randomized to receive either fluocinolone acetonide (Synalar/Medicis) or white petrolatum in addition to efalizumab.

The patients (n = 339) who entered this study had experienced psoriasis for a mean of 17.9 years, and over half (57.5%) of them had a history of prior systemic therapy. Patients in this study population also had significant disease: the mean baseline PASI score was 19.8 and the mean BSA affected by psoriasis was 31.5%.

At FT Week 12, patients who had at least a 50% improved PASI score or an sPGA grading of "mild", "minimal" or "clear" could enter the maintenance treatment period, which was analyzed in 12-week treatment segments. During the maintenance period, patients received 1.0 mg kg^{-1} efalizumab weekly. In the case of relapse – which was defined as the loss of 50% of the PASI improvement achieved at Week 12 – the patient ended participation in the current 12-week segment and started the next segment at 2.0 mg kg^{-1} per week. During months 4 to 15, the dose for relapsed patients could be increased to a maximum of 4 mg kg^{-1} per week, if clinically indicated. After month 15, further dose escalation was not allowed.

In this study the impact of efalizumab on PASI was analyzed in two ways. The primary analysis was based on the intent-to-treat (ITT), and comprised all 339 patients who entered the study. Patients who discontinued treatment during the first 3 months were classified as PASI nonresponders at 3 months and for subsequent periods. For patients who discontinued treatment during the maintenance treatment, the last available PASI assessment was carried forward to the end of the current 12-week segment, after which the patient was considered to be a nonresponder. As a secondary analysis, an analysis of the "as-treated" population (i.e., the number of patients decreased over time due to study discontinuations) was also performed. Use of these two populations provides a reliable estimate of the range of efficacy.

At Week 12, 86% (290/339) of the patients showed a 50% improved PASI score or an sPGA grading of "mild", "minimal", or "clear" and were eligible to enter the maintenance phase; 41% and 13% of patients achieved PASI 75 and PASI 90, respectively. At month 36, the PASI 75 responses were 45% in the ITT population and 73% in the as-treated population (n = 113). PASI 90 responses at month 36 were 25% for ITT, and 40% for as-treated patients (n = 115). The PASI 75 and PASI 90 results of the ITT population are illustrated graphically in Fig. 5.4. More than 50% of patients maintained or improved their response during efalizumab therapy. Moreover, more than half of the patients who achieved a PASI 75 response at 33 and 36 months had a PASI 90 response and were therefore virtually clear of their disease.

5.5.2
Safety and Tolerability

The most frequent symptomatic adverse drug reactions (ADRs) observed during efalizumab therapy were mild-to-moderate acute flu-like symptoms, including headache, fever, chills, nausea, and myalgia. In the efalizumab group, the incidence of acute adverse events was highest following the first dose (27.4% versus 21.2% of patients), and decreased with each subsequent dose (Papp et al. 2006). These reactions were generally less frequent from the third and subsequent

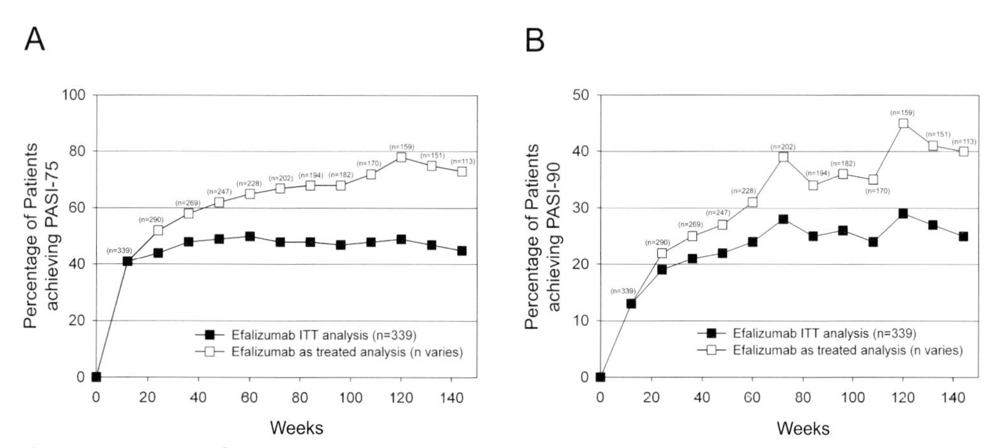

Fig. 5.4 Percentages of patients with PASI 75 and PASI 90 during efalizumab long-term treatment. Data are based on "Intent-to-treat" (ITT) and "As-treated" analyses.

weekly injections, and occurred at similar rates to those seen in the placebo group (3.7% and 3.9%, respectively). The incidence of acute adverse events in other placebo-controlled trials was similar.

Antibodies to efalizumab were detected in 2–6% of patients (Gordon et al. 2003; Lebwohl et al. 2003; Leonardi et al. 2005b). There were no differences in pharmacokinetics, clinically noteworthy adverse events or clinical efficacy between patients with or without antibody responses.

About 40% of patients developed sustained asymptomatic lymphocytosis (this was related to the drug's mechanism of action) during efalizumab therapy (Papp et al. 2005). Typical values were less than threefold the ULN, and lymphocyte counts returned to baseline after therapy discontinuation.

Eight patients (0.3%) developed thrombocytopenia, with a cell count of $<52 \times 10^3 \mu L^{-1}$ reported (Leonardi 2004). One patient was lost to follow-up, but in the remaining seven patients the symptoms were resolved with conservative management of patients. Causality between the administration of efalizumab and thrombocytopenia is likely. Platelet monitoring is recommended. The rate of psoriasis-related ADRs was 2.2% in the efalizumab-treated patients and 0.8% in the placebo group (Papp et al. 2006). Serious psoriasis-related adverse events included erythrodermic, pustular and guttate subtypes (Papp et al. 2006). Exacerbation or flare-up of psoriatic arthritis was observed in 1.6% of efalizumab-treated patients compared to 1.3% in the placebo group (Papp et al. 2006). A recent analysis of pooled data from phase II, III, and IV clinical trials showed that both efalizumab-treated patients and placebo subjects showed a similar incidence of arthropathy during the first 12-week treatment. However, there was no increase in the incidence of arthropathy among efalizumab-treated patients during an extended long-term treatment of up to 3 years (Pincelli and Casset-Semanez 2005).

As with other protein products, efalizumab is potentially immunogenic. During the placebo-controlled clinical studies, the percentage of patients experiencing an adverse event suggestive of hypersensitivity – including urticaria, rash and allergic reactions – was comparable between the efalizumab and placebo groups (Gordon et al. 2003).

Approximately 4.5% of patients developed sustained elevation of alkaline phosphatase throughout efalizumab therapy, compared with 1% of those receiving placebo. All values were between 1.5- and 3-times the ULN, and returned to baseline levels after therapy discontinuation (Serono Europe 2004). About 5.7% of patients developed elevated levels of alanine aminotransferase during efalizumab therapy compared to 3.5% after placebo. However, all such occurrences were asymptomatic, and values above 2.5 ULN were no more frequent in the efalizumab group than in the placebo group. All values returned to baseline levels upon therapy discontinuation (Serono Europe 2004).

Therapies that alter T-lymphocyte function have been associated with an increased risk of developing serious infections. In clinical trials, infection rates were similar between efalizumab- and placebo-treated patients. High rates of malignancy have been associated with therapies that affect the immune system, but in these clinical trials the overall incidences of malignancy were similar among efalizumab- and placebo-treated patients. In addition, the incidences of specific tumors in patients treated with efalizumab were in line with those observed in control psoriasis populations. Among psoriasis patients who received efalizumab at any dose level, the overall incidence of malignancies of any type was 1.8 per 100 patient-years after efalizumab treatment, compared to 1.6 per 100 patient-years for placebo-treated patients. There appears to be no evidence of a risk of developing malignancy exceeding that expected among the psoriasis population (Leonardi 2004).

Continuous treatment with efalizumab for 36 months was generally well tolerated, and no new adverse events were reported (Gottlieb et al. 2005). The adverse event profiles associated with long-term efalizumab therapy showed no evidence for end-organ damage or cumulative toxicity. The incidence of infection-related adverse events observed during this trial (Fig. 5.5) appeared to be similar to that observed during placebo-controlled trials (28.9% and 26.3% of efalizumab-treated and placebo-treated patients, respectively; see Fig. 5.5). There were no opportunistic infections, no tuberculosis, no demyelination, no congestive heart failure, and no deaths during the 3-year study. There was also no evidence of increased risk of opportunistic infections (Langley et al. 2005).

During continuous therapy, the incidence of psoriasis adverse events and of malignancies did not increase over time (Gottlieb et al. 2005) (Fig. 5.6). Whilst the majority of malignancies were nonmelanoma skin carcinomas, other malignancy events observed included one lymphoma, two carcinomas of the colon, two lung carcinomas, and one melanoma.

Based on clinical development data, there seems to be no evidence of risk of developing malignancy. Further data from post-marketing exposure, however, are still needed to confirm this observation.

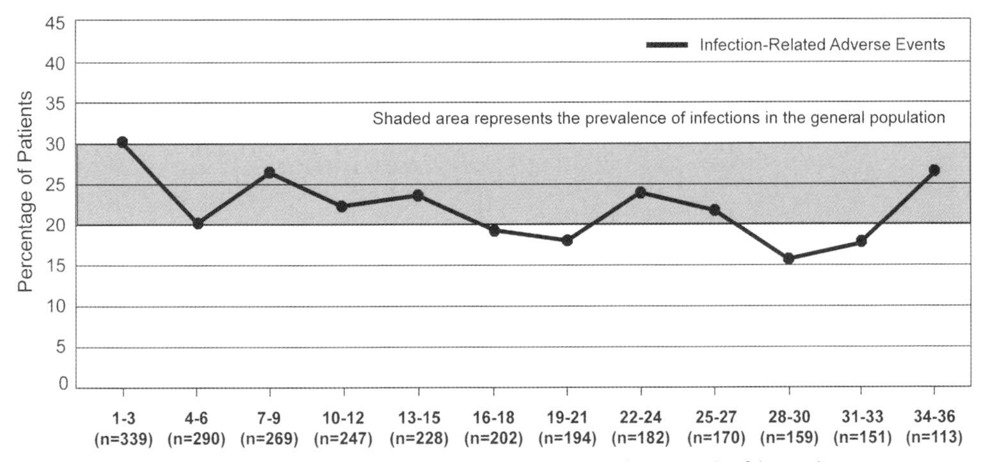

Fig. 5.5 Infection-related adverse events during 3 years' continuous therapy with efalizumab.

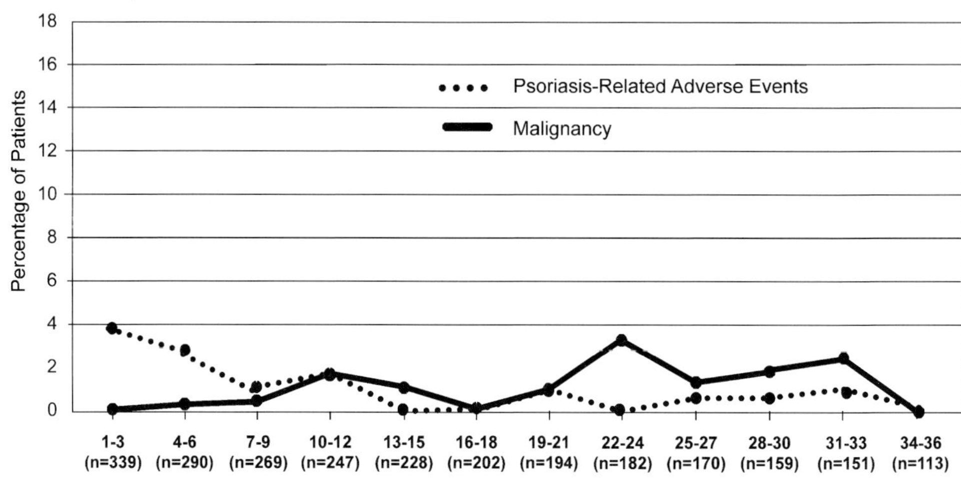

Fig. 5.6 Psoriasis-related adverse events and malignancies during 3 years' continuous therapy with efalizumab.

5.5.3 Health-Related Quality of Life (HRQoL)

It is well known that psoriasis may lead to a reduced quality of life, psychosocial distress, and even disability. The impact of psoriasis on the patients' physical and psychological function has been shown to exceed that of other diseases, such as heart disease, cancer, diabetes, or arthritis (Rapp et al. 1999).

In order to evaluate the effect of efalizumab treatment on HRQoL and disease burden, the results of patient-reported outcomes of three phase III, randomized double-blind, placebo-controlled, multicenter clinical trials were pooled and analyzed (Menter et al. 2004a). A total of 1242 patients with moderate-to-severe plaque psoriasis, receiving either efalizumab ($1\,mg\,kg^{-1}$) or placebo for 12 weeks was included in the analyses. The methods used were dermatology-related quality of life (DLQI), PSA, and an itch scale. Scores were assessed prior to treatment and at the end of the 12-week treatment period, comparing results of the active treatment group with those of the placebo group. Further score assessments were made at the end of an extended treatment period at Week 24. The extension period was performed open label, with the former placebo patients receiving active treatment of $1\,mg\,kg^{-1}$ efalizumab for Weeks 13 to 24.

Baseline data on the disease burden of moderate-to-severe psoriasis indicate a substantial burden across all items of the DLQI. More than 60% of the patients in both the active treatment and placebo groups reported that they were bothered "a lot" or "very much" by itchy, sore painful skin, or by being embarrassed or self-conscious. Over one-third of patients (36%) reported problems with their social life, and about 20% had problems with their partner or had sexual difficulties. In the PSA, more than 80% of patients reported suffering from itching or scaling skin.

Twelve-week treatment with efalizumab led to significant improvements in patient-reported outcomes compared with placebo. Across all measures – DLQI, PSA severity, PSA frequency and itch – the proportion of patients with improvement was at least twofold greater in the active treatment group than in placebo-treated patients (Menter et al. 2004a). Moreover, improvements were maintained during the additional 12-week extended treatment period. Patients randomized to placebo in the FT course and switched to efalizumab for Weeks 13 to 24 showed similar improvements in DLQI and symptom assessments as the efalizumab group during the first treatment phase (Menter et al. 2004a).

5.6 Practical Considerations for Therapy with Efalizumab

Efalizumab gained regulatory approval in the US in 2003 for the continuous treatment of adult patients with moderate-to-severe plaque psoriasis who are eligible for systemic therapy. In 2004, the European Commission approved the drug for adult patients with moderate-to-severe chronic plaque psoriasis who have failed to respond to, or have contraindications to, or are intolerant of other systemic therapies, including cyclosporine, methothrexate, and PUVA (psoralen + UV-A).

Efalizumab is provided as a lyophilized powder in vials containing 150 mg of the drug substance; before injection it is reconstituted in water for injections. After an initial single 0.7 mg kg^{-1} dose, weekly injections of 1.0 mg kg^{-1} should be given (maximum single dose not to exceed a total of 200 mg). Continuous therapy can be given to those patients in Europe who have responded to treatment (PGA of "good" or better) after 12 weeks of initial treatment.

5.6.1 Managing Patients during Long-Term Efalizumab Treatment

Although efalizumab is well tolerated by the majority of patients, some aspects require special attention (Hamilton 2005; Leonardi et al. 2005b).

Patients may develop transient localized papular eruptions, typically within 4 to 8 weeks after the start of therapy. These 2- to 4-mm inflammatory papules are observed at sites not previously affected by the disease – usually the neck, torso, or flexural areas. Discontinuation of efalizumab is not required in those cases. "Treating through" with efalizumab – alternately combining efalizumab with a topical corticosteroid – will resolve the issue in most cases.

Patients may develop generalized inflammatory exacerbations. This is an infrequent complication, generally observed in nonresponders at 6 to 10 weeks after the initiation of therapy. Patients experiencing such an exacerbation should be transferred to alternative systemic therapies to control the flare.

Patients may develop rebound after discontinuation of efalizumab therapy (defined as PASI >125% of baseline or new widespread pustular, erythrodermic or more inflammatory psoriasis occurring within 8 weeks of discontinuing treatment in responding patients). As efalizumab is approved for continuous therapy, this should normally not occur in patients who were responders during the first 12 weeks of treatment. However, should discontinuation be necessary (for any reason), patients are advised to consult their physician at the first signs of a relapse. Efalizumab re-treatment should then be initiated. Patients who did not respond within the first 12 weeks of treatment should be given alternative systemic therapies to prevent rebound. Therapies with the highest success rates at preventing rebound are cyclosporine, methotrexate and systemic corticosteroids (Menter et al. 2004b; Papp et al. 2005).

5.6.2 Other Concerns

Efalizumab is contraindicated in patients with a history of malignancies. As for other anti-psoriatic therapies for systemic use that alter T-lymphocyte functions, efalizumab may affect host defenses against infections. As a consequence, efalizumab should not be used in patients with active tuberculosis or other severe infections, as well as in immune-deficient patients. Patients developing an infection during treatment with efalizumab should be monitored, and efalizumab – according to the severity of the infection – should be discontinued.

Pregnant and breastfeeding women should not use efalizumab; women of childbearing potential should be advised to use appropriate contraception. Immunoglobulins are known to cross the placental barrier. Although preclinical studies have shown no evidence of efalizumab-associated adverse events during the course of pregnancies, or of any harm to fetuses or infants, animal studies indicate a reversible impairment of immune function of offspring exposed to CD11a. Additionally, immunoglobulins are known to be secreted in human milk.

In a trial characterizing immune responses to the neoantigen phiX174, single doses of efalizumab significantly attenuated the secondary immune response at doses recommended for treating psoriasis (Gottlieb et al. 2002). Therefore, in cases of necessary vaccination, efalizumab treatment should be stopped 8 weeks before and be re-initiated 2 weeks after vaccination.

As several cases of thrombocytopenia have occurred during efalizumab treatment, periodic assessment of platelet counts is recommended upon initiating and during efalizumab treatment. It is recommended that assessments be performed monthly during the first 12 weeks of treatment, and every 3 months during continued treatment. In cases of clinical symptoms such as echymoses, spontaneous bruising or bleeding from mucocutaneous tissues, efalizumab should be stopped immediately, a platelet count should be performed, and appropriate symptomatic treatment should be instituted immediately.

5.7 Summary

Psoriasis is a chronic, debilitating disease that may affect patients for several decades. Until recently, no satisfactory systemic treatment was available that showed both efficacy and long-term safety, especially for those patients suffering from moderate-to-severe forms of the condition.

Developments in understanding the pathophysiology of inflammatory diseases such as psoriasis, together with the progress made in biotechnology in designing compounds that specifically target pathogenic molecules within the immune system, has opened a new horizon for the treatment of psoriasis. One of these compounds, efalizumab – a humanized, monoclonal antibody against the CD11a subunit of the lymphocyte function associated antigen-1 (LFA-1) – has been shown in clinical trials to be effective and well tolerated in the treatment of patients with moderate-to-severe psoriasis. Although more data are required to determine the safety profile of continuous therapy, such a strategy appears to be successful for maintaining tight control of psoriasis symptoms, as demonstrated in a recent phase III study in which patients received efalizumab continuously over 3 years (Gottlieb et al. 2005).

References

Bauer, R.J., Dedrick, R.L., White, M.L., Murray, M.J., Garovoy, M.R. (1999) Population pharmacokinetics and pharmacodynamics of the anti-CD11a antibody hu1124 in human subjects with psoriasis. *J Pharmacokinet Biopharm* 27: 397–400.

Bos, J.D., De Rie, M.A. (1999) The pathogenesis of psoriasis: Immunological facts and speculations. *Immunol Today* 20: 40–46.

Bos, J.D., Hulsebosch, H.J., Krieg, S.R., Bakker, P.M., Cormane, R.H. (1983) Immunocompetent cells in psoriasis: In situ immunophenotyping by monoclonal antibodies. *Arch Dermatol Res* 275: 181–189.

Boyman, O., Hefti H., Conrad, C., Nickoloff, B.J., Suter M., Nestle, F.O. (2004) Spontaneous development of psoriasis in a new animal model shows an essential role for resident T cells and tumor necrosis factor-α. *J Exp Med* 199: 731–736.

Bradley, L.M., Watson, S.R. (1996) Lymphocyte migration into tissue: The paradigm derived from CD4 subsets. *Curr Opin Immunol* 8: 312–320.

Carlin, C.S., Feldman, S.R., Krueger, J.G., Menter, A., Krueger, G. (2004) A 50% reduction in the Psoriasis Area and Severity Index (PASI 50) is a clinically significant endpoint in the assessment of psoriasis. *J Am Acad Dermatol* 50: 859–866.

Coffey, G.P., Stefanich, E., Palmieri, S., Eckert, R., Padilla-Eager, J., Fielder, P., Pippig, S. (2004) In vitro internalization, intracellular transport and clearance of efalizumab (Raptiva) by humanT-cells. *J Pharmacol Exp Ther* 310: 896–904.

de Fougerolles, A.R., Stacker, S.A., Schwalting, R., Springer, T.A. (1991) Characterization of ICAM-2 and evidence for a third counter-receptor for LFA- I. *J Exp Med* 174: 253–267.

de Fougerolles, A.R., Qin, X., Springer, T.A. (1994) Characterization of the function of intercellular adhesion molecule (ICAM)-3 and comparison with ICAM-I and ICAM-2 in immune responses. *J Exp Med* 179: 619–629.

Dougherty, G.J., Hogg, N. (1987) The role of monocyte lymphocyte function-associated antigen I (LFA-1) in accessory cell function. *Eur J Immunol* 17: 943–947.

Dustin, M.L., Rothlein, R., Bhan, A.K., Dinarello, C.A., Springer, T.A. (1986) Induction of IL-1 and IFN-γ: tissue distribution, biochemistry, and function of a natural adherence molecule (ICAM-I). *J Immunol* 137: 245–254.

Ferenczi, K., Burack. L., Pope, M., Krueger, J.G., Austin, L.M. (2000) CD69, HLA-DR and the IL-2R identify persistently activated T cells in psoriasis vulgaris lesional skin: Blood and skin comparisons by flow cytometry. *J Autoimmun* 14: 63–78.

Fuhlbrigge, R.C., King, S.L., Dimitroff, C.J., Kupper, T.S., Sackstein, R. (2002) Direct real-time observation of E- and P-selectin-mediated rolling on cutaneous lymphocyte-associated antigen immobilized on Western blots. *J Immunol* 168: 5645–5651.

Gilhar, A., David, M., Ullmann, Y., Berkutski, T., Kalish, R.S. (1997) T-lymphocyte dependence of psoriatic pathology in human psoriatic skin grafted to SCID mice. *J Invest Dermatol* 109: 283–288.

Gollnick, H., Bonnekoh, B. (2001) *Psoriasis – Pathogenese, Klinik und Therapie.* Uni-Med Verlag AG, Bremen.

Gordon, K.B., Papp, K.A., Hamilton, T.K., Walicke, P.A., Dummer, W., Li, N., Bresnahan, B.W., Menter, A. (2003) Efalizumab for patients with moderate to severe plaque psoriasis. A randomised controlled trial. *JAMA* 290: 3073–3080.

Gottlieb, S.L., Gilleaudeau, P., Johnson, R., Estes, L., Woodworth, T.G., Gottlieb, A.B., Krueger, J.G. (1995) Response of psoriasis to a lymphocyte-selective toxin (DAB389IL-2) suggests a primary immune, but not keratinocyte, pathogenic basis. *Nat Med* 1: 442–447.

Gottlieb, A., Krueger, J.G., Bright, R., Ling, M., Lebwohl, M., Kang, S., Feldman, S., Spellman, M., Wittkowski, K., Ochs, H.D. (2000) Effects of administration of a single dose of a humanized monoclonal antibody to CD11a on the immunobiology and clinical activity of psoriasis. *J Am Acad Dermatol* 42: 428–435.

Gottlieb, A.B., Krueger, J.B., Wittkowski, K. (2002) Psoriasis as a model for T-cell mediated disease: immunobiologic and clinical effects of treatment with multiple doses of efalizumab, an anti CD 11a antibody. *Arch Dermatol* 138: 591–600.

Gottlieb, A.B., Hamilton, T.K., Walicke, P.A., Li, N., Joshi, A., Garovoy, M., Gordon, K.B. (2003) Efficacy, safety, and pharmacokinetic outcomes observed in patients with plaque psoriasis during long-term efalizumab therapy: results from an open-label trial. International Investigative Dermatology Meeting April 30–May 4, 2003 Miami Beach Florida, Poster 379.

Gottlieb, A.B., Gordon, K.B., Lebwohl, M.G., Caro, I., Walicke, P.A., Li, N., Leonardi, G. (2004) Extended efalizumab therapy sustains efficacy without increasing toxicity in patients with moderate to severe chronic plaque psoriasis. *J Drugs Dermatol* 3: 614–624.

Gottlieb, A.B., Gordon, K.B., Hamilton, T.K., Leonardi, C. (2005) Maintenance of efficacy and safety with continuous efalizumab therapy in patients with moderate to severe plaque psoriasis: final phase IIIb study results. *J Am Acad Dermatol* 52: P4.

Grakoui, A., Bromley, S.K., Sumen, C., Davis, M.M., Shaw, A.S., Allen, P.M., Dustin, M.L. (1999) The immunological synapse: a molecular machine controlling T cell activation. *Science* 285: 221–227.

Hamilton, T.K. (2005) Clinical considerations of efalizumab therapy in patients with psoriasis. *Semin Cutan Med Surg* 24: 19–27.

Hildreth, J.E.K., August, J.T. (1985) The human lymphocyte function-associated (HLFA) antigen and a related macrophage differentiation antigen (HMac-1): functional effects of subunit-specific monoclonal antibodies. *J Immunol* 134: 3272–3280.

Hildreth, J.E.K., Gotch, F.M., Hildreth, P.D.K., McMichael, A.J. (1983) A human lymphocyte-associated antigen involved in cell-mediated lympholysis. *Eur J Immunol* 13: 202–208.

Homey, B., Alenius, H., Muller, A., Soto, H., Bowman, E.P., Yuan, W., McEnvoy, L., Lauerman, A.I., Assmann, T., Bunemann, E., Lehto, N., Wolff, H., Yen, D., Marxhausen, H., To, W., Sedgwick, J., Ruzicka, T., Lehmann, P., Zlotnik, A. (2002) CCL27-CCR10 interactions regulate T cell-mediated skin inflammation. *Nat Med* 8: 157–165.

Janeway, C.A., Travers, P., Walport, M., Shlomchick, M. (2001) *Immunobiology. The Immune System in Health and Disease.* 5th edn. Garland Publishing.

Jullien, D., Prinz J.C., Langley, R.G.B., Caro, I., Dummer, W., Joshi, A., Dedrick, R., Natta, P. (2004) T-cell modulation for the treatment of chronic plaque psoriasis with efalizumab (Raptiva): Mechanisms of action. *Dermatology* 208: 297–306.

Jung, E.G., Moll, I. (2003) *Dermatologie.* Georg Thieme Verlag, Stuttgart.

Krueger, J.G. (2002) The immunologic basis for the treatment of psoriasis with new biologic agents. *J Am Acad Dermatol* 46: 1–23.

Kuypers, T., Roos, D. (1989) Leukocyte membrane adhesion proteins LFA-I, CR3 and p150.95: a review of functional and regulatory aspects. *Res Immunol* 140: 461–486.

Langley, R.G., Ellis, C.N. (2004) Evaluating psoriasis with Psoriasis Area and Severity Index, Psoriasis Global Assessment, and Lattice System Physician's Global Assessment. *J Am Acad Dermatol* 51: 563–569.

Langley, R.G., Carey, W.P., Rafal, E.S., Tyring, S.K., Caro, I., Wang, X., Wetherill, G., Gordon, K.B. (2005) Incidence of infection during efalizumab therapy for psoriasis: analysis of the clinical trial experience. *Clin Ther* 27: 1317–1328.

Lebwohl, M., Tyring, S.K., Hamilton, T.K., Toth, D., Glazer, S., Tawfik, N.H., Walicke, P., Dummer, W., Wang, X., Garovoy, M.R., Pariser, D. (2003) A novel targeted T-cell modulator, efalizumab, for plaque psoriasis. *N Engl J Med* 349: 2004–2013.

Leonardi, C.L. (2004) Efalizumab in the treatment of psoriasis. *Derm Ther* 17: 393–400.

Leonardi, C.L., Papp, K.A., Gordon, K.B., Menter, A., Feldman, S.R., Caro, I., Walicke, P.A., Compton, P.G., Gottlieb, A.B. (2005a) Extended efalizumab therapy improves chronic plaque psoriasis: Results from a randomized phase III trial. *J Am Acad Dermatol* 52: 425–433.

Leonardi, C.L., Menter, A., Sterry, W., Bos, J.D., Papp, K.A. (2005b) Long-term management of chronic plaque psoriasis: guidelines for continuous therapy with efalizumab. 14th EADV Fall Symposium, London, UK, 12–16 October, 2005.

Lo, S., van Seventer, G, Levin, S., Wright S. (1989) Two leukocyte receptors (CD11a/CD18 and CD11b/CD18) mediate transient adhesion to endothelium by binding to different ligands. *J Immunol* 143: 3325–3329.

Menter, A., Griffiths, C. (2004) *Efalizumab and psoriasis.* Science Press Ltd., London.

Menter, A., Kosinski, M., Bresnahan, B.W., Papp, K.A., Ware, J.E. (2004a) Impact of efalizumab on psoriasis-specific patient reported outcomes. Results from three randomized, placebo-controlled clinical trials of moderate to severe plaque psoriasis. *J Drugs Dermatol* 3: 27–38.

Menter, A., Kardatzke, D., Rundle, A.C., Garovoy, M.R., Leonardi, C.L. (2004b) Incidence and prevention of rebound upon efalizumab discontinuation. Poster Presentation, International Psoriasis Symposium (IPS), Toronto, 2004.

Menter, A., Gordon, K., Carey, W., Hamilton, T., Glazer, S., Caro, I., Li, N., Gulliver, W. (2005) Efficacy and safety observed during 24 weeks of efalizumab therapy in patients with moderate to severe plaque psoriasis. *Arch Dermatol* 141: 31–38.

Mortensen, D.L., Walicke, P.A., Wang, X., Kwon, P., Kuebler, P., Gottlieb, A.B., Krueger, J.G., Leonardi, C.L., Miller, B., Joshi, A. (2005) Pharmacokinetics and pharmacodynamics of multiple weekly subcutaneous efalizumab doses in patients with plaque psoriasis. *J Clin Pharmacol* 45: 286–298.

Papp, K.A., Toth, D., Rosoph, L. (2005) Investigational study comparing different transitioning therapies for patients discontinuing treatment with efalizumab. 14th EADV Fall Symposium, London, UK, 12–16 October, 2005.

Papp, K.A., Bressinck, R., Fretzin, S., Goffe, B., Kempers, S., Gordon, K.B., Caro, I.,

Walicke, P.A., Wang, X., Menter, A. (2006) The safety of efalizumab in adults with chronic moderate to severe plaque psoriasis: a phase IIIb, randomized, controlled trial. *Int J Dermatol* 45: 605–614.

Pincelli, P., Casset-Semanez, F. (2005) Efalizumab therapy and incidence of arthropathy adverse events: analysis of pooled data from phase II, III, IV clinical trials. Psoriasis from Gene to Clinic, 4th International Congress, London, UK, 1–3 December, 2005.

Prinz, J.C., Braun-Falco, O., Meurer, M., Daddona, P., Reiter, C., Rieber, P., Riethmuller, G. (1991) Chimaeric CD4 monoclonal antibody in treatment of generalised pustular psoriasis. *Lancet* 338: 320–321.

Prinz, J.C., Gross, B., Vollmer, S., Trommler, P., Strobel, I., Meurer, M., Plewig, G. (1994) T cell clones from psoriasis skin lesions can promote keratinocyte proliferation in vitro via secreted products. *Eur J Immunol* 24: 593–598.

Rapp, S.R., Feldman, S.R., Exum, M.L. (1999) Psoriasis causes as much disability as other major medical diseases. *J Am Acad Dermatol* 41: 410–417.

Ring, J., Dubertret, L., May, T., Chimenti, S., Sterry, W. (2005) The safety and efficacy of efalizumab patients with chronic plaque psoriasis. Results from retreatment and extended treatment phases following a placebo-controlled trial, 14th EADV Fall Symposium, London, UK 12–16 October, 2005, Poster P06.87.

Schön, M.P., Boehncke, W.-H. (2005) Medical progress: Psoriasis. *N Engl J Med* 352: 1899–1912.

Serono Europe Ltd. (2004) Raptiva (Efalizumab). Summary of Product Characteristics.

Sterry, W., Dubertret, L., Papp, K., Chimenti, S., Larsen, C.G. (2004) Efalizumab for patients with moderate to severe chronic plaque psoriasis: results of the international, randomized, controlled phase III Clinical Experience Acquired with Raptiva (CLEAR) trial. *J Invest Dermatol* 123: A64.

Sun, Y., Yu, J., Joshi, A., Compton, P., Kwon, P., Bruno, R.A. (2005) Population pharmacokinetics of Efalizumab (humanized monoclonal CD 11 antibody) following long term subcutaneous weekly dosing in psoriasis subjects. *J Clin Pharmacol* 45: 468–476.

Valdimarsson, H., Baker, B.S., Jonsdottir, I., Powles, A., Fry, L. (1995) Psoriasis: A T-cell-mediated autoimmune disease induced by streptococcal superantigens? *Immunol Today* 16: 145–149.

Vugmeyster, Y., Kikuch, T., Lowes, M.A., Chamian, F., Kagen, M., Gilleaudeau, P., Lee, E., Howell, K., Bodary, S., Dummer, W., Krueger, J.G. (2004) Efalizumab (anti-CD11a)-induced increase in peripheral blood leukocytes in psoriasis patients is preferentially mediated by altered trafficking of memory CD8+ T cells into lesional skin. *Clin Immunol* 113: 38–46.

Weinstein, G.D., Menter A. (2003) An overview of psoriasis. In: Weinstein, G.D., Gottlieb, A.B. (Eds.), *Therapy of Moderate to Severe Psoriasis*. Marcel Dekker Inc., New York, pp. 1–29.

Werther, W.A., Gonzalez, T.N., O'Connor, S.J., McCabe, S., Chan, B., Hotaling, T., Champe, M., Fox, J.A., Jardieu, P.M., Berman, P.W., Prestal, L.C. (1996) Humanization of an anti-lymphocyte function-associated antigen (LFA)-1 monoclonal antibody and reengineering of the humanized antibody for binding to rhesus LFA-1. *J Immunol* 157: 4986–4995.

Wrone-Smith, T., Nickoloff, B.J. (1996) Dermal injection of immunocytes induces psoriasis. *J Clin Invest* 98: 1878–1887.

6
^{99m}Tc-Fanolesomab (NeutroSpec)

Christopher J. Palestro, Josephine N. Rini, and Charito Love

6.1
Introduction

Nuclear medicine plays an important role in the evaluation of patients suspected of harboring infection. Among the available radionuclide techniques, *in-vitro*-labeled autologous leukocyte imaging currently is the "gold standard" for imaging most infections in the immunocompetent population. There are, unfortunately, several disadvantages to the technique. The *in-vitro* labeling process is labor-intensive, not always available, and involves the direct handling of blood products. For musculoskeletal infection, the need to perform complementary marrow and/or bone imaging adds complexity and expense to the procedure, and is also an inconvenience to patients [1]. A satisfactory *in-vivo* method of labeling leukocytes would overcome many of these problems, and consequently several such methods have been investigated, including peptides and antigranulocyte antibodies/antibody fragments [2–9]. One of these agents is ^{99m}Tc-fanolesomab (NeutroSpec; Palatin Technologies, Cranberry, New Jersey, USA) – an antigranulocyte antibody that binds to CD-15 receptors present on human leukocytes. This radiolabeled antibody is injected directly into patients, labeling leukocytes *in vivo* and eliminating the disadvantages of *in-vitro* labeling [10].

6.2
The Agent

Fanolesomab is a murine monoclonal M class antibody with a molecular weight of approximately 900 kDa. It was originally raised against stage-specific embryonic antigen (SSEA)-1 in mice immunized with murine embryonal carcinoma F9 cells. The antibody exhibits a high affinity (association constant $K_d = 10^{-11}$ M) for the carbohydrate moiety 3-fucosyl-*N*-acetyl lactosamine contained in the CD15 antigen, which is expressed on human neutrophils, eosinophils, and lymphocytes [11,12].

Handbook of Therapeutic Antibodies. Edited by Stefan Dübel
Copyright © 2007 WILEY-VCH Verlag GmbH & Co. KGaA, Weinheim
ISBN 978-3-527-31453-9

6.3
Pharmacokinetics/Dosimetry

Fanolesomab binds to CD 15 receptors expressed on neutrophils in greater proportion than to any other cell type. Less than 1% of activity is bound to lymphocytes and platelets, and only about 4% is associated with erythrocytes. The affinity of fanolesomab for the CD15 receptor on neutrophils is 1.6×10^{-11} M. The concentration bound at 50% of maximum binding is 0.44 to 1.05×10^{-8} M, and the leukocyte labeling efficiency is 80 ± 4%. The activity bound to the recovered granulocyte fraction is 40.7 ± 8.2%. Binding is up-regulated with neutrophil activation, and increases proportionately with increasing numbers of circulating neutrophils [10–13].

Following intravenous injection, ^{99m}Tc-fanolesomab is rapidly distributed with a mean distribution half-life of 0.3 h [13]. The mean linear elimination half-life for blood clearance is 8 h (monoexponential half-life = 5.7 h) [14].

Excretion is primarily through the kidneys, although the radioactivity excreted in the urine is not antibody-bound [13]. The renal excretion fraction varies from 31% to 49% over 26 to 31 h (mean = 39.5 ± 6.5%) [15].

Bone marrow activity peaks shortly after injection, and has a longer elimination time compared to background tissues. Most of the radioactivity accumulates in the liver, followed by the spleen and bladder. Peak liver activity consists of 45 to 50% of the injected dose, decreasing to 25 to 40% of the injected dose by 24 h. Spleen activity peaks at 5 to 12% of the injected dose, and drops by half within 24 h. A low level of activity is seen in the testes. Several tissues known to contain surface receptors identical to SSEA-1/CD15, such as the salivary glands, brain, breasts, eyes and ovaries, are not visualized, and activity does not cross the blood–brain barrier. In contrast to leukocytes that are labeled *in vitro*, transient retention of activity in the lungs following ^{99m}Tc-fanolesomab injection has not been observed [14].

The dose-limiting organ is the spleen, which receives an estimated 0.064 mGy/MBq (0.24 rads/mCi), an amount that is considerably lower than the estimated 90 mGy (9 rads) for the typical dose of 18.5 MBq (0.5 mCi) for ^{111}In-labeled leukocytes [15]. Differences in the absorbed radiation doses between males and females are minimal.

There is a rapid accumulation of radioactivity in target sites as early as 10 min after injection [13]. Uptake of ^{99m}Tc-fanolesomab in infection is governed by two factors. First, the antibody presumably binds to circulating neutrophils, which eventually migrate to the focus of infection. Second, the agent also binds to neutrophils and neutrophil debris which express CD-15 receptors, and already are sequestered at a site of infection [11,16].

6.4
Biodistribution

The biodistribution of ^{99m}Tc-fanolesomab includes the blood pool, which decreases over time, the reticuloendothelial system (liver, spleen, bone marrow), and the kidneys and urinary tract. Gastrointestinal (GI) tract activity is variable. In one investigation of 46 patients, no GI tract activity was identified up to 2h after injection, while in another investigation of 17 patients, none was observed up to 24h after injection [12,16]. In a study of 10 normal volunteers, colonic activity was seen only at 24h after injection [15]. In our experience, the distribution of ^{99m}Tc-fanolesomab in the gastrointestinal tract is similar to that of ^{99m}Tc-HMPAO-labeled leukocytes (unpublished data). Small bowel activity has been observed within 3 to 4h after injection, and colonic activity – the intensity of which is variable – is usually present by 24h after injection. Gallbladder activity has also been observed on occasion, though whether this activity reflects the hepatobiliary excretion of unbound radiolabeled antibody, or perhaps hydrophilic technetium complexes, is uncertain (Figs. 6.1–6.3).

The axial and appendicular bone marrow usually are well seen. In adults, hematopoietically active marrow normally is confined to the axial and proximal appendicular skeletons; however, generalized marrow expansion, which is a response to systemic processes such as anemia, tumor and myelophthisic states, and localized marrow expansion, which is a response to local stimuli such as fracture, surgery, orthopedic hardware, and the neuropathic joint, can alter the usual distribution. As with *in-vitro*-labeled leukocytes, the distribution of marrow activity on ^{99m}Tc-fanolesomab images varies in response to these conditions (Fig. 6.4).

Granulating wounds, such as tracheostomy, colostomy/ileostomy, and percutaneous feeding gastrostomy sites, as well as surgical incisions healing by secondary intention, incite an intense neutrophil response. Uptake of ^{99m}Tc-fanolesomab, as with *in-vitro*-labeled leukocytes, is a normal finding in these conditions and should not be mistaken for infection (Figs. 6.5 and 6.6).

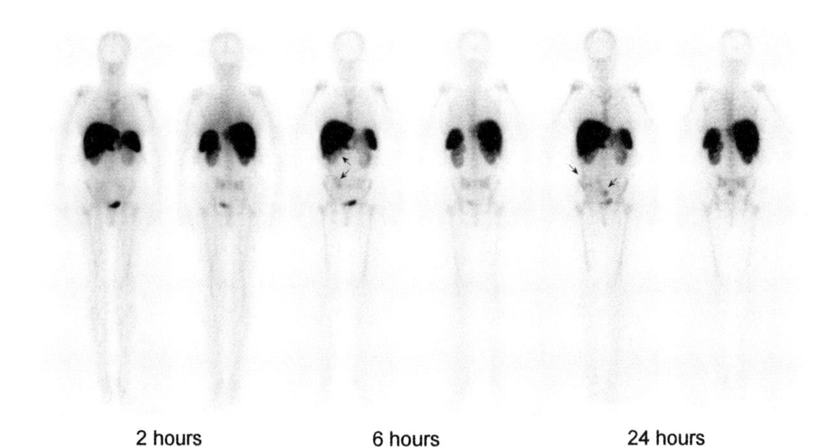

Fig. 6.1 Normal anterior and posterior whole-body images performed at multiple time points after injection of ^{99m}Tc-fanolesomab. At 2h activity is present in the blood pool, genitourinary tract, and reticuloendothelial system. Over time, the blood pool activity decreases, and bone marrow activity becomes more prominent. Colonic activity (arrows) is seen at 6 and 24h, but not at 2h.

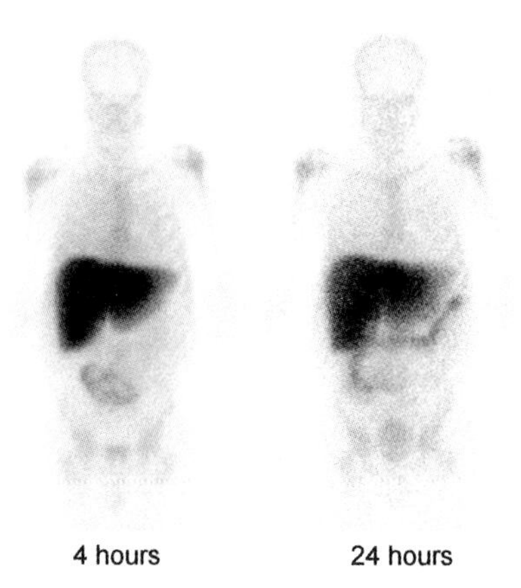

Fig. 6.2 Small bowel activity is present in the right lower quadrant at 4h after injection. By 24h most of this activity has passed into the colon.

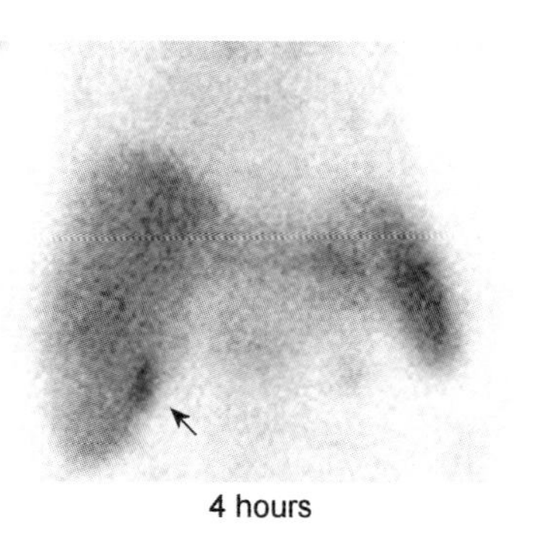

Fig. 6.3 Gallbladder activity (arrow) occasionally can be observed on ^{99m}Tc-fanolesomab studies; its presence cannot automatically be attributed to infection.

6.5 Technique

Fanolesomab is supplied as a lyophilized sterile unit dose kit containing 250 μg of antibody that can be reconstituted and labeled instantly. For the labeling procedure, 40 mCi ^{99m}Tc-pertechnetate ($^{99m}TcO_4^-$) in 0.20–0.35 mL solution is added to the vial and incubated for 30 min at 37°C. It should be noted that a high specific activity is required for satisfactory labeling. If ^{99m}Tc-fanolesomab is prepared using $^{99m}TcO_4^-$ that is more than a few hours old, then free $^{99m}TcO_4^-$ may be observed (Fig. 6.7). In order to stabilize the radiolabeled antibody and to dilute the solution, a sufficient volume of 500 mg mL^{-1} ascorbic acid is added to the vial to bring the final volume to 1.0 mL. Prior to patient administration, the preparation should be tested for radiochemical purity using instant thin-layer chromatography, and should be used only if the percentage of unbound ^{99m}Tc is not more than 10%. The preparation must be kept at room temperature and used within 6 h after reconstitution. The recommended dose to be administered to an adult is 75 to 125 μg of fanolesomab bound to 370–740 MBq (10–20 mCi) $^{99m}TcO_4^-$. For

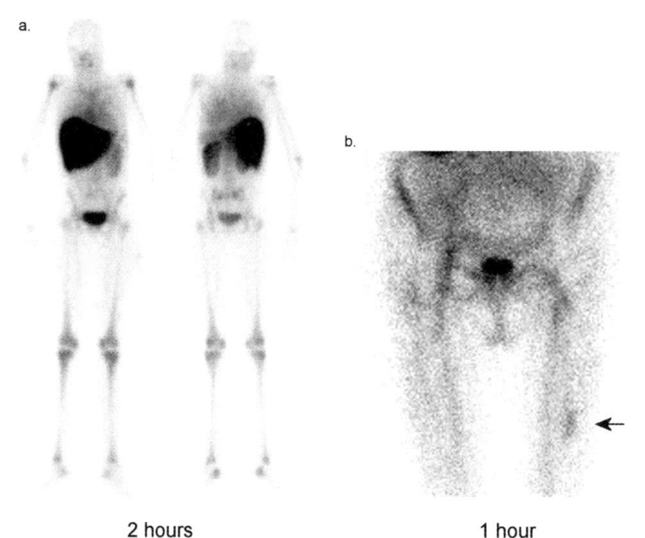

Fig. 6.4 (a) There is diffusely increased activity in the appendicular marrow, which is compatible with generalized marrow expansion, in a patient with sickle-cell disease. Note also the absence of splenic activity, which is secondary to the functional asplenia typical of this entity. (b) There is focally increased activity (arrow) at the distal tip of the femoral component of a left hip prosthesis. For reasons that are not clear, implantation of a prosthetic joint frequently produces localized marrow expansion.

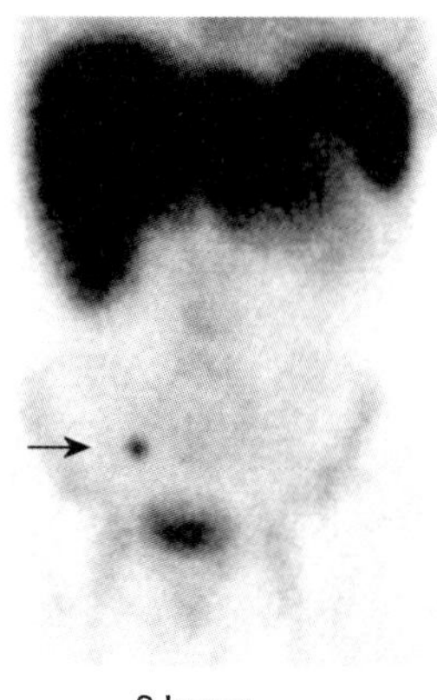

Fig. 6.5 Physiologic accumulation of ^{99m}Tc-fanolesomab at an ileostomy site (arrow) in the right lower quadrant of the abdomen.

children, 3.7 MBq (0.1 mCi) kg^{-1}, to a maximum of 740 MBq (20 mCi), of ^{99m}Tc-fanolesomab is administered [10].

The imaging procedure is well established for patients with suspected appendicitis. Dynamic imaging over the lower abdomen beginning at the time of injection, followed by static images, up to 90 min post-injection, is recommended [17]. Preliminary investigations suggest that for diabetic pedal

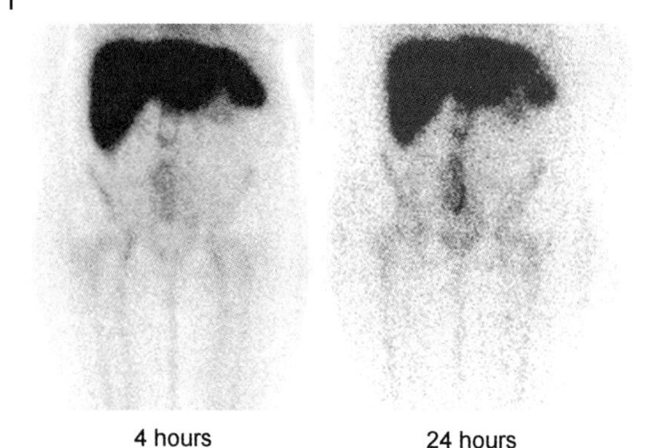

Fig. 6.6 Physiologic accumulation of ^{99m}Tc-fanolesomab in a midline abdominal wall incision healing by secondary intention. Uptake is better delineated at 24 h than at 4h.

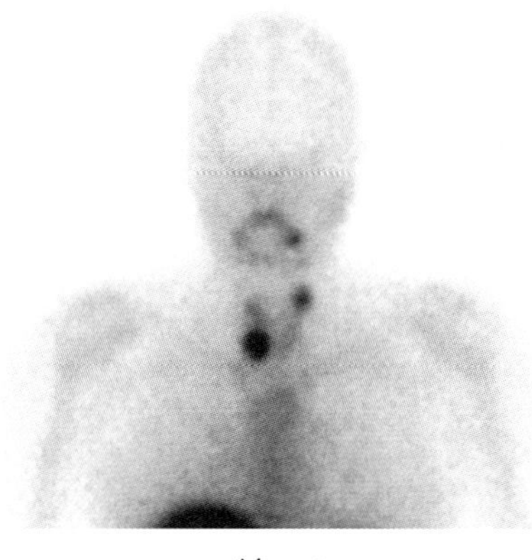

Fig. 6.7 Note the presence of activity in the oral cavity and in the multinodular thyroid gland. High specific activity is required for satisfactory labeling. If fanolesomab is labeled with $^{99m}TcO_4^-$ that is more than a few hours old, free pertechnetate may be seen, as this image illustrates.

osteomyelitis, 1-h fanolesomab images are comparable to 24-h ^{111}In-labeled leukocyte images [18].

Optimal imaging protocols for other sites of infection have not been established, though in some cases dramatic changes have been observed between images obtained at 3 to 4 h and those obtained 24 h after injection of ^{99m}Tc-fanolesomab (Figs. 6.8 and 6.9).

6.6 Indications

6.6.1 Appendicitis

Accurate and timely diagnosis of acute appendicitis can be clinically challenging. The typical presentation, including vague epigastric or periumbilical pain that

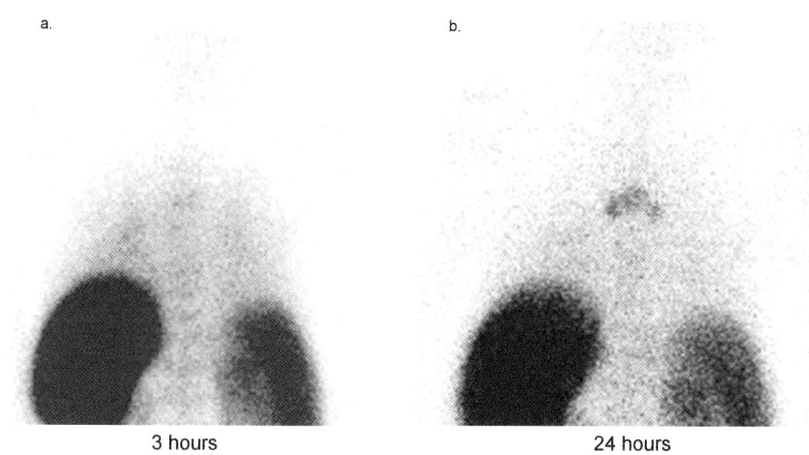

Fig. 6.8 (a) Posterior image of the thorax demonstrates questionable bilateral pulmonary activity, but is otherwise unremarkable. (b) When the image was repeated the next day, however, abnormal activity is clearly seen within the mid-thoracic spine. Osteomyelitis was subsequently confirmed on magnetic resonance imaging.

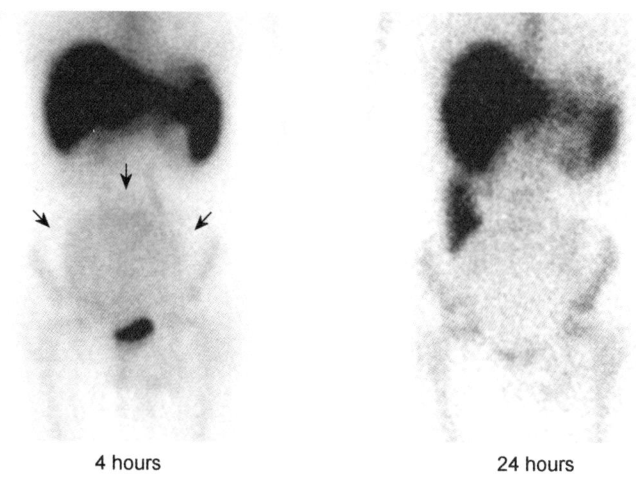

Fig. 6.9 Blood pool activity, present in a large uterine fibroid (arrows) on the initial image, has cleared by 24 h. It would be difficult to exclude infection of this lesion on the basis of the 4-h image alone.

subsequently localizes to the right lower quadrant and is associated with localized right lower quadrant tenderness, anorexia, nausea, fever and leukocytosis, is found in only about 50 to 60% of patients, and is even less common in very young and very old patients, and women of childbearing age. Consequently, appendicitis is one of the most commonly misdiagnosed entities in Emergency Departments.

Almost 30% of pediatric patients ultimately diagnosed with acute appendicitis were originally misdiagnosed, with the result that complications occurred in up to 40% of these individuals [19,20].

Individuals with an atypical or equivocal presentation for acute appendicitis usually are observed in-hospital with frequent examination and imaging studies, or are discharged and advised to return if symptoms worsen. Computed tomography (CT), with an accuracy of approximately 94% when using oral and intravenous contrast is the imaging study of choice for diagnosing appendicitis. The disadvantages of CT include the time delay required for luminal contrast opacification to reach the area of the appendix, lower test sensitivity in patients with low body fat content, and the potential for contrast reactions [10].

Although generally not appreciated by the imaging community, appendicitis can be diagnosed accurately with *in-vitro*-labeled leukocyte imaging. In a study of 100 children with an equivocal presentation, Rypins et al. [21] found that the test was 97% sensitive and 94% specific for diagnosing appendicitis. The negative predictive value of the test was 98%. The negative laparotomy rate in this series was only 4%, compared to the current standard of about 12%. Disadvantages, however, included limited availability, lengthy preparation time (2h or more), hazards of *ex-vivo* leukocyte labeling and, in the case of small children, the amount of blood required for the labeling process. Consequently, despite the favorable results obtained with labeled leukocyte imaging, this test has not gained widespread use for diagnosing appendicitis.

In contrast to the *in-vitro*-labeled leukocyte procedure, ^{99m}Tc-fanolesomab does not require specially trained personnel, is supplied in kit form, and can be formulated in less than 1h. Recent studies have found that this antigranulocyte antibody is a valuable diagnostic adjunct in atypical appendicitis and may, in fact, serve as a screening test for acute appendicitis [17,22]. Its efficacy for diagnosing appendicitis was assessed in a phase II study in 49 patients with equivocal signs and symptoms [22]. The sensitivity, specificity and accuracy of the test were 100%, 83%, and 92%, respectively. The positive predictive value was 87% and the negative predictive value 100%.

A large multicenter phase III trial was conducted to assess the efficacy of ^{99m}Tc-fanolesomab for diagnosing acute appendicitis in patients with an equivocal presentation, to evaluate its safety, and to assess its potential impact on the clinical management of these patients [17]. Fifty-nine of 200 patients enrolled in the trial had histopathologically confirmed acute appendicitis. The sensitivity, specificity, and accuracy of ^{99m}Tc-fanolesomab were found to be 91%, 86%, and 87%, respectively, while the positive and negative predictive values of the test were 73% and 96%, respectively. The diagnosis of appendicitis was made within 90min after injection in all cases; in fact, studies became positive within 8min in 50% of the patients with acute appendicitis. Similar results were obtained in the 48 pediatric patients studied, with sensitivity, specificity, and accuracy of ^{99m}Tc-fanolesomab of 91%, 86%, and 88%, respectively. Positive and negative predictive values were 67% and 97%, respectively. The high negative predictive value is particularly valuable, because a negative result means that acute appendicitis is very unlikely,

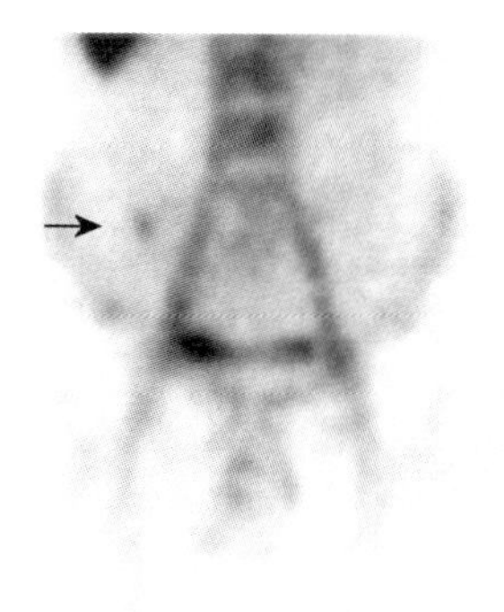

Fig. 6.10 Focal ^{99m}Tc-fanolesomab accumulation in the right lower quadrant of the abdomen (arrow) of a patient with acute appendicitis.

thereby reducing unnecessary time in the hospital for observation, as well as unnecessary surgery. The antibody was well tolerated with no serious adverse events reported. Finally, there was a significant improvement in making the appropriate management decision, both in patients with and in those without appendicitis, after the scan.

The imaging protocol for suspected appendicitis is simple. Planar imaging is carried out over a 90-min period. Single photon emission computed tomography (SPECT) imaging usually is not needed. Abnormal right lower quadrant activity in the "appendicitis zone" that persists over time is the hallmark appearance of appendicitis (Fig. 6.10) [22].

6.6.2 Osteomyelitis

Initial results indicate that ^{99m}Tc-fanolesomab accurately diagnoses osteomyelitis in the appendicular skeleton. A phase II trial was undertaken to assess the accuracy of the antibody for diagnosing osteomyelitis and to compare it with ^{111}In-labeled leukocyte imaging and three-phase bone scintigraphy [23]. A total of 24 patients (10 men, 14 women, aged 48 to 91 years) was enrolled in whom indications included: infected joint replacement (n = 12), diabetic pedal osteomyelitis (n = 8) and long bone osteomyelitis (n = 4). Patients were imaged at multiple time points up to 2 h after tracer injection. There were 11 cases of osteomyelitis. Bone scintigraphy, not surprisingly, was sensitive (100%) but not specific (38%). The 2-h ^{99m}Tc-fanolesomab images were sensitive (91%) and moderately specific (69%), and were comparable in accuracy to 24-h ^{111}In-labeled leukocyte images (91% sensitivity, 62% specificity) (Fig. 6.11). When interpreted together with bone images, the sensitivity and specificity of both the antibody and ^{111}In-labeled leukocytes improved to 100% and 85%, and 100% and 77%, respectively. The performance of the antigranulocyte antibody in this investigation was comparable to that of ^{111}In-labeled leukocytes and, when combined with bone imaging, was more accurate for diagnosing osteomyelitis than any of the other tests.

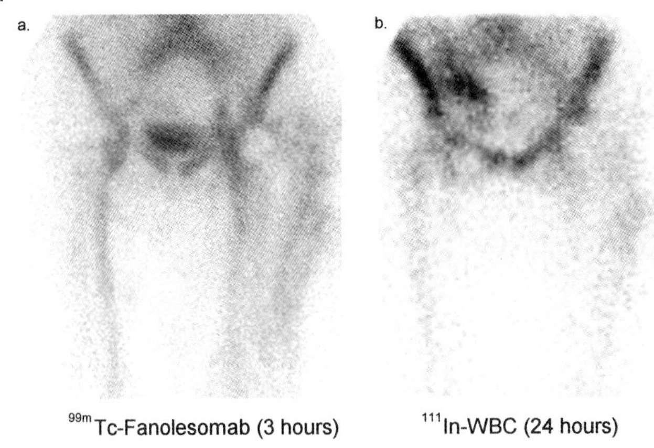

Fig. 6.11 (a) Diffusely increased activity along the lateral aspect of the infected left hip prosthesis at 3 h after injection. (b) A similar, though less obvious, abnormality is present on the ^{111}In-labeled leukocyte image obtained at about 24 h after injection.

The role of ^{99m}Tc-fanolesomab in the diagnosis of osteomyelitis in diabetic patients with pedal ulcers also has been studied [18]. In a phase II investigation, 25 diabetic patients with pedal ulcers (22 in the forefoot, three in the midfoot) underwent ^{99m}Tc-fanolesomab, ^{111}In-labeled leukocyte, and three-phase bone imaging. The 1-h antibody, 24-h labeled leukocyte and three-phase bone images were interpreted separately and classified as either positive or negative for osteomyelitis. Antibody and labeled leukocyte images also were interpreted together with the bone images. The sensitivity, specificity, and accuracy of ^{99m}Tc-fanolesomab alone were 90%, 67%, and 76%, respectively, similar to those obtained with labeled leukocyte imaging alone (80%, 67%, and 72%) (Fig. 6.12). The antibody was as sensitive as – and significantly more specific ($p = 0.004$) than – three-phase bone imaging (90% sensitivity, 38% specificity). Interpreting the antibody together with the bone scan did not improve the results.

Although these initial reports about the value of ^{99m}Tc-fanolesomab for diagnosing osteomyelitis are encouraging, there are nevertheless many issues that are not resolved. In the vast majority of diabetic patients studied, the area of concern involved the forefoot. Patients with open, granulating, surgical incisions were excluded from this investigation, with only those patients receiving antibiotic therapy for less than 7 days being eligible for entry into the study. Consequently, the utility of this agent in the neuropathic joint, in patients with healing surgical incisions, and in patients receiving antibiotics for more than one week is not known.

When performing *in-vitro*-labeled leukocyte studies for osteomyelitis, it is often necessary to perform complementary bone marrow imaging to facilitate the differentiation of labeled leukocyte uptake in bone marrow from uptake in infection [24]. Although the need for marrow imaging has not been confirmed, as the

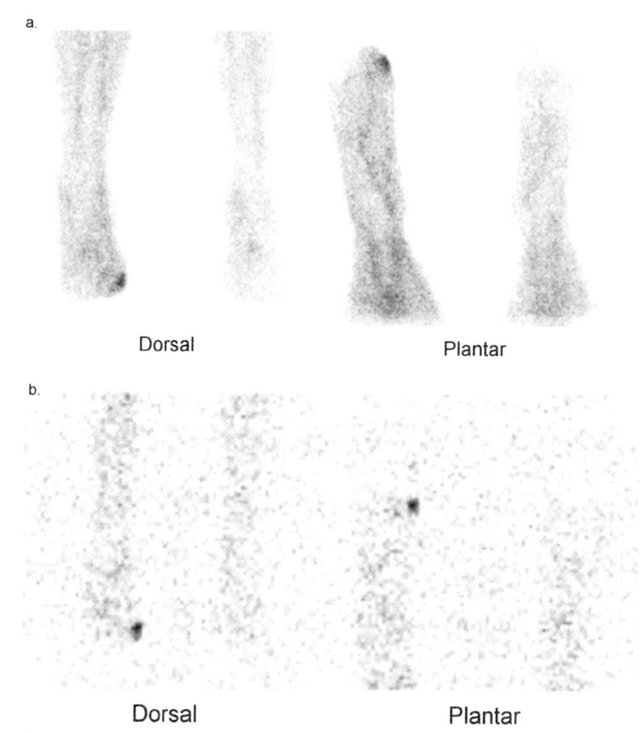

Fig. 6.12 (a) At 1 h after injection, there is focal accumulation of ^{99m}Tc-fanolesomab in the right great toe of a diabetic patient with osteomyelitis of this digit. (b) The uptake pattern is nearly identical on the ^{111}In-labeled leukocyte study of the same patient obtained at about 24 h after injection.

normal distribution of ^{99m}Tc-fanolesomab includes the bone marrow, it is reasonable to conclude that marrow imaging will be necessary (Fig. 6.13).

Finally, because it is essentially another method for performing labeled leukocyte imaging, it is likely that, although data are not available, ^{99m}Tc-fanolesomab will probably not be useful for diagnosing spinal osteomyelitis [25].

6.6.3 Other Infections

Few data are available demonstrating the utility of ^{99m}Tc-fanolesomab for diagnosing infections other than appendicitis or osteomyelitis [12,13,16,26–28]. Thakur et al. [12] studied 12 patients with clinical evidence of active inflammatory processes at the time of imaging, and all 12 had unequivocally positive images within 3 h after injection. In another investigation, 46 patients suspected of having infection underwent fanolesomab imaging at multiple time points up to 2 h after injection [13]. In some patients, additional imaging was performed at various

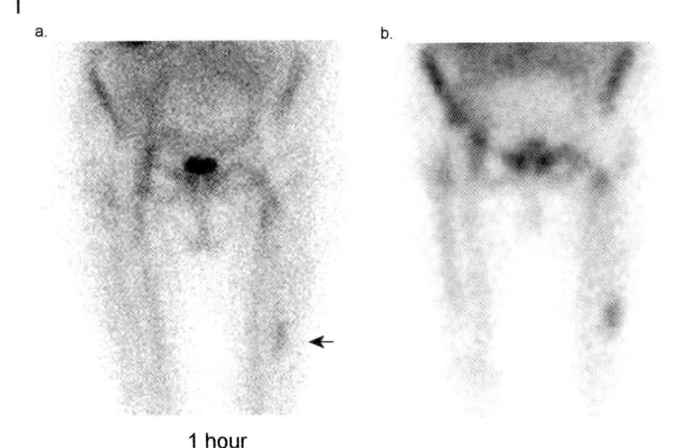

Fig. 6.13 (a) Focally increased activity at the distal tip of the femoral component of a left hip replacement (arrow). It is not possible, based on the ^{99m}Tc-fanolesomab images alone, to determine whether this uptake represents infection or localized marrow expansion. (Same patient as illustrated in Fig. 6.4b). (b) An identical focus is present on the ^{99m}Tc-sulfur colloid marrow image, confirming that the activity on the antibody image represents marrow, not infection.

time points up to 24 h. Indications for imaging included appendicitis (n = 19), osteomyelitis (n = 5), abdominal infection/inflammation (n = 9), pulmonary (n = 5), and miscellaneous (n = 8). The sensitivity, specificity, and accuracy of the test in this population were 95%, 100%, and 96%, respectively, and the positive and negative predictive values were 100% and 75%, respectively. Thirty-three scans were found to be diagnostic within 10 min, and the remainder within 2 h, after injection. Gratz et al. [16] compared ^{99m}Tc-fanolesomab to ^{99m}Tc-HMPAO-labeled leukocytes in 17 patients, with 23 foci of infection, all within soft tissue or bone. The sensitivity and specificity of the antibody were 95% and 96%, respectively, compared to 91% and 82%, respectively, for ^{99m}Tc-HMPAO-labeled leukocyte imaging.

The present authors' clinical experience with ^{99m}Tc-fanolesomab, in a variety of conditions, has been favorable. This agent has been used, in lieu of ^{111}In-labeled leukocytes, to diagnose abscesses, prosthetic vascular graft infections, and to identify the source of an "occult" infection. Generally, the results have been satisfactory (Figs. 6.14 and 6.15) [26–28], although under certain circumstances ^{111}In-labeled leukocytes may have been preferable to ^{99m}Tc-fanolesomab. For example, unlike ^{111}In-labeled leukocytes, fanolesomab normally accumulates in the genitourinary and GI tracts, making detection of infection in these systems more difficult to diagnose with this agent. Therefore, in the setting of occult infection, or when genitourinary or GI tract infections are suspected, ^{111}In-labeled leukocytes, rather than ^{99m}Tc-fanolesomab, should most likely be used (Fig. 6.16).

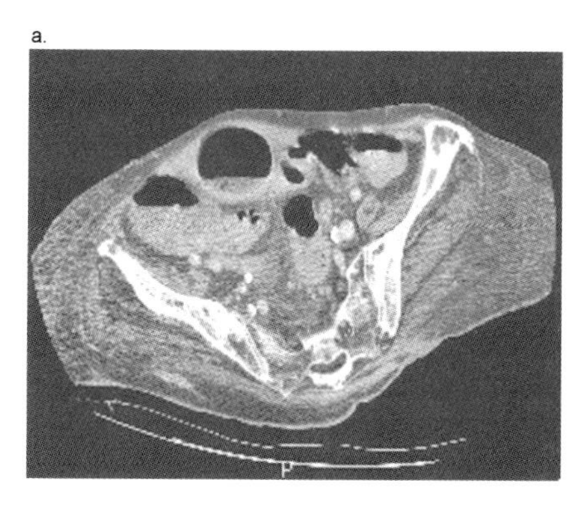

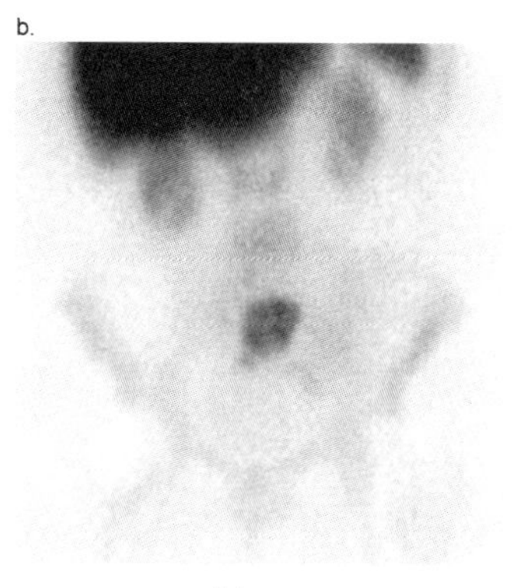

Fig. 6.14 (a) A large cystic mass with an air/fluid level in the lower abdomen was identified on a CT scan performed on a patient with fever and abdominal pain. (b) The intense accumulation of ^{99m}Tc-fanolesomab within this lesion confirmed the presence of infection.

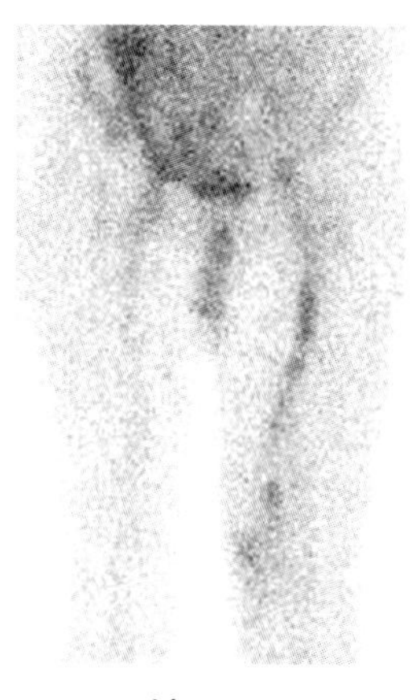

Fig. 6.15 A linear area of activity extending along the medial aspect of the left thigh can be appreciated in a patient with an infected left lower-extremity prosthetic vascular graft.

6.7
Adverse Side Effects and Safety

The chemotactic function of neutrophils is not affected by the degree of receptor occupancy. At 10% receptor occupancy by the antibody, the phagocytic and adherence functions are substantially diminished, while at 4–5% occupancy no effect on neutrophil function is observed. At least 1mg of fanolesomab is needed to saturate 4% of the estimated 5.1×10^5 CD15 surface receptors present on a neutrophil [11,12]. The usual dose injected contains 75 to 125 μg of antibody, which is considerably less than the amount needed to produce pharmacological effects [12,29].

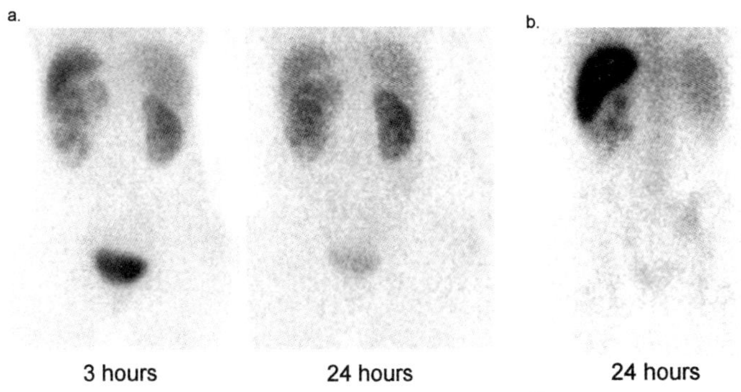

Fig. 6.16 (a) A febrile 56-year-old woman underwent ^{99m}Tc-fanolesomab imaging for occult infection. Posterior images showed only a large left kidney. (b) A ^{111}In-labeled autologous leukocyte study performed 2 days (24 h post injection) later clearly demonstrated abnormal labeled leukocyte accumulation in an infected left kidney. Urine cultures grew *Candida albicans*.

There is a transient decrease in the number of circulating white blood cells within 20 min after injection of fanolesomab; typically, the relative number of granulocytes decreases from 57.2 ± 10.5 to 35.3 ± 17.7% of total circulating leukocytes. The effects on the other blood cells are qualitatively similar, but less pronounced. Recovery is rapid, averaging about 20 min, and there have been no clinical complaints associated with this phenomenon [14,16].

Murine monoclonal antibodies may result in the development of human anti-mouse antibody (HAMA) in individuals to whom these agents are administered. One study to assess formation of HAMA and safety of fanolesomab administration was undertaken in 30 healthy subjects [13]. Each subject received 125 μg antibody labeled with decayed $^{99m}TcO_4^-$, with blood samples being obtained before and at 3–4 weeks and 3–4 months after antibody injection. There was no statistically significant elevation of HAMA titers following injection; neither were any adverse reactions or changes in vital signs observed.

An investigation of 30 healthy volunteers was undertaken to evaluate the safety and the extent of induction of HAMA response following a second injection of fanolesomab [30]. Subjects were injected on two separate occasions, 3 weeks apart, with 125 μg of fanolesomab labeled with decayed $^{99m}TcO_4^-$. HAMA assays were performed on blood samples drawn prior to each injection, and at 1 and 4 weeks after the second injection. Five subjects (17%) exhibited induction of HAMA; two of these were considered marginal and three moderate responses. Seven subjects (23%) experienced adverse events, most of which were coincidental to fanolesomab administration. None of the adverse events was serious or severe, and the investigators concluded that repeated fanolesomab injections, at clinically useful doses, did not appear to induce a strong HAMA response or to present a risk for serious adverse events.

In pre-market studies submitted to the United States Food and Drug Administration (FDA), there were relatively few safety concerns. Among 523 patients and normal volunteers enrolled in various clinical trials, only 49 adverse events were reported by 37 patients (7%). Flushing (2%) and dyspnea (1%) were the most frequently reported adverse events; less-common adverse events included syncope, dizziness, hypotension, chest pressure, paresthesia, nausea, and burning at the injection site. Four of the 49 adverse events were severe: hypotension, worsening of sepsis, chest pressure and decreased O_2 saturation. Two patients enrolled in clinical trials of post-surgical infection or abscess died following injection, although no relationship could be determined between ^{99m}Tc-fanolesomab and these fatalities [31].

At the end of 2005, approximately 18 months after its introduction, ^{99m}Tc-fanoleosmab was withdrawn from the United States market as a result of post-marketing reports of serious and life-threatening cardiopulmonary events following its administration. The onset of these events generally occurred within minutes after injection, and included two deaths attributed to cardiopulmonary failure within 30 min after injection. Additional cases of serious cardiopulmonary events, including cardiac arrest, hypoxia, dyspnea and hypotension, required resuscitation with fluids, vasopressors, and oxygen. However, there is no evidence that patients who already safely received the drug face any long-term risks.

6.8 Summary

The antigranulocyte antibody ^{99m}Tc-fanolesomab labels human leukocytes *in vivo*, rapidly, and accurately diagnoses appendicitis in patients with an equivocal presentation. Preliminary investigations suggest that ^{99m}Tc-fanolesomab is comparable to *in-vitro*-^{111}In-labeled leukocytes for diagnosing diabetic pedal osteomyelitis. Although at present the data are limited, it is likely that ^{99m}Tc-fanolesomab may also be used to accurately diagnose other infections. Based on results of clinical trials conducted to date, the drug appeared to be safe, but as a result of post-marketing reports of untoward (and sometimes severe) adverse events following its administration, ^{99m}Tc-fanolesomab was withdrawn from the US market, and its future remains uncertain.

References

1 Love, C., Palestro, C.J. (2004) Radionuclide imaging of infection. J Nucl Med Tech 32: 47–57.

2 Palestro, C.J., Weiland, F.L., Seabold, J.E., et al. (2001) Localizing infection with a technetium labeled peptide: initial results. Nucl Med Commun 22: 695–701.

3 Becker, W., Goldenberg, D.M., Wolf, F. (1994) The use of monoclonal antibodies and antibody fragments in the imaging of infectious lesions. Semin Nucl Med 24: 142–153.

4 Becker, W., Blair, J., Behr, T., et al. (1994) Detection of soft tissue infections and

osteomyelitis using a technetium-99m-labeled anti-granulocyte monoclonal antibody fragment. J Nucl Med 35: 1436–1443.
5 Becker, W., Saptogino, A., Wolf, F.G. (1992) The single late Tc-99m-granulocyte antibody scan in inflammatory diseases. Nucl Med Commun 13: 186–1920.
6 Gratz, S., Schipper, M.L., Dorner, J., et al. (2003) Leukoscan for imaging infection in different clinical settings: A retrospective evaluation and extended review of the literature. Clin Nucl Med 28: 267–276.
7 Becker, W., Palestro, C.J., Winship, J., et al. (1996) Rapid diagnosis of infections with a monoclonal antibody fragment (Leukoscan). Clin Orthop 329: 263–272.
8 Hakki, S., Harwood, S.J., Morrissey, M.A., Camblin, J.G., Laven, D.L., Webster, Jr. W.B. (1997) Comparative study of monoclonal antibody scan in diagnosing orthopaedic infection. Clin Orthop 335: 275–285.
9 Deveillers, A., Garin, E., Polard, J.L., et al. (2000) Comparison of Tc-99m-labelled antileukocyte fragment Fab' and Tc-99m-HMPAO leukocyte scintigraphy in the diagnosis of bone and joint infections: a prospective study. Nucl Med Commun 21: 747–753.
10 Love, C., Palestro, C.J. (2003) 99mTc-fanolesomab. IDrugs 6: 1079–1085.
11 Thakur, M.L., Richard, M.D., White, F.W., III. (1988) Monoclonal antibodies as agents for selective radiolabeling of human neutrophils. J Nucl Med 29: 1817–1825.
12 Thakur, M.L., Marcus, C.S., Henneman, P., et al. (1996) Imaging inflammatory diseases with neutrophil-specific technetium-99m-labeled monoclonal antibody anti-SSEA-1B. J Nucl Med 37: 1789–1795.
13 Thakur, M.L., Marcus, C.S., Kipper, S.L., et al. (2001) Imaging infection with LeuTech. Nucl Med Commun 22: 513–519.
14 Mozley, P.D., Thakur, M.L., Alavi, A., et al. (1999) Effects of a 99mTc-labeled murine immunoglobulin M antibody to CD15 antigens on human granulocyte membranes in healthy volunteers. J Nucl Med 40: 2107–2114.
15 Mozley, P.D., Stubbs, J.B., Dresel, S.H., et al. (1999) Radiation dosimetry of a 99mTc-labeled IgM murine antibody to CD15 antigens on human granulocytes. J Nucl Med 40: 625–630.
16 Gratz, S., Behr, T., Herrmann, A., Dresing, K. (1998) Intraindividual comparison of 99mTc-labelled anti-SSEA-1 antigranulocyte antibody and 99mTc-HMPAO labeled white blood cells for the imaging of infection. Eur J Nucl Med 25: 386–393.
17 Rypins, E.B., Kipper, S.L., Weiland, F., et al. (2002) 99mTc anti-CD15 monoclonal antibody (LeuTech) imaging improves diagnostic accuracy and clinical management in patients with equivocal presentation of appendicitis. Ann Surg 235: 232–239.
18 Palestro, C.J., Caprioli, R., Love, C., et al. (2003) Rapid diagnosis of pedal osteomyelitis in diabetics with a technetium-99m labeled monoclonal antigranulocyte antibody. J Foot Ankle Surg 42: 2–8.
19 Sarosi, Jr., G.A., Turnage, R.H. (2002) Appendicitis. In: Feldman, M., Tschumy, Jr., W.O., Friedman, L.S., Sleisenger, M.H. (Eds.), Sleisenger & Fordtran's Gastrointestinal and Liver Disease. 7th edition. St Louis: W.B. Saunders, pp. 2089–2099.
20 Kipper, S.L. (1999) The role of radiolabeled leukocyte imaging in the management of patients with acute appendicitis. Q J Nucl Med 43: 83–92.
21 Rypins, E.B., Kipper, S.L. (1997) 99mTc-hexamethylpropyleneamine oxime (Tc-WBC) scan for diagnosing acute appendicitis in children. Am Surg 63: 878–881.
22 Kipper, S.L., Rypins, E.B., Evans, D.G., et al. (2000) Neutrophil-specific 99mTc-labeled anti-CD15 monoclonal antibody imaging for diagnosis of equivocal appendicitis. J Nucl Med 41: 449–455.
23 Palestro, C.J., Kipper, S.L., Weiland, F.L., et al. (2002) Osteomyelitis: Diagnosis with 99mTc-labeled antigranulocyte antibodies compared with diagnosis with indium-111-labeled leukocytes-

initial experience. Radiology 223: 758–764.

24 Palestro, C.J., Love, C., Tronco, G.G., et al. (2006) Combined labeled leukocyte and technetium-99m sulfur colloid marrow imaging for diagnosing musculoskeletal infection: principles, technique, interpretation, indications and limitations. RadioGraphics 26: 859–870.

25 Palestro, C.J., Kim, C.K., Swyer, A.J., Vallabhajosula, S., Goldsmith, S.J. (1991) Radionuclide diagnosis of vertebral osteomyelitis: Indium-111-leukocyte and technetium-99m-methylene diphosphonate bone scintigraphy. J Nucl Med 32: 1861–1865.

26 Rini, J.N., Love, C., Zia, N., et al. (2005) Initial clinical experience with Tc-99m-fanolesomab in patients with suspected infection. Radiological Society of North America 91st Scientific Assembly and Annual Meeting Program; (P)312 [abstract].

27 Rini, J.N., Love, C., Zia, N., et al. (2005) Tc99m-fanolesomab: a new diagnostic agent for the detection of infection. Radiological Society of North America 91st Scientific Assembly and Annual Meeting Program; (P)806 [abstract].

28 Tronco, G.G., Love, C., Rini, J.N., et al. Diagnosing prosthetic vascular graft infection with the antigranulocyte antibody ^{99m}Tc-fanolesomab. Nucl Med Commun (in press).

29 Thakur, M.L., Lee, J., DeFulvio, M.D., et al. (1990) Human neutrophils: Evaluation of adherence, chemotaxis, and phagocytosis, following interaction with radiolabeled antibodies. Nucl Med Commun 11: 37–43.

30 Line, B.R., Breyer, R.J., McElvany, K.D., et al. (2004) Evaluation of human anti-mouse antibody response in normal volunteers following repeated injections of fanolesomab (NeutroSpec), a murine anti-CD15 IgM monoclonal antibody for imaging infection. Nucl Med Commun 25: 807–811.

31 Package insert: NeutroSpec [Kit for the preparation of technetium (99m Tc) fanolesomab].

7
Gemtuzumab Ozogamicin (Mylotarg)

Matthias Peipp and Martin Gramatzki

7.1
Introduction

Monoclonal antibodies (mAbs) have proven their usefulness as biotherapeutics across a spectrum of diseases, including cancer, infection and immune disorders, as shown by the growing list of mAbs that have been approved by the US Food and Drug Administration (FDA) and by the European Medicines Agency (EMEA) [1,2]. Oncology has been a major area of interest for mAb-based therapeutics, because various antigens have been identified that are overexpressed on certain types of cancer cells compared with normal tissues. Even so, translation into clinically effective therapeutics required that many challenges be overcome. The initial application of murine antibodies was limited by immunogenicity, short serum half-lives, and a lack of efficient interaction with human immune effector cells [3]. Today, however, the introduction of chimeric, humanized and fully human antibodies has overcome these limitations, and such agents can now be produced on a routine basis through protein engineering (humanization), display technologies (phage-, ribosomal-display), or by immunization of transgenic animals carrying human immunoglobulin genes [4–6]. Several agents such as Rituximab, Trastuzumab and Cetuximab have been approved for clinical applications, and demonstrate varying activity in different tumor types [2]. Although these unmodified mAbs show some therapeutic potency, the effects are often not curative. Present knowledge about the clinically relevant mechanisms of action for mAbs is rather limited, but several lines of evidence point to an important role of Fc receptors [7,8]. Therefore, different approaches to improve the therapeutic efficacy of antibody therapeutics by enhancing Fc/FcR interaction have been followed (see Volume I: Fc engineering).

A conceptionally different approach to enhance the efficacy of antibody therapeutics represents arming mAbs with drugs, toxins, or radionuclides [1]. Although such agents are simple in principle, deriving activities from them has been a major challenge because the mAb-conjugated drugs can decompose or decay before being delivered to the target site, the pharmacokinetics of the mAb carriers

Handbook of Therapeutic Antibodies. Edited by Stefan Dübel
Copyright © 2007 WILEY-VCH Verlag GmbH & Co. KGaA, Weinheim
ISBN 978-3-527-31453-9

may be suboptimal, the conjugation process can disturb the binding characteristics, the chemical linkers used may have inappropriate stability, and the drugs may therefore not be released in active states or in quantities needed to achieve therapeutic efficacy. Despite these complex requirements, progress has now been made, and the FDA has recently approved three mAb conjugates for cancer therapy [1]. Whilst two of these mAb conjugates are murine radiolabeled antibodies, the third mAb conjugate is a humanized, CD33-specific IgG4 mAb conjugated to a calicheamicin derivative – gemtuzumab ozogamicin – which is used for the treatment of leukemia [9], and is reviewed in this chapter.

7.2 CD33 as a Target Antigen in Acute Leukemia Therapy

One important factor that contributes to the success of antibodies and antibody conjugates as drugs represents the choice of an appropriate target antigen for intervention in the pathophysiology of disease [2,3,10].

Approximately 85 to 90% of adult and pediatric acute myeloid leukemia (AML) cases and 15 to 25% of acute lymphoblastic leukemia (ALL) cases are considered CD33 positive, as defined by the presence of antigen on at least 20 to 25% of leukemic blasts. Importantly, AML blasts express large amounts of CD33 compared to normal tissue [11,12]. The presence of CD33 on AML blasts motivated the development of mAbs against this myeloid cell-surface antigen for the treatment of AML [13,14]. CD33 is a 67 kDa type 1 transmembrane sialoglycoprotein, and a founding member of a rapidly evolving immunoglobulin superfamily subset of sialic acid-binding immunoglobulin-related lectins (siglecs; siglec-3) [15,16]. It was identified by the murine monoclonal antibody anti-MY9 [17], and the human CD33 gene has been mapped to chromosome 19q13.3 and most closely resembles the genes for two adhesion receptors, the myelin-associated glycoprotein and the B-cell antigen CD22 [18,19].

Physiologically, CD33 expression is restricted to early multilineage hematopoietic progenitors, myelomonocytic precursors, and to more mature myeloid cells, macrophages, monocytes and dendritic cells. CD33 is highly expressed on granulocyte precursors, but its expression decreases with maturation and differentiation. It is also expressed on liver cells, but importantly it is absent from normal pluripotent hematopoietic stem cells [20–26].

The natural ligand of CD33 and its biologic functions are unknown. The cytoplasmic tail of CD33 has two tyrosine residues in sequences that closely resemble immunoreceptor tyrosine-based inhibitory motifs (ITIMs). When phosphorylated, these tyrosine motifs provide docking sites for Src homology-2 (SH2) domain-containing tyrosine phosphatases (SHP-1 and SHP-2) [27–29]. CD33 activation and association with SHP-1 may result in inhibitory signals, and may affect the function of neighboring membrane receptors. Recent *in-vitro* studies using genetically modified myeloid cell lines demonstrated that mutations in the ITIM-like motifs impair internalization of antibody-bound CD33, suggesting that structural and functional variations altering this pathway might affect the susceptibility to CD33-targeted therapies [30]. The expression pattern, its overexpression on tumor cells, and the efficient internalization make CD33 an ideal target for immunoconjugates.

7.3
Gemtuzumab Ozogamicin

Gemtuzumab ozogamicin (GO; Mylotarg, CMA-676; Wyeth Laboratories, Philadelphia, PA, USA) consists of a recombinant humanized immunoglobulin G4 (IgG4) CD33 monoclonal antibody (hP67.6) conjugated to the antitumor antibiotic calicheamicin-γ1, *N*-acetyl-γ1-calicheamicin dimethyl hydrazide (NAc-gamma calicheamicin DMH; Fig. 7.1). GO was very potent and selectively cytotoxic to HL-60 leukemia cells in tissue culture, and low doses given to mice bearing HL-60 xenografts resulted in long-term, tumor-free survivors. In addition, GO selectively inhibited leukemia colony formation by marrow cells from a significant proportion of AML patients [31,32].

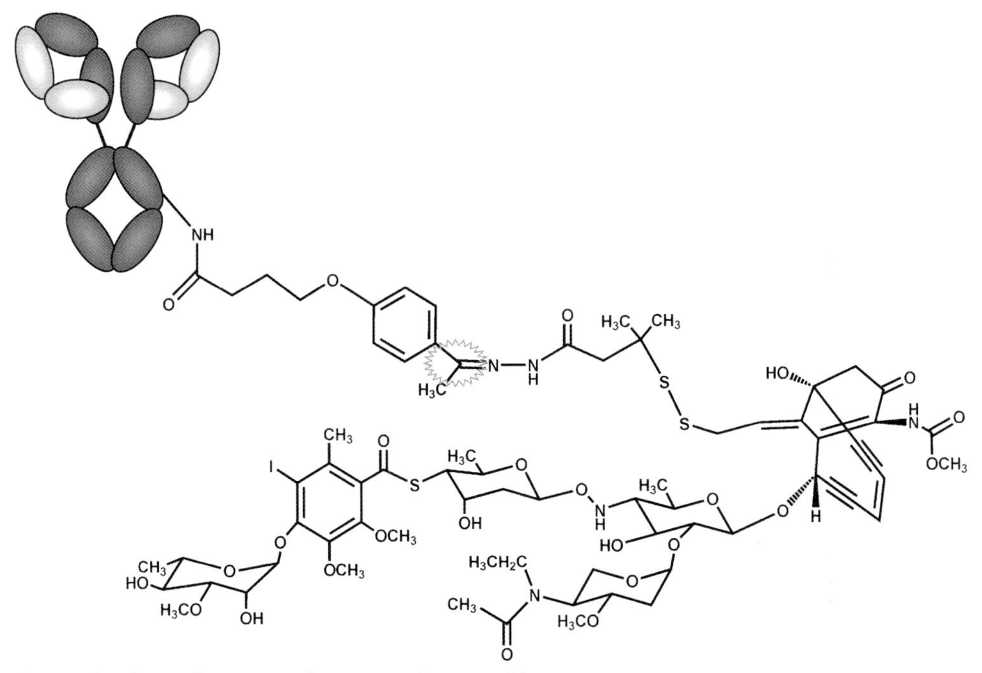

Fig. 7.1 The chemical structure of gemtuzumab ozogamicin GO. The designed hybrid conjugate contains both a lysine attachment to the antibody through amide formation, and a hydrazone linkage (indicated by the gray star-shaped motif). Hydrazones release the drug under acidic conditions within the lysosomes of target cells. The drug loading is 2–3 mol mol^{-1}.

7.3.1
The IgG4 Moiety

The human IgG4 component of GO is not cytotoxic *in vitro*, nor does it induce complement-mediated (CDC) or antibody-dependent cellular cytotoxicity (ADCC; See Chapter 8, Vol I, FC engineering). Therefore, the antibody moiety is utilized primarily as vehicle to transport the cytotoxic drug to CD33-positive leukemia cells.

7.3.2
Calicheamicin

The calicheamicins are highly potent anti-tumor antibiotics of the enediyne family. Originally isolated from a broth extract of the soil microorganism *Micromonospora echinospora* ssp. *calichensis*, the calicheamicins were detected in a screen for potent DNA-damaging agents [33]. The parent of this family of xenobiotics, chalicheamicin γI^1, has been shown to bind to double-stranded DNA in a sequence-specific manner [33–36]. Upon binding, it cleaves the sugar backbone of the DNA, even at sub-picomolar concentrations. The major DNA contact surface of the antibiotic is an aryltetrasaccharide moiety that interacts with the minor groove of the DNA duplex (Fig. 7.2) [34,37–39]. The enediyne portion of the antibiotic appears to contribute to the drug–DNA interaction by increasing binding

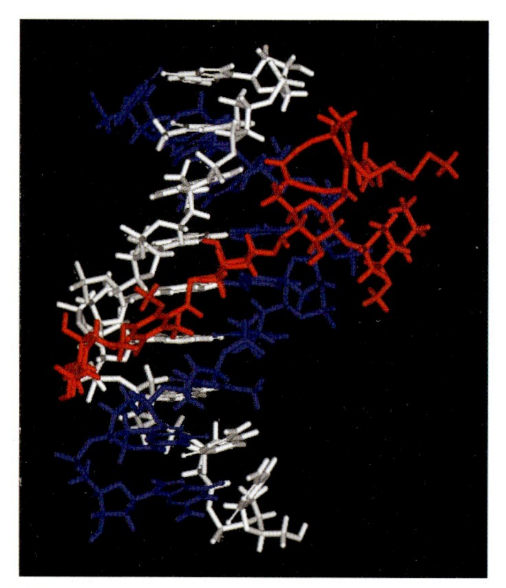

Fig. 7.2 Calicheamicin binding to the minor groove of DNA. Solution structure of the calicheamicin $\gamma 1^I$-DNA hairpin duplex complex containing a central (T-C-C-T)·(A-G-G-A) segment based on a combined analysis of NMR and molecular dynamics calculations. The DNA backbone and base pairs are shown in white and blue. Calicheamicin is shown in red [92,93]. [RSCB Protein Data Bank Accession Number 2PIK].

energy and widening the spectrum of target sequences [40]. One of the preferred target sequences is the tetranucleotide TCCT [36]; others include TCTC and TTTT [34,41,42]. Structural and chemical analyses have shown that the carbohydrate tail of chalicheamycin is orientated towards the 3′ end of the TCCT surface, and that the iodine of the aryltetrasaccharide plays a critical role in the sequence-selective recognition of DNA [34,41]. In addition, the chalicheamycin oligosaccharide was demonstrated to inhibit transcription factor binding to DNA [43,44]. Due to their unique mode of action and potency, several analogs of the calicheamicins have been tested in preclinical models as potential anti-tumor agents. Since the therapeutic range is limited by the high toxicity, their potency makes them ideal for antibody-targeted chemotherapy. Calicheamicin-conjugates may offer significant advantages over previous antibody-targeted agents, because they are approximately 1000-fold more cytotoxic than other clinically used anti-cancer agents [31], although they may have much lower immunogenicity than for example protein toxins such as ricin or pseudomonas exotoxin A.

7.3.3 The Design of Antibody-Chalicheamicin Conjugates: Humanization and Choice of Linker

Earlier studies with chalicheamicin conjugates of the murine, CD33 antibody P67.6 demonstrated that a site of hydrolytic release, as supplied by a hydrazone, was necessary for potent and selective cytotoxicity in tissue culture, and for a maximum efficacy in xenograft models of AML in athymic mice [45]. A carbohydrate-based conjugate was chosen that was made by oxidizing the naturally occurring carbohydrate residues in murine P67.6 with periodate, and reacting the resultant aldehydes with a hydrazide derivative of calicheamicin, NAc-gamma calicheamicin DMH [46]. In order to minimize the potential for immune responses in human trials, the P67.6 antibody has been humanized by CDR-grafting [31]. However, the periodate oxidation required when preparing a humanized hP67.6 calicheamicin carbohydrate conjugate resulted in a virtually total loss of immunoreactivity. Subsequent experiments indicated that an oxidatively sensitive methionine residue near the CDR that originated with the human sequences might be the reason for this sensitivity to oxidation. This necessitated the design of a new class of conjugates in which the designed hybrid conjugates contain both lysine attachment to the antibody through amide formation, which obviates the need for oxidation of the carbohydrates, and a hydrazone linkage, which allows for the necessary hydrolytic release. The most optimal conjugates were made with unsubstituted acetophenones, and the 4-(4-acetylphenoxy)butanoic acid) linker, which appeared to afford the most favorable balance between hydrolytic stability in physiological buffers (pH 7.4) and efficient drug release at the pH of lysosomes (~4). The final product, gemtuzumab ozogamicin, consists of a 1:1 mixture of unconjugated hP67.6 and hP67.6 conjugated to four to six moles of NAc-gamma calicheamicin, thus providing an average drug loading ratio of two to three drug molecules to one antibody molecule [47].

7.4 Mechanisms of Action

After antibody binding, CD33-antibody complexes are rapidly internalized and translocated into lysosomes [12,48,49]. In the acidic environment of the lysosome, the acid-hydrolyzable linker is efficiently cleaved [31], and the chalicheamicin moiety is released and subsequently reduced to a highly reactive 1,4-dehydrobenzene-diradical species, through the action of glutathione [36,50]. This diradical species binds within the minor groove of DNA and abstracts hydrogen atoms from the deoxyribose backbone. The resulting radicals scavenge oxygen and initiate a sequence of events that lead to site-specific single- and double-stranded DNA scission [12,51,52]. The mitochondrial pathway of apoptosis appears to be predominantly utilized for GO-induced cell death. GO treatment of primary AML samples and AML cell lines causes the loss of mitochondrial membrane potential and activation of caspase 3, while overexpression of Bcl-2 or Bcl-X_L inhibits GO-induced cytotoxicity [53,54]. *In-vitro* exposure to GO also causes G_2/M cell cycle arrest in susceptible myeloid leukemia cell lines, with activation of cyclin B1 and phosphorylation of checkpoint kinases 1 and 2 (Chk1; Chk2) [53,55]. In summary, calicheamicin-induced DNA damage stimulates a strong DNA damage response in the cell [50,56,57], resulting in permanent growth arrest, cell death, or temporary arrest followed by DNA repair. Thus, multiple pathways appear to interact during GO-induced DNA damage, resulting in either cytotoxicity or cell survival [12].

7.5 Potential Mechanisms of Resistance

Several potential mechanisms of resistance have been postulated from clinical and *in-vitro* observations. Different members of the ATP-binding cassette (ABC) superfamily of transporter proteins such as Pgp (MDR1; ABCB1) and multidrug resistance protein 1 (MRP1; ABCC1) have been demonstrated to mediate *in-vitro* resistance to GO [12,55,58–60], whereas the ABC transporter breast cancer resistance protein (BCRP; ABCG2), which is variably expressed in AML, does not confer resistance to GO [12,61,62]. In clinical phase II trials, Pgp expression on AML blasts correlated with treatment failure [9,58]. Although multidrug resistance inhibitors, such as cyclosporine A (CsA) and PK11195 have been shown to increase *in-vitro* GO sensitivity in AML cells [54], incorporating the Pgp inhibitor CsA in GO-containing regimes as induction, salvage or post-remission therapy in AML did not appear to increase rates of response and survival [63,64].

Alternative resistance mechanisms include reduced GO-binding capacity to leukemic blasts [65], altered pharmacokinetics, high levels of tumor load in the peripheral blood [65,66], overexpression of anti-apoptotic proteins, such as bcl-2 or bcl-X_L [54], or resting state of cell cycle [67,68]. Therefore, it is postulated that combinations of GO with myeloid cytokines that stimulate AML blast proliferation or agents that down-modulate anti-apoptotic proteins (e.g., Bcl-2 antisense) might enhance drug sensitivity and clinical efficacy [12,69–71].

7.6 Clinical Trials: The Data of GO

CD 33 has a wide expression on myeloid cells [72], and thus may serve as a target in many patients with AML. In contrast, hematopoietic stem cells are CD 33-negative. The group of Scheinberg [13] used lintuzumab (HuM195), a humanized (CDR-grafted) antibody, to treat various AML forms and achieved limited success. Lintuzumab could be added safely to intensive induction chemotherapy, and has also been applied as an immunoconjugate with alpha-emitting bismuth [73]. However, the first – and until now only FDA-approved targeted therapy for CD 33-positive leukemic cells is GO (Mylotarg). Following numerous unsuccessful attempts to develop immunoconjugates with mitoxantrone, American Cynamide/Wyeth succeeded in coupling the carbohydrate toxin calicheamicine to CD33. Maytansinoids such as calicheamicine have been evaluated since the 1970s for their cancer therapy properties, but due to the narrow therapeutic range [74] they did not find any role in cancer treatment. Thus, the possibility of delivering calicheamicine directly to the myeloid blast by coupling to a CD33 antibody proved to be an exciting event.

GO – which originally was named CMA-676 – had first shown efficacy in xenotransplanted nude mice [31]. In a preliminary phase I study, 40 patients with relapsed or refractory CD33-positive AML were treated with GO [32], and a promising high rate of clinical response was observed. At the highest dose level of $9\,mg\,m^{-2}$, long-term neutropenia was observed. Whilst GO was well tolerated, a syndrome of fever and chills after infusion was usually observed, though it is believed that this limited toxicity might have been due to rapid internalization and degradation of the chemical linker being confined to the lysosome.

In phase II evaluations, following a first dose of $9\,mg\,m^{-2}$, a second 2-h infusion was given 14 days later after appropriate premedication which included antihistamines [9]. A total of 142 patients with AML in first relapse and a median age of 61 years was treated in this international study, and 30% achieved remission. Due to a favorable toxicity profile, 27 patients went on to hematopoietic stem cell transplantation. Not surprisingly, GO led to grade IV myelosuppression (neutropenia and thrombocytopenia), though 28% of patients developed grade III and IV infectious complications. Approximately 20% of these patients showed liver abnormalities that might have contributed to death in two cases. Specificity of the drug was documented by minimal mucositis and complete lack of alopecia (Fig. 7.3). No antibody responses to the immunoconjugate were found. From a practical viewpoint, this therapy could in part even be given in an outpatient setting.

These data have recently been updated by Larson and co-workers for 277 patients participating in phase III trials on both sides of the Atlantic [75]. These clinical

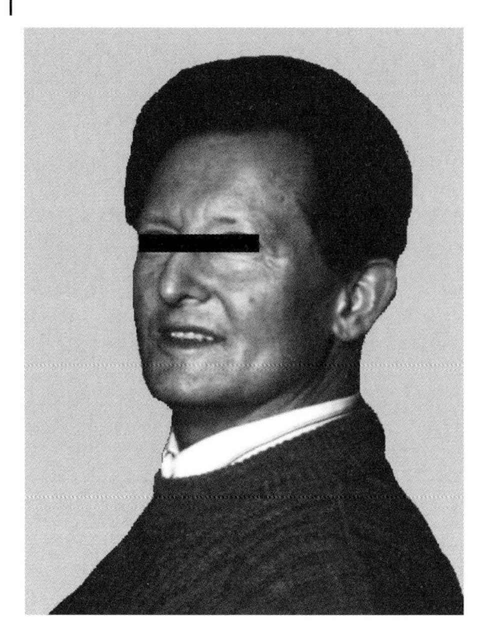

Fig. 7.3 A patient after gemtuzumab ozogamicin (GO) treatment. The specificity of GO is documented by a complete lack of alopecia. This patient received GO infusions three times for relapsed acute myeloid leukemia. No hair loss inevitable to conventional chemotherapy was seen, the patient achieved complete remission, and was latter successfully transplanted with peripheral blood stem cells from his sister.

studies clearly showed that, if remissions were achieved, they were rather short-lived and were of approximately 12 months' duration. Notably, similar clinical responses were obtained in younger and older individuals. In 2000, GO was approved by the FDA as the third antibody-based drug for cancer therapy, and the first immunoconjugate ever to be used in this manner.

The Italian GIMEMA group assessed GO in elderly patients who in general poorly tolerate the toxicity associated with chemotherapy [76]. These frail patients, who had untreated AML and were aged between 61 and 89 years, were to receive two courses of GO as induction therapy, followed by two additional infusions if complete remission (CR) was achieved. For CD33-positive patients aged less than 75 years, a median survival of 11.4 months was quite remarkable, but those patients aged over 75 years or who were CD33-negative had a dismal outcome. Whilst almost no mucositis was observed, the problems associated with neutropenia were severe. In addition, hepatotoxicity was seen. These findings confirmed the observations made by Larson and co-workers [77] in relapsed elderly patients who had little mucositis, cardiotoxicity, cerebellar toxicity or alopecia. However, when administered to elderly patients with newly diagnosed AML or myelodysplasia, GO did not seem superior to standard idarubicin combined with cytosine-arabinoside [78].

Recently, several studies conducted in the US and in Europe also documented activity of GO in children [79–81]. Among children with relapsed or refractory AML, GO treatment led to only limited toxicity of organs other than bone marrow, but significant hepatotoxicity occurred. Subsequently, among 13 patients reported by Arceci et al. who proceeded to stem cell transplantation after GO therapy, 40% developed hepatic veno-occlusive disease/sinusoidal obstruction syndrome (VOD/SOS) [79].

Not surprisingly, when trying to integrate an immunoconjugate into chemotherapy, both the scheduling and dose were difficult to determine. In monotherapy, lower doses of GO such as $6\,mg\,m^{-2}$ had shown activity. Moreover, colleagues from the MD Anderson Cancer Center combined GO at $4.5\,mg\,m^{-2}$ with fludarabine, cytarabine, and CsA and showed feasibility [82]. Currently, several multicenter trials are under way to identify the correct chemotherapy partner and schedule for GO in AML therapy (for reviews, see [68,83,84]).

Clearly, GO alone will not achieve longlasting remissions in the majority of patients, and therefore allogeneic bone marrow transplantation is a logical choice for consolidation therapy. Many patients have received such treatment in early pivotal studies [9,75], with the median overall survival time exceeding 18.3 months, compared to 16.5 months for autologous stem cell transplantation, 12.2 months for chemotherapy, and 11.2 months for supportive care. As CD33-positive cells are found in the sinusoids of the liver [85] and VOD/SOS is seen in patients receiving GO outside the transplant setting [86], liver toxicity was of particular concern. Wadleigh and co-workers from the Dana Faber Cancer Institute, when retrospectively comparing patients with and without GO prior to human stem cell transplantation, found VOD rates of 64% and 8%, respectively [87]. In particular, a short interval between GO treatment and stem cell transplantation appeared to be a major risk factor. Versluys et al. [88] reported on a group of children who received prophylactic defibrotide ($40\,mg\,m^{-2}$ per day) which may have helped to prevent graft-versus-host disease (GvHD), an experience which we have shared in adults at our center. Taking this into consideration, it appears justified to conduct additional studies to include GO into conditioning regimens such as that proposed by Bornhäuser and colleagues [89], with split doses of GO administered 21 and 14 days before allogeneic transplantation. In fact, defibrotide treatment may have positive effects even on GO-induced VOD seen outside the transplant situation [90].

7.7 Summary and Conclusions

Gemtuzumab ozogamicin was the first immunoconjugate to be approved for cancer therapy. The carbohydrate toxin calicheamicin is efficiently delivered by specific monoclonal antibodies to CD33-positive leukemic cells and, after internalization, deliberates from the acid-sensitive linker in the lysosome. Clinical studies in patients with AML have demonstrated remarkable single-agent efficacy for GO, with few adverse side effects that were mainly caused by hepatic toxicity and the inevitable targeting of normal CD33-positive hematopoiesis. Due to the overall favorable side-effect profile, GO may be particularly helpful in elderly AML patients, or may be administered in conjunction with bone marrow transplantation. In the latter situation, however, it is vital that the occurrence of VOD/SOS is avoided.

To date, the clinical experiences with GO have stimulated the development of similarly designed immunoconjugates. For example, an additional calicheamicin immunoconjugate targeting CD22 has demonstrated potent and selective inhibition of CD22-positive B-cell lymphoma cell lines both *in vitro* and in xenografted tumor models [91]. This immunoconjugate is currently undergoing clinical trials for the treatment of lymphoma patients. Nonetheless, it remains to be seen how many immunoconjugates against other internalizing antigens can be successfully developed, and what the ultimate role of these therapeutics will be in the clinic.

References

1 Wu, A.M., Senter, P.D. (2005) Arming antibodies: prospects and challenges for immunoconjugates. Nat Biotechnol 23: 1137–1146.

2 Carter, P.J. (2006) Potent antibody therapeutics by design. Nat Rev Immunol 6: 343–357.

3 Glennie, M.J., van de Winkel, J.G. (2003) Renaissance of cancer therapeutic antibodies. Drug Discov Today 8: 503–510.

4 Rondot, S., Koch, J., Breitling, F., Dubel, S. (2001) A helper phage to improve single-chain antibody presentation in phage display. Nat Biotechnol 19: 75–78.

5 Lonberg, N. (2005) Human antibodies from transgenic animals. Nat Biotechnol 23: 1117–1125.

6 Hanes, J., Schaffitzel, C., Knappik, A., Pluckthun, A. (2000) Picomolar affinity antibodies from a fully synthetic naive library selected and evolved by ribosome display. Nat Biotechnol 18: 1287–1292.

7 Nimmerjahn, F., Ravetch, J.V. (2005) Divergent immunoglobulin g subclass activity through selective Fc receptor binding. Science 310: 1510–1512.

8 Weng, W.K., Levy, R. (2003) Two immunoglobulin G fragment C receptor polymorphisms independently predict response to rituximab in patients with follicular lymphoma. J Clin Oncol 21: 3940–3947.

9 Sievers, E.L., Larson, R.A., Stadtmauer, E.A., et al. (2001) Efficacy and safety of gemtuzumab ozogamicin in patients with CD33-positive acute myeloid leukemia in first relapse. J Clin Oncol 19: 3244–3254.

10 Waldmann, T.A. (2003) Immunotherapy: past, present and future. Nat Med 9: 269–277.

11 Jilani, I., Estey, E., Huh, Y., et al. (2002) Differences in CD33 intensity between various myeloid neoplasms. Am J Clin Pathol 118: 560–566.

12 Linenberger, M.L. (2005) CD33-directed therapy with gemtuzumab ozogamicin in acute myeloid leukemia: progress in understanding cytotoxicity and potential mechanisms of drug resistance. Leukemia 19: 176–182.

13 Feldman, E.J., Brandwein, J., Stone, R., et al. (2005) Phase III randomized multicenter study of a humanized anti-CD33 monoclonal antibody, lintuzumab, in combination with chemotherapy, versus chemotherapy alone in patients with refractory or first-relapsed acute myeloid leukemia. J Clin Oncol 23: 4110–4116.

14 Caron, P.C., Dumont, L., Scheinberg, D.A. (1998) Supersaturating infusional humanized anti-CD33 monoclonal antibody HuM195 in myelogenous leukemia. Clin Cancer Res 4: 1421–1428.

15 Crocker, P.R. (2002) Siglecs: sialic-acid-binding immunoglobulin-like lectins in cell-cell interactions and signalling. Curr Opin Struct Biol 12: 609–615.

16 Freeman, S.D., Kelm, S., Barber, E.K., Crocker, P.R. (1995) Characterization of CD33 as a new member of the sialoadhesin family of cellular interaction molecules. Blood 85: 2005–2012.

17 Griffin, J.D., Linch, D., Sabbath, K., Larcom, P., Schlossman, S.F. (1984) A monoclonal antibody reactive with normal and leukemic human myeloid progenitor cells. Leuk Res 8: 521–534.

18 Peiper, S.C., Ashmun, R.A., Look, A.T. (1988) Molecular cloning, expression, and chromosomal localization of a human gene encoding the CD33 myeloid differentiation antigen. Blood 72: 314–321.

19 Simmons, D., Seed, B. (1988) Isolation of a cDNA encoding CD33, a differentiation antigen of myeloid progenitor cells. J Immunol 141: 2797–2800.

20 Andrews, R.G., Singer, J.W., Bernstein, I.D. (1989) Precursors of colony-forming cells in humans can be distinguished from colony-forming cells by expression of the CD33 and CD34 antigens and light scatter properties. J Exp Med 169: 1721–1731.

21 Brashem-Stein, C., Flowers, D.A., Smith, F.O., Staats, S.J., Andrews, R.G., Bernstein, I.D. (1993) Ontogeny of hematopoietic stem cell development: reciprocal expression of CD33 and a novel molecule by maturing myeloid and erythroid progenitors. Blood 82: 792–799.

22 Andrews, R.G., Takahashi, M., Segal, G.M., Powell, J.S., Bernstein, I.D., Singer, J.W. (1986) The L4F3 antigen is expressed by unipotent and multipotent colony-forming cells but not by their precursors. Blood 68: 1030–1035.

23 Andrews, R.G., Torok-Storb, B., Bernstein, I.D. (1983) Myeloid-associated differentiation antigens on stem cells and their progeny identified by monoclonal antibodies. Blood 62: 124–132.

24 Gao, Z., McAlister, V.C., Williams, G.M. (2001) Repopulation of liver endothelium by bone-marrow-derived cells. Lancet 357: 932–933.

25 Tchilian, E.Z., Beverley, P.C., Young, B.D., Watt, S.M. (1994) Molecular cloning of two isoforms of the murine homolog of the myeloid CD33 antigen. Blood 83: 3188–3198.

26 Robertson, M.J., Soiffer, R.J., Freedman, A.S., et al. (1992) Human bone marrow depleted of CD33-positive cells mediates delayed but durable reconstitution of hematopoiesis: clinical trial of MY9 monoclonal antibody-purged autografts for the treatment of acute myeloid leukemia. Blood 79: 2229–2236.

27 Paul, S.P., Taylor, L.S., Stansbury, E.K., McVicar, D.W. (2000) Myeloid specific human CD33 is an inhibitory receptor with differential ITIM function in recruiting the phosphatases SHP-1 and SHP-2. Blood 96: 483–490.

28 Taylor, V.C., Buckley, C.D., Douglas, M., Cody, A.J., Simmons, D.L., Freeman, S.D. (1999) The myeloid-specific sialic acid-binding receptor, CD33, associates with the protein-tyrosine phosphatases, SHP-1 and SHP-2. J Biol Chem 274: 11505–11512.

29 Ulyanova, T., Blasioli, J., Woodford-Thomas, T.A., Thomas, M.L. (1999) The sialoadhesin CD33 is a myeloid-specific inhibitory receptor. Eur J Immunol 29: 3440–3449.

30 Walter, R.B., Raden, B.W., Kamikura, D.M., Cooper, J.A., Bernstein, I.D. (2005) Influence of CD33 expression levels and ITIM-dependent internalization on gemtuzumab ozogamicin-induced cytotoxicity. Blood 105: 1295–1302.

31 Hamann, P.R., Hinman, L.M., Hollander, I., et al. (2002) Gemtuzumab ozogamicin, a potent and selective anti-CD33 antibody-calicheamicin conjugate for treatment of acute myeloid leukemia. Bioconj Chem 13: 47–58.

32 Sievers, E.L., Appelbaum, F.R., Spielberger, R.T., et al. (1999) Selective ablation of acute myeloid leukemia using antibody-targeted chemotherapy: a phase I study of an anti-CD33 calicheamicin immunoconjugate. Blood 93: 3678–3684.

33 Lee, M.D., Dunne, T.S., Siegel, M.M., Chang, C.C., Morton, G.O., Borders, D.B. (1987) Calichemicins, a novel family of antitumor antibiotics. 1. Chemistry and partial structure of calichemicin. gamma.1I. J Am Chem Soc 109: 3464–3466.

34 Nicolaou, K.C., Tsay, S.C., Suzuki, T., Joyce, G.F. (1992) DNA-carbohydrate interactions. Specific binding of the calicheamicin.gamma.1I oligosaccharide with duplex DNA. J Am Chem Soc 114: 7555–7557.

35 Zein, N., Poncin, M., Nilakantan, R., Ellestad, G.A. (1989) Calicheamicin gamma 1I and DNA: molecular recognition process responsible for site-specificity. Science 244: 697–699.

36 Zein, N., Sinha, A.M., McGahren, W.J., Ellestad, G.A. (1988) Calicheamicin gamma 1I: an antitumor antibiotic that cleaves double-stranded DNA site specifically. Science 240: 1198–1201.

37 Aiyar, J., Danishefsky, S.J., Crothers, D.M. (1992) Interaction of the aryl tetrasaccharide domain of calicheamicin. gamma.1I with DNA: influence on aglycon and methidiumpropyl-EDTA. cntdot.iron(II)-mediated DNA cleavage. J Am Chem Soc 114: 7552–7554.

38 Walker, S., Gupta, V., Kahne, D., Gange, D. (1994) Analysis of hydroxylamine glycosidic linkages: Structural consequences of the NO bond in calicheamicin. J Am Chem Soc 116: 3197–3206.

39 Walker, S., Valentine, K.G., Kahne, D. (1990) Sugars as DNA binders: a comment on the calicheamicin oligosaccharide. J Am Chem Soc 112: 6428–6429.

40 Uesugi, M., Sugiura, Y. (1993) New insights into sequence recognition process of esperamicin A1 and calicheamicin gamma 1I.: origin of their selectivities and 'induced fit' mechanism. Biochemistry 32: 4622–4627.

41 Gomez-Paloma, L., Smith, J.A., Chazin, W.J., Nicolaou, K.C. (1994) Interaction of calicheamicin with duplex DNA: Role of the oligosaccharide domain and identification of multiple binding modes. J Am Chem Soc 116: 3697–3708.

42 Li, T., Zeng, Z., Estevez, V.A., Baldenius, K.U., Nicolaou, K.C., Joyce, G.F. (1994) Carbohydrate-minor groove interactions in the binding of calicheamicin. gamma.1I to duplex DNA. J Am Chem Soc 116: 3709–3715.

43 Ho, S.N., Boyer, S.H., Schreiber, S.L., Danishefsky, S.J., Crabtree, G.R. (1994) Specific inhibition of formation of transcription complexes by a calicheamicin oligosaccharide: a paradigm for the development of transcriptional antagonists. Proc Natl Acad Sci USA 91: 9203–9207.

44 Liu, C., Smith, B.M., Ajito, K., et al. (1996) Sequence-selective carbohydrate-DNA interaction: dimeric and monomeric forms of the calicheamicin oligosaccharide interfere with transcription factor function. Proc Natl Acad Sci USA 93: 940–944.

45 Hamann, P.R., Hinman, L.M., Beyer, C.F., et al. (2002) An anti-CD33 antibody-calicheamicin conjugate for treatment of acute myeloid leukemia. Choice of linker. Bioconj Chem 13: 40–46.

46 Rodwell, J.D., Alvarez, V.L., Lee, C., et al. (1986) Site-specific covalent modification of monoclonal antibodies: in vitro and in vivo evaluations. Proc Natl Acad Sci USA 83: 2632–2636.

47 Bross, P.F., Beitz, J., Chen, G., et al. (2001) Approval summary: gemtuzumab ozogamicin in relapsed acute myeloid leukemia. Clin Cancer Res 7: 1490–1496.

48 McGrath, M.S., Rosenblum, M.G., Philips, M.R., Scheinberg, D.A. (2003) Immunotoxin resistance in multidrug resistant cells. Cancer Res 63: 72–79.

49 Press, O.W., Shan, D., Howell-Clark, J., et al. (1996) Comparative metabolism and retention of iodine-125, yttrium-90, and indium-111 radioimmunoconjugates by cancer cells. Cancer Res 56: 2123–2129.

50 Elmroth, K., Nygren, J., Martensson, S., Ismail, I.H., Hammarsten, O. (2003) Cleavage of cellular DNA by calicheamicin gamma1. DNA Repair (Amst.) 2: 363–374.

51 Krishnamurthy, G., Brenowitz, M.D., Ellestad, G.A. (1995) Salt dependence of calicheamicin-DNA site-specific interactions. Biochemistry 34: 1001–1010.

52 Walker, S., Landovitz, R., Ding, W.D., Ellestad, G.A., Kahne, D. (1992) Cleavage behavior of calicheamicin gamma 1 and calicheamicin T. Proc Natl Acad Sci USA 89: 4608–4612.

53 Amico, D., Barbui, A.M., Erba, E., Rambaldi, A., Introna, M., Golay, J. (2003) Differential response of human acute myeloid leukemia cells to gemtuzumab ozogamicin in vitro: role of Chk1 and Chk2 phosphorylation and caspase 3. Blood 101: 4589–4597.

54 Walter, R.B., Raden, B.W., Cronk, M.R., Bernstein, I.D., Appelbaum, F.R., Banker, D.E. (2004) The peripheral benzodiazepine receptor ligand PK11195 overcomes different resistance mechanisms to sensitize AML cells to gemtuzumab ozogamicin. Blood 103: 4276–4284.

55 Naito, K., Takeshita, A., Shigeno, K., et al. (2000) Calicheamicin-conjugated humanized anti-CD33 monoclonal antibody (gemtuzumab zogamicin, CMA-676) shows cytocidal effect on CD33-positive leukemia cell lines, but is inactive on P-glycoprotein-expressing sublines. Leukemia 14: 1436–1443.

56 Martensson, S., Nygren, J., Osheroff, N., Hammarsten, O. (2003) Activation of the DNA-dependent protein kinase by drug-induced and radiation-induced DNA strand breaks. Radiat Res 160: 291–301.

57 Zhao, B., Konno, S., Wu, J.M., Oronsky, A.L. (1990) Modulation of nicotinamide adenine dinucleotide and poly(adenosine diphosphoribose) metabolism by calicheamicin gamma 1 in human HL-60 cells. Cancer Lett 50: 141–147.

58 Linenberger, M.L., Hong, T., Flowers, D., et al. (2001) Multidrug-resistance phenotype and clinical responses to gemtuzumab ozogamicin. Blood 98: 988–994.

59 Matsui, H., Takeshita, A., Naito, K., et al. (2002) Reduced effect of gemtuzumab ozogamicin (CMA-676) on P-glycoprotein and/or CD34-positive leukemia cells and its restoration by multidrug resistance modifiers. Leukemia 16: 813–819.

60 Walter, R.B., Raden, B.W., Hong, T.C., Flowers, D.A., Bernstein, I.D., Linenberger, M.L. (2003) Multidrug resistance protein attenuates gemtuzumab ozogamicin-induced cytotoxicity in acute myeloid leukemia cells. Blood 102: 1466–1473.

61 Abbott, B.L. (2003) ABCG2 (BCRP) expression in normal and malignant hematopoietic cells. Hematol Oncol 21: 115–130.

62 Walter, R.B., Raden, B.W., Thompson, J., et al. (2004) Breast cancer resistance protein (BCRP/ABCG2) does not confer resistance to gemtuzumab ozogamicin and calicheamicin-gamma1 in acute myeloid leukemia cells. Leukemia 18: 1914–1917.

63 Tsimberidou, A., Cortes, J., Thomas, D., et al. (2003) Gemtuzumab ozogamicin, fludarabine, cytarabine and cyclosporine combination regimen in patients with CD33+ primary resistant or relapsed acute myeloid leukemia. Leuk Res 27: 893–897.

64 Tsimberidou, A., Estey, E., Cortes, J., et al. (2003) Gemtuzumab, fludarabine, cytarabine, and cyclosporine in patients with newly diagnosed acute myelogenous leukemia or high-risk myelodysplastic syndromes. Cancer 97: 1481–1487.

65 van Der Velden, V.H., te Marvelde, J.G., Hoogeveen, P.G., et al. (2001) Targeting of the CD33-calicheamicin immunoconjugate Mylotarg (CMA-676) in acute myeloid leukemia: in vivo and in vitro saturation and internalization by leukemic and normal myeloid cells. Blood 97: 3197–3204.

66 van der Velden, V.H., Boeckx, N., Jedema, I., et al. (2004) High CD33-antigen loads in peripheral blood limit the efficacy of gemtuzumab ozogamicin (Mylotarg) treatment in acute myeloid leukemia patients. Leukemia 18: 983–988.

67 Jedema, I., Barge, R.M., van der Velden, V.H., et al. (2004) Internalization and cell cycle-dependent killing of leukemic cells by Gemtuzumab Ozogamicin: rationale for efficacy in CD33-negative malignancies with endocytic capacity. Leukemia 18: 316–325.

68 Tsimberidou, A.M., Giles, F.J., Estey, E., O'Brien, S., Keating, M.J., Kantarjian, H. M. (2006) The role of gemtuzumab ozogamicin in acute leukaemia therapy. Br J Haematol 132: 398–409.

69 Leone, G., Rutella, S., Voso, M.T., Fianchi, L., Scardocci, A., Pagano, L. (2004) In vivo priming with granulocyte colony-stimulating factor possibly enhances the effect of gemtuzumab-ozogamicin in acute myeloid leukemia: results of a pilot study. Haematologica 89: 634–636.

70 Moore, J., Seiter, K., Kolitz, J., et al. (2006) A phase II study of Bcl-2 antisense (oblimersen sodium) combined with gemtuzumab ozogamicin in older patients with acute myeloid leukemia in first relapse. Leuk Res 30: 777–783.

71 Rutella, S., Bonanno, G., Procoli, A., et al. (2006) Granulocyte colony-stimulating factor enhances the in vitro cytotoxicity of gemtuzumab ozogamicin against acute myeloid leukemia cell lines and primary blast cells. Exp Hematol 34: 54–65.

72 Repp, R., Schaekel, U., Helm, G., et al. (2003) Immunophenotyping is an independent factor for risk stratification in AML. Cytometry B Clin Cytom 53: 11–19.

73 Jurcic, J.G., Larson, S.M., Sgouros, G., et al. (2002) Targeted alpha particle immunotherapy for myeloid leukemia. Blood 100: 1233–1239.

74 Chabner, B.A., Levine, A.S., Johnson, B.L., Young, R.C. (1978) Initial clinical trials of maytansine, an antitumor plant alkaloid. Cancer Treat Rep 62: 429–433.

75 Larson, R.A., Sievers, E.L., Stadtmauer, E.A., et al. (2005) Final report of the efficacy and safety of gemtuzumab ozogamicin (Mylotarg) in patients with CD33-positive acute myeloid leukemia in first recurrence. Cancer 104: 1442–1452.

76 Amadori, S., Suciu, S., Stasi, R., et al. (2005) Gemtuzumab ozogamicin (Mylotarg) as single-agent treatment for frail patients 61 years of age and older with acute myeloid leukemia: final results of AML-15B, a phase 2 study of the European Organisation for Research and Treatment of Cancer and Gruppo Italiano Malattie Ematologiche dell'Adulto Leukemia Groups. Leukemia 19: 1768–1773.

77 Larson, R.A., Boogaerts, M., Estey, E., et al. (2002) Antibody-targeted chemotherapy of older patients with acute myeloid leukemia in first relapse using Mylotarg (gemtuzumab ozogamicin). Leukemia 16: 1627–1636.

78 Estey, E.H., Thall, P.F., Giles, F.J., et al. (2002) Gemtuzumab ozogamicin with or without interleukin 11 in patients 65 years of age or older with untreated acute myeloid leukemia and high-risk myelodysplastic syndrome: comparison with idarubicin plus continuous-infusion, high-dose cytosine arabinoside. Blood 99: 4343–4349.

79 Arceci, R.J., Sande, J., Lange, B., et al. (2005) Safety and efficacy of gemtuzumab ozogamicin in pediatric patients with advanced CD33+ acute myeloid leukemia. Blood 106: 1183–1188.

80 Brethon, B., Auvrignon, A., Galambrun, C., et al. (2006) Efficacy and tolerability of gemtuzumab ozogamicin (anti-CD33 monoclonal antibody, CMA-676, Mylotarg) in children with relapsed/refractory myeloid leukemia. BMC Cancer 6: 172.

81 Reinhardt, D., Diekamp, S., Fleischhack, G., et al. (2004) Gemtuzumab ozogamicin (Mylotarg) in children with refractory or relapsed acute myeloid leukemia. Onkologie 27: 269–272.

82 Tsimberidou, A.M., Estey, E., Cortes, J.E., et al. (2003) Mylotarg, fludarabine, cytarabine (ara-C), and cyclosporine (MFAC) regimen as post-remission therapy in acute myelogenous leukemia. Cancer Chemother Pharmacol 52: 449–452.

83 Burnett, A.K., Mohite, U. (2006) Treatment of older patients with acute myeloid leukemia – new agents. Semin Hematol 43: 96–106.

84 Fenton, C., Perry, C.M. (2005) Gemtuzumab ozogamicin: a review of its use in acute myeloid leukaemia. Drugs 65: 2405–2427.

85 Rajvanshi, P., Shulman, H.M., Sievers, E.L., McDonald, G.B. (2002) Hepatic sinusoidal obstruction after gemtuzumab ozogamicin (Mylotarg) therapy. Blood 99: 2310–2314.

86 Giles, F.J., Kantarjian, H.M., Kornblau, S.M., et al. (2001) Mylotarg (gemtuzumab

ozogamicin) therapy is associated with hepatic venoocclusive disease in patients who have not
received stem cell transplantation. Cancer 92: 406–413.

87 Wadleigh, M., Richardson, P.G., Zahrieh, D., et al. (2003) Prior gemtuzumab ozogamicin exposure significantly increases the risk of veno-occlusive disease in patients who undergo myeloablative allogeneic stem cell transplantation. Blood 102: 1578–1582.

88 Versluys, B., Bhattacharaya, R., Steward, C., Cornish, J., Oakhill, A., Goulden, N. (2004) Prophylaxis with defibrotide prevents veno-occlusive disease in stem cell transplantation after gemtuzumab ozogamicin exposure. Blood 103: 1968.

89 Hanel, M., Thiede, C., Helwig, A., et al. (2003) Successful combination of anti-CD33 antibody (gemtuzumab ozogamicin) and minimal conditioning before second allografting in recurrent acute myeloid leukaemia. Br J Haematol 120: 1093–1094.

90 Lannoy, D., Decaudin, B., Grozieux de Laguerenne, A., et al. (2006) Gemtuzumab ozogamicin-induced sinusoidal obstructive syndrome treated with defibrotide: a case report. J Clin Pharmacol Ther 31: 389–392.

91 Di Joseph, J.F., Armellino, D.C., Boghaert, E.R., et al. (2004) Antibody-targeted chemotherapy with CMC-544: a CD22-targeted immunoconjugate of calicheamicin for the treatment of B-lymphoid malignancies. Blood 103: 1807–1814.

92 Ikemoto, N., Kumar, R.A., Ling, T.T., Ellestad, G.A., Danishefsky, S.J., Patel, D. J. (1995) Calicheamicin-DNA complexes: warhead alignment and saccharide recognition of the minor groove. Proc Natl Acad Sci USA 92: 10506–10510.

93 Kumar, R.A., Ikemoto, N., Patel, D.J. (1997) Solution structure of the calicheamicin gamma 1I-DNA complex. J Mol Biol 265: 187–201.

8
Infliximab (Remicade)

Maria Wiekowski and Christian Antoni

8.1
Antibody Characteristics

Infliximab was developed as a potential therapeutic agent for various inflammatory chronic diseases that are believed to be driven by the proinflammatory cytokine tumor necrosis factor alpha (TNF-α). Infliximab (cA2) is a human–murine chimeric monoclonal antibody that potently binds and neutralizes the soluble TNF-α homotrimer and its membrane-bound precursor. It does not neutralize TNF-β (lymphotoxin α), a related cytokine that utilizes the same receptors as TNF-α.

Infliximab was genetically engineered by fusing the variable region of the murine antibody A2 that binds with high selectivity and specificity human TNF-α to the constant region of human IgG1 κ immunoglobulin using recombinant DNA technology. It had been expected that the chimeric antibody had a better immunogenic and pharmacokinetic profile than the murine antibody [1].

The chimeric product cA2 binds soluble TNF-α in its monomeric [2] and trimeric form with an affinity of K_d 100 pM for the latter [3]. The binding of TNF monomers might slow down or even prevent the formation of bioactive trimeric TNF. The binding affinity for the membrane-bound form is about twofold higher (K_d 46 pM), as determined from a cell line expressing membrane-bound recombinant human TNF-α [3]. Comparison of binding affinities of cA2 FAB fragments demonstrated 50-fold higher binding affinity of the dimeric $F(ab')_2$ fragment over the monomeric FAB fragment, indicating bivalent interaction with the ligand [3]. The stability of the infliximab–TNF-α complex is further demonstrated by its slow dissociation rate – in fact, in an *in-vitro* assay no dissociation was observed within 4 h from soluble or within 2 h from membrane-bound TNF-α [2].

Handbook of Therapeutic Antibodies. Edited by Stefan Dübel
Copyright © 2007 WILEY-VCH Verlag GmbH & Co. KGaA, Weinheim
ISBN 978-3-527-31453-9

8.2
Preclinical Characterization

In vitro, cA2 inhibited TNF-induced mitogenesis and interleukin (IL)-6 secretion by human fibroblasts, and blocked expression of adhesion molecules on endothelial cells [4].

The *in-vivo* activity of infliximab was demonstrated in a transgenic mouse that overexpresses human TNF-α and dies prematurely of wasting. In this model, twice-weekly injections of infliximab ($2\,mg\,kg^{-1}$) allowed 93% of the animals to survive by reversing the TNF-induced lethal wasting syndrome [4].

The effect of infliximab on apoptosis was also tested in an animal model. Severely compromised immunodeficient (SCID) mice, which lack T and B lymphocytes as well as NK cells, were reconstituted with a human monocytic cell line (THP-1) or a human T-cell line (Jurkat). Following intraperitoneal injection of infliximab, apoptotic cell death was shown to occur within 1h of injection [3,5]. Similarly, the induction of apoptosis has been reported for synovial macrophages from rheumatoid patients treated with infliximab [6]. Currently it is unknown which pathway of apoptosis infliximab engages.

8.3
Pharmacokinetics

Infliximab is administered to patients by intravenous (IV) infusion as an induction therapy (infusions at Weeks 0, 2, and 6), and then every 8 weeks thereafter. The pharmacokinetics of infliximab were determined in clinical trials for various indications; those from a phase II clinical trial in psoriasis are shown graphically in Fig. 8.1 [7]. The highest serum levels of infliximab were observed in patients

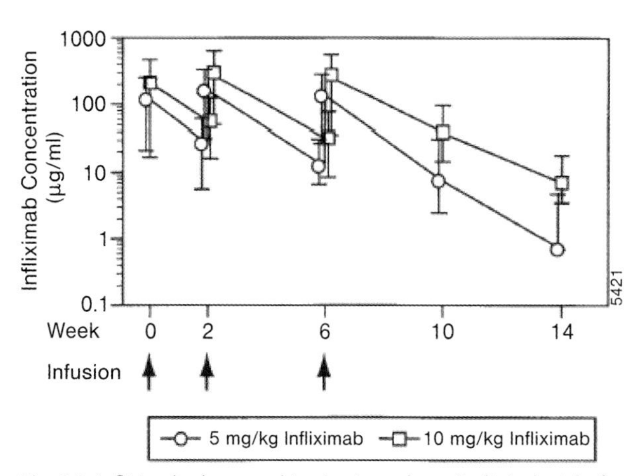

Fig. 8.1 Infliximab pharmacokinetics in a phase II clinical trial of psoriasis.

at Week 2, immediately after the second dose, with median concentrations of 158.14 and 298.89 mg mL^{-1} for the 5 and 10 mg kg^{-1} dose groups, respectively. The serum infliximab concentrations were directly proportional to the administered dose. The lowest serum concentrations (0.67 mg mL^{-1}) were observed at Week 14 for the 5 mg kg^{-1} group (range from <0.1 to 6.09 mg mL^{-1}), while the mean serum level for the 10 mg kg^{-1} group was 7.11 mg mL^{-1} at Week 14.

Infliximab is distributed primarily in the intravascular space; the median elimination half-life of infliximab was 7.62 days (interquartile range 6.62 to 10.15 days) for the 5 mg kg^{-1} dose group and 9.97 (interquartile range 6.17 to 10.14 days) for the 10 mg kg^{-1} dose group [7]. In patients with Crohn's disease, infliximab was found to have a median half-life of ~9.5 days [8].

The relationship between serum infliximab concentrations and clinical improvement was monitored in the ATTRACT trial for rheumatoid arthritis. Doses of 3 or 10 mg kg^{-1} infliximab resulted in maximal serum concentrations that were directly proportional to the intravenous dose, with serum concentrations of 68.8 and 219.1 μg mL^{-1} identified at 1 h post infusion [9]. The trough concentrations were also found to depend on the amount and frequency of dosing; in the ATTRACT trial, median trough levels were comparable in patients receiving 10 mg kg^{-1} every 8 weeks and those receiving 3 mg kg^{-1} every 4 weeks [9].

When the trough serum levels of infliximab at Week 54 were correlated with clinical response it became apparent that a higher ACR-N response was significantly associated with a higher trough serum concentration of infliximab [9]. In general, in rheumatoid arthritis (RA), a trough concentration of >1.0 μg mL^{-1} is associated with a good therapeutic response [9], and the clinical response declined rapidly when serum infliximab levels fell below this threshold [10].

8.4 Clinical Response

8.4.1 Therapeutic Indications

In Europe, infliximab is currently (2006) approved for the treatment of six chronic inflammatory conditions:

- For the reduction of signs and symptoms, as well as improvement in physical function, in patients with RA.
- Severe and active Crohn's disease in patients who do not respond to corticosteroid and/or immunosuppressant therapy, or those with fistulating, active Crohn's disease.
- In ulcerative colitis, in patients who do not respond to conventional therapy.
- In patients with ankylosing spondylitis who have severe axial symptoms, elevated serological markers of

inflammatory activity, and who have responded inadequately to conventional therapy.

- In patients with active and progressive psoriatic arthritis, alone or in combination with methotrexate (MTX); in patients who have responded inadequately to disease-modifying anti-rheumatic drugs.
- In patients with moderate to severe plaque psoriasis who are intolerant of other systemic therapies.

8.4.1.1 Crohn's Disease

Crohn's disease is a chronic inflammatory disorder of the gastrointestinal (GI) tract. Originally, the condition was treated with corticosteroids, until infliximab as a TNF-α blocker presented a significant advancement in the treatment of this disease.

In the ACCENT I trial, Crohn's disease patients experienced a clinical response following one infusion of infliximab, and were then randomized at Week 2 to receive either placebo or infliximab at Weeks 2 and 6, and at every 8 weeks thereafter. Evaluation at Week 30 demonstrated that 39% and 45% of patients achieved remission after infliximab doses of 5 and 10 $mg\,kg^{-1}$, respectively, while only 21% of those in the placebo group were in remission. Evaluation at Week 54 confirmed the sustained efficacy of infliximab. In addition, the time to loss of response was significantly longer among infliximab-treated patients than in the placebo group. In addition, the number of patients receiving infliximab who had been able to discontinue corticosteroid treatment at Week 54 was threefold higher than in the placebo group [11]. The benefit of disease control through infliximab therapy was associated with reduced hospitalizations and surgeries, increased employment, and a normalized quality of life [12].

The efficacy of infliximab in inducing response and remission in Crohn's disease triggered the question of whether the disease would be better controlled by continuous infliximab exposure (maintenance treatment; infusions every 8 weeks) or by treatment only upon relapse of the disease (episodic treatment). Consequently, a clinical study showed that maintenance treatment resulted in better clinical responses, higher rates of mucosal healing and lower hospitalization rates, and also fewer incidences of development of antibodies to infliximab [13].

An important advancement in the management of Crohn's disease was achieved with a clinical study that compared the efficacy of the widely used immunosuppressors azathioprine (AZA) or 6-mercaptorpurine (6-MP) to the efficacy of the immunosuppressors in combination with infliximab. In steroid-dependent Crohn's patients, almost twice as many receiving the combination therapy were in clinical remission without steroids at Weeks 12, 24, and 54 as compared to the AZA/6-MP monotherapy group [14].

In addition, infliximab has been shown to be effective in inducing and maintaining complete and durable closures of draining fistulas (ACCENT II) [15] which occur in 17 to 43% of patients with Crohn's disease. This effect also resulted in reduced hospitalizations and surgeries when compared with placebo [16].

Thus, while infliximab offered a new and highly efficacious treatment entity for Crohn's disease, there was concern originally that such treatment might result in a higher incidence of intestinal strictures or obstructions. In order to address these and other safety concerns, registries were implemented that followed Crohn's disease patients receiving infliximab or other therapies besides TNF-α blockers for several years.

Indeed, the analysis of data obtained from the observational TREAT (the Crohn's Therapy, Resource, Evaluation and Assessment Tool) registry and also ACCENT I, have indicated that symptoms such as intestinal strictures are associated with other factors of Crohn's disease, including disease duration, ileal disease, and new corticosteroid use [17].

8.4.1.2 **Rheumatoid Arthritis**

The effect of repeated doses of infliximab in rheumatoid arthritis was first assessed in a double-blind, placebo-controlled study conducted during the late 1990s and involving 101 patients with active RA, despite treatment with low doses of MTX. Infliximab was administered in the described induction scheme (infusions at Weeks 0, 2, 6 followed by 8-week infusions) up to Week 14 at doses of 1, 3, and $10\,mg\,kg^{-1}$ body weight, with or without combined MTX treatment. The clinical response was evaluated at Week 26. Independent of concomitant treatment with MTX, 60% of patients receiving either 3 or $10\,mg\,kg^{-1}$ responded to the treatment with a median duration of over 18 weeks, while the lowest dose of $1\,mg\,kg^{-1}$ induced a response only when given in combination with MTX.

Pharmocokinetic analysis of antibody concentrations in serum showed that the response to infliximab treatment correlated with sustained antibody levels in the circulation [18]. The rapid induction of response to infliximab in combination with MTX compared to MTX alone was demonstrated in a phase III trial with 428 patients who had active RA despite MTX treatment. A response to 3 or $10\,mg\,kg^{-1}$ infliximab treatment was detected as early as 2 weeks after the first infusion. Maintenance treatment with 8-week infusions maintained the response for over 30 weeks [19]. During this trial, two fatal infections of tuberculosis and coccidiomyosis occurred, which then led to an awareness of the risk that patients might encounter when treated with TNF-α blockers (see below). However, the benefits still greatly outweighed the risk for patients, as infliximab treatment not only greatly improved the quality of life but also preserved the structure of the joint. While at Week 54 joint damage had increased in the MTX group (as determined by radiography), no radiographic progression was observed in the infliximab group [20]. Infliximab was found to have a significant benefit when erosions and joint-space narrowing were examined, and when the hands and feet were examined separately [20]. The deceleration of joint damage was even reduced in patients who did not show a clinical response.

The benefit of infliximab treatment was sustained, when treatment was extended to 102 weeks. Patients experienced significant, clinically relevant improvements in physical function and quality of life, inhibition of progressive joint damage and sustained improvement in the signs and symptoms of RA [21,22]. A subanalysis of the patient cohort in this study highlighted the importance for early intervention to slow the progression of physical damage, as greater joint damage at baseline was associated with poorer physical function at baseline and also less improvement in physical function after treatment [21]. In addition, patients treated with infliximab in combination with MTX showed minimal disease progression on radiologic assessments, and improvements in physical function as measured by the health assessment questionnaire (HAQ) score. In contrast, those patients treated only with MTX experienced worsening of these measures [9].

Interestingly, in some patients, an inhibition of joint destruction was even observed on treatment with infliximab when no improvement of clinical variable was seen (ACR20 nonresponders). These results suggest a significant effect of infliximab on joint damage, implying that TNF-α inhibition in RA has a much higher impact on joint destruction and cartilage degradation than on inflammation [23]. In fact, TNF-α synergizes with the receptor activator of NF-κB ligand (RANKL) in osteoclastogenesis, which results in an increase in the numbers of joint and circulating osteoclast precursors in patients with RA. A similar situation occurs in psoriatic arthritis, which might explain the highly beneficial effect of TNF-α blockade on the bone erosion seen in these patients [24].

The infliximab-induced attenuation of joint destruction suggested that the initiation of infliximab treatment early in the disease might result in a greater benefit to the patient. Thus, in the ASPIRE trial patients were enrolled who had been diagnosed with RA less than 3 years prior to the study. In combination with MTX, infliximab (3 and 6 $mg\,kg^{-1}$ bodyweight) caused greater improvement in signs and symptoms at Week 54 than MTX treatment alone [25]. While high-dose MTX monotherapy showed efficacy in RA, infliximab in combination with MTX proved to be more efficacious, especially in early RA patients with the highest swollen joint counts and an acute-phase response [26]. Indeed, the radiologic progression of structural damage over 54 weeks was significantly attenuated in patients treated with infliximab in combination with MTX. This indicated that intervention with infliximab in RA early in the disease preserves the joint before destruction and the progression of physical disability [27].

While infliximab is recommended in combination with MTX for the treatment of RA, the long-term follow-up of RA patients captured by registries has demonstrated that, in the real-life setting, both monotherapy and combination therapy are used [28].

8.4.1.3 **Ankylosing Spondylitis**

Ankylosing spondylitis (AS) is a chronic, inflammatory rheumatic disease that most frequently involves the sacroiliac joints, enthuses, and spine. A hallmark of this disease is chronic inflammatory back pain. Progressive ankylosis of the spine results in restricted mobility, disability, and a decreased quality of life which is often accompanied by unemployment.

A randomized, placebo-controlled 3-month trial in patients with AS showed that infliximab is superior to placebo in improving disease activity, function, spine mobility, peripheral arthritis, enthesitis, and quality of life [29]. This response was maintained at Weeks 54 and 102 in about 50% of patients receiving infliximab treatment [30,31]. When patients who had received infliximab for 2 years were compared to randomly selected patients on standard therapy that did not include TNF-α blockade, it became apparent that infliximab treatment reduced the number of the inflammatory spinal lesions in AS patients by 50 to 70% [32], and also decelerated radiographic progression [33]. Patients with AS receiving infliximab for 3 years showed a durable clinical response without loss of efficacy [34]. The discontinuation of treatment resulted in relapse of disease in almost all patients (41/42), but retreatment was safe and restored the response [35]. These results were confirmed in the ASSERT trial, where at Week 24 some 61% of AS patients showed an ASAS (Assessment in Ankylosing Spondylitis) 20 response compared to only 19% of placebo-treated patients [36].

8.4.1.4 Psoriatic Arthritis

Psoriatic arthritis (PsA) is a chronic and inflammatory arthritis that occurs in association with skin psoriasis. T cells and proinflammatory cytokines have been identified as important components of the pathogenesis of both psoriasis and PsA.

In psoriatic skin lesions and in synovial membranes of patients with PsA, the concentrations of proinflammatory cytokines such as TNF-α or IL-1 are increased [37,38]. Likewise, expression of the nuclear factor κB (NF-κB), which is regulated by proinflammatory cytokines such as TNF-α, is up-regulated in psoriatic skin [39] and synovial membranes of patients with PsA [40], which suggests that, in PsA, TNF-α triggers the inflammation in skin as well as in the synovial membrane.

Thus, it was postulated that anti-TNF-α treatment would show efficacy in PsA, as it does in RA [40]. The success of anti-TNF-α treatment in RA and the scientific rationale of the role of proinflammatory cytokines in PsA led to open and double-blind trials to investigate the efficacy of infliximab treatment in PsA.

In a small study involving six PsA patients, infusions with infliximab at Weeks 0, 2, and 6 resulted in a drastic improvement in the psoriasis area and severity index (PASI) score at Week 10; moreover, arthritis symptoms as well as skin lesions were improved [41].

In an open-label follow-up study, ten PsA patients were treated with infliximab at 0, 2, and 6 weeks and subsequently at 8-week intervals up to Week 54. Infliximab treatment resulted in an improvement of all global and peripheral assessments of arthritis [42]. Furthermore, in patients who achieved an ACR50 response at Week 6 the response was sustained until Week 54, despite lowering the dose or discontinuing infliximab treatment due to remission in some patients. The sizes of the psoriatic plaques were also visibly reduced. Magnetic resonance imaging of the joints at Weeks 0 and 10 demonstrated a more than 80% mean reduction in inflammation compared to baseline [42].

These results were confirmed in a placebo-controlled, randomized, double-blind study for infliximab (IMPACT) in 102 PsA patients. These patients had failed at least one prior therapy with a disease-modifying anti-rheumatic drug (DMARD). Again, the ACR20 at Week 16 was 65% for infliximab-treated patients, and 10% for the placebo arm. While no placebo-treated patients achieved ACR50 or ACR70, 46% and 29% of infliximab-treated patients achieved ACR50 and ACR70, respectively. Improvements in the arthritis score occurred concomitantly with the improvements in skin lesions, and >68% of infliximab-treated patients achieved at least 75% improvement in the PASI score at Week 16 [43]. Continued therapy with infliximab resulted in a sustained improvement in articular and dermatologic manifestations of PsA through Week 50 [44].

In the IMPACT 2 study, the efficacy of infliximab in combination with MTX was confirmed in 200 PsA patients. Infliximab treatment caused a significant improvement in the ACR 20, 50, and 70 responses over placebo, and concomitant treatment with MTX did not appear to further improve the efficacy. In addition, two distinctive and common clinical manifestations of PsA – dactylitis and enthesopathy, which are not commonly included as outcomes in PsA clinical studies – were present in a substantial proportion of patients at baseline and improved significantly at Weeks 14 and 24 in those receiving infliximab [44].

An assessment of quality of life in these patients demonstrated a significant improvement of the HAQ score as early as Week 2, and this was maintained through Week 24. In addition, an improvement in the physical component summary score and the mental component summary score, as well as the raw score of the quality of life short form-36 (SF-36) assessment, and a greater improvement in both scores, correlated with a greater improvement in the ACR and PASI scores [45].

Analyses of serum samples obtained from PsA patients before and after infliximab treatment indicated that there were no changes in serum levels of TNF, but significant reductions occurred in IL-6, E-selectin, vascular endothelial growth factor (VEGF) and matrix metalloproteinase (MMP)-2 expression [46]. Infliximab treatment also significantly reduced the expression of angiogenic growth factors in synovial tissue in patients with PsA, in parallel with dramatic clinical skin and joint responses; this suggested that vascular regression is one potential mechanism of infliximab therapy [47].

8.4.1.5 **Psoriasis**

The improvement of skin lesions in PsA patients treated with infliximab triggered clinical studies to test the efficacy of infliximab in psoriasis. Psoriasis is a chronic, inflammatory skin disease of which plaque psoriasis is the most common form. Psoriatic skin lesions can cause physical discomfort and emotional debilitation, as this disease can significantly affect a patient's perception of general health and social functioning.

Proof-of-concept for the efficacy of infliximab in psoriasis was obtained in a phase II trial where patients received 5 or 10 $mg\,kg^{-1}$ infliximab or placebo at Weeks 0, 2, and 6. A significant improvement in the PASI score was observed as early as Week 2, and by Week 10 this was manifested in 93% and 95% improvements for the 5 and 10 $mg\,kg^{-1}$ groups, respectively, compared to 11% in the placebo group [7].

Histological analysis of skin biopsies at Week 10 showed a significant decrease in epidermal T cells and a significant decrease in epidermal thickness. Similarly, keratin K16 and ICAM expression by keratinocytes was decreased to normal levels at Week 10. This study demonstrated, for the first time, the important role of TNF-α in the pathogenesis of psoriasis [7].

These results were confirmed in a phase III trial involving 249 patients; subsequently, 72% and 88% of patients treated with 3 or 5 $mg\,kg^{-1}$ infliximab monotherapy, respectively, achieved 75% improvement of their baseline PASI score (PASI 75) at Week 10, compared with 6% of placebo patients. The response began to decline at 4 to 8 weeks after the last infusion, but could be regained in 38% and 64% of the patients after retreatment with 3 or 5 $mg\,kg^{-1}$ infliximab, respectively, compared to 18% of placebo-treated patients [48]. Patients responding to infliximab treatment also experienced a significant improvement in their quality of life [49].

An additional study showed that continued infliximab treatment resulted in long-term maintenance of the therapeutic response for skin as well as nail lesions. When treatment with infliximab was continued every 8 weeks following the induction therapy, 82% of infliximab-treated patients achieved PASI 75 at Week 24 (compared to 4% for placebo) and 58% a PASI 90 at Week 24 (compared to 1% placebo). At Week 50, the PASI 75 response was maintained for 61% of patients, and PASI 90 for 45% [50]. Similarly, the nail psoriasis severity index (NAPSI) score decreased by 56% in infliximab-treated patients at Week 54 and was maintained through Week 50. This demonstrated that infliximab had high efficacy in improving skin as well as nail lesions, and furthermore caused a rapid onset of response in psoriasis.

8.4.1.6 **Ulcerative Colitis**

Ulcerative colitis is an inflammatory disease characterized by mucosal ulceration of the colon, rectal bleeding, diarrhea, and abdominal pain. TNF-α has been found in increased concentrations in the blood, colonic tissue and stools of patients with ulcerative colitis. In the ACT1 and ACT2 trials, the efficacy of infliximab was evaluated in patients who had not responded to corticosteroid or AZA/6-MP. About twofold more patients in the infliximab group showed a clinical response, as defined by a decrease in the Mayo score of at least 3 points, and no rectal bleeding at Weeks 8, 30, and 54 than in the placebo group. Mucosal healing also occurred in about twofold more patients, while remission rates were almost threefold higher in the infliximab group than in the placebo group [51].

Infliximab is also a safe and effective rescue therapy in patients experiencing an acute severe or moderately severe attack of ulcerative colitis and do not respond to conventional corticosteroid treatment. Only 29% of these patients had colectomies following one infusion of infliximab compared to 67% who had received a sham infusion [52].

8.5 Safety

TNF-α is a central mediator in inflammation and immunity, and plays a crucial role in host defense. Thus, certain serious adverse events such as bacterial infections, tuberculosis and certain opportunistic infections, as well as demyelinating syndromes, have been observed with all TNF-α antagonists.

8.5.1 Serious Infections

In clinical studies, 36% of infliximab-treated patients experienced infections compared with 28% of placebo-treated patients; respiratory and urinary tract infections were the most commonly reported problems. No increased risk of serious infections was observed when infliximab was compared to placebo in the Crohn's disease studies. However, in the RA trials the incidence of serious infections, including pneumonia, was higher in patients treated with infliximab in combination with MTX than in those treated with MTX alone, especially at infliximab doses of $6\,mg\,kg^{-1}$ or greater. In the psoriasis studies, 1.5% of patients (average 41.9 weeks follow-up) receiving infliximab and 0.6% of patients (average 18.1 weeks follow-up) receiving placebo developed serious infections.

In order to determine the risk of developing serious infections, approximately 1000 patients with active RA were followed for 22 months. The risk of developing serious infections was similar in patients who received the approved infliximab dose of $3\,mg\,kg^{-1}$ in combination with MTX to those receiving MTX alone, although the risk was increased in patients receiving high doses of infliximab ($10\,mg\,kg^{-1}$) combined with MTX [53].

Opportunistic infections, such as tuberculosis (TB), atypical mycobacteria, pneumocystosis, histoplasmosis, coccidioidomycosis, cryptococcosis, aspergillosis, listeriosis and candidiasis, have also been reported to occur in patients treated with other TNF-α antagonists.

8.5.1.1 Tuberculosis

The first case of TB in an infliximab-treated patient occurred during the ATTRACT trial in RA, and since then the risk of developing TB infection or reactivation of latent TB has been recognized as a class-effect risk for TNF-α-blocking agents. Even in a country such as Sweden, where the incidence of TB is very low (5 cases in 100000 population), the risk of TB in RA patients is about twofold higher than in the normal population, and RA patients treated with TNF antagonists have an approximately fourfold higher risk than RA patients not treated with TNF antagonists for developing TB [54].

Thus, before, during, and after infliximab treatment it is recommended that patients are closely monitored for signs and symptoms of active TB and other serious infections. Before starting infliximab treatment the patient should be evaluated for a personal history of TB and appropriately screened using a skin test and chest X-radiography (SPC). In the case of apparent or suspected TB, patients should be treated with isoniazid (INH) before infliximab treatment can be initiated. An analysis of the Spanish registry BIOBADASER (Society of Rheumatology Database on Biologic Products) on the incidence ratio of TB indicated that the rate of TB cases in patients taking TNF-α antagonists fell by 78% after TB screening had been recommended [55]. In addition, all severe adverse events are collected in a central database by the market authorization holder, and safety updates are reported periodically to the health authorities. In addition, the TB education of rheumatologists, dermatologists and gastroenterologists is targeted to raise awareness of the risk of TB infections in patients receiving TNF-α blockers.

While serious infection is considered a risk for patients using TNF-α blockers, the long-term follow-up of Crohn's disease patients (over 5000 patient-years) has shown no correlation between the incidences of serious infections and infliximab treatment. Rather, prednisone use, narcotic analgesic use and disease activity correlated with the occurrence of infections [17].

8.5.2 Antibody Formation against Infliximab

The detection or interpretation of the analyses of antibodies to infliximab can be hindered by the presence of infliximab in the serum. Taking into account that infliximab remains in the circulation for at least 4 to 12 weeks after infusion, serum samples are usually collected for analysis at 12 weeks or later following the last infusion. When the formation of antibodies against infliximab (human antichimeric antibodies; HACA) was determined at 12 weeks after the last infusion in RA patients, HACA were detected in about 17% of all treated patients. The rate of HACA responses was inversely proportional to the dosage; thus, HACA formation occurred in 53%, 21% and 7% of patients treated with 1, 3 and 10 mg kg^{-1} infliximab, respectively. Concomitant treatment with MTX reduced the rates to 15%, 7% and 3% for the three respective dosages. These data suggest that anti-TNF-α treatment induces a phenomenon resembling tolerance [18].

A significantly higher incidence of antibodies to infliximab was detected in Crohn's disease patients who had received episodic treatment – that is, one dose of infliximab at the start of the trial, but then only placebo until Week 14 or later until the disease worsened. Whilst 30% of patients with this treatment schedule developed HACA, only 8% who received infliximab infusions every 8 weeks did so. The incidence of antibodies to infliximab was higher in patients who did not receive immunomodulators compared to those who did (18% versus 10%, $p = 0.02$) [56]. The rate of infusion reactions correlated positively with the development of HACA, but the antibody results proved to be poorly predictive of these events [56,57]. The development of antibodies to infliximab is also associated with a reduced duration of response to treatment [57]. Taken together, the results of these studies suggest that scheduled treatment in combination with an immunosuppressive drug carries the least risk of antibody formation against infliximab, and offers the best probability to maintain the response.

8.5.3 Infusion Reactions/Delayed Hypersensitivity Reactions

The highest frequency of adverse events in response to infliximab treatment are infusion reactions occurring within 2 h of the infusion. Typically, these consist of fever, chills, nausea, dyspnea, and headaches; the symptoms can be controlled with drug treatment (e.g., antihistamines). Infusion reactions led to the discontinuation of treatment in approximately 3% of patients, and were considered serious in <1% [15,20,25,48]. Delayed reactions such as myalgias, arthralgias, fever, rash, pruritis, facial, hand or lip edema, dysphagia, urticaria, sore throat and headache may occur within 3 to 12 days following infliximab infusion [20].

8.5.4 Auto-Antibody Formation

The formation of auto-antibody has been noted in patients treated with infliximab for refractory spondyloarthropathy [58], RA [59], and Crohn's disease [60]. In these studies, 50 to 60% of patients developed antinuclear antibodies (ANA), and 17 to 30% developed antibodies to double-stranded (ds)-DNA. These auto-antibodies are generally not associated with clinical autoimmunity. Infliximab does not affect the formation of other non-organ-specific antibodies [60], although antibodies to antiphospholipid/anticardiolipin have been detected in RA patients upon infliximab treatment [61]. Co-medication with MTX or systemic maintenance therapy with infliximab appears to reduce the risk of auto-antibody induction or systemic maintenance therapy [58,60]. The immunopathological mechanism of auto-antibody induction after infliximab treatment is currently unknown.

8.5.5 Neurological Disorders/Demyelinating Disease

Rare cases of optic neuritis, seizures and new-onset or exacerbation of demyelinating disorders, including multiple sclerosis, have been reported after treatment with TNF-α antagonists. The discontinuation of anti-TNF-therapy resulted in complete or partial improvement of symptoms [62]. While the association of TNF-α blockade with demyelination remains unclear, the treating physician should carefully evaluate risks and benefits before recommending infliximab treatment in patients with demyelination disorders.

8.5.6 Malignancies/Lymphoma

No apparent increase in the risk of developing malignancies has been shown for RA patients receiving anti-TNF-α therapy [63]. Current data are insufficient to establish a causal relationship between RA treatment and the development of lymphoma; indeed, an analysis of the National Data Bank for Rheumatic Diseases identified only 29 cases of lymphoma in 18572 patients. The standardized incidence ratio (SIR) for RA patients for lymphoma was 1.9 (955 confidence interval [95% CI] 1.3–2.7), while the SIR for infliximab (with or without etanercept in patients who had received multiple TNF-α blockers) was 2.6 (95% CI 1.4–4.5). However, it remains to be seen whether the increased SIR with anti-TNF treatment is reflective of channeling bias, as patients with the highest risk of lymphoma preferentially receive anti-TNF therapy [64]. Similarly, five cases of lymphoma were identified in 1603 person-years at risk in anti-TNF-treated RA patients compared to two lymphoma cases in 3948 person-years treated with conventional anti-rheumatic therapies. This translated to a lymphoma risk of 11.5(95% CI 3.7–26.9) and 1.2 (95% CI 0.2–4.5). Thus, the apparent increase of lymphoma risk in anti-TNF-treated patients is based on only a few cases [63].

Long-term data from the TREAT registry that follows Crohn's disease patients for up to 5 years while they are being treated with standard therapy or infliximab, do not show any increased risk of malignancies. However, Crohn's disease patients treated with the immunosuppressive drugs AZA or 6-MP have a fourfold higher risk of developing lymphomas compared to the general population [65].

8.5.7 Congestive Heart Failure

In a randomized, double-blind trial in patients with moderate-to-severe heart failure (New York Heart Association class III to IV) the risk of death from any cause or hospitalization for heart failure was significantly higher in infliximab-treated patients [66]. However, post-marketing analysis of approximately 1000 Swedish RA patients of whom about 50% were treated with TNF-α blockers, suggests that treatment with TNF-α antagonists actually lowers the risk of developing cardiovascular disease [67]. Patients starting on TNF-α therapy have most likely more severe RA and might be predisposed to developing ischemic heart disease. However, the risk of developing cardiovascular disease in these patients appears to be reduced by aggressive antirheumatic therapy that includes TNF-α blockers.

8.5.8 Other Adverse Events

8.5.8.1 Hepatic Events

Hepatic events, most often seen as mild to moderate increases in alanine aminotransferase (ALT) and aspartate aminotransferase (AST), have occurred in patients treated with infliximab, but a relationship with infliximab has not been established.

8.5.8.2 Pregnancy Outcome

An analysis of the worldwide infliximab safety database (Centocor) on data collected between 1998 and 2001 has identified 96 women who were directly exposed to infliximab during pregnancy, and for whom outcome data could be collected. Live births occurred in 67% of cases, miscarriages in 15% and therapeutic termination in 19%. This outcome does not differ from that reported for the general US population [68]. While no increased risk is apparent from these data, the follow-up of larger numbers of pregnancies occurring during infliximab treatment will need to be analyzed.

8.6 Summary

Infliximab therapy has demonstrated efficacy in a wide variety of chronic inflammatory diseases, thus highlighting the central role of TNF-α in inflammation. The preservation of joint structure in RA and PsA, as well as the rapid clearance of skin lesions in psoriasis and PsA, has led to an important and impressive improvement in the quality of life in these patients. Similarly, in Crohn's disease and ulcerative colitis infliximab treatment has resulted in a reduced number of surgeries and hospitalization of patients, thus contributing to the improved quality of life in these cases. While infliximab treatment thus provides definite benefits to patients, its mechanism of action as a TNF-α blocker carries some degree of risk, which is especially apparent in the increased susceptibility of patients receiving TNF-α antagonists to serious infections. However, an improved awareness of this risk and the employment of preventive measures can in fact reduce the number of patients encountering infections. Other major concerns associated with TNF-α blockers, such as increased malignancies and lymphoma rates, are currently being addressed by the long-term follow-up of patients, although at present TNF-α blockade is considered to be a treatment paradigm for chronic inflammatory disease with a very positive benefit/risk profile.

References

1 Knight, D.M., Trinh, H., Le, J., et al. (1993) Construction and initial characterization of a mouse-human chimeric anti-TNF antibody. Mol Immunol 30: 1443–1453.

2 Scallon, B., Cai, A., Solowski, N., Rosenberg, A., Song, X.Y., Shealy, D., Wagner, C. (2002) Binding and functional comparisons of two types of tumor necrosis factor antagonists. J Pharmacol Exp Ther 301: 418–426.

3 Scallon, B.J., Moore, M.A., Trinh, H., Knight, D.M., Ghrayeb, J. (1995) Chimeric anti-TNF-alpha monoclonal antibody cA2 binds recombinant transmembrane TNF-alpha and activates immune effector functions. Cytokine 7: 251–259.

4 Siegel, S.A., Shealy, D.J., Nakada, M.T., Le, J., Woulfe, D.S., Probert, L., Kollias, G., Ghrayeb, J., Vilcek, J., Daddona, P.E. (1995) The mouse/human chimeric monoclonal antibody cA2 neutralizes TNF in vitro and protects transgenic mice from cachexia and TNF lethality in vivo. Cytokine 7: 15–25.

5 Shen, C., Maerten, P., Geboes, K., Van Assche, G., Rutgeerts, P., Ceuppens, J.L. (2005) Infliximab induces apoptosis of monocytes and T lymphocytes in a human-mouse chimeric model. Clin Immunol 115: 250–259.

6 Catrina, A.I., Trollmo, C., af Klint, E., Engstrom, M., Lampa, J., Hermansson, Y., Klareskog, L., Ulfgren, A.K. (2005) Evidence that anti-tumor necrosis factor therapy with both etanercept and infliximab induces apoptosis in macrophages, but not lymphocytes, in rheumatoid arthritis joints: extended report. Arthritis Rheum 52: 61–72.

7 Gottlieb, A.B., Masud, S., Ramamurthi, R., Abdulghani, A., Romano, P., Chaudhari, U., Dooley, L.T., Fasanmade, A.A., Wagner, C.L. (2003) Pharmacodynamic and pharmacokinetic response to anti-tumor necrosis factor-alpha monoclonal antibody (infliximab) treatment of moderate to severe psoriasis vulgaris. J Am Acad Dermatol 48: 68–75.

8 Cornillie, F., Shealy, D., D'Haens, G., Geboes, K., Van Assche, G., Ceuppens, J., Wagner, C., Schaible, T., Plevy, S.E., Targan, S.R., Rutgeerts, P. (2001) Infliximab induces potent anti-inflammatory and local immunomodulatory activity but no systemic immune suppression in patients with Crohn's disease. Aliment Pharmacol Ther 15: 463–473.

9 St Clair, E.W., Wagner, C.L., Fasanmade, A.A., Wang, B., Schaible, T., Kavanaugh, A., Keystone, E.C. (2002) The relationship of serum infliximab concentrations to clinical improvement in rheumatoid arthritis: results from ATTRACT, a multicenter, randomized, double-blind, placebo-controlled trial. Arthritis Rheum 46: 1451–1459.

10 Markham, A., Lamb, H.M. (2000) Infliximab: a review of its use in the management of rheumatoid arthritis. Drugs 59: 1341–1359.

11 Hanauer, S.B., Feagan, B.G., Lichtenstein, G.R., Mayer, L.F., Schreiber, S., Colombel, J.F., Rachmilewitz, D., Wolf, D.C., Olson, A., Bao, W., Rutgeerts, P. (2002) Maintenance infliximab for Crohn's disease: the ACCENT I randomised trial. Lancet 359: 1541–1549.

12 Lichtenstein, G.R., Yan, S., Bala, M., Hanauer, S. (2004) Remission in patients with Crohn's disease is associated with improvement in employment and quality of life and a decrease in hospitalizations and surgeries. Am J Gastroenterol 99: 91–96.

13 Rutgeerts, P., Feagan, B.G., Lichtenstein, G.R., Mayer, L.F., Schreiber, S., Colombel, J.F., Rachmilewitz, D., Wolf, D.C., Olson, A., Bao, W., Hanauer, S.B. (2004) Comparison of scheduled and episodic treatment strategies of infliximab in Crohn's disease. Gastroenterology 126: 402–413.

14 Lemann, M., Mary, J.Y., Duclos, B., Veyrac, M., Dupas, J.L., Delchier, J.C., Laharie, D., Moreau, J., Cadiot, G., Picon, L., Bourreille, A., Sobahni, I., Colombel, J.F. (2006) Infliximab plus azathioprine

for steroid-dependent Crohn's disease patients: a randomized placebo-controlled trial. Gastroenterology 130: 1054–1061.

15 Sands, B.E., Anderson, F.H., Bernstein, C.N., et al. (2004) Infliximab maintenance therapy for fistulizing Crohn's disease. N Engl J Med 350: 876–885.

16 Lichtenstein, G.R., Yan, S., Bala, M., Blank, M., Sands, B.E. (2005) Infliximab maintenance treatment reduces hospitalizations, surgeries, and procedures in fistulizing Crohn's disease. Gastroenterology 128: 862–869.

17 Lichtenstein, G.R., Feagan, B.G., Cohen, R.D., Salzberg, B.A., Diamond, R.H., Chen, D.M., Pritchard, M.L., Sandborn, W.J. (2006) Serious infections and mortality in association with therapies for Crohn's disease: TREAT registry. Clin Gastroenterol Hepatol 4: 621–630.

18 Maini, R.N., Breedveld, F.C., Kalden, J.R., Smolen, J.S., Davis, D., Macfarlane, J.D., Antoni, C., Leeb, B., Elliott, M.J., Woody, J.N., Schaible, T.F., Feldmann, M. (1998) Therapeutic efficacy of multiple intravenous infusions of anti-tumor necrosis factor alpha monoclonal antibody combined with low-dose weekly methotrexate in rheumatoid arthritis. Arthritis Rheum 41: 1552–1563.

19 Maini, R., St Clair, E.W., Breedveld, F., Furst, D., Kalden, J., Weisman, M., Smolen, J., Emery, P., Harriman, G., Feldmann, M., Lipsky, P. (1999) Infliximab (chimeric anti-tumour necrosis factor alpha monoclonal antibody) versus placebo in rheumatoid arthritis patients receiving concomitant methotrexate: a randomised phase III trial. ATTRACT Study Group. Lancet 354: 1932–1939.

20 Lipsky, P.E., van der Heijde, D.M., St Clair, E.W., Furst, D.E., Breedveld, F.C., Kalden, J.R., Smolen, J.S., Weisman, M., Emery, P., Feldmann, M., Harriman, G. R., Maini, R.N. (2000) Infliximab and methotrexate in the treatment of rheumatoid arthritis. Anti-Tumor Necrosis Factor Trial in Rheumatoid Arthritis with Concomitant Therapy Study Group. N Engl J Med 343: 1594–1602.

21 Breedveld, F.C., Emery, P., Keystone, E., Patel, K., Furst, D.E., Kalden, J.R., St Clair, E.W., Weisman, M., Smolen, J., Lipsky, P.E., Maini, R.N. (2004) Infliximab in active early rheumatoid arthritis. Ann Rheum Dis 63: 149–155.

22 Maini, R.N., Breedveld, F.C., Kalden, J.R., Smolen, J.S., Furst, D., Weisman, M.H., St Clair, E.W., Keenan, G.F., van der Heijde, D., Marsters, P.A., Lipsky, P.E. (2004) Sustained improvement over two years in physical function, structural damage, and signs and symptoms among patients with rheumatoid arthritis treated with infliximab and methotrexate. Arthritis Rheum 50: 1051–1065.

23 Smolen, J.S., Han, C., Bala, M., Maini, R.N., Kalden, J.R., van der Heijde, D., Breedveld, F.C., Furst, D.E., Lipsky, P.E. (2005) Evidence of radiographic benefit of treatment with infliximab plus methotrexate in rheumatoid arthritis patients who had no clinical improvement: a detailed subanalysis of data from the anti-tumor necrosis factor trial in rheumatoid arthritis with concomitant therapy study. Arthritis Rheum 52: 1020–1030.

24 Ritchlin, C.T., Schwarz, E.M., O'Keefe, R.J., Looney, R.J. (2004) RANK, RANKL and OPG in inflammatory arthritis and periprosthetic osteolysis. J Musculoskelet Neuronal Interact 4: 276–284.

25 St Clair, E.W., van der Heijde, D.M., Smolen, J.S., Maini, R.N., Bathon, J.M., Emery, P., Keystone, E., Schiff, M., Kalden, J.R., Wang, B., Dewoody, K., Weiss, R., Baker, D. (2004) Combination of infliximab and methotrexate therapy for early rheumatoid arthritis: a randomized, controlled trial. Arthritis Rheum 50: 3432–3443.

26 Smolen, J.S., Van Der Heijde, D.M., St Clair, E.W., Emery, P., Bathon, J.M., Keystone, E., Maini, R.N., Kalden, J.R., Schiff, M., Baker, D., Han, C., Han, J., Bala, M. (2006) Predictors of joint damage in patients with early rheumatoid arthritis treated with high-dose methotrexate with or without concomitant infliximab: results from the ASPIRE trial. Arthritis Rheum 54: 702–710.

27 Taylor, P.C., Steuer, A., Gruber, J., Cosgrove, D.O., Blomley, M.J., Marsters, P.A., Wagner, C.L., McClinton, C., Maini, R.N. (2004) Comparison of ultrasonographic assessment of synovitis and joint vascularity with radiographic evaluation in a randomized, placebo-controlled study of infliximab therapy in early rheumatoid arthritis. Arthritis Rheum 50: 1107–1116.

28 Hyrich, K.L., Symmons, D.P., Watson, K.D., Silman, A.J. (2006) Comparison of the response to infliximab or etanercept monotherapy with the response to cotherapy with methotrexate or another disease-modifying antirheumatic drug in patients with rheumatoid arthritis: Results from the British Society for Rheumatology Biologics Register. Arthritis Rheum 54: 1786–1794.

29 Braun, J., Brandt, J., Listing, J., Zink, A., Alten, R., Golder, W., Gromnica-Ihle, E., Kellner, H., Krause, A., Schneider, M., Sorensen, H., Zeidler, H., Thriene, W., Sieper, J. (2002) Treatment of active ankylosing spondylitis with infliximab: a randomised controlled multicentre trial. Lancet 359: 1187–1193.

30 Braun, J., Brandt, J., Listing, J., Zink, A., Alten, R., Burmester, G., Golder, W., Gromnica-Ihle, E., Kellner, H., Schneider, M., Sorensen, H., Zeidler, H., Reddig, J., Sieper, J. (2003) Long-term efficacy and safety of infliximab in the treatment of ankylosing spondylitis: an open, observational, extension study of a three-month, randomized, placebo-controlled trial. Arthritis Rheum 48: 2224–2233.

31 Braun, J., Brandt, J., Listing, J., Zink, A., Alten, R., Burmester, G., Gromnica-Ihle, E., Kellner, H., Schneider, M., Sorensen, H., Zeidler, H., Sieper, J. (2005) Two year maintenance of efficacy and safety of infliximab in the treatment of ankylosing spondylitis. Ann Rheum Dis 64: 229–234.

32 Sieper, J., Baraliakos, X., Listing, J., Brandt, J., Haibel, H., Rudwaleit, M., Braun, J. (2005) Persistent reduction of spinal inflammation as assessed by magnetic resonance imaging in patients with ankylosing spondylitis after 2 yrs of treatment with the anti-tumour necrosis factor agent infliximab. Rheumatology (Oxford) 44: 1525–1530.

33 Baraliakos, X., Listing, J., Rudwaleit, M., Brandt, J., Sieper, J., Braun, J. (2005) Radiographic progression in patients with ankylosing spondylitis after 2 years of treatment with the tumour necrosis factor alpha antibody infliximab. Ann Rheum Dis 64: 1462–1466.

34 Braun, J., Baraliakos, X., Brandt, J., Listing, J., Zink, A., Alten, R., Burmester, G., Gromnica-Ihle, E., Kellner, H., Schneider, M., Sorensen, H., Zeidler, H., Sieper, J. (2005) Persistent clinical response to the anti-TNF-alpha antibody infliximab in patients with ankylosing spondylitis over 3 years. Rheumatology (Oxford) 44: 670–676.

35 Baraliakos, X., Listing, J., Brandt, J., Zink, A., Alten, R., Burmester, G., Gromnica-Ihle, E., Kellner, H., Schneider, M., Sorensen, H., Zeidler, H., Rudwaleit, M., Sieper, J., Braun, J. (2005) Clinical response to discontinuation of anti-TNF therapy in patients with ankylosing spondylitis after 3 years of continuous treatment with infliximab. Arthritis Res Ther 7: R439–R444.

36 van der Heijde, D., Dijkmans, B., Geusens, P., Sieper, J., DeWoody, K., Williamson, P., Braun, J. (2005) Efficacy and safety of infliximab in patients with ankylosing spondylitis: results of a randomized, placebo-controlled trial (ASSERT). Arthritis Rheum 52: 582–591.

37 van Kuijk, A.W., Reinders-Blankert, P., Smeets, T.J., Dijkmans, B.A., Tak, P.P. (2006) Detailed analysis of the cell infiltrate and the expression of mediators of synovial inflammation and joint destruction in the synovium of patients with psoriatic arthritis: implications for therapy. Ann Rheum Dis. (in press).

38 Olaniran, A.K., et al. (1996) Cytokine expression in psoriatic skin lesions during PUVA therapy. Arch Dermatol Res 288(8): 421–425.

39 Lizzul, P.F., Aphale, A., Malaviya, R., Sun, Y., Masud, S., Dombrovskiy, V., Gottlieb, A.B. (2005) Differential expression of phosphorylated NF-kappaB/RelA in normal and psoriatic epidermis and downregulation of NF-kappaB in response to treatment with etanercept. J Invest Dermatol 124: 1275–1283.

40 Danning, C.L., Illei, G.G., Hitchon, C., Greer, M.R., Boumpas, D.T., McInnes, I.B. (2000) Macrophage-derived cytokine and nuclear factor kappaB p65 expression in synovial membrane and skin of patients with psoriatic arthritis. Arthritis Rheum 43(6): 1244–1256.

41 Ogilvie, A.L., Antoni, C., Dechant, C., Manger, B., Kalden, J.R., Schuler, G., Luftl, M. (2001) Treatment of psoriatic arthritis with antitumour necrosis factor-alpha antibody clears skin lesions of psoriasis resistant to treatment with methotrexate. Br J Dermatol 144: 587–589.

42 Antoni, C., Dechant, C., Hanns-Martin Lorenz, P.D., Wendler, J., Ogilvie, A., Lueftl, M., Kalden-Nemeth, D., Kalden, J.R., Manger, B. (2002) Open-label study of infliximab treatment for psoriatic arthritis: clinical and magnetic resonance imaging measurements of reduction of inflammation. Arthritis Rheum 47: 506–512.

43 Antoni, C.E., Kavanaugh, A., Kirkham, B., et al. (2005) Sustained benefits of infliximab therapy for dermatologic and articular manifestations of psoriatic arthritis: results from the infliximab multinational psoriatic arthritis controlled trial (IMPACT). Arthritis Rheum 52: 1227–1236.

44 Antoni, C., Krueger, G.G., de Vlam, K., Birbara, C., Beutler, A., Guzzo, C., Zhou, B., Dooley, L.T., Kavanaugh, A. (2005) Infliximab improves signs and symptoms of psoriatic arthritis: results of the IMPACT 2 trial. Ann Rheum Dis 64: 1150–1157.

45 Kavanaugh, A., Antoni, C.E., Gladman, D., Wassenberg, S., Zhou, B., Beutler, A., Keenan, G., Burmester, G., Furst, D.E., Weisman, M.H., Kalden, J.R., Smolen, J., van der Heijde D. (2006) The Infliximab Multinational Psoriatic Arthritis Controlled Trial (IMPACT): results of radiographic analyses after 1 year. Ann Rheum Dis 65: 1038–1043.

46 Mastroianni, A., et al. (2005) Cytokine profiles during infliximab monotherapy in psoriatic arthritis. Br J Dermatol 153(3): 531–536.

47 Canete, J.D., Pablos, J.L., Sanmarti, R., Mallofre, C., Marsal, S., Maymo, J., Gratacos, J., Mezquita, J., Mezquita, C., Cid, M.C. (2004) Antiangiogenic effects of anti-tumor necrosis factor alpha therapy with infliximab in psoriatic arthritis. Arthritis Rheum 50: 1636–1641.

48 Gottlieb, A.B., Evans, R., Li, S., Dooley, L.T., Guzzo, C.A., Baker, D., Bala, M., Marano, C.W., Menter, A. (2004) Infliximab induction therapy for patients with severe plaque-type psoriasis: a randomized, double-blind, placebo-controlled trial. J Am Acad Dermatol 51: 534–542.

49 Feldman, S.R., Gordon, K.B., Bala, M., Evans, R., Li, S., Dooley, L.T., Guzzo, C., Patel, K., Menter, A., Gottlieb, A.B. (2005) Infliximab treatment results in significant improvement in the quality of life of patients with severe psoriasis: a double-blind placebo-controlled trial. Br J Dermatol 152: 954–960.

50 Reich, K., Nestle, F.O., Papp, K., Ortonne, J.P., Evans, R., Guzzo, C., Li, S., Dooley, L.T., Griffiths, C.E. (2005) Infliximab induction and maintenance therapy for moderate-to-severe psoriasis: a phase III, multicentre, double-blind trial. Lancet 366: 1367–1374.

51 Rutgeerts, P., Sandborn, W.J., Feagan, B.G., Reinisch, W., Olson, A., Johanns, J., Travers, S., Rachmilewitz, D., Hanauer, S.B., Lichtenstein, G.R., de Villiers, W.J., Present, D., Sands, B.E., Colombel, J.F. (2005) Infliximab for induction and maintenance therapy for ulcerative colitis. N Engl J Med 353: 2462–2476.

52 Jarnerot, G., Hertervig, E., Friis-Liby, I., Blomquist, L., Karlen, P., Granno, C., Vilien, M., Strom, M., Danielsson, A., Verbaan, H., Hellstrom, P.M., Magnuson, A., Curman, B. (2005) Infliximab as rescue therapy in severe to moderately severe ulcerative colitis: a randomized, placebo-controlled study. Gastroenterology 128: 1805–1811.

53 Westhovens, R., Yocum, D., Han, J., Berman, A., Strusberg, I., Geusens, P., Rahman, M.U. (2006) The safety of infliximab, combined with background treatments, among patients with

rheumatoid arthritis and various comorbidities: a large, randomized, placebo-controlled trial. Arthritis Rheum 54: 1075–1086.

54 Askling, J., Fored, C.M., Brandt, L., Baecklund, E., Bertilsson, L., Coster, L., Geborek, P., Jacobsson, L.T., Lindblad, S., Lysholm, J., Rantapaa-Dahlqvist, S., Saxne, T., Romanus, V., Klareskog, L., Feltelius, N. (2005) Risk and case characteristics of tuberculosis in rheumatoid arthritis associated with tumor necrosis factor antagonists in Sweden. Arthritis Rheum 52: 1986–1992.

55 Carmona, L., Gomez-Reino, J.J., Rodriguez-Valverde, V., Montero, D., Pascual-Gomez, E., Mola, E.M., Carreno, L., Figueroa, M. (2005) Effectiveness of recommendations to prevent reactivation of latent tuberculosis infection in patients treated with tumor necrosis factor antagonists. Arthritis Rheum 52: 1766–1772.

56 Hanauer, S.B., Wagner, C.L., Bala, M., Mayer, L., Travers, S., Diamond, R.H., Olson, A., Bao, W., Rutgeerts, P. (2004) Incidence and importance of antibody responses to infliximab after maintenance or episodic treatment in Crohn's disease. Clin Gastroenterol Hepatol 2: 542–553.

57 Baert, F., Noman, M., Vermeire, S., Van Assche, G., Carbonez, A., Rutgeerts, P. (2003) Influence of immunogenicity on the long-term efficacy of infliximab in Crohn's disease. N Engl J Med 348: 601–608.

58 Sellam, J., Allanore, Y., Batteux, F., Deslandre, C.J., Weill, B., Kahan, A. (2005) Autoantibody induction in patients with refractory spondyloarthropathy treated with infliximab and methotrexate. Joint Bone Spine 72: 48–52.

59 Eriksson, C., Engstrand, S., Sundqvist, K.G., Rantapaa-Dahlqvist, S. (2005) Autoantibody formation in patients with rheumatoid arthritis treated with anti-TNF alpha. Ann Rheum Dis 64: 403–407.

60 Nancey, S., Blanvillain, E., Parmentier, B., Flourie, B., Bayet, C., Bienvenu, J., Fabien, N. (2005) Infliximab treatment does not induce organ-specific or nonorgan-specific autoantibodies other than antinuclear and anti-double-stranded DNA autoantibodies in Crohn's disease. Inflamm Bowel Dis 11: 986–991.

61 Jonsdottir, T., Forslid, J., van Vollenhoven, A., Harju, A., Brannemark, S., Klareskog, L., van Vollenhoven, R.F. (2004) Treatment with tumour necrosis factor alpha antagonists in patients with rheumatoid arthritis induces anticardiolipin antibodies. Ann Rheum Dis 63: 1075–1078.

62 Mohan, N., Edwards, E.T., Cupps, T.R., Oliverio, P.J., Sandberg, G., Crayton, H., Richert, J.R., Siegel, J.N. (2001) Demyelination occurring during anti-tumor necrosis factor alpha therapy for inflammatory arthritides. Arthritis Rheum 44: 2862–2869.

63 Geborek, P., Bladstrom, A., Turesson, C., Gulfe, A., Petersson, I.F., Saxne, T., Olsson, H., Jacobsson, L.T. (2005) Tumour necrosis factor blockers do not increase overall tumour risk in patients with rheumatoid arthritis, but may be associated with an increased risk of lymphomas. Ann Rheum Dis 64: 699–703.

64 Wolfe, F., Michaud, K. (2004) Lymphoma in rheumatoid arthritis: the effect of methotrexate and anti-tumor necrosis factor therapy in 18,572 patients. Arthritis Rheum 50: 1740–1751.

65 Kandiel, A., Fraser, A.G., Korelitz, B.I., Brensinger, C., Lewis, J.D. (2005) Increased risk of lymphoma among inflammatory bowel disease patients treated with azathioprine and 6-mercaptopurine. Gut 54: 1121–1125.

66 Chung, E.S., Packer, M., Lo, K.H., Fasanmade, A.A., Willerson, J.T. (2003) Randomized, double-blind, placebo-controlled, pilot trial of infliximab, a chimeric monoclonal antibody to tumor necrosis factor-alpha, in patients with moderate-to-severe heart failure: results of the anti-TNF Therapy Against Congestive Heart Failure (ATTACH) trial. Circulation 107: 3133–3140.

67 Jacobsson, L.T., Turesson, C., Gulfe, A., Kapetanovic, M.C., Petersson, I.F., Saxne, T., Geborek, P. (2005) Treatment with tumor necrosis factor blockers is associated with a lower incidence of first cardiovascular events in patients with rheumatoid arthritis. J Rheumatol 32: 1213–1218.

68 Katz, J.A., Antoni, C., Keenan, G.F., Smith, D.E., Jacobs, S.J., Lichtenstein, G.R. (2004) Outcome of pregnancy in women receiving infliximab for the treatment of Crohn's disease and rheumatoid arthritis. Am J Gastroenterol 99: 2385–2392.

9 Muromonab-CD3 (Orthoclone OKT3)

Harald Becker

9.1 Introduction

Prior to the availability of cyclosporine (CsA), acute allograft rejection was a major cause of graft loss following transplantation. Often, acute renal allograft rejections in patients receiving CsA maintenance immunosuppression were not reversible with high-dose corticosteroids (Thistlethwaite et al. 1987), and consequently a more effective agent was needed. The introduction of muromonab CD3 (Orthoclone OKT3) into clinical trials – and its subsequent approval by the US Food and Drug Administration as well as by the European authorities in 1985 and 1986 for use as an anti-rejection agent for renal transplantation – was a landmark in the field of clinical transplantation of solid organs. Hence, muromonab-CD3 was the first monoclonal antibody (mAb) to be approved for clinical use in humans.

Because it is a mAb preparation, muromonab-CD3 is a homogeneous, reproducible antibody product with consistent, measurable reactivity to human T cells. Each 5-mL ampoule of Orthoclone OKT3 sterile solution contains 5 mg ($1\,mg\,mL^{-1}$) of muromonab-CD3 in a clear, colorless buffered solution (pH 7.0 ± 0.5) of monobasic sodium phosphate (2.25 mg), dibasic sodium phosphate (9.0 mg), sodium chloride (43 mg), and polysorbate 80 (1.0 mg) in water for injection.

The recognition of foreign antigens forms the physiological basis of acute allograft rejection, and proteins located on the cell surfaces play a major role in this immune response. The genes that encode these proteins are termed "histocompatibility" genes, and in every species this set of genes – known as the major histocompatibility complex (MHC) class II antigens – plays a critical role in the T-cell-dependent immune responses that initiate acute allograft rejection. Host recognition of donor antigen depends on the expression of recipient T-cell antigen receptors (TcRs). The TcR is closely associated with polypeptide chains known as clusters of differentiation (CD3), and together this unit is referred to as the TcR–CD3 complex, located on the surfaces of the T cells. The TcR–CD3 (Fig. 9.1)

Handbook of Therapeutic Antibodies. Edited by Stefan Dübel
Copyright © 2007 WILEY-VCH Verlag GmbH & Co. KGaA, Weinheim
ISBN 978-3-527-31453-9

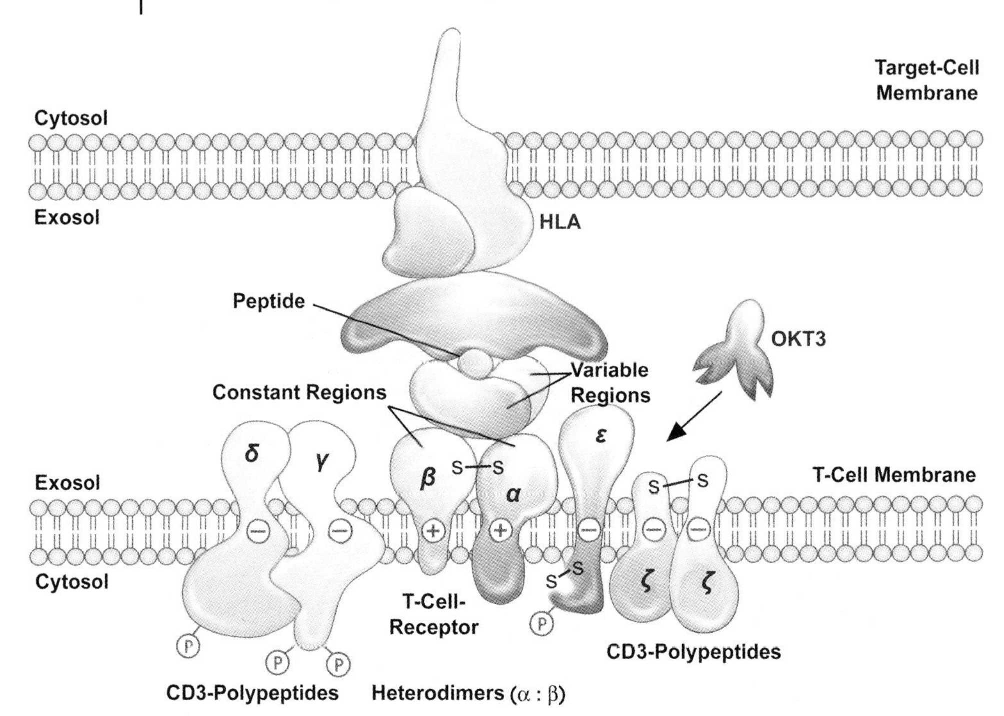

Fig. 9.1 Diagrammatic representation of the CD3 complex, T-cell antigen receptor (TCR) and site of muromonab-CD3 binding (epsilon chain of the CD3 complex). (Modified from Krensky et al. 1990.)

complex triggers the recognition of foreign antigens bound to MHC class II antigens on antigen-presenting cells (APCs), and transduces signals that subsequently induce lymphocyte activation against the allograft (Smith 1996).

9.2
Production of the Monoclonal Antibody

A mAb is an antibody derived from a single clone that is active against a single target antigen. The development of mAbs was derived from the knowledge that a single B cell (and its expanded clone) would produce a single specific antibody. The means of translating this knowledge into a reagent with practical applications in clinical transplantation occurred when Kohler and Milstein first successfully fused individual antibody-producing cells with myeloma cells, thus establishing permanent cell lines capable of secreting a single defined antibody (Kohler and Milstein 1976). In 1984, Kohler and Milstein were awarded the Nobel Prize in Medicine for their findings in this area. In theory, a mAb can be produced for any antigen against which an antibody can be produced. In the case of transplantation, however, the key was to produce an antibody against the cell surface molecules that mediate acute allograft rejection.

Muromonab-CD3 was first identified by Kung and colleagues in 1979 as part of their efforts to describe T-cell subsets in humans (Kung et al. 1979). Muromonab-CD3 is a murine monoclonal antibody to the TcR–CD3 complex (CD3) on the surface of circulating human T cells which functions as an immunosuppressant. It is administered only via the intravenous route. The antibody is biochemically purified IgG_{2a} immunoglobulin with two heavy chains each of approximately 50000 Da, and two light chains each of approximately 25000 Da. It is directed to a glycoprotein (the 20000-Da epsilon chain) in the human T-cell surface, which is essential for T-cell functions.

In order to produce immortal hybridomas, mice are immunized with an appropriate antigen complex such as a human T cell or a specific antigen complex such as CD 3; this increases the number of cells responding to the desired antigen. Approximately 1 month later, splenocytes are harvested and fused to murine myeloma cells, after which the hybridomas are grown in a medium which selects for their survival. The secreted antibody is tested for both, survival and specificity, and the clones producing the desired antibody are maintained and expanded (Fig. 9.2). Expansion is generally accomplished by passage through the peritoneal cavities of mice, or by sequential bulk tissue culture techniques (Chatenoud 1995).

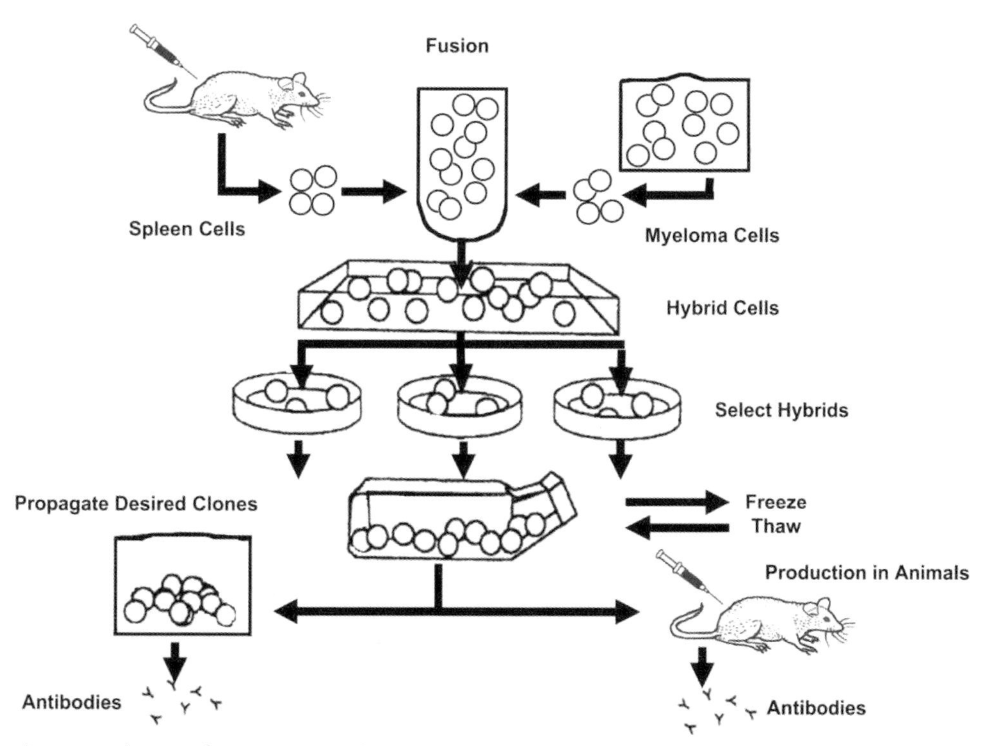

Fig. 9.2 Production of a murine monoclonal antibody (modified from Smith 1996).

The specific mAb is purified from the ascitic fluid or culture supernatant and tested to ensure the absence of microbial pathogens. The nature of the immunization process and the specificity screening employed prior to clonal expansion preclude the development of antibodies specific to platelet or red blood cell (RBC) determinants. New anti-lymphoid monoclonal antibodies considered for clinical use must be carefully tested to rule out cross-reactivity with nonlymphoid cells, as some functional surface proteins are shared by several cell types (Latham et al. 1970). Between 90 and 95% of the protein in commercially produced mAb preparations is specific antibody (Chatenoud 1995). The generation of hybridoma cell lines and the production of mAbs is described in detail elsewhere in this book (see Chapter 2, Vol I). Muromonab-CD3 solution is a homogenous, reproducible antibody product with consistent measurable reactivity to human T cells.

9.3 The Pharmacology of Muromonab-CD3

The murine mAb IgG2a specifically reacts with the TcR–CD3 complex on the surface of circulating human T cells. Muromonab-CD3 binds to a glycoprotein (the 20-kDa epsilon chain) on the CD3 complex to activate circulating T cells, and this interaction results in a transient activation of T cells with release of cytokines, and the blocking of T-cell proliferation and differentiation. As a result, almost all functional T cells are eliminated transiently from the peripheral circulation. Although T cells reappear in the circulation during the course of treatment, these cells are CD3-negative and incapable of T-cell activation. T-cell function usually returns to normal within approximately 48 h of the discontinuation of therapy (Smith 1996).

Although the mechanism by which muromonab-CD3 blocks the function of circulating T cells is incompletely understood, the results of *in-vitro* and *in-vivo* studies indicate that the primary modes of action are steric inhibition, stimulation of T-cell activation (induction of cytokine release), peripheral T-cell opsonization and depletion, and modulation of the CD3 complex (Roitt 1993). The initial effect is blockade of the TcR engagement with MHC class II antigens by steric inhibition of APCs. In addition, T-cell activation, or the expression of surface-activation markers such as interleukin (IL)-2 receptors and cellular adhesion counter-receptors, followed by profound T-cell dysfunction, occurs. The T cells are opsonized by macrophages in the reticuloendothelial system (RES), resulting in the margination of T cells into the intravascular spaces and redistribution to the lymph nodes. A rapid and concomitant decrease in the number of circulating CD3+ cells, including those that are CD2+, CD4+ or CD8+, has been observed in patients studied within minutes after the intravenous administration of muromonab-CD3 (Table 9.1). These cells are therefore unavailable to recognize transplant antigens (Kreis et al. 1991). Finally, any remaining T cells undergo TcR–CD3 modulation or internalization, during which TcR with CD3 antigens are removed

Table 9.1 Definition of surface antigens (from Bernard and Boumsell 1984).

OKT monoclonal antibody	*Recognized antigen*			*Cell population*
	CD-class	*T-class*	*Alternative*	
OKT1	CD5	T1		All T cells
OKT3	CD3	T3	LEU4	All matured peripheral T cells
OKT4	CD4	T4	LEU3a	Helper/inducer T cells
OKT6	CD1	T6		Thymocytes
OKT8	CD8	T8	LEU2a	Suppressor cytotoxic T cells
OKT11	CD2	T11	LEU5	T cells with receptor of sheep-erythrocytes

CD = cluster of differentiation; LEU = leucocytes.

from the T-cell surfaces, rendering the T cells CD3-negative and immunologically inactive (Caillat-Zucman et al. 1981).

9.3.1 Pharmacokinetic Properties of Muromonab-CD3

Plasma muromonab-CD3 concentrations vary according to the muromonab-CD3 antibody status, transplanted organ, and patient age. In renal transplant recipients receiving muromonab-CD3 5 mg once daily for 10 to 14 days, mean plasma concentrations were 996 $\mu g\,L^{-1}$ after 1 h and 104 $\mu g\,L^{-1}$ at 24 h. The mean trough steady-state serum concentrations range from 500 to 1000 $\mu g\,L^{-1}$ after 2 to 4 days. A serum level of approximately 1000 $\mu g\,L^{-1}$ is required to block cytotoxic T-cell function *in vitro*. Moreover, steady-state serum muromonab-CD3 concentrations are achieved earlier with prophylactic administration than with administration to treat transplant rejection.

There is evidence of drug accumulation after repeated doses, with muromonab-CD3 plasma elimination half-lives of approximately 18 h having been reported after administration to treat rejection, and of 36 h after prophylactic administration (Wilde et al. 1996).

Low plasma levels of muromonab-CD3 have been associated with a failure of muromonab-CD3 prophylaxis (Abramowicz et al. 1994; McDiarmid et al. 1990; Schroeder et al. 1991). The CD3+ cell level alone is not a reliable indicator for muromonab-CD3 concentrations or early sensitization (Broughan et al. 1994; Gebel et al. 1989a; McCarthy et al. 1993; Shaefer et al. 1990), and therefore the achievement of optimal dosage adjustments and efficacy requires continued surveillance for clinical signs of rejection during muromonab-CD3 prophylaxis in addition to monitoring the drug's serum levels, CD3+ cell levels, and anti-muromonab-CD3 antibody titers, as appropriate (Abramowicz et al. 1994; Gebel et al. 1989a,b; Hammond et al. 1993; Henell et al. 1993; Moore et al. 1991; Ohman et al. 1993; Ryckman et al. 1991; Schroeder et al. 1991; Toyoda et al. 1995).

9.3.2
Pharmacodynamics of Muromonab-CD3

Muromonab-CD3 reverses graft rejection, most probably by blocking all T-cell functions and is, therefore, classified as a pan-T-cell suppressive mAb. This blockade plays a major role in acute allograft rejection, as muromonab-CD3 reacts with and blocks the function of a 20000-Da molecule (CD3) in the membrane of human T-cells. *In vivo*, muromonab-CD3 reacts with most peripheral blood T cells and T cells in body tissues, but has not been found to react with other hematopoietic elements or other tissues of the body. In all patients studied, a rapid and concomitant decrease in the number of circulating CD3+, CD4+ and CD8+ T cells was observed within minutes after the administration of muromonab-CD3 (Table 9.2). Between Days 2 and 7 after administration, increasing numbers of circulating CD4+ and CD8+ cells have been observed in patients, although CD3+ cells are not detectable. Due to their rapid clearance from the peripheral circulation, the T cells are unavailable to recognize transplant antigens. Finally, any remaining T cells undergo TcR–CD3 modulation or internalization, during which TcR with CD3 antigens are removed from the T-cell surfaces, thus rendering the T cells CD3-negative and immunologically inactive (Caillat-Zucman et al. 1981).

T-cell receptor stimulation by muromonab-CD3 *in vitro* led to a strong increase in tyrosine phosphorylation in all specimens of kidney transplant recipients and healthy controls (Muller et al. 1999). A significant release of IL-10, which plays a role in the down-modulation of the muromonab-CD3-triggered T-cell activation cascade, was observed after muromonab-CD3 administration to human renal allograft recipients (Herbelin et al. 1999). The pretreatment of renal allograft

Table 9.2 Possible mechanisms of action of muromonab-CD3. Modified from Wilde MI et al. (1996).

Accepted mechanisms	*Supposed mechanisms*
• Antigenic modulation of the CD3/T-cell receptor complex on peripheral T cells resulting in failure of antigen recognition. • Opsonization of circulating T cells and subsequent removal by the reticuloendothelial system.	• Immunomodulation of graft-infiltrating lymphocytes. • Elimination of activated CD3+ cells by induction of apoptosis. • Modulation of CD3 complex by shedding CD3 antigens of the CD3 complex. • Increasing lymphocyte adhesion molecule expression on peripheral blood lymphocytes. • Induction of cell-mediated cytolysis.

patients with doses of up to 1 μg kg^{-1} bodyweight recombinant human (rh) IL-10 was found safe and reduced the release of tumor necrosis factor alpha (TNF-α) induced by muromonab-CD3. Higher doses however, seemed to promote early sensitization to muromonab-CD3 (Wissing et al. 1997).

CD3+ cells reappear rapidly and reach pretreatment levels within a week after termination of muromonab-CD3 therapy. Increasing numbers of CD3+ cells have been observed in some patients during the second week of muromonab-CD3 therapy, possibly as a result of the development of neutralizing antibodies to the compound. Antibodies to muromonab-CD3 have been observed, occurring with an incidence of 21% (n = 43) for IgM, 86% (n = 43) for IgG, and 29% (n = 35) for IgE. The mean (± SD) time of appearance of IgG antibodies was 20 ± 2 days, while early IgG antibodies appeared by the end of the second week of treatment in 3% (n = 86) of the patients. Serum levels of muromonab-CD3 are measurable using an enzyme-linked immunosorbent assay (ELISA). During treatment with 5 mg per day for 14 days, mean serum-trough levels of muromonab-CD3 were increased during the first 3 days of administration, but then remained at steady state (mean 0.9 μg mL^{-1}) on Days 3 to 14. The levels obtained during therapy have been shown to block T-cell effector functions *in vitro*. Following *in-vivo* administration, leukocytes have been observed in cerebrospinal and peritoneal fluids, though the mechanism of this effect is not understood.

9.3.3 Activation of Human T Cells

The cytokines TNF-α, IL-2, -3, -6 and -10 and interferon gamma (IFN-γ) are all released following the administration of muromonab-CD3, associated with an acute-phase reaction involving C-reactive protein (CRP), neopterin, endothelin-1, complement, transferrin, alpha-1-proteinase inhibitor and neutrophilic granulocytes (Abramowicz et al. 1989; Chatenoud 1994; Ellenhorn et al. 1990; Gaston et al. 1991; Goumy et al. 1990; Kreis 1993; Raasveld et al. 1993). This response usually occurs after the first and possibly second and third dose(s). A similar response, particularly the release of IL-6 (Bloemena et al. 1990), may also occur later during the course of treatment if CD3+ cell levels are increased substantially. This may account for some of the late adverse events observed with muromonab-CD3. The T-cell-activating properties of muromonab-CD3 manifest clinically as first-dose events (Cytokine Release Syndrome; CRS), while T-cell receptor stimulation by muromonab-CD3 *in vitro* led to a strong increase in tyrosine phosphorylation in all specimens of kidney transplant recipients and healthy controls (Muller et al. 1999). Monocyte-dependent Fc receptor-mediated cell activation appears to be the main mechanism underlying these events (Ellenhorn et al. 1990; Gaston et al. 1991; Hoffman et al. 1992; Vossen et al. 1995; Woodle et al. 1992).

The administration of muromonab-CD3 to human renal allograft recipients led to a significant release of IL-10 that may play a major role in the down-modulation of the muromonab-CD3-triggered T-cell activation cascade (Herbelin et al. 1999).

The release of IL-6 and IL-10 may also be involved in the pathogenesis of Epstein–Barr virus (EBV)-associated lymphoproliferative disorders in transplant recipients receiving muromonab-CD3 (Goldman et al. 1992; Swinnen et al. 1993). Anti-CD3 monoclonal antibodies have also been shown to trigger T-cell mitogenesis *in vitro* (Chatenoud 1994).

Several agents, including corticosteroids and pentoxifylline, reduce muromonab-CD3-induced cytokine release (Chatenoud et al. 1990, 1991; Ferran et al. 1990; Leimenstoll et al. 1993). However, although corticosteroids can reduce first-dose cytokine-related adverse events, pretreatment with pentoxifylline does not appear to reduce the CRS associated with prophylactic muromonab-CD3 (Alegre et al. 1991; De Vault et al. 1994).

9.3.4 Immunogenicity

The time to first appearance of anti-muromonab-CD3 antibodies in heart transplant patients who received muromonab-CD3-based quadruple sequential immunosuppression varied from Day 0 to Day 35. Anti-muromonab-CD3 antibodies decreased to negative levels in 11 of 12 patients who were initially positive following treatment by 1 year post transplant. Anti-idiotype and anti-isotype antibodies occurred at similar times and titers in patients who were anti-muromonab-CD3-antibody-positive (Bhat et al. 1997).

Stimulation of the T cells of healthy persons by muromonab-CD3 has led to a biphasic increase in phosphotyrosine levels, with the first peak occurring after 15 s and the absolute maximum occurring after 3 to 5 min. Levels remained high up to 30 min and subsequently returned to baseline. Using muromonab-CD3 to stimulate the T cells of bone marrow transplant (BMT) patients led to strong increases in phosphotyrosine levels comparable to those of controls. In contrast, the response of T cells of HIV-infected patients with AIDS-syndrome was severely impaired ($p = 0.01$) (Muller et al. 1998).

The selective depletion of activated type 1-like T cells by muromonab-CD3 resulted in longlasting immune deviation that might explain the long-term effects of muromonab-CD3 treatment (Reinke et al. 1997). Complement activation – an early event after muromonab-CD3 administration – was associated with the increased expression of adhesion molecules on neutrophils, and with pulmonary hemodynamic changes (Vallhonrat et al. 1999). T-cell depletion using muromonab-CD3 resulted in a higher T-cell content and higher rates of acute graft-versus-host disease (GvHD) and posttransplant lymphoproliferative disease (PTLD) compared with T10B9 (anti-TCR). A decreased risk of relapse for patients with high-risk disease was seen with muromonab-CD3-treated grafts (Keever-Taylor et al. 2001).

Following the infusion of muromonab-CD3 (three daily doses), a statistically significant increase in the binding of the mAb anti-factor V/Va to platelets (2.2% versus 12.8%, $p = 0.04$) was seen at 15 min after the second dose (Lozano et al. 2001). In uremic patients receiving prophylactic muromonab-CD3, an increase in the binding of anti-factor V/Va was noted, denoting an increased exposure of anionic phospholipids in platelets.

High antibody titers (≥1:1000) were detected in 5.8% of 12133 serum samples from patients who received muromonab-CD3 for the treatment or prevention of transplant rejection (Carey et al. 1995). These antibodies may result in decreased muromonab-CD3 plasma concentrations and increased circulating CD3+ levels, and may preclude re-use of the agent in some patients (Chatenoud 1993a).

The two types of anti-muromonab-CD3 antibodies that may be induced are anti-idiotypic and anti-isotypic (Chatenoud 1993a, 1994; Todd and Brogden 1989). Anti-idiotypic antibodies compete with muromonab-CD3 for binding to the CD3 complex, and can neutralize the activity of muromonab-CD3. IgG – but not IgM – anti-muromonab-CD3 antibodies are able to reduce the activity of the agent. Although anti-isotypic antibodies bind to the constant portion of the muromonab-CD3 antibody molecule, they do not block the effects of the drug (Wilde et al. 1996). Anti-idiotypic antibodies to the muromonab-CD3 do not cross-react with murine antibodies of similar or different isotypes, and cross-react with only 10% of other anti-CD3 mAbs (Norman 1992).

Although variation in antibody test results between centers is significant, the incidence of anti-muromonab-CD3 antibodies and the percentage of high antibody titers (≥1:1000) appears to be greatest in liver or kidney transplant recipients, and least in cardiac transplant recipients (Carey et al. 1995; Kimball et al. 1993; O'Connell et al. 1991; Schroeder et al. 1994). The risk of high anti-muromonab-CD3 antibody titers also appears to be greatest in patients aged <30 years, in those who have undergone previous transplantation or muromonab-CD3 courses, and in those receiving muromonab-CD3 for rescue treatment (versus prophylaxis or first-line treatment of rejection) (Carey et al. 1995). The administration of concomitant immunosuppressants reduces the likelihood of anti-muromonab-CD3 antibody formation (Chatenoud 1993a, 1994; Kreis 1993; Norman et al. 1993; Schroeder et al. 1990; Taylor et al. 1994).

9.3.5 Interactions

Concomitant medications (azathioprine, corticosteroids, CsA) may have contributed to the neuropsychiatric, infectious, nephrotoxic, thrombotic, and/or neoplastic events reported in patients treated with muromonab-CD3. A study by Vasquez and Pollack (1997) presented clinical evidence for an interaction between muromonab-CD3 and CsA, a CYP3A4 substrate and p-glycoprotein inhibitor. The authors postulated that cytokine release by muromonab-CD3 and subsequent cytokine-induced inhibition of CYP3A4 enzyme activity might be responsible for elevations in CsA blood levels observed during muromonab-CD3 therapy. The study was designed as a single-center retrospective analysis in 33 subjects (17 subjects received combined muromonab CD3/CsA versus 16 subjects receiving CsA/anti-lymphocyte globulin [ALG]). Generally, CsA is known to demonstrate high inter-patient variability (Mendez et al. 1999), and changes in plasma protein binding may impact its clearance and thus its plasma or blood concentrations (Strong et al. 1997). CsA is also known to have saturable blood and tissue distribution (Tanaka et al. 1999), which may cause higher variability in the estimations of blood concentrations. As these authors did not standardize their dose adjustment algorithm, the adjustments made might be arbitrary and further strengthen the fact that that there was a large variability in the CsA blood concentrations. In view of the above-mentioned limitations, it is difficult to ascertain whether the described effect is indeed true.

A summary report by Mignat (1997) provided clinically significant drug interactions with immunosuppressive agents, although there was no mention of interactions of muromonab-CD3 with either tacrolimus or CsA. The use of indomethacin by some patients who were simultaneously receiving therapy with muromonab-CD3 may have contributed to some encephalopathic and other CNS adverse events, but the mechanisms of these effects are unknown.

9.4 Therapeutic Use

Muromonab-CD3 is indicated for the treatment of acute allograft rejection in renal transplant patients, and for the treatment of steroid-resistant acute allograft rejection in cardiac and hepatic transplant patients. The effectiveness of muromonab-CD3 for prophylaxis of renal allograft rejection has not been established. The dosage of other immunosuppressive agents used in conjunction with muromonab-CD3 should be reduced to the lowest level compatible with an effective therapeutic response. The recommended dose of muromonab-CD3 to treat acute renal, steroid-resistant cardiac or steroid-resistant hepatic allograft rejection in adults is 5 mg per day in a single (bolus) IV injection given in less than 1 min, over a period of 10 to 14 days. Lower initial doses of muromonab-CD3 (<5 mg day^{-1}) may be effective and better tolerated than higher doses (Norman et al. 1991, 1994). Rejection incidences with low-dose muromonab-CD3-based induction therapy in renal transplant recipients were similar to or lower than those with higher dosages (Norman et al. 1991, 1994).

High muromonab-CD3 doses (10 mg day^{-1}) have been associated with intragraft thromboses, whilst single doses (30, 40 or 50 mg) have provided no advantages in terms of efficacy over a 5 mg day^{-1} dose, and were in fact associated with a high incidence of adverse events (Welter et al. 1990).

In pediatric patients, the initial recommended dose is 2.5 mg day^{-1} for those weighing $\leq$30 kg, and 5 mg day^{-1} for those weighing $>$30 kg; all daily doses are given as a single (bolus) IV injection in less than 1 min, and for 10 to 14 days. Daily increases in muromonab-CD3 doses (i.e., 2.5-mg increments) may be required to achieve depletion of CD3+ cells ($<$25 cells mm^{-3}) and ensure therapeutic muromonab-CD3 serum concentrations ($>$800 ng mL^{-1}). Pediatric patients may require augmentation of the muromonab-CD3 dose. In all patients with acute renal rejection, treatment should begin upon diagnosis, but for steroid-resistant cardiac or hepatic allograft rejection the treatment should begin when a rejection has not been reversed by an adequate course of corticosteroid therapy.

Laboratory screening should be carried out prior to the administration of muromonab-CD3. Likewise, the patient's volume (fluid) status and a chest X-radiograph should be assessed to rule out volume overload, uncontrolled hypertension, or uncompensated heart failure. Patients should not weigh >3% above their minimum weight during the week prior to injection. The following investigations should be conducted prior to and during muromonab-CD3 therapy:

- Renal: Blood urea nitrogen (BUN), serum creatinine, etc.;
- Hepatic: transaminases, alkaline phosphatase, bilirubin;
- Hematopoietic: WBCs and differential, platelet count, etc.;
- Chest X-radiograph within 24 h before initiating muromonab-CD3 treatment to rule out heart failure or fluid overload.
- Blood tests: Periodic assessment of organ system functions (renal, hepatic, and hematopoietic).

In addition, during therapy with muromonab-CD3 the following parameters should be followed:

- Plasma muromonab-CD3 levels to be $\geq$800 ng mL^{-1}; and
- T-cell clearance (CD3+ T cells <25 cells mm^{-3}).

In adults, these should be monitored *periodically*; in pediatric patients, they should be monitored *daily*.

The end-points of immunosuppressive trials should include rejection incidence and severity, time to first rejection episode, and effects on organ function as well as graft and patient survival. Two types of induction therapy to prevent rejection are used in transplantation:

- Administration of an anti-lymphocyte preparation, azathioprine and corticosteroids from the time of transplantation with CsA withheld until renal function is established (sequential therapy).
- Administration of an anti-lymphocyte preparation, azathioprine, corticosteroids and low-dose CsA followed by maintenance therapy.

Muromonab-CD3 was usually administered during the immediate postoperative period, although some patients received the drug intraoperatively (Wilde et al. 1996).

Most clinical trials were randomized, but the majority were not double-blind because the first-dose reaction to muromonab-CD3 and different modes of administration of anti-lymphocyte antibodies precluded double-blinding. Therefore, small numbers of patients were involved in the studies. Dosage regimens of concomitant immunosuppressants were varied and complex, and the administration of muromonab-CD3 was preceded by a variety of prophylactic agents, making between-study comparisons difficult. Most recipients were undergoing cadaveric transplantation. Rejection episodes were documented histologically or with appropriate laboratory tests of organ function (Wilde et al. 1996).

Factors influencing the outcome of rejection prophylaxis include:

- Primary disease and severity including baseline organ function.
- The age and gender of the recipient and donor (Wechsler et al. 1995).
- Pre-existing anti-muromonab-CD3 antibody titers.
- Delayed graft function (Howard et al. 1993; Troppmann et al. 1995).
- ABO blood group compatibility.
- Previous transplantation or blood transfusions.
- Donor/recipient cytomegalovirus (CMV) status.
- Histocompatibility antigen matching and preformed reactive anti-HLA antibodies, although the potential beneficial effects of HLA-matching, especially in those receiving anti-lymphocyte agents, are controversial (Costanco-Nordin et al. 1993; Kermann et al. 1994).
- Duration of organ cold ischemia time.
- Surgical technique.
- Postoperative patient management, including treatment of rejection.

9.4.1
Renal and or Renal-Pancreas Transplant Recipients

Some studies with muromonab-CD3-based immunosuppression and triple therapy or regimens containing other lymphocyte preparations are summarized in Table 9.3.

Muromonab-CD3 was successfully used by Chkotua et al. (2003) as effective treatment for steroid-resistant rejections in renal transplant patients with severe rejection who did not respond to the first few doses of antithymocyte globulin (ATG). In a prospective, randomized trial with thymoglobulin/daclizumab induction, tacrolimus and steroid maintenance, comparing rapamycin with mycophenolate mofetil (MMF) as immunosuppression in simultaneous pancreas renal-pancreas transplantation, five acute rejection episodes in the MMF group were steroid-resistant, but responsive to muromonab-CD3 or thymoglobulin (Ciancio et al. 2004). Another study compared the outcomes of renal transplantation with two distinct induction protocols, basiliximab versus muromonab-CD3, in the setting of CsA-based immunosuppression. Post-induction protocols included either total prednisone avoidance or prednisone sparing versus standard prednisone dosing. A total of 245 adult patients receiving kidney transplantation between 1995 and 2000 was included in the study. Treatment in group 1 was

Table 9.3 Efficacy of muromonab-CD3 (M-CD3) immunosuppression (delayed cyclosporine A) as prophylaxis of renal allograft rejection versus standard triple therapy (immediate cyclosporine A) or other anti-lymphocyte-based regimens. A selection of clinical studies. Modified from Wilde et al. (1996).

Reference	*Study design*	*No. of evaluable patients (characteristics)*	*Treatment*	*Time to initial rejection [days]*	*Actuarial graft survival [%]*	*Rejection incidence [%]*	*Actuarial patient survival [%]*	*Overall efficacy*	*Comments*
Abramowicz et al. (1994)	Randomized	56	M-CD3[a]	23 (first 3 months)	83 (overall; 3 years) 92 graft survival	32 (3 years)	94.5 (3 years)	M-CD3 ≥ triple therapy	Fewer rejections with M-CD3 corticosteroid-resistant (15 vs. 30%)
		52	Triple therapy[c]	11 (first 3 months)	75 (overall; 3 years) 79 graft survival	22 (3 years)	93 (3 years)		
Benvenisty et al. (1990)	Non-blinded	34 (delayed graft function)	M-CD3[a]		80 (1 year) 74 (2 years)	44	89	M-CD3 ≥ triple therapy	M-CD3 decreased the duration of graft non-function (9.4 vs. 14.9 days)
		40 (delayed graft function)	Triple therapy[c]		55 (1 year) 47 (2 years)	82	89		
Bock et al. (1995)	Randomized, non-blinded	51	M-CD3[a]		78 (1 year)	45	92 (1 year)	ATG > M-CD3	
		53	ATG[d]		91 (1 year)	26	96 (1 year)		
Broyer et al. (1993)	Randomized	77	M-CD3[a]		79 (1 year) 71 (2 years) 68 (3 years)	11	96	M-CD3 = ALG	
		71	ALG[b]		80 (1 year) 77 (2 years) 73 (3 years)	11	99		

Table 9.3 *Continued*

Reference	*Study design*	*No. of evaluable patients (characteristics)*	*Treatment*	*Time to initial rejection [days]*	*Actuarial graft survival [%]*	*Rejection incidence [%]*	*Actuarial patient survival [%]*	*Overall efficacy*	*Comments*
Cole et al. (1994)	Randomized	83	M-CD3[a]		81 (1 year)	31 (patients with no rejection)	95 (1 year)	ATG ≥ M-CD3	More steroid-resistant rejections with M-CD3 (25 vs. 12)
		83	ATG[d]		78 (1 year)	57 (patients with no rejection)	89.5 (1 year)		
Frey et al. (1992)	Randomized	67	M-CD3[a]	M-CD3 = ALG	87 (1 year) 83 (2 years)	36	96 (1 year) 96 (2 years)	M-CD3 = ALG	
		71	ALG[b]		84 (1 year) 80 (2 years)	45	93 (1 year) 91 (2 years)		
Hanto et al. (1994)	Randomized	59	M-CD3[a]	35	84 (1 year) 79 (2 years) 79 (2 years)	33 (37)	98 (1, 2, and 3 years)	M-CD3 = ALG	
		58	ALG[b]	29	81 (1 year) 78 (2 years) 78 (3 years)	27 (31)	96 (1, 2, and 3 years)		
Steinmuller et al. (1991)	Randomized	25	M-CD3[a]	67.5	84 (<6 months)	64	99 (<6 months)		
		26	ALG[b]	95	85 (<6 months)	35	99 (<6 months)		

a MCD3-based regimen: MCD3 + Az + MPr + Pr + CsA.
b ALG + MPr + Az + CsA.
c CsA + Az + MPr + Pr.
d ATG + MPr + Az + CsA.
Efficacy status: = indicates equivalence; > indicates significantly ($p < 0.05$) more effective than comparator; > indicates significantly ($p < 0.05$) greater efficacy than comparator in terms of at least one efficacy parameter; *$p < 0.05$ compared with comparator.
ALG = anti-lymphocyte globulin (Minnesota; horse antibody); ATG = anti-thymocyte globulin; Az = azathioprine; CsA = cyclosporine A; MPr = methylprednisolone; Pr = prednisone.

muromonab CD3 + CsA + adjunct + standard prednisone; group 2 received basiliximab + CsA + adjunct + steroid sparing; group 3 received basiliximab + CsA + adjunct + no prednisone. The demographics between all groups were similar. The incidences of acute rejection within 1 year in the respective groups were 28% versus 15% versus 16%. Thus, the authors concluded that the use of basiliximab in transplant recipients resulted in long-term patient and graft survival similar to those achieved with muromonab-CD3 (Kung et al. 2004).

9.4.2
Liver Transplant Recipients

In addition to the toxic effects of early CsA use, hepatic transplant recipients are at an increased risk of renal impairment during the early postoperative period because of intraoperative hemodynamic instability, hepatorenal syndrome and pre-existing renal or hepatic dysfunction. Studies comparing muromonab-CD3-based immunosuppression with triple therapy or regimens containing other anti-lymphocyte preparations or anti-IL-2 antibody are summarized in Table 9.4.

In a retrospective study of 156 patients with recurrent primary biliary cirrhosis after liver transplantation, 10.9% of patients experienced recurrence (Sanchez et al. 2003). Muromonab-CD3 was used for steroid-resistant rejection in 41% of recurrent primary biliary cirrhosis patients, and in 23.7% of nonrecurrent primary biliary cirrhosis patients. There was no significant difference in the development of recurrent primary biliary cirrhosis between the groups ($p = 0.36$). Arenas et al. (2003) evaluated the influence of rejection treatment (methylprednisolone) on hepatitis C virus (HCV) quasispecies after liver transplantation, with muromonab-CD3 being used in two patients for steroid-resistant rejection. Those liver transplant recipients treated for rejection had a decrease in HCV quasispecies diversity after transplantation.

9.4.3
Cardiac Transplant Recipients

The prevention of rejection is particularly important for cardiac transplant recipients because, unlike renal transplant failure, cardiac graft failure is usually fatal. Although direct comparisons are few in number, on the available evidence muromonab-CD3-based therapy appears to be similar to CsA plus prednisone or triple therapy as assessed by rejection incidence, time to first rejection episode, and graft and patient survival (see Table 9.5). The time to rejection was significantly longer with muromonab-CD3-based induction therapy than triple therapy, but there was no significant between-treatment difference in rejection incidence (Barr et al. 1990). The results of selected studies are summarized in Table 9.5. Data relating to muromonab-CD3-based therapy compared with other anti-lymphocyte-based regimens are conflicting. For example, some investigators have reported a longer time to first rejection and/or a lower incidence of rejection with muromonab-CD3 (Costanco-Nordin et al. 1990; Renlund et al. 1989), whereas

Table 9.4 Efficacy of muromonab-CD3 (M-CD3) and delayed cyclosporine A (CsA) as prophylaxis of hepatic allcgraft rejection versus standard triple therapy (Tr) (immediate CsA), other anti-lymphocyte-based regimens or anti-interleukin-2 monoclonal antibody-based immunosuppression. Modified from Wilde et al. (1996).

Reference	*Study design*	*No. of evaluable patients (charac-teristics)*	*Treatment*	*Time to initial rejection [days]*	*Actuarial graft survival [%]*	*Rejection incidence [%]*	*Actuarial patient survival [%]*	*Overall efficacy*	*Comments*
Farges et al. (1994)	Randomized, multicenter	44	M-CD3[a]	38	61 (4 years)	34 (2 weeks) 67 (1 year)	82 (1 year) 69 (4 years)	M-CD3 ≥ triple therapy	
		50	Triple therapy[b]	9	54 (4 years)	61 (2 weeks) 75 (1 year)	78 (1 year) 62 (4 years)		
McDiarmid et al. (1991a,b)	Randomized	46	M-CD3[a]	12	63 (>3 months)	28 (postop. day 0–14) 46 (<1 months) 91 (>3 months)	67 (>3 months)	M-CD3 ≥ triple therapy	Mean duration of follow-up 648 days for M-CD3 and 682 days for triple therapy
		39	Triple therapy[b]	9.5	73 (>3 months)	67 (postop. day 0–14) 31 (rejection-free < 1 month) 99 (rejection-free >3 months)	84 (>3 months)		
Mühlbacher et al. (1989)		30	M-CD3[a]			56	57 (1 year)	M-CD3 > triple therapy	
		58	Triple therapy[b]			79.5	45 (1 year)		

Pons et al. (1993)		25	M-CD3[a]		25	3.5 (mortal. 2 months)		
		19	Triple therapy[b]		69	26 (mortal. 2 months)		
Reding et al. (1993)	Randomized	37 (adults + children)	M-CD3[a]	86 (1 year)	81 (first 3 months)	86 (1 year)		
		35	Anti-IL-2 receptor[d]	97 (1 year)	91 (first 3 months)	100 (1 year)		
		28	Triple therapy[b]	75 (1 year)	96 (first 3 months)	79 (1 Year)		
Steininger et al. (1991)	Retrospective	63	CsA+Pred.		67 (first 3) weeks		ATG > CsA + pred./ M-CD3	Not standard regimens. Rejection advantage of ATG abolished when severe rejection observed
		32	M-CD3[a]		44			
		49	ATG[c]		24			
Cosimi AB, et al. 1990	Randomized	38	M-CD3[a]		13/39 (1/2 weeks) 68 (1 year)	84	M-CD3 ≥ triple therapy	Duration of the initial rejection-free period longer with M-CD3
		41	Triple therapy[b]		46/71 (1/2 weeks) 78 (1 year)	73		

For details of superscripts, abbreviations, etc., see Tab. 9.3.

Table 9.5 Selection of clinical studies concerning efficacy of muromonab-CD3 (M-CD3) and delayed cyclosporine A (CsA) as prophylaxis of cardiac allograft rejection versus standard triple therapy and immediate CsA or anti-lymphocyte regimens. Modified from Wilde et al. (1996).

Reference	*Study design*	*No. of evaluable patients (characteristics)*	*Treatment*	*Time to initial rejection (days)*	*Actuarial graft survival (%)*	*Rejection Incidence (%)*	*Actuarial patient survival (%)*	*Overall efficacy*
Balk et al. (1991)	Randomized	33	M-CD3[a]		91	1.33 (1 year) 80/31/28 (1, 3, 6 months)	91 (2 years)	M-CD3 = CysA + prednisone
		33	CsA + Pr		94	1.36 (1 year) 66, 33, 27 (1, 3, 6 months)	94 (2 years)	
Barr et al. (1990)	retro-spective	26	M-CD3[a]	42		0.003	88/81 (6/18 months)	M-CD3 = triple therapy
		26	Triple therapy[b]	21		0.003	92/87 (6/18 months)	
Costanzo-Nordin et al. (1990)	Randomized	12	M-CD3[a]	32		3.4		M-CD3 ≥ ATG
		11	ATG[c]	15		5		
Griffith et al. (1990)	Randomized	43	M-CD3[a]	33		0.58/pat.; 0.43/pat.; 0.2 (1; 2; >2 months)	98 (1 year)	M-CD3 ≤ ATG
		39	ATG[c]	67		0.08/pat.; 0.24/pat.; 0.2 (1; 2; >2 months)	95 (1 year)	

Ippoliti et al. (1991)	Randomized	15	M-CD3[a]			80 (6 months)	ATG ≥ M-CD3
		15	ATG[c]			93 (6 months)	
Ladowski et al. (1993)	Randomized	30	M-CD3[a]		0.8/pat. (3 months) 1.03/pat. (1 year)	83 (1 year)	ATG ≥ ALG and M-CD3
		34	ATG[c]		0.24/pat. (3 months) 0.26/pat. (1 year)	82 (1 year)	
		15	ALG[d]		1.14/pat. (3 months) 1.27/pat. (1 year)	80 (1 year)	
Macdonald et al. (1993)	Randomized	20	M-CD3[a]	33	90	83 (1 year)	M-CD3 = ATG
		21	ATG[c]	27	100	81 (1 year)	
Menkis et al. (1992)	Randomized	20	M-CD3[a]	5.6 weeks	2.1/pat. (6 months)	92 (2 years)	M-CD3 = ALG
		19	ALG	5.3 weeks	1.4/pat. (6 months)	84 (2 years)	
Renlund et al. (1989)	Randomized	26	M-CD3[a]	76	1.5/pat. (6 months) 0.2/pat. (1 year)		
		25	ATG[c]	36	2.2/pat. (6 months) 0.8/pat. (1 year)		

For details of superscripts, abbreviations, etc., see Tab. 9.3.

others have not observed any differences (Kirklin et al. 1990; Macdonald et al. 1993; Menkis et al. 1992; Wollenek et al. 1989) or have reported a lower incidence of rejection (Griffith et al. 1990; Ladowski et al. 1993) and/or a longer time to first rejection episode (Griffith et al. 1990; Ippoliti et al. 1991) with ATG-based therapy. No between-treatment differences in graft or patient survival have been reported (Wilde et al. 1996).

9.5 Cytokine Release Syndrome

Cytokines are small peptides that regulate cell growth and tissue homeostasis. Proinflammatory cytokines, such as IL-1, -2, and -6, INF-γ, and TNFs play an important role in allograft rejection. Understanding the cytokine response is crucial to the effective management of the transplant patient. Cytokines are multifunctional, are involved in virtually all organ systems, and have effects that are sometimes stimulatory and sometimes inhibitory. The cell-mediated event of acute rejection is stimulated by cytokines following the introduction of allogeneic organs to the transplant recipient. An enhanced production of IL-1, IL-6 and TNF-α results when APCs detect the presence of a foreign body. T-cell activation also increases the production of IL-2, which in turn produces an antigen-specific T-cell response. Other stimulated cytokines that extend the immune response include T lymphocyte-derived IL-2, IL-3, INF-γ, TNF-α, and TNF-β. The presence of adhesion molecules also promotes the proinflammatory response. The result of this cascade of events is the destruction and failure of the allograft tissue. Monoclonal antibodies such as muromonab-CD3 disrupt the cytokine process, thereby allowing for greater allograft transplantation success (Gaston 1994).

Muromonab-CD3 is directed to a glycoprotein in the human T-cell surface that is essential for T-cell function. In *in-vitro* cytolytic assays, muromonab-CD3 blocks both the generation and function of effector cells. Binding of muromonab-CD3 to T lymphocytes results in the early activation of T cells, which leads to cytokine release, followed by blocking of T-cell functions. T-cell activation results in the release of numerous cytokines/lymphokines, which are felt to be responsible for many of the acute clinical manifestations seen following muromonab-CD3 administration.

9.5.1 The Pathophysiology of the Cytokine Release Syndrome

Early investigators first attributed the fever, chills and other adverse reactions typically experienced with a dose of muromonab-CD3 to an idiosyncratic or allergic-type reaction, or to massive lymphocytolysis (Gaston 1994; Rossi et al. 1993). However, the observation that symptoms resolved after repeated doses of muromonab-CD3 challenged these theories. The current accepted etiology of the CRS involves a massive and transient release of proinflammatory cytokines into the circulation, attributed to the mitogenic properties of muromonab-CD3 (Chatenoud et al. 1991; First et al. 1993; Rossi et al. 1993). The first dose of muromonab-CD3 induces a surge of cytokine secretion that is probably a response to the transient T-cell activation by the mAb (Gaston 1994). Several cytokines are seen in the circulation, including IL-1, IL-2, IL-3, IL-6, INF-γ, TNF, and granulocyte-macrophage colony-stimulating factor (GM-CSF) (First et al. 1993; Jeyarajah et al. 1993). As soon as 1 h following administration of the first muromonab-CD3 dose, the release of TNF is observed in the circulation (Chatenoud 1993b; Gaston 1994). TNF-α is believed to be the key mediator of the CRS, causing the most severe symptoms of the first-dose response as well as meningeal inflammation, encephalopathy, nephropathy, and cardiopulmonary deterioration (Eason et al. 1996; Gaston, 1994; Jeyarajah et al. 1993; Rossi et al. 1993; Vincenti et al. 1993). IL-6 production also increases in the circulation approximately 6 to 48 h after muromonab-CD3 dosing (Chatenoud 1993b; Gaston 1994), while INF-γ and IL-2 peak at approximately 2 to 4 h after administration. Elevations decrease significantly with subsequent therapy. All cytokines return to values seen at baseline approximately 12 to 24 h following the first muromonab-CD3 treatment (Chatenoud 1993b).

Abramowicz et al. (1989) evaluated the presence of IL-2, INF-γ and TNF-α in the serum of renal transplant patients who had been treated prophylactically with muromonab-CD3. Beginning on the day of transplantation, nine patients received IV muromonab-CD3 (5 mg day^{-1}), azathioprine (2 mg kg^{-1} per day), and prednisolone (0.3 mg kg^{-1} per day) for 14 days, and six patients received orally on the day of transplantation CsA (6 mg kg^{-1}) and azathioprine (1 mg kg^{-1}), as well as methylprednisolone (2 mg kg^{-1}, intravenously). Muromonab-CD3 patients also received a 1 mg kg^{-1} bolus of methylprednisolone immediately preceding therapy, and 100 mg hydrocortisone 30 min later. No increase in TNF was found in patients treated with CsA, and these patients had no detectable levels of IL-2 or INF-γ. However, all nine patients treated with muromonab-CD3 experienced a significant rise in the serum levels of TNF ($p < 0.01$), IL-2 ($p < 0.01$) and INF-γ ($p < 0.01$). The TNF levels peaked at 1 h following muromonab-CD3 administration, and remained elevated compared to those in CsA-treated patients ($p < 0.05$) for 4 h. IL-2 and INF-γ first became evident in the serum at 1 h after dosing, and peaked after 2 h in eight of the nine patients. IL-2 and INF-γ were not detectable in the serum after 24 h. All patients developed a fever ($p < 0.01$ compared with CsA-treated patients). The second and third injections of muromonab-CD3 induced no significant release of cytokines, with the exception of one patient who exhibited a low level of INF at 2 h after the second injection.

Several authors have suggested that the CRS appears only after the first few injections of muromonab-CD3 (Abramowicz et al. 1991; Chatenoud 1993b; Chatenoud et al. 1991; First et al. 1993; Rossi et al. 1993; Todd et al. 1989). However, Vasquez et al. (1995) conducted a retrospective review which suggested that many patients continue to experience this syndrome subsequent to the first two doses. In a review of 83 evaluable renal transplant patients, the investigators discovered that 50% of patients experienced fever and chills at the first dose.

However, over one-third of patients developed reactions that were apparently mediated by muromonab-CD3 after doses beyond the second. Those patients who experienced the later effects were usually being treated for rejection rather than to induce immunosuppression. The authors concluded that the extension of adverse reactions to the later doses suggests that other mechanisms and/or mediators to this phenomenon might exist that have yet to be identified.

9.5.2
Symptoms of the Cytokine Release Syndrome

Most patients develop this acute clinical syndrome associated with the first few doses of muromonab-CD3 (particularly the first two to three doses). Symptoms generally resolve after 4 to 6 h, and cessation of therapy is usually not necessary (Rossi et al. 1993). The CRS has ranged from a more frequently reported mild, self-limited, "flu-like" illness to a less frequently reported severe, life-threatening shock-like reaction, which may include serious cardiovascular and CNS manifestations. The syndrome typically begins approximately 30 to 60 min after administration of a dose of muromonab-CD3 (though it may also occur later) and may persist for several hours. The frequency and severity of this symptom complex are usually greatest with the first dose. With each successive dose of muromonab-CD3, both the frequency and severity of the CRS tend to diminish. Increasing the amount of muromonab-CD3 or resuming treatment after a hiatus may result in a reappearance of the CRS.

The CRS displays a distinctive group of clinical signs and symptoms (Jeyarajah et al. 1993). Common clinical manifestations may include high fever (often spiking >40 °C), chills/rigors, headache, tremor, nausea/vomiting, diarrhea, abdominal pain, malaise, muscle/joint aches and pains, and generalized weakness. Less frequently reported adverse experiences include minor dermatologic reactions (e.g., rash, pruritus, etc.) and a spectrum of often serious, occasionally fatal, cardiorespiratory and CNS adverse experiences. A higher frequency of cardiac, gastrointestinal, and neurologic adverse events is usually observed in patients treated prophylactically than is observed in those treated for actual acute rejection (Jeyarajah et al. 1993).

Cardiorespiratory findings may include dyspnea, shortness of breath, bronchospasm/wheezing, tachypnea, respiratory arrest/distress, cardiovascular collapse, cardiac arrest, angina/myocardial infarction, chest pain/tightness, tachycardia (including ventricular), hypertension, hemodynamic instability, hypotension including profound shock, heart failure, pulmonary edema (cardiogenic and noncardiogenic), adult respiratory distress syndrome, hypoxemia, apnea, and arrhythmias.

In the initial studies of renal allograft rejection, potentially fatal severe pulmonary edema occurred in 5% of the initial 107 patients. Fluid overload was present before treatment in all of these cases. The reaction did not occur in any of the subsequent 311 patients treated with first-dose volume/weight restrictions. In subsequent trials, severe pulmonary edema has also occurred in patients who appeared to be euvolemic. The pathogenesis of pulmonary edema may involve all or some of the following: volume overload, increased pulmonary vascular permeability, and/or reduced left ventricular compliance/contractility. During the first 1 to 3 days of muromonab-CD3 therapy, some patients have experienced an acute and transient decline in the glomerular filtration rate (GFR), there being a diminished urine output with a resultant increase in the level of serum creatinine. A massive release of cytokines appears to lead to reversible renal functional impairment and/or delayed renal allograft function. Similarly, transient elevations in hepatic transaminases have been reported following administration of the first few doses of muromonab-CD3.

In controlled clinical trials for the treatment of acute renal allograft rejection, patients treated with muromonab-CD3 plus concomitant low-dose immunosuppressive therapy (primarily azathioprine and corticosteroids) were observed to have an increased incidence of adverse experiences during the first 2 days of treatment, as compared with the group of patients receiving azathioprine and high-dose steroid therapy. During this period the majority of patients experienced pyrexia (90%; 40.0 °C or above in 19% of cases) and chills (59%). In addition, other adverse experiences occurring in 8% or more of the patients during the first 2 days of muromonab-CD3 therapy included dyspnea (21%), nausea (19%), vomiting (19%), chest pain (14%), diarrhea (14%), tremor (13%), wheezing (13%), headache (11%), tachycardia (10%), rigor (8%), and hypertension (8%). A similar spectrum of clinical manifestations has been observed in open clinical studies and in post-marketing experience involving patients treated with muromonab-CD3 for rejection following renal, cardiac, and hepatic transplantation.

9.5.3 Muromonab-CD3 and the Cytokine Release Syndrome

Thistlethwaite et al. (1988) reported the complications associated with muromonab-CD3 therapy in 122 renal transplant recipients. Muromonab-CD3 was administered to 83 patients for rejection treatment, and to 39 patients for prophylactic immunosuppression induction. Therapy for all patients was supplemented with CsA, and most received triple therapy that included CsA, prednisone, and azathioprine. Muromonab-CD3 (5 mg day^{-1}, 2.5 mg day^{-1} for children under 30 kg) was administered to 39 patients treated for induction, along with azathioprine 3 mg kg^{-1} per day and prednisone 0.25 mg kg^{-1} per day. Treatment of muromonab-CD3 continued for 7 or 14 days. Two days prior to the last muromonab-CD3 dose, CsA (10 mg kg^{-1} per day) was given, and the dose of azathioprine was reduced to 2 mg kg^{-1} per day. In the 83 patients treated for rejection therapy, methylprednisolone (5–10 mg kg^{-1}) was given up to three times a day in addition to baseline immunosuppression. A percutaneous needle biopsy of the renal allograft was performed to confirm a rejection diagnosis if adequate response was not achieved. If rejection was confirmed, muromonab-CD3 (5 mg day^{-1}, 2.5 mg day^{-1} for children under 30 kg) was administered. Adult patients received acetaminophen (paracetamol) 600 mg and diphenhydramine 50 mg prior to the first muromonab-

CD3 dose, and hydrocortisone 100 mg 30 min later. CsA was discontinued while azathioprine was reduced to 25 mg day^{-1} on the day that muromonab-CD3 was commenced. Peripheral blood CD3+ cell counts were monitored prior to muromonab-CD3 therapy, and then three times weekly.

The adverse event experienced with highest incidence in both the rejection treatment and immunosuppression groups during the 122 courses of muromonab-CD3 treatment was hyperpyrexia (89% and 92%, respectively). Most patients experienced this event during the first treatment only; however, some continued to develop fevers at later doses, and a few experienced post-dose fever spikes throughout the course of therapy. Fluid-overload contributed to hypertension and/or mild dyspnea, and tachycardia and/or hypotension were credited to relative volume depletion due to planned dialysis or forced diuresis prior to therapy. There were no reports of serious pulmonary complications. Headache (35% and 54%, respectively) and rigors (29% and 31%, respectively) were also frequently reported events. Self-limiting events included gastrointestinal symptoms (nausea/vomiting/diarrhea) and CNS effects (headache, and rare cases of aseptic meningitis or seizures). Most of these events occurred between Days 2 and 5 of treatment, and they generally resolved after 2 to 4 days. Aseptic meningitis developed in four patients (3%), three of whom also had leukocytosis revealed by lumbar puncture. Fifty-two patients (44%) experienced headaches without any other CNS sequelae. Generalized seizures were reported in eight patients (6%); these patients presented with multiple risk factors, including fever and hypercalcemia. The use of muromonab-CD3 in uremic patients during the early postoperative period tended to predispose patients to seizure activity. There were only rare reports of late adverse events to muromonab-CD3 therapy. Infectious complications were an identified late event of muromonab-CD3 therapy, and two-thirds of patients who received two or three doses experienced infections, compared to one-sixth of patients who received only one prophylactic dose.

A randomized, prospective, double-blind study was conducted by Norman et al. (1993) to compare the severity of CRS between two dosage regimens of muromonab-CD3. A total of 26 renal transplant patients was enrolled. Patients in the high-dose arm received 5 mg muromonab-CD3 for 12 days (60 mg total), while those in the low-dose arm received 1 mg muromonab-CD3 for 2 days, followed by 2 mg daily for 10 days (22 mg total). All patients also received a standard immunosuppression regimen of CsA, azathioprine, methylprednisolone, and prednisone. There were no differences in demographics or medical history between the groups, except that more men were randomly assigned to the low-dose group ($p < 0.05$). CRS side effects, including anorexia, nausea, headache, vomiting, diarrhea, weakness, dyspnea, pulmonary edema, and hallucinations, were observed with equivalent frequency between the two groups. The high-dose group also had no significantly greater degree and occurrence of pyrexia than the low-dose group. There were significantly fewer reports of hypotension in the low-dose group ($p < 0.02$). In this study, both protocols were effective, but the lesser degree of hypotension seen with the low-dose protocol improved its safety profile over the high-dose regimen.

9.5.4
Management of the Cytokine Release Syndrome

Clinical management of the CRS involves precautionary measures to prevent the serious sequelae of first-dose reactions. No treatment modality is known completely to prevent this syndrome, but the severity can be greatly reduced with careful patient management (Jeyarajah et al. 1993). Several pharmaceutical agents have been used to decrease the severity of CRS related to muromonab-CD3 therapy. Acetaminophen has been administered 30 min prior to therapy for fever reduction, and diphenhydramine is often administered in an attempt to ward off any allergic-type reaction. Methylprednisolone is used for both antipyretic and anti-inflammatory properties. Patients with fluid overload benefit from forced diuresis, hemodialysis, or hemofiltration to prevent pulmonary edema. Muromonab-CD3 should be avoided in patients who are more than 3% over their baseline bodyweight, or who demonstrate hemodynamic instability. Together with chest radiography to rule out fluid overload, these measures can dramatically reduce the likelihood of life-threatening complications. Finally, bronchospasm may be controlled by nasal oxygen or inhaled bronchodilators (Gaston 1994; Rossi et al. 1993).

9.5.4.1 Methylprednisolone

Corticosteroids have nonspecific immunosuppressive and anti-inflammatory activity, which makes this class of drug a logical choice for use in transplant patients. Research has shown that increased doses of these agents are also useful in inhibiting cytokine synthesis (Rossi et al. 1993). Goldman et al. (1989) conducted a study to evaluate the effect of high-dose methylprednisolone prior to the first muromonab-CD3 injection on levels of cytokine release. Peak levels of TNF, INF-γ, and IL-2 were compared in patients receiving an IV bolus of either 1 or 8 $mg\,kg^{-1}$ methylprednisolone 4 h prior to the administration of muromonab-CD3. The extent of TNF and INF-γ release was significantly reduced in patients receiving high-dose methylprednisolone ($p < 0.05$) compared to those receiving low-dose therapy. Additionally, only five of 15 patients pretreated at high doses required postoperative dialysis, in contrast to 14 of the 21 patients treated with low-dose methylprednisolone ($p = 0.04$). These authors recommended that high-dose methylprednisolone be administered by IV bolus preceding muromonab-CD3 therapy in kidney transplant patients.

Chatenoud et al. (1991) conducted a prospective, randomized trial to determine the optimal timeframe for administration of high-dose corticosteroids with muromonab-CD3 therapy. Three groups of patients participated in this study. Group 1 consisted of 27 renal transplant recipients who received 5 $mg\,day^{-1}$ muromonab-CD3 for 14 days for rejection prophylaxis. In addition, patients received corticosteroids (0.25 $mg\,kg^{-1}$ per day) and azathioprine (3 $mg\,kg^{-1}$ per day). All patients in Group 1 were given high-dose (1 g) methylprednisolone without any precise schedule, either at the same time as the first muromonab-CD3 injection or between 30 min and 4 h later. The six patients assigned to Group 2 received intrafamilial renal transplant, and consequently did not receive daily corticosteroids. These patients received 5 mg day^{-1} muromonab-CD3 for 20 days and azathioprine (3 mg kg^{-1} per day). Three patients did not receive the 1-g dose of methylprednisolone, while three other patients received the same pretreatment as patients in Group 1 due to the severity of their illness. Again, no dosing schedule was followed for corticosteroid administration. Data from 12 renal transplant patients in Group 3 were included in an open randomized analysis, where the regimen was identical to that of Group 1. However, six of the patients were assigned to receive high-dose methylprednisolone concomitantly while six were assigned to receive it 1 h preceding muromonab-CD3 administration.

In Group 1, a marked increase in TNF plasma levels was observed. Levels started to decrease at 4 h after the injection of muromonab-CD3, but remained higher than pretreatment values. Subsequent injections did not produce any further increases in TNF levels. At 1 h after injection, INF-γ levels increased, and then peaked by 4 h. Although the nonprospective collection approach made the clinical data difficult to assess, a trend was recognizable: corticosteroids given 30 to 60 min prior to muromonab-CD3 decreased the CRS symptoms and decreased cytokine levels. Group 2 exhibited the highest TNF and INF-γ levels, associated with serious acute clinical syndrome. The results from patients in Group 3 confirmed that the methylprednisolone administration schedule markedly affects the activity of TNF and INF-γ. With high-dose methylprednisolone given 1 h before muromonab-CD3 therapy, TNF levels were significantly lower ($p < 0.007$) than with concomitant administration. The same effect holds true for levels of INF-γ measured at 4 h after muromonab-CD3 administration ($p < 0.003$). Although these data are encouraging, the authors state that anesthesia was given to all patients and may have contributed to the improved symptomatology. The authors recommend administering high-dose methylprednisolone 1 h prior to muromonab-CD3 therapy to decrease cytokine production and resultant CRS.

Bemelman et al. (1994) attempted to reduce the severity of CRS symptoms by manipulating the time interval between steroid and muromonab-CD3 administration in renal allograft recipients treated for acute cellular rejection. Three groups of patients were treated with 5 mg kg^{-1} muromonab-CD3 for 10 days. Group 1 ($n = 10$) received 500 mg methylprednisolone at 6 h before the first muromonab CD3 dose; Group 2 ($n = 10$) received the same dose of methylprednisolone at 1 h prior; and Group 3 ($n = 6$) received two equal doses of methylprednisolone 250 mg, one given at 6 h and one at 1 h prior to muromonab-CD3 bolus injection. All methylprednisolone doses were administered as a 20-min infusion. Patients also received basic immunosuppressive therapy consisting of CsA and prednisolone. Clinical side effects, lymphocyte counts, and plasma TNF and IL-6 levels were monitored at frequent intervals. The patients' body temperature was measured prior to the muromonab-CD3 bolus and again at 2, 4, 8, 12, 24, 36, and 48 h. Other adverse side effects were monitored by observation and questioning; patient blood pressure was assessed twice daily. In Group 3, which received steroids administered in a divided dose, the peak temperature levels were significantly lower than in Groups 1 and 2 ($p = 0.0459$ and 0.0157, respectively), and patients experienced significantly fewer side effects ($p = 0.0392$ and 0.016, respectively). There were no differences in cumulative side effect scores or body temperature between Groups 1 and 2 during the 24-h period following muromonab-CD3 administration. The blood pressures of the three groups remained unchanged. In Group 3, the median plasma level of IL-6 was 14 pg mL^{-1}, compared to 180 pg mL^{-1} in Group 1 and 43 pg mL^{-1} in Group 2; in addition, peak levels were significantly decreased compared to Group 1 ($p = 0.0152$). Peak IL-6 levels occurred at 3 h in Groups 2 and 3, and were apparently postponed by 3 h in Group 1. Group 2 patients displayed significantly higher peak levels of TNF than did the other groups ($p = 0.0321$). The authors concluded that an additive effect was observed when methylprednisolone was administered in two divided doses, at 6 h and at 1 h prior to the first dose of muromonab-CD3. Although the prevention was not complete, administration of methylprednisolone in this manner aided in preventing the first-dose side effects of muromonab-CD3.

9.5.4.2 **Pentoxifylline**

A prospective study was conducted by Vincenti et al. (1993) to assess the effects of pretreatment with pentoxifylline on the cytokine-induced first-dose side effects of muromonab-CD3 in 20 renal transplant patients. Ten patients were historical controls and were also treated with muromonab-CD3. Patients were treated with 5 mg day^{-1} muromonab-CD3 for 10 days. All patients received 7 mg kg^{-1} prednisolone (IV) at 1 h before the first muromonab-CD3 dose; pretreatment consisted of 2 mg kg^{-1} before the second dose and 0.5 mg kg^{-1} before subsequent doses. Additionally, patients were administered acetaminophen 650 mg and diphenhydramine 50 mg every 6 h for 2 days. Patients were given pentoxifylline 600 mg every 8 h for the 24-h period prior to muromonab-CD3 therapy and throughout the course of treatment. Muromonab-CD3-related adverse events (fever, chills, headache, dyspnea) were rated for severity on a scale of 0 to 3 (0 = absent, 3 = severe); gastrointestinal events were graded as either +1 (present) or 0 (absent). Compared to historic controls, patients treated with pentoxifylline had significantly lower reaction scores ($p < 0.0001$). It has been hypothesized that pentoxifylline might be successful in reducing the severity of the CRS because of its ability to inhibit transcription of TNF mRNA and thus prevent the accumulation of TNF mRNA in activated monocytes (Rossi et al. 1993; Vincenti et al. 1993). In this study the first-dose reactions stimulated by muromonab-CD3 were significantly decreased by pentoxifylline. Rossi et al. (1993) have cautioned that in the study conducted by Vincenti and colleagues, pentoxifylline was used in conjunction with high-dose corticosteroids, thus making it difficult to ascertain the effects of pentoxifylline alone. Additionally, pentoxifylline has shown no direct effect on end-organ cytokine response, and does not affect TNF already in circulation. Consequently, Rossi et al. (1993) concluded that the clinical evidence for a beneficial effect remains controversial, and recommended starting pentoxifylline prior to treatment with muromonab-CD3.

9.5.4.3 Indomethacin

Various studies have shown that inhibition of the cyclooxygenase pathway decreases cytokine release, thus initiating attention to the use of indomethacin for its potential role in reducing CRS symptoms (Rossi et al. 1993). Shield et al. (1992) conducted a randomized, placebo-controlled, multicenter study to evaluate the possible effects of indomethacin on the incidence and severity of CRS symptoms during the first 48 h after muromonab-CD3 injection. Sixty patients being treated with muromonab-CD3 as either first-line or rescue therapy for renal rejection were randomly assigned to receive either indomethacin 100 mg ($n = 29$) or placebo ($n = 31$) at 1 to 2 h before muromonab-CD3, then indomethacin 50 mg or placebo respectively at 8-h intervals for a total of 48 h. Patients also received methylprednisolone and diphenhydramine at 1 to 2 h before the initial muromonab-CD3 dose. The indomethacin-treated group reported significantly fewer episodes of body pain than the placebo group (0% versus 16%, $p = 0.053$). Additionally, the indomethacin group experienced a milder febrile response ($p = 0.02$). The incidence of chills in the indomethacin group was also significantly lower (45% versus 71%, $p = 0.066$). Significantly fewer patients in the indomethacin group developed pyrexia after the second muromonab-CD3 dose than did placebo-treated patients (28% and 87%, $p = 0.001$). The severity of pyrexia ($p = 0.001$) and tachycardia ($p = 0.024$) and overall reported adverse experiences ($p = 0.05$) were all significantly lower in the indomethacin group. Although these findings are encouraging, the use of indomethacin still induced significant increases in circulating IL-1β ($p = 0.01$), TNF ($p = 0.001$) and IL-6 ($p = 0.025$) after the first dose of muromonab-CD3. The placebo group had significant increases only in TNF and GM-CSF levels ($p = 0.001$ for both). The authors concluded that, while indomethacin did not inhibit the production and circulation of cytokines, the prevalence and severity of CRS symptoms associated with muromonab-CD3 were greatly reduced during the first 48 h post dose.

9.5.4.4 Recombinant Human Soluble Tumor Necrosis Factor Receptor (TNFR:Fc)

TNFR:Fc, a dimer of the p80 TNF receptor, binds to both TNF-α and lymphotoxin (LT). This product may be more appealing as preventive therapy for muromonab-CD3-related CRS than anti-TNF mAb as it has a higher affinity for TNF, has minimal immunogenicity, and is physiologically closer to a human receptor. Eason et al. (1996) performed a prospective, randomized study in 12 renal graft recipients who were diagnosed with steroid-resistant rejection to evaluate the ability of TNFR:Fc to decrease the symptoms of CRS and promote renal function. All patients received a bolus dose of methylprednisolone 500 mg, oral acetaminophen 350 mg, and oral diphenhydramine 25 mg before the first muromonab-CD3 dose. The first two doses of muromonab-CD3 were received either alone or with TNFR:Fc (0.05 to 0.15 $mg\,kg^{-1}$), administered 30 min prior. There was a significant decrease in the frequency of chills ($p < 0.05$) in the group treated with TNFR:Fc compared to the control group. The TNFR:Fc group also had fewer symptoms ($p = 0.032$) by Day 2. TNFR:Fc returned patients to normal renal function beginning at 24 h compared to 48 h in the control group, and in addition was associated with lower serum creatinine levels during the first 6 days of muromonab-CD3 therapy ($p = 0.032$).

9.5.4.5 Anti-TNF Monoclonal Antibodies

Studies are under way to develop new treatments to control CRS, such as nonactivating anti-CD3 antibodies. The use of the $F(ab')_2$ fragment of muromonab-CD3 antibody may show promise in reducing cytokine activation (Jeyarajah et al. 1993). Current research has demonstrated the potential of anti-TNF IgG, which may bind to TNF already released into the circulation of allograft recipients. The results from murine models of this process have demonstrated significant TNF inactivation and consequent prevention of the synthesis of other cytokines (Rossi et al. 1993).

9.6 The Consequences of Immunosuppression

9.6.1 Infections

Muromonab-CD3 is usually added to immunosuppressive therapeutic regimens, thereby augmenting the degree of immunosuppression. This increase in the total amount of immunosuppression may alter the spectrum of infections observed and increase the risk, the severity, and the morbidity of infectious complications. During the first month post transplant, patients are at greatest risk for the following infections:

- those present prior to transplant, perhaps exacerbated by posttransplant immunosuppression;
- infection conveyed by the donor organ; and
- the usual postoperative urinary tract, intravenous line-related, wound, or pulmonary infections due to bacterial pathogens.

At approximately 1 to 6 months post transplant, patients are at risk for viral infections [e.g., CMV, EBV, herpes simplex virus (HSV), etc.] which produce serious systemic disease and which also increase the overall state of immunosuppression.

Reactivation (at 1 to 4 months post transplant) of EBV and CMV has been reported. When administration of an anti-lymphocyte antibody, including Orthoclone OKT3, is followed by an immunosuppressive regimen including CsA, there is an increased risk of reactivating CMV and an impaired ability to limit its proliferation, resulting in symptomatic and disseminated disease. EBV infection – either primary or reactivated – may play an important role in the development of posttransplant lymphoproliferative disorders.

In the pediatric transplant population, viral infections often include pathogens uncommon in adults, such as varicella zoster virus (VZV), adenovirus, and respiratory syncytial virus (RSV). A large proportion of pediatric patients have not been infected with the herpes viruses prior to transplantation and, therefore, are susceptible to developing primary infections from the grafted organ and/or blood products.

Anti-infective prophylaxis may reduce the morbidity associated with certain potential pathogens, and should be considered for pediatric and other high-risk patients. The judicious use of immunosuppressive drugs, including type, dosage, and duration, may limit the risk and seriousness of some opportunistic infections. It is also possible to reduce the risk of serious CMV or EBV infection by avoiding transplantation of a CMV-seropositive (donor) and/or EBV-seropositive (donor) organ into a seronegative patient.

9.6.2
Neoplasia

As a result of depressed cell-mediated immunity from immunosuppressive agents, organ transplant patients have an increased risk of developing malignancies. This risk is evidenced almost exclusively by the occurrence of lymphoproliferative disorders, squamous cell carcinomas of the skin and lip, and sarcomas. In immunosuppressed patients, T-cell cytotoxicity is impaired allowing for transformation and proliferation of EBV-infected B lymphocytes. Transformed B lymphocytes are thought to initiate oncogenesis, which ultimately culminates in the development of most posttransplant lymphoproliferative disorders. Patients – especially pediatric patients – with primary EBV infection may be at a higher risk for the development of EBV-associated lymphoproliferative disorders. Current data support an association between the development of lymphoproliferative disorders at the time of active EBV infection and muromonab-CD3 administration in pediatric liver allograft recipients.

Following the initiation of muromonab-CD3 therapy, patients should be continuously monitored for evidence of lymphoproliferative disorders through physical examination and histological evaluation of any suspect lymphoid tissue. Close surveillance is advised, as early detection with subsequent reduction of total immunosuppression may result in the regression of some of these lymphoproliferative disorders. As the potential for the development of lymphoproliferative disorders is related to the duration and extent (intensity) of total immunosuppression, physicians are advised:

- to adhere to the recommended dosage and duration of muromonab-CD3 therapy;
- to limit the number of courses of muromonab-CD3 and other anti-T-lymphocyte antibody preparations administered within a short period of time; and
- if appropriate, to reduce the dosage(s) of immunosuppressive drugs used concomitantly to the lowest level compatible with an effective therapeutic response.

A recent study examined the incidence of non-Hodgkin's lymphoma (NHL) among 45 000 kidney transplant recipients and over 7500 heart transplant recipients. This study suggested that all transplant patients, regardless of the immunosuppressive regimen employed, are at increased risk of NHL over the general population. The relative risk was highest among those receiving the most aggressive regimens. The long-term risk of neoplastic events in patients being treated with muromonab-CD3 has not been determined.

References

Abramowicz, D., Schandene, L., Goldman, M., et al. (1989) Release of tumor necrosis factor, interleukin-2, and gamma-interferon in serum after injection of OKT3 monoclonal antibody in kidney transplant recipients. *Transplantation* 47: 606–608.

Abramowicz, D., Goldman, M., Mat, O., et al. (1994) OKT3 serum levels as a guide for prophylactic therapy: a pilot study in kidney transplant recipients. *Transpl Int* 7: 258–263.

Alegre, M.-L., Gastadello, K., Abramowicz, D., et al. (1991) Evidence that pentoxifylline reduces anti-CD3 monoclonal antibody-induced cytokine release syndrome. *Transplantation* 52: 674–679.

Arenas, J.I., Laskus, T., Wilkinson, J., et al. (2003) Rejection treatment modifies HCV quasispecies evolution after transplantation. *Hepatology* 38(Suppl.1): 532.

Balk, A.H.M.M., Simoons, M.L., Jutte, N.H.P.M., et al. (1991) Sequential OKT3 and cyclosporine after heart transplantation. A randomized study with single and cyclic OKT3. *Clin Transpl* 5: 301–305.

Barr, M.L., Sanchez, J.A., Seche, L.A., et al. (1990) Anti-CD3 monoclonal antibody induction therapy. Immunological equivalence with triple-drug therapy in heart transplantation. *Circulation* 82(Suppl. IV): 291–294.

Bemelman, F.J., Buysmann, S., Surachno, J., et al. (1994) Pretreatment with divided dose of steroids strongly decreases side effects of OKT3. *Kidney Int* 46: 1674–1679.

Benvenisty, A.I., Cohen, D., Stegall, M.D., et al. (1990) Improved results using OKT3 as induction immunosuppression in renal allograft recipients with delayed graft function. *Transplantation* 49: 321–327.

Bernard, A., Boumsell, L. (1984) The clusters of differentiation (CD) defined by the First International Workshop on Human Leucocyte Differentiation Antigens. *Hum Immunol* 11(1): 1–10.

Bhat, G., Schroeder, T.J. (1997) Clinical role of immunologic monitoring during OKT3 treatment. *Transpl Proc* 29(Suppl.8): 21–26.

Bloemena, E., ten Berge, I.J.M., Surachno, J., et al. (1990) Kinetics of interleukin 6 during OKT3 treatment in renal allograft recipients. *Transplantation* 50: 330–331.

Bock, H.A., Gallati, H., Zürcher, R.M., et al. (1995) A randomized prospective trial of prophylactic immunosuppression with ATG-Fresenius versus OKT3 after renal transplantation. *Transplantation* 59: 830–840.

Broughan, T.A., Valenzuela, R., Escorcia, E., et al. (1994) Mouse antibody-coated lymphocytes during OKT3 therapy in liver transplantation. *Clin Transpl* 8: 488–491.

Broyer, M., Gagnandoux, M.-F., Guest, G., et al. (1993) Prophylactic OKT3 monoclonal antibody versus antilymphocyte globulins: a prospective, randomised study in 148 first cadaver kidney grafts. *Transplant Proc* 25: 570–571.

Caillat-Zucman, S., Blumenfield, N., Legendre, C., et al. (1981) The OKT3 immunosuppressive effect: in situ antigenic modulation of human graft-infiltrating T cells. *Transplantation* 32: 535–539.

Carey, G., Lisi, P.J., Schroeder, T.J. (1995) The incidence of antibody formation to OKT3 consequent to its use in organ transplantation. *Transplantation* 60: 151–158.

Chatenoud, L. (1993a) Humoral immune response against OKT3. *Transplant Proc* 25(Suppl.1): 68–73.

Chatenoud, L. (1993b) OKT3-induced cytokine-release syndrome: preventive effect of anti-tumor necrosis factor monoclonal antibody. *Transplant Proc* 25(2): 47–51.

Chatenoud, L. (1994) Use of CD3 antibodies in transplantation and autoimmune disease. *Tansplant Proc* 26: 3191–3193.

Chatenoud, L. (1995) *Monoclonal antibodies in transplantation.* Medical Intelligence Unit. RG Landes Company, p. 3.

Chatenoud, L., Legendre, C., Ferran, C., et al. (1990) In vivo cell activation following OKT3 administration. Systemic cytokine release and modulation by corticosteroids. *Transplantation* 49: 697–702.

Chatenoud, L., Legendre, C., Ferran, C., et al. (1991) Corticosteroid inhibition of the OKT3-induced cytokine-related syndrome – dosage and kinetics prerequisites. *Transplantation* 51: 334–338.

Chkotua, A.B., Klein, T., Shabtai, E., et al. (2003) Kidney transplantation from living-unrelated donors: comparison of outcome with living-related and cadaveric transplants under current immunosuppressive protocols. *Urology* 62: 1002–1006.

Ciancio, G., Mattiazzi, A., Vaidya, A., et al. (2004) A prospective study with thymoglobulin/daclizumab induction, tacrolimus and steroid maintenance, comparing rapamycin with mycophenolate mofetil as adjunctive immunosuppression in simultaneous pancreas-kidney (SPK) transplantation. *J Urol* 171(4): 489–490.

Cole, E.H., Cattran, D.C., Farewell, V.T., et al. (1994) A comparison of rabbit antithymocyte serum and OKT3 as prophylaxis against renal allograft rejection. *Transplantation* 57: 60–67.

Cosimi, A.B., Jenkins, R.L., Rohrer, R.J., et al. (1990) A randomized clinical trial of prophylactic OKT3 monoclonal antibody in liver allograft recipients. *Arch Surg* 125: 781–785.

Costanzo-Nordin, M.R., Sullivan, E.J., Johnson, M.R., et al. (1990) Prospective randomized trial of OKT3 versus horse antithymocyte globulin based immunosuppressive prophylaxis in heart transplantation. *J Heart Transplant* 9: 306–315.

Costanzo-Nordin, M.R., Fisher, S.G., O'Sullivan, E.J., et al. (1993) HLA-DR incompatibility predicts heart transplant rejection independent of immunosuppressive prophylaxis. *J Heart Lung Transpl* 12: 779–789.

De Vault, Jr., G.A., Kohan, D.E., Nelson, E. W., et al. (1994) The effect of oral pentoxifylline on the cytokine release syndrome during inductive OKT3. *Transplantation* 57: 532–540.

Eason, J.D., Pasqual, M., Wee, S., et al. (1996) Evaluation of recombinant human soluble dimeric tumor necrosis factor receptor for prevention of OKT3-associated acute clinical syndrome. *Transplantation* 61(2): 224–228.

Ellenhorn, J.D.I., Woodle, E.S., Ghobreal, I., et al. (1990) Activation of human T cells in vivo following treatment of transplant recipients with OKT3. *Transplantation* 50: 608–612.

Farges, O., Ericzon, B.-G., Bresson-Hadni, S., et al. (1994) A randomized trial of OKT3-based versus cyclosporine-based immunoprophylaxis after liver-transplantation: long-term results of a European and Australian multicenter study. *Transplantation* 58: 891–898.

Ferran, C., Dy, M., Merite, S., et al. (1990) Reduction of morbidity and cytokine release in anti-CD3 MoAb-treated mice by corticosteroids. *Transplantation* 50: 642–648.

First, M.R., Schroeder, T.J., Hariharan, S. (1993) OKT-3 induced cytokine-release syndrome: renal effects (cytokine nephropathy). *Transplant Proc* 25(Suppl.1): 25–26.

Frey, D.J., Matas, A.J., Gillingham, K.J., et al. (1992) Sequential therapy – a prospective randomized trial of malg versus OKT3 for prophylactic immunosuppression in cadaver renal allograft recipients. *Transplantation* 54: 50–56.

Gaston, R.S. (1994) Cytokines and transplantation: a clinical perspective. *Transplant Sci* 4(1): 9–19.

Gaston, R.S., Deierhoi, M.H., Patterson, T., et al. (1991) OKT3 first-dose reaction: association with T-cell subsets and cytokine release. *Kidney Int* 39: 141–148.

Gebel, H.M., Lebeck, L.L., Jensik, S.C., et al. (1989a) Discordant expression of CD3 and T cell receptor antigens on lymphocytes from patients treated with OKT3. *Transplant Proc* 21: 1745–1746.

Gebel, H.M., Lebeck, L.L., Jensik, S.C., et al. (1989b) T cells from patients successfully treated with OKT3 do not react with the T-cell receptor antibody. *Hum Immunol* 26: 123–130.

Goldman, M., Gerard, C., Abramowicz, D., et al. (1992) Induction of interleukin-6 and interleukin-10 by the OKT3 monoclonal antibody: possible relevance to posttransplant lymphoproliferative disorders. *Clin Transpl* 6: 265–268.

Goumy, L., Ferran, C., Merite, S., et al. (1990) In vivo anti-CD3-driven cell activation. Cellular source of induced tumor necrosis factor, interleukin-1β, and interleukin 6. *Transplantation* 61: 83–87.

Griffith, B.P., Kormos, R.L., Armitage, J.M., et al. (1990) Comparative trial of immunoprophylaxis with RATG versus OKT3. *J Heart Transplant* 9: 301–305.

Hammond, E.A., Yowell, R.L., Greenwood, J., Hartung, L., Renlund, D., Wittwer, C. (1993) Prevention of adverse clinical outcome by monitoring of cardiac transplant patients for murine monoclonal CD3 antibody (OKT3) sensitization. *Transplantation* 55(5): 1061–1063.

Hanto, D.W., Jendrisak, M.D., So, S.K.S., et al. (1994) Induction immunosuppression with antilymphocyte globulin or OKT3 in cadaver kidney transplantation. Results of a single institution prospective randomized trial. *Transplantation* 57: 377–384.

Henell, K.R., Norman, D.J. (1993) Monitoring OKT3 treatment: pharmacodynamic and pharmacokinetic measures. *Transplant Proc* 25(Suppl.1): 83–85.

Herbelin, A., Abramowicz, D., de Groote, D., et al. (1999) CD3 antibody-induced IL-10 in renal allograft recipients: an in vivo and in vitro analysis. *Transplantation* 68(5): 616–622.

Hoffman, T., Tripathi, A.K., Lee, Y.L., et al. (1992) Stimulation of human monocytes by anti-CD3 monoclonal antibody: induction of inflammatory mediator release via immobilization of Fc receptor by adsorbed immunoglobulin and T-lymphocytes. *Inflammation* 16: 571–585.

Howard, R.J., Pfaff, W.W., Brunson, M.E., et al. (1993) Delayed graft function is associated with an increased incidence of occult rejection and results in poorer graft survival. *Transplant Proc* 25: 884.

Ippoliti, G., Negri, M., Abelli, P., et al. (1991) Preoperative prophylactic OKT3 vs RATG. A randomized clinical study in heart transplant patients. *Transplant Proc* 23: 2272–2274.

Jeyarajah, D.R., Thistlethwaite, J.R. (1993) General aspects of cytokine-release syndrome: timing and incidence of symptoms. *Transplant Proc* 25(2): 16–20.

Keever-Taylor, C.A., Craig, A., Molter, M., et al. (2001) Complement-mediated T-cell depletion of bone marrow: comparison of T10B9. 1A-31 and Muromonab-Orthoclone OKT3. *Cytotherapy* 3(6): 467–481.

Kermann, R.H., Sullivan, K., Tejani, A. (1994) Impact of HLA matching, type of crossmatch and immunosuppressive therapy on primary pediatric cadaver renal allograft survival [Abstract]. *Hum Immunol* 40(Suppl.1): 17.

Kimball, J.A., Norman, D.J., Shield, C.F., et al. (1993) OKT3 antibody response study: comparative testing of human anti-mouse antibody. *Transplant Proc* 25(Suppl.1): 74–76.

Kirklin, J.K., Bourge, R.C., White-Williams, C., et al. (1990) Prophylactic therapy for rejection after cardiac transplantation – a comparison of rabbit antithymocyte globulin and OKT3. *J Thorac Cardiovasc Surg* 99: 716–724.

Kohler, G., Milstein, C. (1976) Derivation of specific antibody producing tissue culture and tumor lines by cell fusion. *Eur J Immunol* 6: 171–178.

Kreis, H. (1993) Adverse events associated with OKT3 immunosuppression in the prevention or treatment of allograft rejection. *Clin Transpl* 7: 431–446.

Kreis, H., Legendre, C., Chatenoud, L. (1991) OKT3 in organ transplantation. *Transplant Rev* 5: 181–189.

Krensky, A.M., Weiss, A., Crabtree, G., Davis, M.M., Parham, P. (1990) T-lymphocyte-antigen interactions in transplant rejection. *N Engl J Med* 322(8): 510–517.

Kung, G., Goldstein, G., Reinherz, E.L., Schlossman, S.F. (1979) Monoclonal antibodies defining distinctive T-cell surface antigens. *Science* 206: 347.

Kung, S.C., Parikh, M., Fyfe, B., et al. (2004) Simulect induction facilitates Neoral-based steroid-free immunosuppression in primary kidney transplant recipients. *Transplant Proc* 36(Suppl.2): 475–477.

Ladowski, J.S., Dillon, T., Schatzlein, M.H., et al. (1993) Prophylaxis of heart transplant rejection with either antithymocyte globulin-, Minnesota antilymphocyte globulin-, or OKT3-based protocol. *J Cardiovasc Surg* 34: 135–140.

Latham, W.C., Cooney, R.M., Brown, K.J., et al. (1970) Preparation of purified antilymphocyte serum on an immuno-absorbent column. In proceedings of a symposium on standardization of antilymphocyte serum. 16: 171–178.

Leimenstoll, G., Zabel, P., Schroeder, P., et al. (1993) Suppression of OKT3-induced tumor necrosis factor alpha formation by pentoxifylline in renal transplant recipients. *Transplant Proc* 25: 561–563.

Lozano, M., Oppenheimer, F., Cofan, F., et al. (2001) Platelet procoagulant activity induced in vivo by muromonab CD3 infusion in uremic patients. *Thrombosis Res* 104(6): 405–411.

Macdonald, P.S., Mundy, J., Keogh, A.M., et al. (1993) A prospective randomized study of prophylactic OKT3 versus equine antithymocyte globulin after heart transplantation – increased morbidity with OKT3. *Transplantation* 55: 110–116.

McCarthy, C., Light, J.A., Aquino, A., et al. (1993) Correlation of CD3+ lymphocyte depletion with rejection and infection in renal transplants. *Transplant Proc* 25: 2477–2478.

McDiarmid, S.V., Millis, M., Terashita, G., et al. (1990) Low serum OKT3 levels correlate with failure to prevent rejection in orthotopic liver transplant patients. *Transplant Proc* 22: 1774–1776.

McDiarmid, S.V., Millis, M.J., Terasaki, P.I. (1991a) OKT3 prophylaxis in liver transplantation. *Dig Dis Sci* 36: 1418–1426.

McDiamid, S.V., Busuttil, R.W., Levy, P., et al. (1991b) The long-term outcome of OKT3 compared with cyclosporine prophylaxis after liver transplantation. *Transplantation* 52: 91–97.

Mendez, R., Abboud, H., Burdick, J., et al. (1999) Reduced intrapatient variability of cyclosporin pharmacokinetics in renal transplant recipients switched from oral Sandimmune to Neoral. *Clin Ther* 21(1): 161–171.

Menkis, A.H., Powel, A.-M., Novick, R.J., et al. (1992) A prospective randomized controlled trial of initial immunosuppression with ALG versus OKT3 in recipients of cardiac allografts. *J Heart Lung Transplant* 11: 569–576.

Mignat, C. (1997) Clinically significant drug interactions with new immunosuppressive agents. *Drug Safety* 16(4): 267–278.

Moore, C.K., O'Connell, J.B., Renlund, D.G., et al. (1991) Cardiac allograft cellular rejection during OKT3 prophylaxis in the absence of sensitization. *Transplant Proc* 23: 1055–1058.

Mühlbacher, F., Steininger, R., Längle, F., et al. (1989) OKT3 immunoprophylaxis in human liver transplantation. *Transplant Proc* 21: 2253–2254.

Muller, C., Patzke, J., Bonmann, M., et al. (1998) Alterations of protein tyrosine phosphorylation in T-cells of immunocompromised patients. *Scand J Immunol* 47(2): 101–105.

Muller, C., Bonmann, M., Heidenreich, S., et al. (1999) Tyrosine phosphorylation in peripheral T-cells of kidney transplant recipients: analyses of baseline levels and response to T-cell receptor stimulation. *Int J Mol Med* 4(2): 141–144.

Norman, D.J. (1992) Antilymphocyte antibodies in the treatment of allograft rejection: targets, mechanisms of action, monitoring, and efficacy. *Semin Nephrol* 12: 315–324.

Norman, D.J., Barry, J.M., Bennett, W.M., et al. (1991) OKT3 for induction immunosuppression in renal transplantation: a comparative study of high versus low doses. *Transplant Proc* 23: 1052–1054.

Norman, D.J., Chatenoud, L., Cohen, D., et al. (1993) Consensus statement regarding OKT3-induced cytokine-release syndrome and human anti-mouse antibodies. *Transplant Proc* 25(Suppl.1): 89–92.

Norman, D.J., Kimball, J.A., Bennett, W.M., et al. (1994) A prospective, double-blind, randomized study of high- versus low-dose OKT3 induction immunosuppression in cadaveric renal transplantation. *Transpl Int* 7: 356–361.

O'Connell, J.B., Bristow, M.R., Hammond, E.H., et al. (1991) Antimurine antibody to OKT3 in cardiac transplantation: implications for prophylaxis and retreatment of rejection. *Transplant Proc* 23: 1157–1159.

Ohman, M., Kolb, M., Leathers, L.K., et al. (1993) Multiparameter monitoring of efficacy of OKT3-induced immune suppression in immune renal allograft recipients (abstract). *Hum Immunol* 37(Suppl.1): 95.

Pons, J.A., Bueno, F., Parilla, P., et al. (1993) Cyclosporine vs. OKT3 prophylaxis after orthotopic liver transplantation. *Transplant Proc* 25: 1949.

Raasveld, M.H.M., Bemelman, F.J., Schellekens, P.A., et al. (1993) Compliment activation during OKT3 treatment: a possible explanation for respiratory side effects. *Kidney Int* 43: 1140–1149.

Reding, R., Vraux, H., de Goyet, J. de V., et al. (1993) Monoclonal antibodies in prophylactic immunosuppression after liver transplantation. A randomized controlled trial comparing OKT3 and anti-IL-2 receptor monoclonal antibody LO-Tact-1. *Transplantation* 55: 534–541.

Reinke, P., Schwinzer, H., Hoflich, C., et al. (1997) Selective in vivo deletion of alloactivated TH1 cells by OKT3 monoclonal antibody in acute rejection. *Immunol Lett* 57(1–3): 151–153.

Renlund, D.G., O'Connell, J.B., Gilbert, E. M., et al. (1989) A prospective comparison of murine monoclonal CD-3 OKT3 antibody-based and equine antithymocyte globulin-based rejection prophylaxis in cardiac transplantation decreased rejection and less corticosteroid use with OKT3. *Transplantation* 47: 599–605.

Roitt, I.M. (1993) OKT3: immunology, production, purification and pharmacokinetics. *Clin Transplant* 7: 367–373.

Rossi, S.J., Schroeder, T.J., Hariharan, S., et al. (1993) Prevention and management of the adverse effects associated with immunosuppressive therapy. *Drug Safety* 9(2): 104–131.

Ryckman, F.C., Schroeder, T.J., Pedersen, S.H., et al. (1991) Use of monoclonal antibody immunosuppressive therapy in paediatric renal and liver transplantation. *Clin Transplant* 5: 186–190.

Sanchez, R.Q., Levy, M.F., Goldstein, R.M., et al. (2003) The changing clinical presentation of recurrent primary biliary cirrhosis after liver transplantation. *Transplantation* 76(11): 1583–1588.

Schroeder, T.J., First, M.R., Mansour, M.E., et al. (1990) Antimurine antibody formation following OKT3 therapy. *Transplantation* 49: 48–51.

Schroeder, T.J., Ryckman, F.C., Hurtubise, P.E., et al. (1991) Immunological monitoring during and following OKT3 therapy in children. *Clin Transpl* 5: 191–196.

Schroeder, T.J., Michael, A.T., First, M.R., et al. (1994) Variations in serum OKT3 concentration based upon age, sex, transplanted organ, treatment regimen, and anti-OKT3 antibody status. *Ther Drug Monit* 16: 361–367.

Shaefer, M.S., Stratta, R.J., Pirruccello, S.J., et al. (1990) Peripheral lymphocyte monitoring of liver transplant recipients being treated with OKT3 for rejection or induction immunosuppression (abstract). *Pharmacotherapy* 10(3): 248.

Shield, C.F., Kahana, L., Pirsh, J., et al. (1992) Use of indomethacin to minimize the adverse reactions associated with Orthoclone OKT3 treatment of kidney allograft rejection. *Transplantation* 54(1): 164–166.

Smith, S.L. (1996) Ten years of Orthoclone OKT3 (muromonab-CD3): a review. *J Transpl Coord* 6(3): 109–119.

Steininger, R., Mühlbacher, F., Hamilton, G., et al. (1991) Comparison of CyA, OKT3, and ATG immunoprophylaxis in human liver transplantation. *Transplant Proc* 23: 2269–2271.

Steinmuller, D.R., Hayes, J.M., Novick, A.C., et al. (1991) Comparison of OKT3 with ALG for prophylaxis for patients with acute renal failure after cadaveric renal transplantation. *Transplantation* 52: 67–71.

Strong, M.L., Ueda, C.T. (1997) Effects of low and high density lipoproteins on renal cyclosporin A and cyclosporin G disposition in the isolated perfused rat kidney. *Pharm Res* 14(10): 1466–1471.

Swinnen, L.J., Fisher, R.I. (1993) OKT3 monoclonal antibodies induce interleukin-6 and interleukin-10: a possible cause of lymphoproliferative disorders associated with transplantation. *Curr Opin Nephrol Hypertens* 2: 670–678.

Tanaka, C., Kawai, R., Rowland, M. (1999) Physiologically based pharmacokinetics of cyclosporin A: reevaluation of dose-nonlinear kinetics in rats. *J Pharmacokinet Biopharm* 27(6): 597–623.

Taylor, D.O., Bristow, M.R., O'Connell, J.B., et al. (1994) A prospective, randomized comparison of cyclophosphamide and azathioprine for early rejection prophylaxis after cardiac transplantation: decreased sensitization to OKT3. *Transplantation* 58: 645–649.

Thistlethwaite, J.R., Haag, B.W., Gaber, A., et al. (1987) The use of OKT3 to treat steroid-resistant renal allograft rejection in patients receiving cyclosporine. *Transplant Proc* 19: 1901–1904.

Todd, P.A., Brogden, R.N. (1989) Muromonab CD3: a review of its pharmacologically and therapeutic potential. *Drugs* 37: 871–899.

Toyoda, M., Galfayan, K., Wachs, K., et al. (1995) Immunologic monitoring of OKT3 induction therapy in cardiac allograft recipients. *Clin Transplant* 9: 472–480.

Troppmann, C., Gillingham, K.J., Benedetti, C., et al. (1995) Delayed graft function, acute rejection, and outcome after cadaver renal transplantation: a multivariate analysis. *Transplantation* 59: 962–968.

Vallhonrat, H., Williams, W.W., Cosimi, A. B., et al. (1999) In vivo generation of C4d, Bb, iC3b, and SC5b-9 after OKT3 administration in kidney and lung transplant recipients. *Transplantation* 67(2): 153–258.

Vasquez, E.M., Pollak, R. (1997) OKT3 therapy increases cyclosporine blood levels. *Clin Transpl* 11(1): 38–41.

Vasquez, E.M., Fabrega, A.J., Pollak, R. (1995) OKT3-induced cytokine release syndrome: occurrence beyond the second dose and association with rejection therapy. *Transplant Proc* 27(1):873–874.

Vincenti, F.G., Vasconcelos, M., Birnbaum, J.L., et al. (1993) Pentoxifylline reduces the first-dose reactions following OKT3. *Transplant Proc* 25(2): 57–59.

Vosssen, A.C.T.M., Tibbe, G.J.M., Kroos, M. J., et al. (1995) Fc receptor binding of anti-CD3 monoclonal antibodies is not essential for immunosuppression, but triggers cytokine-related side effects. *Eur J Immunol* 25: 1492–1496.

Wechsler, M.E., Giardina, E.-G.V., Sciacca, R.R., et al. (1995) Increased early mortality in women undergoing cardiac transplantation. *Circulation* 91: 1029–1035.

Welter, H.F., Illner, W.-D., Schleibner, S., et al. (1990) Pilot study on induction treatment with high-dose OKT3: preliminary observations in kidney transplantation. *Transplant Proc* 22: 2272.

Wilde, M.I., Goa, L.G. (1996) Muromonab CD3. A reappraisal of its pharmacology and use as prophylaxis of solid organ transplant rejection. *Drugs* 51(5): 865–894.

Wissing, K.M., Morelon, E., Legendre, C., et al. A pilot trial of recombinant human interleukin-10 in kidney transplant recipients receiving OKT3 induction therapy. *Transplantation* 64(7): 999–1006.

Wollenek, G., Laufer, G., Laczkovics, A., et al. (1989) Comparison of a monoclonal anti-T cell antibody vs ATG as prophylaxis after heart transplantation. *Transplant Proc* 21: 2499–2501.

Woodle, E.S., Thistlethwaite, J.R., Jolliffe, L. K., Zivin, R.A., Collins, A., Adair, J.R., Bodmer, M., Athwal, D., Alegre, M.L., Bluestone J.A. (1992) Humanized OKT3 antibodies: successful transfer of immune modulating properties and idiotype expression. *J Immunol* 148(9): 2756–2763.

10
Natalizumab (Tysabri)

Sebastian Schimrigk and Ralf Gold

10.1
Introduction

Multiple sclerosis (MS) is a chronic disease of the central nervous system (CNS) where multiple foci of inflammation accompanied by demyelination and axonal transection cause various symptoms, including weakness, ataxia and other troublesome complications. During the early 1990s, different disease-modifying drugs (DMDs), including interferon (INF)-β1a (Avonex, BiogenIdec; Rebif, Serono), IFN-β1b (Betaseron, Schering), glatiramer acetate (Copaxone, Teva Pharmaceutical Industries) and mitoxantrone (Ralenova or Novantrone, Lederle-Wyeth), were found to be effective to a certain extent in influencing the course of MS. However, additional effective treatments are warranted to slow disease progression, reduce disability, and limit lesion evolution and irreversible tissue destruction.

During the past decade, substantial progress has been made in the understanding of some of the key mechanisms underlying the pathogenesis of neuroimmunological diseases. In parallel, the development of new biologicals is under way to improve treatment. A main focus here is set on monoclonal antibodies (mAbs), where natalizumab is the very first selective immunmodulatory drug with an orphan drug status for the treatment of MS.

Natalizumab is a recombinant humanized mAb that blocks the mechanism of endothelial adhesion of activated T cells in inflammatory diseases such as MS. Natalizumab (Elan Corp. plc.; Biogen Idec.) binds to the α4-chain of the heterodimeric VLA-4 receptor which is expressed predominantly on lymphocytes. Thus, natalizumab is an α_4-integrin antagonist in the class of selective adhesion molecule (SAM) inhibitors.

A rapid translation of *in-vitro* findings with VLA-4 blockade into experiments with experimental models for MS (experimental autoimmune encephalomyelitis; EAE) (Yednock et al. 1992) led to the first clinical investigations (Miller 2003) with natalizumab, followed by phase III studies where it finally proved to be a very effective drug to treat MS (Polmann et al. 2006). Approximately 3000 patients

Handbook of Therapeutic Antibodies. Edited by Stefan Dübel
Copyright © 2007 WILEY-VCH Verlag GmbH & Co. KGaA, Weinheim
ISBN 978-3-527-31453-9

were enrolled in clinical trials evaluating the efficacy of natalizumab, not only in MS but also in Crohn's disease and rheumatoid arthritis.

Nevertheless, the clinical trials have shown that this drug has potentially serious adverse side effects. On February 28, 2005 natalizumab was taken off the market after reports of three patients with progressive multifocal leukencephalopathy (PML), two of whom unfortunately died. Two subjects were enrolled in the SENTINEL-Trail where they received combination treatment of natalizumab and IFN-beta (Avonex). The third patient participated in the Crohn's disease study. Following an outstanding and meticulous safety investigation, no further PML cases were found and natalizumab has been reintroduced into MS treatment in the US and Europe in June 2006.

10.2 Basic Principles

The adhesion of leukocytes to the microvascular endothelium is essential for their migration into inflamed tissue. This response is mediated by the interaction of adhesion molecules expressed on the cell surface of leukocytes and venular endothelial cells (Springer 1990). The very-late-antigen-4 (VLA-4; $\alpha_4\beta_1$ integrin, CD49d/CD29) (Fig. 10.1) and its counter-receptor on brain endothelium, vascular adhesion molecule-1 (VCAM-1), are members of the adhesion molecule family. Natalizumab binds the α_4-chain irrespective of its associated β-chain (β_1 or β_7) so it can also block the interaction of the $\alpha_4\beta_7$-integrin expressed on leukocytes interacting with the mucosal addressin cell adhesion molecule-1 (MadCAM-1), which is expressed on intestinal endothelium. This was the basis for studies in Crohn's disease.

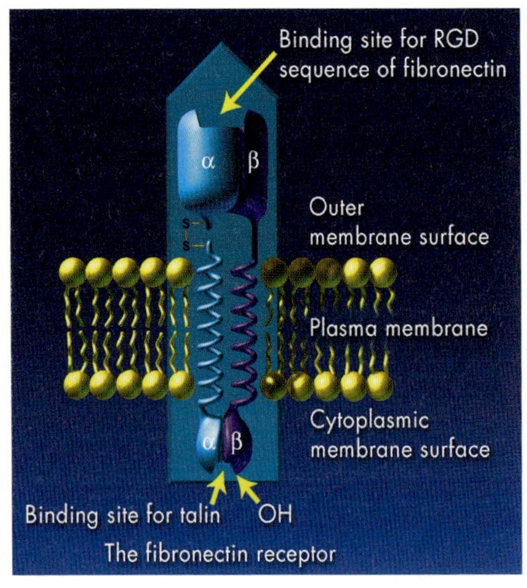

Fig. 10.1 A model of the cell-surface adhesion molecule VLA-4.

The VLA-4 receptor is a member of a large family of cell-surface adhesion molecules (integrins) that mediate both cell–cell and cell–extracellular matrix (ECM) interactions. Adhesion molecules stabilize interactions between migrating cells (e.g., lymphocytes) and their tissue standing partners. In turn, these receptors are then able to induce intracellular signal transduction. VLA-4 is built like a classic integrin molecule with a noncovalently linked α- and a β-chain which is predominantly present on T-lymphocyte membranes, but is also found on monocytes, eosinophilic granulocytes and, to a lesser degree, on neutrophils.

10.3
Mode of Action

The complex adhesion process of activated lymphocytes on inflamed brain endothelium (Fig. 10.2) is necessary for the extravasation of leukocytes through the blood–brain barrier (BBB) as a first step in the inflammation cascade of CNS tissue, as in MS (Alter et al. 2003; Baron et al. 1993). Leukocytes exhibit an initial adhesive contact to the activated endothelium which allows them to slow down and adhere to the vascular wall. This step is mediated by different selectins and integrins, and is still reversible. Activation signals provided by chemokines, matrix metalloproteinases and proinflammatory cytokines switch the integrins to a high affinity state and allow the cell to arrest and finally to transmigrate through the vascular endothelium into the brain (von Andrian and Mackay 2000). Integrins and their counter-receptors have a unique role in this multistep cascade as they modulate the process of *rolling* and *arresting* of lymphocytes.

Various modes of action of anti-VLA4 mAbs have been postulated:

- Blockade of T-cell migration by blocking adhesion to endothelial cells and interaction with ECM proteins (Gonzales-Amaro et al. 2005; Vajkoczy et al. 2001).

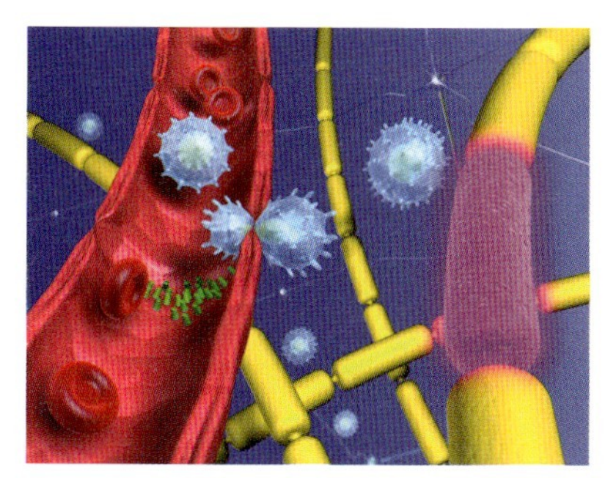

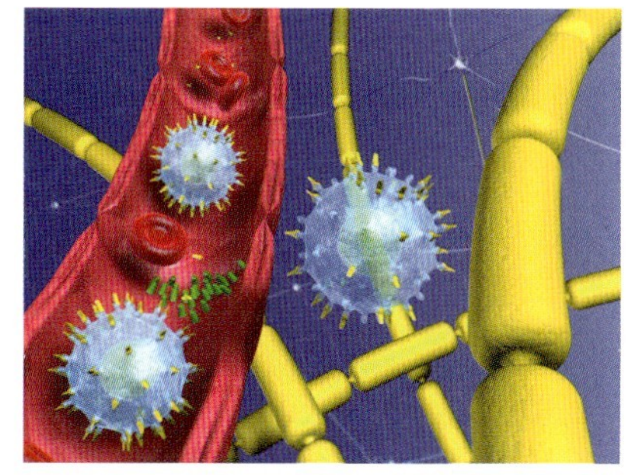

Fig. 10.2 Left: Process of extravasation of activated lymphocytes (blue) via complex binding to adhesion molecules (green). Right: blocking the adhesion process via natalizumab (yellow).

- Circulating B cells and monocytes express significant amounts of VLA-4, which highlights the possible role for VLA-4 in modulating the migration of these cell types across the BBB endothelium (Niino et al. 2006).
- Induction of apoptosis by blocking interaction of α_4-integrin-bearing leukocytes with ECM proteins such as fibronectin (Leussink et al. 2002).

A well-established hypothesis of MS pathogenesis suggests that peripherally activated T-lymphocytes migrate through the BBB and cause brain tissue destruction (Weiner 2004). VLA-4 has been found to be a most interesting target for MS treatment. During the early 1990s, Yednock discovered that the ligand pair VCAM-1 and VLA-4 was of key importance for the migration of inflammatory T cells into the brain parenchyma. Antibodies to α_4-chains of integrins reduce cellular infiltration, inhibit the development of EAE, and halt the progression of disease – or even reverse existing symptoms – by preventing inflammatory cells from crossing the BBB (von Andrian and Engelhardt 2003). Natalizumab has shown to inhibit the trafficking of leukocytes across the BBB and rapidly to reverse EAE (Kent et al. 1995; Léger et al. 1997; Yednock et al. 1992). In addition, natalizumab inhibits inflammation in experimental autoimmune neuritis (EAN) due to almost complete blockade by VLA-4 (Enders et al. 1998). The expression of VCAM-1, the major ligand for $\alpha_4\beta_1$-integrin, is increased in active CNS plaques. Blockade of this adhesion mechanism results in a reduction in proinflammatory cytokines and subsequently in a sustained decrease of inflammation (Wolinsky 2003).

10.3.1 Pharmacodynamic Profile

Natalizumab consists of the recombinant humanized α_4-integrin antibody and stabilizing salts and agents (i.e., sodium chloride, sodium dibasic phosphate heptahydrate, sodium monobasic phosphate monohydrate and polysorbate 80). The current formulation of natalizumab is provided in a buffered solution at a concentration of $20\,mg\,mL^{-1}$. The product is stable for 30 months when stored at 2 to 8 °C.

10.3.2 Pharmacokinetics

Single natalizumab doses of 1 to $6\,mg\,kg^{-1}$ produce maximal saturation (defined as >80%) of α_4-integrin receptors on the surface of lymphocytes within 24 h after administration. Approximately 90% of patients achieve over 80% saturation of α_4-integrin receptors after natalizumab doses of 3 and $6\,mg\,kg^{-1}$. Receptor saturation is maintained for 1, 3 to 4 and 6 weeks with natalizumab doses of 1, 3, and $6\,mg\,kg^{-1}$, respectively. Serum concentrations exhibit a biexponential decline, with a relatively rapid distribution phase followed by a prolonged terminal phase (Sheremata et al. 1999). The pharmacodynamic effects and therapeutic response observed in the phase II clinical studies supported a fixed dose (300 mg) and a fixed interval (every 4 weeks) for the phase III trials.

10.4 Technology

Studies with the murine form of the antibody (AN100226m) prior to humanization had demonstrated that it was a potent inhibitor of a specific adhesion molecule important in the pathogenesis of inflammatory CNS diseases such as MS. The humanization of AN100226m was undertaken in order to reduce potential immunogenicity, to increase the *in-vivo* half-life, and to allow repeated administration for increased therapeutic benefit (Léger et al. 1997). Grafting of the complementarity-determining regions (CDRs) of AN100226 onto a human immunoglobulin (Ig) G_4 framework resulted in an antibody, natalizumab, that was approximately 99% human-derived (Rice et al. 2005; Rudick and Sandrock 2004). Natalizumab is a recombinant humanized antibody that is produced in mouse NSO murine myeloma cells and binds specifically to the α_4-subunit of the VLA-4 receptor. It is expressed at high levels on all circulating leukocytes, except for polymorphonuclear leukocytes. For principal details, see Volume I, Chapter 1.

10.5 Clinical Findings

10.5.1 Phase I Clinical Trials

Four phase I studies were completed. One study was conducted in healthy volunteers, while the other three were conducted to evaluate natalizumab in patients with either relapsing-remitting or secondary progressive MS. Different intravenous doses of natalizumab ranging between 0.03 and 6 $mg\,kg^{-1}$ were assessed. Overall, the results of the phase I trials showed natalizumab to be relatively well tolerated, with an overall adverse event profile that did not differ significantly from that of placebo (Sheremata et al. 1999, 2005).

10.5.2 Phase II Clinical Trials

Two phase II clinical trials were completed demonstrating significant reductions in inflammatory lesions, as visualized by magnet resonance imaging (MRI), and fewer patients with relapse during natalizumab treatment compared with placebo (Miller et al. 2003; Tubridy et al. 1999). The studies were conducted with relapsing-remitting MS patients in a modern randomized, double-blind, placebo-controlled setting. Different dosage schemes and dosing time points were chosen. Pharmacokinetic and pharmacodynamic results from these studies also helped to identify the most appropriate natalizumab dose and application scheme for use in the subsequent phase III MS trials (Rudick and Sandrock 2004).

10.5.3 Phase III Clinical Trials

Two large randomized, double-blinded, placebo-controlled trials were commenced in MS patients who had experienced at least one clinical relapse during the year before entry. The AFFIRM (natalizumab versus placebo) and the SENTINEL (natalizumab plus IFN-β1a [Avonex] versus IFN-β1a plus placebo) studies were published in March 2006 (Polmann et al. 2006; Rudick et al. 2006). The level of efficacy observed in the earlier phase II trails was at least maintained in the two 2-year trials mentioned above. Overall, more than 2000 patients were included in these MS trials. In addition to the results confirmed from the phase II trial, these studies clearly showed that the progression of disability (defined as a change in the extended disability status scale, EDSS) was reduced by 42% within 2 years. The annualized rate of clinical relapse was reduced by 66% (from 0.78 to 0.27) and the number of new or enlarging brain lesions on MRI was reduced by 83%. By way of perspective, the currently used drugs, β-interferons and glatiramer acetate, diminish acute relapses by approximately 35%. Yet, only head-to-head trials of natalizumab and IFN will allow for a direct comparison (Gold et al. 2006).

Because natalizumab also blocks the interaction of the $\alpha_4\beta_7$-integrin with its ligand, MadCAM-1 (which is involved in leukocyte migration through intestinal endothelium), this treatment has also been investigated in Crohn's disease (Gordon et al. 2001). However, the results in the phase III trial have not shown any clinical significance.

Based on these 1-year phase III trial data, the FDA licensed natalizumab for reduction of relapse rate in November 2004 in the USA. Natalizumab was finally marketed under the name of Tysabri (Elan Pharmaceuticals Inc., San Francisco, CA, USA), but was suspended from the market and all clinical trials on February 28, 2005. This decision was based on two patients being diagnosed with progressive multifocal leukoencephalopathy (PML) participating in the SENTINEL trail. Retrospectively, a third patient who was enrolled in the Crohn's disease trial was shown by brain biopsy to have PML. His neurological symptoms were originally thought to be associated to an astrocytoma.

PML is a rare infection caused by the JC polyomavirus. It usually occurs in people infected with the human immunodeficiency virus (HIV), but it has also been reported in immunocompromised patients receiving prolonged treatment with immunosuppressants. The JC virus is ubiquitous (ca. 80% of the European population) and is usually acquired in childhood. The virus remains dormant in the bone marrow, kidney epithelia, and spleen, but can enter the CNS directly during periods of viremia, such as those occurring during prolonged immunosuppression. The etiology of PML in patients treated with Tysabri is unclear. The influence of natalizumab on B-cell function and the specific manner of altering the adhesion process at the BBB, probably in synergy with β-interferon (and potentially all other immunomodulators or immunosuppressors), are possible reasons for JC virus activation. It should be noted, however, that blockade of VLA-4 also exerts widespread effects in the hematopoietic system. VLA-4 is involved in mobilizing leukocytes from bone marrow, and these effects on bone marrow may be directly relevant for the subsequent development of PML (Ransohoff 2005).

Elan pharmaceuticals and the BiogenIdec company voluntarily withdrew Tysabri from the market, and an investigation of all patients who ever received the drug was initiated. No new case of PML has since been observed (Yousry et al. 2006). An FDA advisory committee recommended that Tysabri could be reintroduced for relapsing-remitting MS as a monotherapy in June 2006 in the USA, and finally the EMEA decided to reintroduce it in July 2006 in Europe.

10.5.4 Adverse Side Effects

Apart from the above-mentioned issue, the overall tolerability of natalizumab treatment is very good. In both clinical trails most adverse side effects did not differ significantly between placebo and natalizumab. Of note, hypersensitivity reactions in more than 3.5% and anaphylactoid reactions in more than 1.4% in the natalizumab group must be mentioned. Natalizumab treatment produces increased levels of circulating white blood cells (i.e., lymphocytes, monocytes and eosinophils), although the mean values did not exceed the normal range (Miller et al. 2003). In this respect, attention must be paid to infections, cholelithiasis, urticaria (hives), irregular menstruation, and other diseases which can be associated with the application of natalizumab.

10.5.5 Neutralizing Antibodies

Persistent high-titer neutralizing anti-natalizumab antibodies (nAbs) developed in 6% of patients in the monotherapy study, and led to a loss of therapeutic efficacy and an increase in adverse side effects (Polman et al. 2006). Therefore nAbs can be evaluated for safety and efficacy after 3 to 4 months of treatment.

10.6 Indications for Tysabri

In June 2006, Tysabri received a license for the treatment of patients with relapsing forms of multiple sclerosis who:

- currently take β-interferons, for an appropriate period with ongoing relapse activity (at least one relapse in the past 12 months *and* at least nine T2 lesions *or* one gadolinium-enhancing lesion in the MRI); or
- with a high relapse rate of naive patients defined by two or more relapses during the past year with progression of disease *and* at least one gadolinium-enhancing lesion *or* significant increase of T2 lesion load compared to a previous recent MRI.

10.6.1 Clinical Applications

10.6.1.1 Preparation and Administration of Natalizumab

Aseptic technique must be used when preparing Tysabri solution for intravenous (IV) infusion. Each vial is intended for single use only. Tysabri is a colorless, clear to slightly opalescent concentrate. The vial must be inspected for particulate material prior to dilution and administration. If visible particulates are observed and/or the liquid in the vial is discolored, the vial must not be used. Tysabri must not be used beyond the expiration date stamped on the carton or vial (Fig. 10.3).

- Dose preparation: To prepare the solution, 15 mL of Tysabri concentrate is withdrawn from the vial using a sterile needle and syringe. The concentrate is then injected into 100 mL of 0.9% sodium chloride injection (USP). No other IV diluents may be used to prepare the solution. The

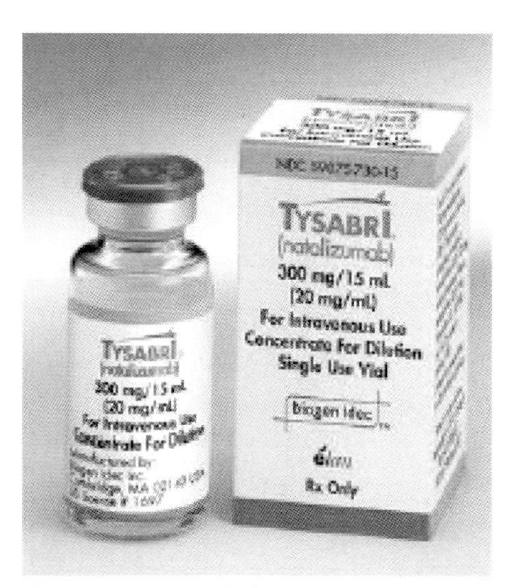

Fig. 10.3 A 15-mL vial of concentrated Tysabri 300 mg.

Tysabri solution is then gently inverted to provide complete mixing; the vial most NOT be shaken. The final solution must then be inspected visually for particulate material prior to administration.

- Infusion: Infuse Tysabri 300 mg in 100 mL 0.9% sodium chloride injection (USP) over a period of approximately 1 h. When the infusion is complete, the infusion system is flushed with 0.9% sodium chloride injection (USP).
- For safety reasons, in order to control any possible allergic reactions, the patient should remain in an outpatient setting for a further 60 min after infusion.

10.7 Outlook

The Advisory Committee of the FDA recommended a risk-minimization program with mandatory patient registration and periodic follow-up. Tysabri is available in the USA only through the Risk Management Plan, referred to as the TOUCH prescribing program.

In Europe, it is emphasized that Tysabri is administer only at centers with experience and availability of MRI. The drug may be prescribed only by neurologists, and all patients receive an "alert card". A prescribing program TYGRIS is also proposed for Europe in the near future.

In order to receive Tysabri, patients must visit their specialist doctor and also understand the risks and benefits associated with Tysabri administration.

References

Alter, A., Duddy, M., Hebert, S., Biernacki, K., Prat, A., Antel, J.P., Yong, V.W., Nuttall, R.K., Pennington, C.J., Edwards, D.R., Bar-Or, A. (2003) Determinants of human B cell migration across brain endothelial cells. *J Immunol* 170: 4497–4505.

Baron, J.L., Madri, J.A., Ruddle, N.H., Hashim, G., Janeaway, C.A. (1993) Surface expression of α4 integrin by CD4 T cells is required for their entry into brain parenchyma. *J Exp Med* 177: 57.

Enders, U., Lobb, R., Pepinsky, R.B., Hartung, H.-P., Toyka, K.V., Gold, R. (1998) The role of the very late antigen-4 (VLA-4) and its counterligand vascular cell adhesion molecule-1 (VCAM-1) in the pathogenesis of experimental autoimmune neuritis (EAN) of the Lewis rat. *Brain* 121: 1257–1266.

Gold, R., Hartung, H.P., Hohlfeld, R. (2006) Monoclonal antibodies improve therapy of relapsing multiple sclerosis. Molecular basis and clinical results of anti-VLA4 (natalizumab) therapy. *Dtsch Med Wochenschr* 131(1–2): 31–34.

Gonzalez-Amaro, R., Mittelbrunn, M., Sanchez-Madrid, F. (2005) Therapeutic anti-integrin (α4 and αL) monoclonal antibodies: two-edged swords? *Immunology* 116: 289–296.

Gordon, F.H., Lai, C.W., Hamilton, M.I., et al. (2001) A randomized placebo-controlled trial of a humanized monoclonal antibody to α4-integrin in active Crohn's disease. *Gastroenterology* 121: 268–274.

Kent, S.J., Karlik, S.J., Cannon, C., et al. (1995) A monoclonal antibody to α4 integrin suppresses and reverses active experimental allergic encephalomyelitis. *J Neuroimmunol* 58: 1–10.

Léger, O.J., Yednock, T.A., Tanner, L., Horner, H.C., Hines, D.K., Keen, S., Saldanha, J., Jones, S.T., Fritz, L.C., Bendig, M.M. (1997) Humanization of a mouse antibody against human alpha-4 integrin: a potential therapeutic for the treatment of multiple sclerosis. *Hum Antibodies* 8(1): 3–16.

Leussink, V.I., Zettl, U.K., Jander, S., et al. (2002) Blockade of signaling via the very late antigen (VLA-4) and its counterligand vascular cell adhesion molecule-1 (VCAM-1) causes increased T cell apoptosis in experimental autoimmune neuritis. *Acta Neuropathol* 103: 131–136.

Miller, D.H., Khan, O.A., Sheremata, W.A., et al. (2003) A controlled trial of natalizumab for relapsing multiple sclerosis. *N Engl J Med* 348: 15–23.

Niino, M., Bodner, C., Simard, M.L., et al. (2006) Natalizumab effects on immune cell responses in multiple sclerosis. *Ann Neurol* 59(5): 748–754.

Polman, C.H., O'Connor, P.W., Havrdova, E., et al. (2006) A randomized, placebo-controlled trial of natalizumab for relapsing multiple sclerosis. *N Engl J Med* 354: 899–910.

Ransohoff, R.M. (2005) Natalizumab and PML. *Nat Neurosci* 8: 1275.

Rice, G.P., Hartung, H.P., Calabresi, P.A. (2005) Anti-α4 integrin therapy for multiple sclerosis: mechanisms and rationale. *Neurology* 64(8): 1336–1342.

Rudick, R.A., Sandrock, A. (2004) Natalizumab: alpha 4-integrin antagonist selective adhesion molecule inhibitors for MS. *Expert Rev Neurother* 4(4): 571–580.

Rudick, R.A., Stuart, W.H., Calabresi, P.A., Confavreux, C., Galetta, S.L., Radue, E.W., Lublin, F.D., Weinstock-Guttman, B., Wynn, D.R., Lynn, F., Panzara, M.A., Sandrock, A.W. and the SENTINEL Investigators. (2006) Natalizumab plus interferon beta-1a for relapsing multiple sclerosis. *N Engl J Med* 354(9): 911–923.

Sheremata, W.A., Vollmer, T.L., Stone, L.A., Willmer-Hulme, A.J., Koller, M. (1999) A safety and pharmacokinetic study of intravenous natalizumab in patients with MS. *Neurology* 52(5): 1072–1074.

Sheremata, W.A., Minagar, A., Alexander, J.S., Vollmer, T. (2005) The role of alpha-4 integrin in the aetiology of multiple sclerosis: current knowledge and therapeutic implications. *CNS Drugs* 19(11): 909–922.

Springer, T.A. (1990) Adhesion receptors of the immune system. *Nature* 346: 425.

Tubridy, N., Behan, P.O., Capildeo, R., et al. (1999) The effect of antiα4 integrin antibody on brain lesion activity in MS. *Neurology* 53(3): 466–472.

Vajkoczy, P., Laschinger, M., Engelhardt, B. (2001) Alpha 4-integrin-VCAM-1 binding mediates G protein-independent capture of encephalitogenic T cell blasts to CNS white matter microvessels. *J Clin Invest* 108: 557–565.

von Andrian, U.H., Mackay, C.R. (2000) T-cell function and migration. *N Engl J Med* 343: 1020–1034.

von Andrian, U.H., Engelhardt, B. (2003) α_4-Integrins as therapeutic targets in autoimmune disease. *N Engl J Med* 348: 68–72.

Weiner, H.L. (2004) Multiple sclerosis is an inflammatory T-cell-mediated autoimmune disease. *Arch Neurol* 61: 1613–1615.

Wolinsky, J.S. (2003) Rational therapy for relapsing multiple sclerosis. *Lancet Neurol* 2(5): 271–272.

Yednock, T.A., Cannon, C., Fritz, L.C., Sanchez-Madrid, F., Steinman, L., Karin, N. (1992) Prevention of experimental autoimmune encephalomyelitis by antibodies against alpha4beta1 integrin. *Nature* 356: 63–66.

Yousry, T.A., Major, E.O., Ryschkewitsch, C., et al. (2006) Evaluation of patients treated with natalizumab for progressive multifocal leukoencephalopathy. *N Engl J Med* 354: 924–933.

11
Omalizumab (Xolair)
Anti-Immunoglobulin E Treatment in Allergic Diseases

Claus Kroegel and Martin Foerster

Summary

Immunoglobulin E (IgE) plays a key role in the induction and maintenance of allergic disorders, making the molecule and its pathways an obvious target for a novel treatment approach. Omalizumab, a recombinant humanized monoclonal antibody, is the first therapeutic agent that specifically targets immunoglobulin E (IgE). Monoclonal anti-IgE Omalizumab (rhuMAb-E25; Xolair) has been approved for the treatment of moderate-to-severe persistent allergic asthma in several nations, including USA, Canada and Brazil, Australia, New Zealand, and a number of European countries.

Monoclonal antibody: Omalizumab (Xolair, Genentech) is a recombinant DNA-derived, humanized IgG1 monoclonal anti-human IgE antibody consisting of a murine mAb MAE11 directed against IgE (Fab terminus) and a human IgG frame (Fc terminus).

Indications and dosing: Omalizumab is approved for the treatment of adults and adolescents (aged > 12 years) having moderate-persistent to severe-persistent asthma with a positive skin test or *in-vitro* reactivity to perennial aeroallergens (e.g., dust mites, cats, dogs, fungi), and whose symptoms are inadequately controlled with inhaled corticosteroids. The recommended dosage of omalizumab is 150–375 mg subcutaneously (s.c.) every 2 or 4 weeks depending on pretreatment total serum IgE level and body weight.

Mode of action: Omalizumab binds to the same domain (Cε3) on the IgE molecule that interacts with the high-affinity IgE receptor (FcεRI), thereby interrupting the proinflammatory signal carried by IgE. The mode of action of rhuMAb-E25 comprises immediate (elimination of free serum IgE), short-term (down-regulation of IgE receptors on circulating basophils and dendritic cells) and long-term effects (reduction the number of tissue dwelling eosinophils and mast cells, T and B cells).

Clinical data: Several large phase III clinical trials have demonstrated that omalizumab is more effective than placebo in controlling moderate to severe allergic asthma with respect to symptom severity scores, lung function, quality of life,

Handbook of Therapeutic Antibodies. Edited by Stefan Dübel
Copyright © 2007 WILEY-VCH Verlag GmbH & Co. KGaA, Weinheim
ISBN 978-3-527-31453-9

and number of exacerbations. In particular, omalizumab provides important benefits in patients with poorly controlled disease or exacerbations despite adequate therapy. In addition, the drug may aid in the treatment of other IgE-related conditions such as allergic rhinities, atopic dermatitis or food intolerance.

Adverse side effects: Omalizumab is generally well tolerated in adults and children with allergic asthma, and the incidence of adverse side effects is low. Adverse events most commonly observed are injection-site reaction, viral infection, upper respiratory tract infection, sinusitis, headache, and pharyngitis; severe side effects (anaphylaxis) are rare, and an association with malignancy is weak.

Summary: Omalizumab provides a novel second-line treatment option for moderate to severe allergic asthma and possibly for other manifestations of atopy, allowing a more individualized therapy. The available data emphasize the fundamental importance of IgE in the pathogenesis of allergic diseases.

11.1 Introduction

Worldwide, allergic diseases continue to increase, affecting more than 100 million individuals (Beasley et al. 2000; ISAAC 1998; Warner et al. 2006). The reasons underlying this worldwide epidemic of allergic diseases are not understood, but are most likely the consequence of environmental changes and improved hygiene, superimposed on a range of genetic susceptibilities.

Immunoglobulin E (IgE) is an important mediator in the pathophysiology of asthma and allergy (Kroegel 2002; Platts-Mills 2001). Approximately two-thirds of asthma is estimated to be allergic (Novak and Bieber 2003), and total IgE levels correlate with asthma rates in adults (Fig. 11.1) (Burrows et al. 1989). Atopy, which has increased over the past 30 years, does not decline with increasing age. Moreover, IgE receptors are expressed on various different cell types, further emphasizing the biological relevance of IgE. The binding of IgE to its receptors primes these cells to respond to allergen. Although the function of mast cells and basophils mainly depends on the binding of IgE, the number of IgE receptor-expressing cells is large, and includes other effector cells (eosinophils, macrophages), regulator cells (B-lymphocytic dendritic cells) as well as structural cells (epithelial cells, smooth muscle cells). Upon repeated exposure to allergens, IgE antibodies bound to FcεRI on mast cells and basophils are crosslinked, and this results in the secretion of preformed mediators (e.g., histamine, tryptase) and the generation of newly formed mediators, such as arachidonic acid metabolites (e.g., prostanoids, leukotrienes) and various cytokines. These secreted mediators mount early- and late-phase allergic reactions via multiple inflammatory effects. The early phase, which occurs within minutes of allergen exposure, is associated with increased vascular permeability, smooth muscle contraction, and local damage caused by vasoactive amines, proteases, and lipid mediators. Cytokines released during the reaction are responsible for the late-phase local inflammatory response observed between 6 and 24 h after allergen exposure. IgE is also believed

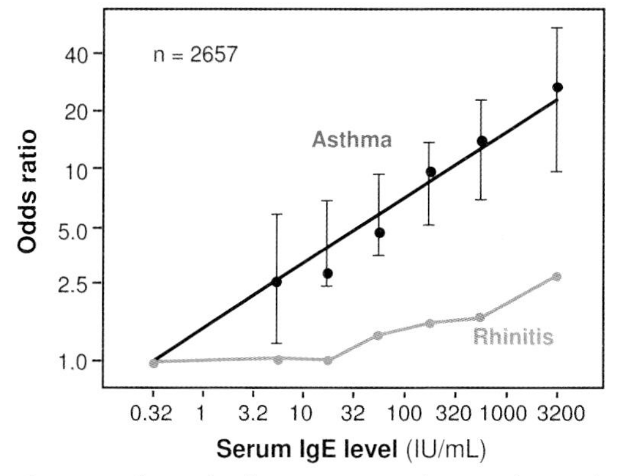

Fig. 11.1 Relationship between IgE, asthma, and rhinitis in adults. The data indicate odds ratio (OR) of having asthma at seven levels of total IgE concentrations after correction for age, gender, smoking habits, and skin-test index in a logistic analysis. The upper line represents the risk of asthma; vertical bars indicate 95% confidence intervals around the regression for each OR corresponding to a given log IgE level. The lower line represents the OR of having rhinitis at the same seven levels of serum IgE following correction of the same factors. (From Burrows et al. 1989.)

to bridge early- and late-phase allergic reactions, both directly via binding to FcεRII on eosinophils, and indirectly via activation and release of cytokines from mast cells and basophils (Fig. 11.2). Other IgE-mediated effects of relevance in asthma include IgE-dependent focusing of antigen to dendritic cells (Kehry and Yamashita 1989) and mobilization of calcium in airway smooth muscle cells (Gounni et al. 2005). Therefore, as the biologic effects of IgE are diverse and far-reaching, the inhibition of IgE can be expected to block multiple allergic mechanisms at different levels of allergic inflammation.

During recent years, several drugs have been developed aimed specifically at interfering with IgE-associated pathways, most notably humanized monoclonal antibodies (mAbs) (e.g., CGP 51901, rhuMAb-E25). Among these mAbs, the most advanced anti-IgE mAb is rhuMAb-E25 (omalizumab; Xolair; Novartis Pharmaceuticals, East Hanover, NJ, USA; Genentech Inc., South San Francisco, CA, USA). This novel drug offers a completely different treatment approach which promises several clinically relevant advantages over current treatment options. This chapter provides an overview of the immunobiology of the IgE molecule and its receptors, followed by a detailed discussion on the construction of rhuMAb-E25 and its use in the treatment of asthma and other IgE-associated diseases.

11.2 The Biology of the IgE Molecule

During the mid-1960s, Ishizaka and colleagues (Ishizaka and Ishizaka 1970) discovered a new antibody isotype, now known as IgE, thereby opening a new era

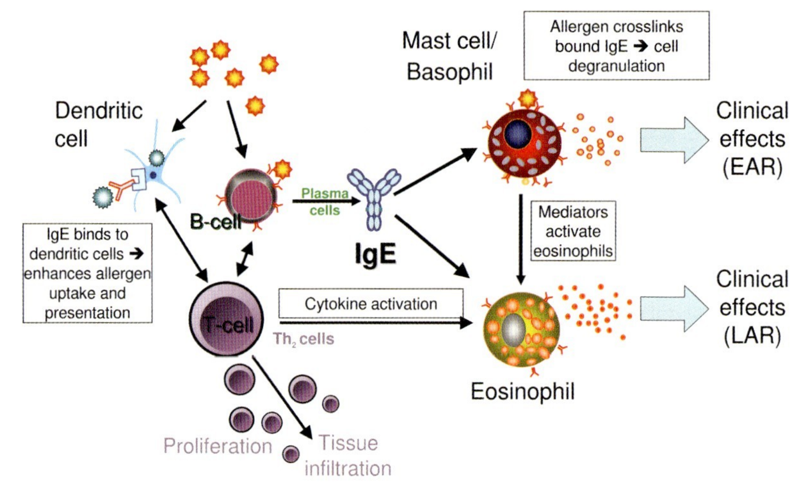

Fig. 11.2 The immunological cascade in allergy and asthma. Following inhalation, the allergen is processed by antigen-presenting cells (APCs) and presented to B cell and Th0 cells; this leads to differentiation of Th2 cells (IL-4, IL-5, IL-9, IL-13) and IgE-producing B cells (plasma cells). Repeated allergen exposure causes the immediate release of mediators from mast cells and produces the typical symptoms of acute asthma (EAR). Mediators released from eosinophils produce asthma symptoms associated with chronic allergic inflammation, in which bronchial hyperresponsiveness predominates (LAR). EAR = early allergic/asthmatic reaction; LAR = late allergic/asthmatic reaction; Th0 = uncommitted T lymphocyte (can differentiate into Th1 or Th2 lymphocytes depending on the cytokines present); Th2 = functional T lymphocyte associated with helper activity for antibody production. (Modified after Holgate et al. 2005b.)

in the pathophysiology of immunologic disorders. Since its discovery, IgE has been shown to be a key mediator in allergic diseases (Maizels 2003; Maizels and Yazdanbakhsh 2003; Yazdanbakhsh et al. 2001). In addition, IgE is associated with a number of other immunologic diseases (Fig. 11.3), where its role is less well-defined.

Like other immunoglobulins, IgE consists of two light chains and two ε-heavy chains (Fig. 11.4), but is distinguished by its ε-heavy chain constant region sequence. Unlike IgG, IgE contains four instead of three heavy chain constant domains (Cε1–Cε4) (Garman et al. 2000), with the additional domain (Cε2) replacing the hinge region (Sutton and Gould 1993). Moreover, as opposed to other immunoglobulin classes, human IgE binds specifically with both high and low affinity to receptors (FcεR) on the surface of several different cell types (Scharenberg and Kinet 1995).

The allergen-binding site on the IgE molecule is located on the variable regions of the heavy and light chains. The Fc fragment of the IgE molecule binds to two types of immunoglobulin Fcε-receptors (see below). The binding site for both FcεRI and FcεRII is located in proximity to, but on different parts of, the Cε3 domain of the IgE molecule (Garman et al. 2000; Nissim et al. 1993; Presta et al.

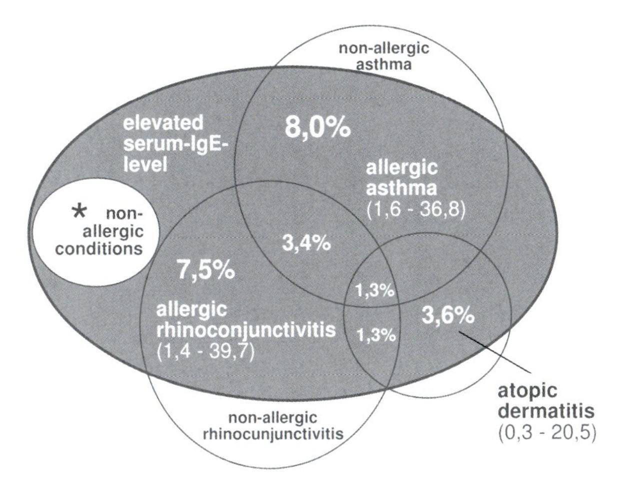

Fig. 11.3 Relationship between elevated serum IgE levels and allergic and non-allergic conditions. Large numbers refer to the prevalence of allergic diseases in children (symptoms during the past 12 months) and are based on data provided by the ISAAC Study (1998). Numbers in brackets indicate ranges. *Non-allergic diseases associated with a raised IgE level are parasitoses, AIDS, Wiskot–Aldrich syndrome, Nezelof syndrome, non-Hodgkin lymphoma, malignant tumors and the hyper-IgE-syndrome (Job syndrome).

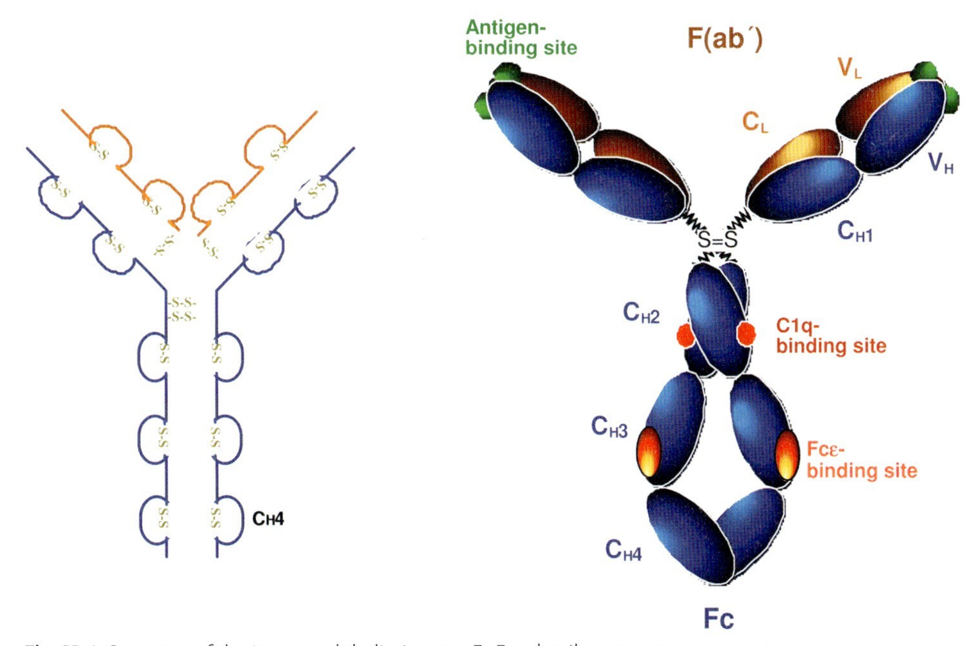

Fig. 11.4 Structure of the immunoglobulin isotype E. For details, see text.

1994; Vercelli et al. 1989). The association of the dimeric Fc fragment of the IgE antibody to its receptors has two significant consequences. First, the IgE receptor adopts the antigenic specificity of the prevalent IgE repertoire (Gaman et al. 2000). Second, a specific interaction with antigens/allergens results in crosslinking of the FcεRI, thereby initiating a signal transduction cascade and downstream intracellular effects (Turner and Kinet 1999). This process ultimately induces the release of a variety of preformed (e.g., histamine) and *de-novo* synthesized chemical mediators (e.g., leukotrienes and prostaglandins), as well as cytokines that exert their immunologic effects by interacting with specific receptors on target organs.

11.2.1 IgE Distribution and Blood Concentration

While being present in very low amounts (ng to $\mu g\,mL^{-1}$ range) in the serum of normal healthy individuals, IgE levels are significantly elevated in allergic subjects. The raised levels of specific IgE in the serum, together with a positive skin test, defines allergic sensitization (or "atopy"). However, even in highly atopic individuals, the level of plasma IgE is less than a 1000-fold that of plasma IgG (Geha et al. 2003). Indeed, in nonallergic individuals, plasma IgE levels may be 10 000- to 50 000-fold lower than that of plasma IgG. However, in contrast to IgG, most of the IgE produced is bound by FcεRI expressed by mast cells and basophils (Geha et al. 2003) and by FcεRII on various other immune cells.

11.2.2 IgE Synthesis and Regulation

IgE molecules are mainly produced by plasma cells in the mucosa-associated lymphoid tissue. The regulation of IgE production is an extremely complex process, and is controlled by various positive as well as negative factors. The major step in IgE synthesis is the regulation of the IgE class-switch recombination (Geha et al. 2003). Common to all isotype switching is the germline transcription of CH genes and the induction of activation-induced cytidine deaminase (AID) expression. IgE class-switch recombination is accomplished either through T-cell-dependent (the classic pathway) or T-cell-independent processes (the alternative pathway) (Geha et al. 2003). In addition, IgE class switching is controlled by several other mechanisms, including cytokines [interferon (IFN)-γ and interleukin (IL)-21], the engagement of surface receptors on B cells (B-cell receptor, CD45, CTLA4, and CD23), as well as transcription factors (B-cell lymphoma 6, inhibitor of DNA binding 2). A detailed review, which also provides an excellent description of the mechanisms underlying the production of IgE, is available (Geha et al. 2003).

11.3 IgE Receptors

The IgE molecule mediates immunologic its effects by coupling with two different binding sites expressed by immune cells (Table 11.1):

- the high-affinity IgE receptor (FcεRI); and
- the low-affinity IgE receptor (FcεRII or CD23).

11.3.1 FcεRI (High-Affinity IgE Receptor)

The FcεRI complex represents a high-affinity cell-surface receptor for the Fc region of antigen-specific IgE molecules. FcεRI is multimeric, and is a member of a family of related antigen/Fc receptors which show conserved structural features and similar roles in initiating intracellular signaling cascades. In humans, FcεRI controls the activation of mast cells and basophils, and participates in IgE-mediated antigen presentation. FcεRI binds to the IgE molecule with high affinity, characterized by an equilibrium dissociation constant (or binding affinity) of 10^{-9} to 10^{-10} M (Garman et al. 2000). The receptor consists of the four transmembrane polypeptides $\alpha\beta\gamma 2$ (Sutton and Gould 1993), and its expression is restricted to mast cells, basophils, and dendritic cells (Allam et al. 2006) (Fig. 11.5).

Table 11.1 Distribution of FcεR, its biologic role and expression in allergic disease (modified according to Nissim et al. 1991).

Cell type	*FcεR-type*	*Biological function*	*Expression in allergy*	*Regulated by IgE-binding*
Mast cells basophils	$\alpha\beta\gamma 2$	Synthesis and secretion of mediators and cytokines, Induction of IgE synthesis, antigen presentation	Increased	Yes
Dendritic cells	$\alpha\beta\gamma 2$	Antigen presentation	Increased	Yes
Langerhans cells	$\alpha\gamma 2$	Antigen presentation, priming of naive T-lymphocytes to IgE-reactive proliferation	Increased	Not known
Monocytes Macrophages	$\alpha\gamma 2$	Antigen presentation, synthesis and secretion of mediators and cytokines, enzymes and oxygen radicals	Increased	Not known
Eosinophils	$\alpha\gamma 2$	Antigen presentation?	Increased	Not known
Platelets	$\alpha\gamma 2$	Unknown	Unchanged	Not known
Epithelial cells	$\alpha\gamma 2$	Secretion of eicosanoids	Increased	Not known

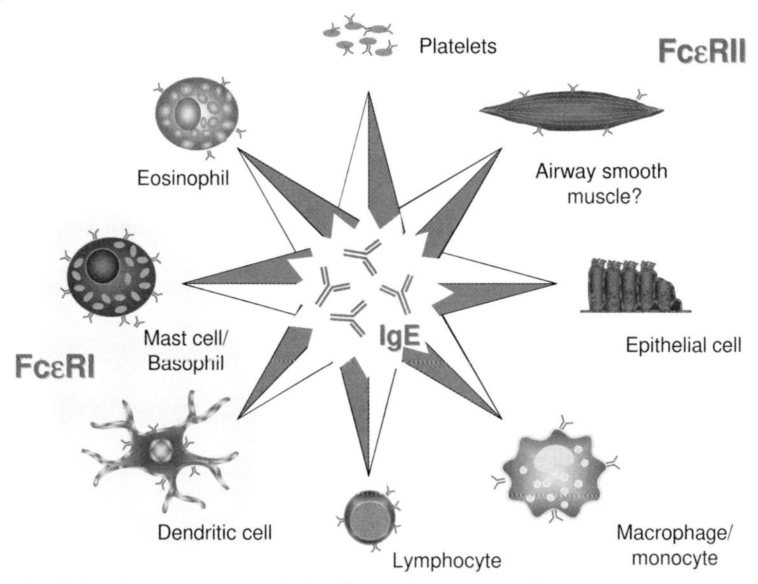

Fig. 11.5 Cells expressing the high-affinity IgE receptor FcεRI (gray shaded area) and the low-affinity IgE receptor FcεRII (the remaining cells).

FcεRI-IgE binding The extracellular portion of the α-chain enables the receptor to interact with the Cε3 of the IgE molecule (Fig. 11.6). The IgE binding site consists of two immunoglobulin-like domains (D1 and D2), and is located on the extracellular portion of the FcεRI α-chain (Turner et al. 1999). The domains are positioned at an acute angle to one another, thus creating a convex surface at the top of the molecule, and enclosing a marked cleft. One receptor binds one dimeric IgE-Fcε molecule asymmetrically through interactions at two sites (Garman et al. 1998), thereby preventing the binding of a second receptor (Turner et al. 1999). In contrast, the β- and γ-chains of the receptor are required for receptor transmembrane insertion and intracellular signal transduction (Garman et al. 2000; Turner and Kinet 1999).

FcεRI activation Whereas other receptor types (hormone or cytokine receptors) dimerize and signal after binding a well-defined ligand, the Fcε receptor types act as adaptors between highly variable antigen-binding sites and the intracellular signal transduction machinery (Garman et al. 2000). Multivalent antigens bind and crosslink IgE molecules held at the cell surface by FcεRI. Crosslinking and receptor aggregation of FcεRI induced by a multivalent antigen, is critical in triggering consecutive cellular events which eventually mount an allergic effector response in various tissues and organs. These include the release of preformed (e.g., histamine, tryptase) and newly generated mediators within minutes (e.g., leukotrienes, prostanoids) and hours [cytokines such as IL-4, IL-6, tumor necrosis

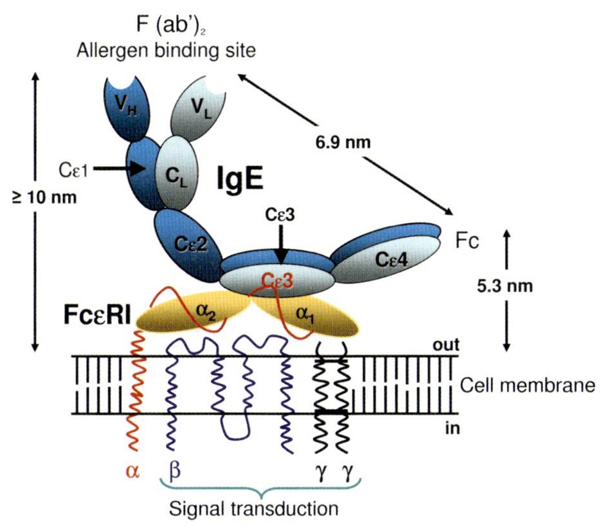

Fig. 11.6 Binding of IgE to the high-affinity receptor (FcεRI). FcεRI consists of four polypeptide chains, γβα2. The α-chain binds to five amino acids (330–335) of the Cε3 domain of Fc segment of IgE in such a manner that it lies on its side with the allergen binding site facing outwards. In contrast, FcεRII is expressed in trimeric form (γα2) on antigen-presenting cells such as monocytes and peripheral blood dendritic cells. (Modified after Holgate 1998.)

factor-alpha (TNF-α)] following repeated antigen/allergen exposure (Garman et al. 2000; Geha et al. 2003; Turner and Kinet 1999) (Fig. 11.7).

Regulation of FcεRI expression Observations made by Malveaux and Lichtenstein in 1978 noted a correlation between cell-surface FcεRI densities on peripheral blood basophils and the serum IgE titer (Malveaux et al. 1978). More recently, several investigations (MacGlashan et al. 1997a,b; Prussin et al. 2003; Saini et al. 1999) have confirmed the relationship between serum IgE concentrations and the magnitude of FcεRI expression on basophils and mast cells (Fig. 11.8). Although the underlying mechanism is not understood, the data lend strong support to the role of total IgE concentrations in controlling the expression of FcεRI (Borkowski et al. 2001). Nevertheless, the observation that blood IgE concentrations modify the expression of FcεRI has important implications both for the immunobiology of atopy and the therapeutic approaches for allergic disease (MacGlashan 2005).

11.3.2
FcεRII (Low-Affinity IgE Receptor, CD23)

The FcεRII receptor is referred to as the low-affinity IgE receptor, although it displays only one log difference in its binding affinity compared with the

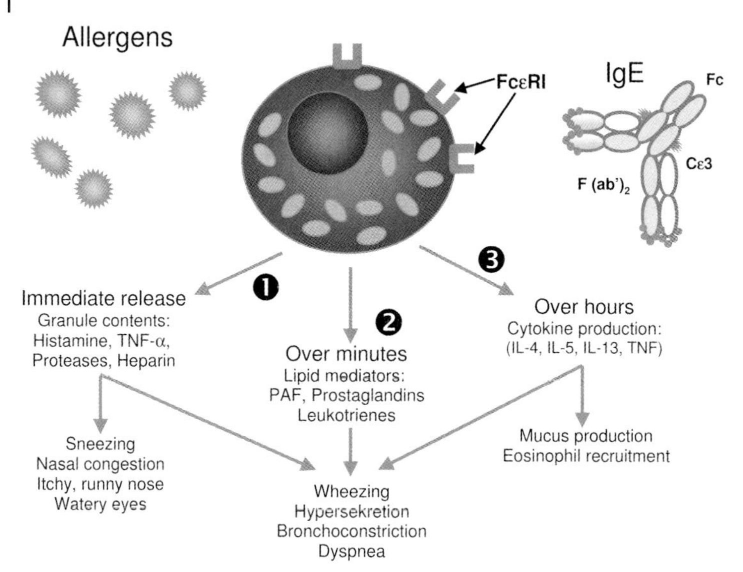

Fig. 11.7 IgE-dependent cell activation. IgE binds to high- and low-affinity receptors (FcεRI or FcεRII) on effector cells. The inflammatory cascade is initiated when IgE bound to effector cells is crosslinked by an allergen. This results in the degranulation of effector cells and the release of a comprehensive array of mediators that are linked to the pathophysiology of asthma. This in turn causes a sequential release of different types of inflammatory mediators.

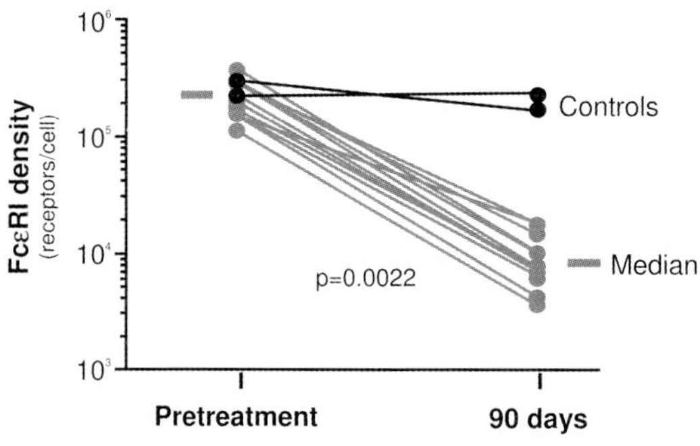

Fig. 11.8 FcεRI density on basophils prior to and 90 days after initiation of rhuMAb-E25 treatment. A series of *in-vitro* studies used circulating basophils taken from 15 subjects with perennial allergic rhinitis receiving rhuMAb-E25 for 3 months. The figure shows the results from 12 rhuMAb-E25-treated patients and two patients who acted as controls. With rhuMAb-E25 treatment, total FcεRI receptor density was markedly and significantly decreased by approximately 97% ($p = 0.0022$). The average IgE receptor numbers were reduced from approximately 240 000 to approximately 8600 receptors per cell at the 3-month measurement. (Modified after MacGlashan et al. 1997.)

high-affinity FcεRI. The equilibrium dissociation constant (binding affinity) is of $10^8 M^{-1}$ for IgE-antigen complex and $10^7 M^{-1}$ for IgE (Kijimoto-Ochiai 2002). In contrast to the four transmembrane polypeptides of FcεRI, the low-affinity form consists of an $\alpha\gamma 2$ trimer (Turner and Kinet 1999), and is expressed on a variety of hematopoietic cell lines, including eosinophils (Gounni et al. 1994), dendritic cells (DCs), activated macrophages, Langerhans cells, platelets, epithelial cells and possibly muscle cells (Bieber et al. 1992; Maurer et al. 1995; Novak and Bieber 2003; Wang et al. 1992) (see below; see also Fig. 11.5).

FcεRII-IgE binding Unlike other Fc receptors that are structurally classified within the immunoglobulin-like superfamily, CD23 is a 45-kDa type II transmembrane protein with an intracytoplasmic C-terminus. It is unique among Fc receptors in its homology to C-type (calcium-dependent) lectins. IgE binds to two CD23 lectin domains (D1-2 and D5-8) within the Cε3 region of the IgE molecule via protein–protein and protein–carbohydrate interactions, respectively (Henchoz et al. 1994; Kijimoto-Ochiai 2002).

FcεRII activation Activation of the trimeric FcεRII also contributes to the pathogenesis of allergic diseases (Saini and MacGlashan 2002). For instance, activation of FcεRI prevents apoptosis of monocytes by induction of the anti-apoptotic ligands *Bcl*-2 and *Bcl*-xL, and also protects from CD95 Fas-mediated apoptosis. Further, activation of FcεRII in atopic donors leads to the downstream activation of the proinflammatory transcription factor “nuclear factor κB” (Kraft et al. 2001). In the presence of IL-4 and granulocyte-macrophage colony-stimulating factor (GM-CSF), crosslinking FcεRI alters monocyte differentiation toward a macrophage species and away from a DC phenotype (Novak et al. 2001). Ligation of IgE to the CD23 receptor expressed on B cells leads to cell differentiation, apoptosis, and regulation of IgE synthesis (Bonnefoy et al. 1996).

FcεRII/CD23 functions CD23 has multiple functions that are controlled by a range of different ligands (Table 11.2). These include IgE (both in its secreted form and on membranes of committed B cells), CD21 (also known as complement receptor 2), CD18/CD11b and CD18/CD11c (complement receptors 3 and 4, respectively), and the vitronectin receptor. Paradoxically, CD23 engages in both the up- and down-regulation of IgE synthesis, thereby constituting a “two-way switch” in IgE homeostasis (Aubrey et al. 1992; Bonnefoy et al. 1996). When IgE binds to membrane CD23, further IgE synthesis is suppressed. In contrast, CD23-deficiency increases the level of circulating IgE by orders of magnitude. Soluble CD23 enhances IgE synthesis on binding to CD21 (Aubrey et al. 1992) and other mechanisms (Henchoz et al. 1994).

The dust mite protease Der p I cleaves CD23 close to the lectin domain, and the resulting monomeric CD23 may be a factor in the high allergenicity of dust mites (Nakamura et al. 2002). Soluble circulating CD23 fragments are found in the blood of healthy human subjects, but are commonly elevated in inflammatory or lymphoproliferative diseases, such as rheumatoid arthritis, asthma, and chronic

Table 11.2 Functional variability of CD23.

Description	*Example*	*Consequence*	*Reference*
Two forms	CD23a expressed constitutively on B cells; CD23b is expressed on monocytes, eosinophils, DCs, Langerhans cells, and platelets	Functional variability on B cells: differentiation, apoptosis, and regulation of IgE synthesis	Novak and Bieber (2003) Bonnefoy et al. (1996)
Non-IgE receptor interactions	CD21 CD11b CD11c	Variety of functions	Bonnefoy et al. (1996) Aubrey et al. (1992)
CD23 is cleaved by proteases	Soluble CD23	Proinflammatory properties, e.g., up-regulation of IgE synthesis	Herbelin et al. (1994) Henchoz et al. (1994)

lymphoblastic leukemia. It has been shown that antibodies to CD23 alleviate all three conditions, and an anti-CD23 antibody, IDEC-152 (lumiliximab), is presently undergoing clinical trials for treatment of asthma (Rosenwasser and Meng 2005).

11.4 Cell Distribution of IgE

11.4.1 Effector Cell-Associated IgE

As mentioned above, IgE is found in various tissues throughout the body. It is mainly bound to cells via either FcεRI or FcεRII on the surface of the effector (e.g., mast cell, basophils eosinophils, macrophages, and monocytes) and regulator cells (e.g., B cells and dendritic cells) (see Fig. 11.5). The complex of allergen, IgE, and FcεRI on the surface triggers a noncytotoxic, energy-dependent release of preformed, granule-associated mediators (histamine and tryptase) and the membrane-derived lipid mediators (leukotrienes, prostaglandins, and platelet-activating factor) which eventually leads to the clinical manifestation seen in allergic disease (see Fig. 11.7).

11.4.2 Antigen-Presenting Cell-Associated IgE

Antigen-presenting cells (APCs) are critical in initiating and controlling allergic inflammation. Dendritic cells and cutaneous Langerhans cells are particularly important in asthma and atopic eczema, respectively. These present antigen to $CD4^+$ Th2 cells in a MHC class II-restricted fashion. The overproduction of GM-CSF in the airway mucosa of patients with asthma enhances antigen presentation and increases the local accumulation of macrophages (Maurer et al. 1997; Stingle and Maurer 1997). Bronchoalveolar lavage macrophages obtained from patients with asthma present allergen to $CD4^+$ T cells and stimulate the production of Th2-type cytokines, whereas alveolar macrophages from control subjects do not.

APCs facilitate antigen/allergen presentation to allergen-specific T-cells and express both FcεRI and FcεRII (or CD23) (Novak and Bieber 2003). The main role of FcεRI on APCs is antigen/allergen focusing. Multivalent allergens activate FcεRI-bound IgE on APCs, initiating a cascade of events that leads to newly synthesized MHC class II molecule for the presentation of FcεRI-targeted agents (Novak and Bieber 2003).

Dendritic cells (DCs) are potent APCs that play a significant role in promoting T-cell responses to antigens/allergens. Classically, myeloid precursor DCs (pDC1s or $CD11c^+$ cells) are high IL-12 producers that induce a Th1 response. In contrast, plasmacytoid DCs (pDC2s or $CD123^+$ cells) are low IL-12 producers that induce a Th2 response (Liu et al. 2001; Rissoan et al. 1999). However, a given DC subset has remarkable plasticity in directing different types of T-cell responses (Kalinski et al. 1999; Liu et al. 2001). Thus, the modulation of surface receptor expression may have important consequences for the development of the downstream immune sequence, including IgE-mediated type 1 hypersensitivity reactions.

11.5 Physiologic and Pathophysiologic Significance of IgE

Despite the fact that IgE was discovered more than 40 years ago, its exact biological function has not been disclosed to date. Because IgE titers are elevated in individuals suffering from helminthic infestations, IgE was originally thought to play a role in the defense against parasitic infestations (Amiri et al. 1994; Hagan et al. 1991). However, the observation that a reduction of IgE levels through anti-IgE antibody treatment of mice infected with *Schistosoma mansoni* or *Nippostrongylus brasiliensis* resulted in accelerated elimination of parasites, a decreased worm burden, and a reduction in the number of worm ova, suggested that high serum IgE levels are independent of the host defense. In fact, it is currently believed that IgE is not involved in the defense processes against parasites, and the increase in concentration may simply represent an epiphenomenon resulting from parasite-induced Th2 cell activation. Thus, to date, except for its pathogenic role in allergic inflammation, a clear biologic function cannot be assigned to IgE.

Nevertheless, because IgE is the central macromolecular mediator involved in several key positions in allergic reactions, interrupting IgE-dependent pathways appears to be a rational approach for the treatment of allergic diseases. If true, this finding has two major – but opposing – consequences. First, due to its

Table 11.3 Requirement for a Anti-IgE mAb applicable in humans (Chang et al. 1990).

- High-affinity binding to IgE,
- No binding to IgE already bound by FcεRI on mast cells and basophils,
- No binding to IgE bound by the low-affinity IgE Fc receptors (FcεRII, CD23) on various other cell types
- Binding to membrane-bound IgE (mIgE) on mIgE-expressing B cells
- Construction as a humanized mAb against IgE
- Binding of circulating IgE regardless of specificity
- Forming only small, biologically inert omalizumab:IgE complexes
- No complement activation

insignificant "physiologic" role, interrupting IgE-related allergic processes is likely to cause few adverse side effects. Second, since its natural biological function during immune reactions remains unclear, it cannot be ruled out that long-term use may eventually reveal side effects related to a so-far hidden physiologic role of IgE.

11.6 The Concept of Anti-IgE-Based Treatment

Immunoglobulin E plays a key role in the induction and maintenance of allergic disorders, and consequently the interruption of IgE-dependent pathways appears to be a rational approach for the treatment of allergic diseases (Barnes 2000; Chang et al. 1990; Davis et al. 1991, 1993). The basic idea was to engineer chimerized or humanized anti-IgE antibodies with a set of unique binding properties which could be used for the isotype-specific neutralization of IgE. However, in order to be applicable in humans, the anti-IgE mAb needed to fulfill a number of the attributes listed in Table 11.3 (Chang et al. 1990; Davis et al. 1993; Heusser and Jardieu 1997).

11.7 Construction of the Monoclonal Anti-IgE Molecule

11.7.1 Antibody Generation

Selective binding of IgE requires that the mAb binds only to free IgE and not to FcεRI-IgE (see above). Thus, the engineering of a safe and effective anti-IgE molecule necessitates identification of the binding site on human immunoglobulin E for FcεRI (Presta et al. 1994). Using a model of the IgE Fcε3 homologous to the second constant domain of IgG, homology scanning mutagenesis and replacement of individual residues were performed to determine the specific amino acids involved in binding of human IgE to FcεRI (Nissim and Eshhar 1992). A total of six key amino acids (Arg408, Ser411, Lys414, Glu452, Arg465, and Met469) were identified. These residues are localized in three loops within the Cε3 domain, forming a ridge on the most exposed portion of the IgE molecule (Presta et al. 1994).

In order to create a novel specific inhibitor capable of blocking IgE-binding to FcεRI but lacking the capacity to stimulate degranulation of mast cells and other cell types, a strategy employing a murine mAb directed against IgE, which would bind IgE at the same site as the high-affinity receptor was developed. It was essential that this antibody lacked the unwanted side effects caused by receptor crosslinking. Thus, FcεRI would already occupy the immunoglobulin epitope preventing anti-IgE binding to cellular IgE (see Table 11.1). By virtue of binding to this epitope, the antibody would have the inherent ability to interfere with IgE responses by blocking binding of IgE to FcεRI.

Using the technique of scanning mutagenesis, a murine mAb (MAE11) directed against IgE (Saban et al. 1994; Shields et al. 1995), which had all of the properties required (see above), was identified. The antibody was selected on the basis of its ability to bind circulating IgE at the same site as the high-affinity receptor, thus blocking binding of IgE to mast cells and basophils.

In order to allow for possible chronic administration, as well as to avoid the problems of antigenicity, MAE11 was humanized with a human IgG1 framework (Fig. 11.9). The best of several humanized variants – version 25 (rhuMAb-E25) – was selected as it possessed IgE binding affinity and biological activity compared to that of the murine antibody MAE11. Only minimal changes restricted to five residues (Presta et al. 1993), corresponding to approximately 5% nonhuman residues located at the complementary-determining regions, were necessary (Fig. 11.10). Several *in-vitro* assays confirmed that rhuMAb-E25 not only prevented IgE-mediated activation of FcεRI-bearing cells but also inhibited IgE binding to human lung mast cells (Saban et al. 1994).

11.7.2
Complex Formation and Tissue Distribution

The immune complexes detected *in vitro* with IgE by both rhuMAb-E25 and a distinct monoclonal, chimeric murine-human antihuman IgE antibody were shown to be of limited size by both size-exclusion chromatography (SEC) and analytical ultracentrifugation (Fox et al. 1996; Liu et al. 1995). These studies showed that the size of rhuMAb-E25:IgE complexes formed *in vitro* were dependent on the molar ratio of rhuMAb-E25 and IgE. At molar equivalence (1:1 ratio), the largest complex with a molecular weight of approximately 1000 kDa predominates (Fox et al. 1996). The data are best described by a cyclic hexamer structure (Fig. 11.11), which consists of three molecules of each immunoglobulin. This cyclic structure accounts for the limited size of the complexes. At antibody or antigen excess, small heterotrimers are formed, with little detectable hexamer present at molar ratios of 3 or larger.

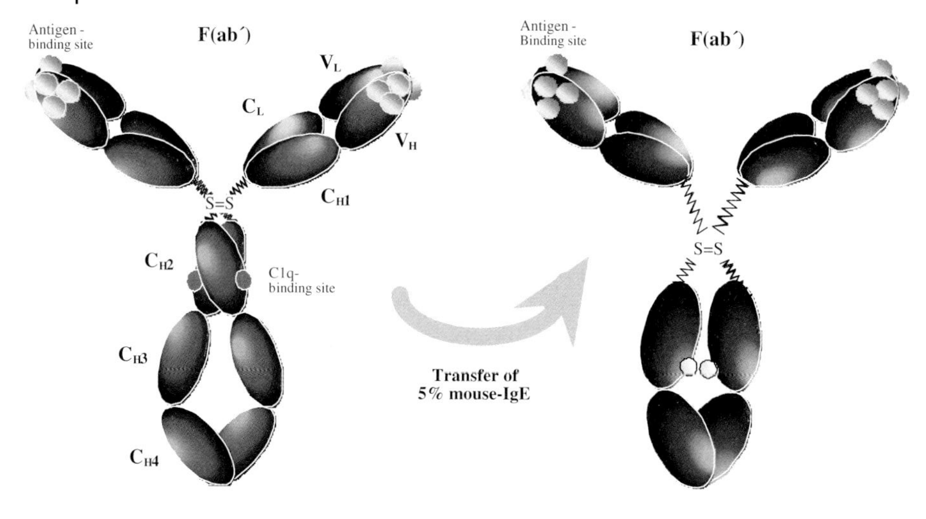

Fig. 11.9 Humanization of the monoclonal anti-IgE-antibody for omalizumab by fusion of the variable region of a mouse-anti-IgE-antibody with a human IgG1-frame. For details, see text. (Modified after Kroegel 2002.)

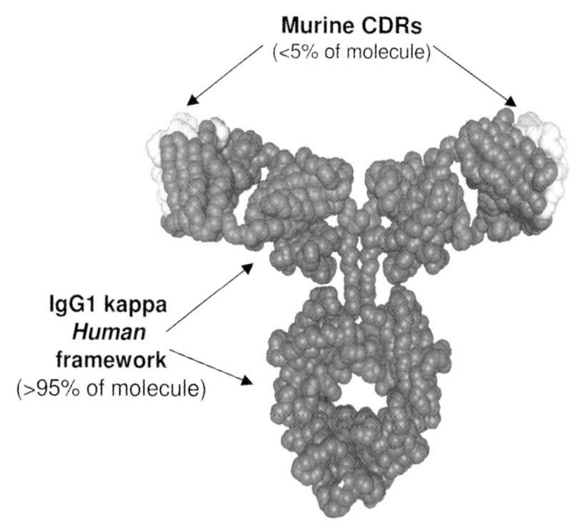

Fig. 11.10 Molecular model of rhuMAb-E25 (omalizumab). CDR = complementarity-determining region. (Adapted from Boushey 2001.)

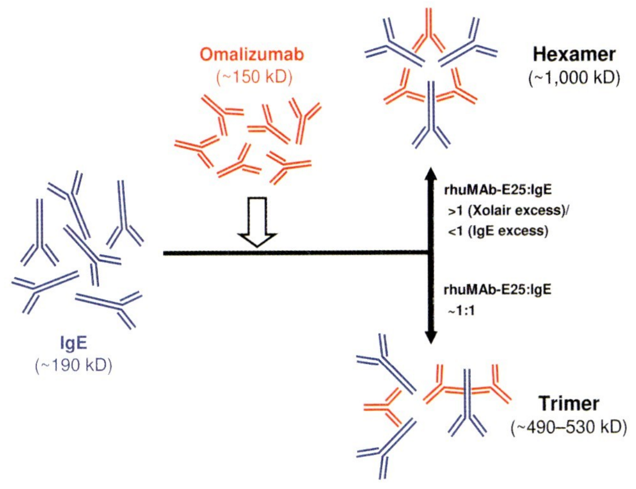

Fig. 11.11 Formation of rhuMAb-E25:IgE complexes in relation to their molar ratio.

Tissue distribution and rhuMAb-E25:IgE complex formation with IgE and rhuMAb-E25 following intravenous (IV) administration was initially studied in cynomolgus monkeys (Fox et al. 1996). ^{125}I-rhuMAb-E25 was administered as an intravenous bolus dose to wild cynomolgus monkeys that had high levels of IgE. Subsequent SEC of serum samples showed that the rhuMAb-E25:IgE complexes were of limited size and were similar to the small complexes formed *in vitro* with human IgE at antigen excess. No specific uptake of radioactivity was seen in any of the tissues collected from the animals at 1 and 96 h post administration.

Pharmacokinetic analysis revealed that both rhuMAb-E25 and rhuMAb-E25: IgE immune complexes cleared the serum compartment (Fig. 11.12), with urinary excretion as the primary route of elimination (Fox et al. 1996). Elimination of the immune complex was mediated by the interaction with FcγR (Fig. 11.13) (Lanier et al. 2003; Shields et al. 1995). Immune complex clearance was slower than IgE clearance, with a half-life of 3 weeks; therefore, although free IgE levels decrease, the total IgE level (IgE complexes with omalizumab and free IgE) is usually increased during omalizumab treatment.

11.7.3
Preclinical Results

The binding of rhuMAb-E25 to human peripheral blood basophils was assessed when basophils from 12 normal donors sensitized with ragweed-specific IgE were challenged with antigen. Only ragweed antigen induced histamine release, whereas rhuMAb-E25 failed to elicit histamine release from any donor. In other studies, the ability of rhuMAb-E25 to block IgE binding to human lung mast cells was analyzed by using strips of normal human lung perfused overnight with

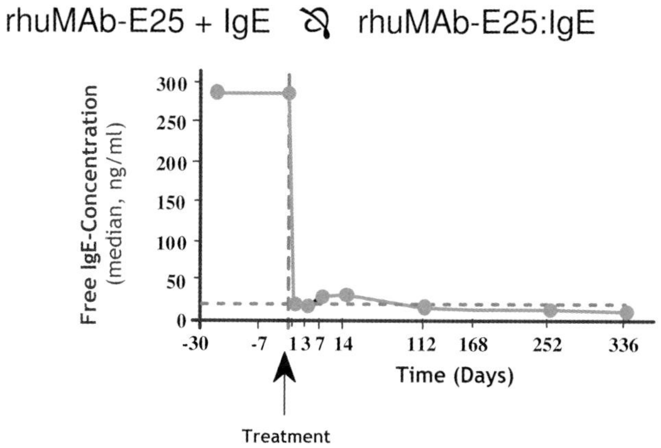

rhuMAb-E25 + IgE ⇌ rhuMAb-E25:IgE

Free IgE-Concentration (median, ng/ml)

300 250 200 150 100 50 0

-30 -7 1 3 7 14 112 168 252 336

Time (Days)

Treatment

Fig. 11.12 Free serum IgE concentration prior to and following application of rhuMAb-E25. Formation of rhuMAb-E25:IgE complexes occurs within several hours and causes almost complete elimination of IgE.

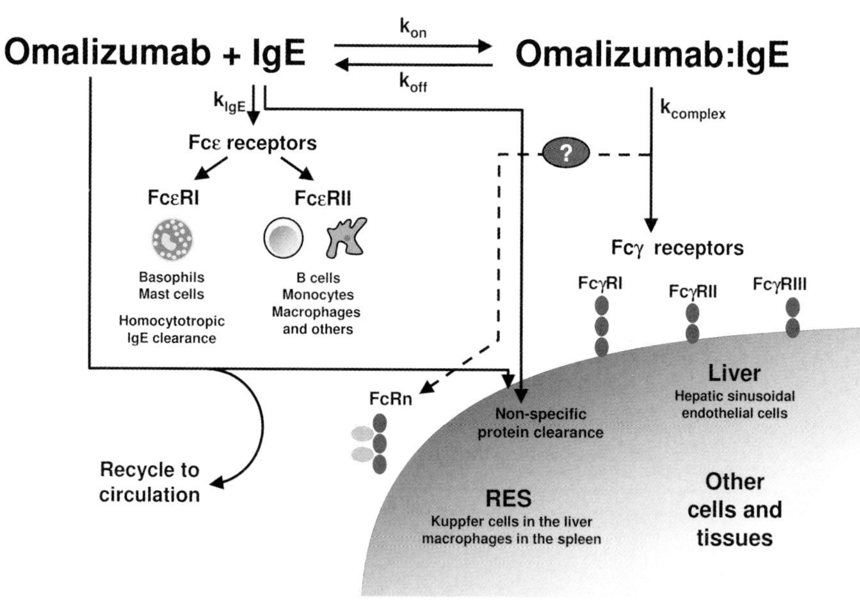

Fig. 11.13 Clearance pathways of rhuMAb-E25 (omalizumab). Clearance of omalizumab: IgE complexes occur via typical IgG pathways, including receptor-bearing cells and nonspecific mechanisms. Complexes are cleared from the body via interaction with the Fcγ receptors of platelets, leucocytes, and the reticuloendothelial system (RES).

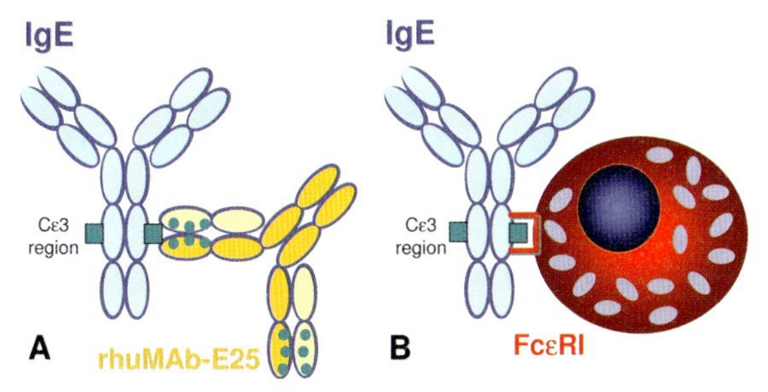

Fig. 11.14 Interaction between IgE rhuMAb-E25 and the mast cell (A) and with the humanized monoclonal anti-IgE antibody rhuMAb-E25 (omalizumab; Xolair). rhuMAb-E25 binds to the Cε3 region of the IgE molecule, which is the same part of the molecule that interacts with IgE receptors. Consequently, rhuMAb-E25 can *not* bind to receptor-bound IgE and is unlikely to generate anaphylactoid events.

ragweed-specific human IgE to sensitize the lung mast cells. Challenge with ragweed antigen induced mast cell degranulation, as measured by histamine release and muscle contraction. In contrast, rhuMAb-E25 completely inhibited this response. These data (Shields et al. 1995) confirm that rhuMAb-E25 is effective in blocking degranulation and a subsequent mediator release (Fig. 11.14).

Studies were then undertaken to examine the effects of rhuMAb-E25 *in vivo*. The ability of rhuMAb-E25 to effect IgE responses was measured in cynomolgus monkeys as rhuMAb-E25 has near-equivalent affinity for IgE purified from cynomolgus monkey serum (3×10^{-10} M) as for human IgE (1.7×10^{-10} M). To determine the ability of rhuMAb-E25 to activate cutaneous mast cells *in vivo*, 1 μg of the antibody was injected into the skin of normal cynomolgus monkeys, but failed to elicit the wheal and flare reaction indicative of mast cell degranulation. In contrast, a positive response was elicited with as little as 1 ng of a crosslinking murine mAb. Furthermore, rhuMAb-E25 failed to induce hive formation in monkey skin presensitized with 27 ng of human ragweed-specific IgE. These data were identical to the results obtained following systemic administration. Even at doses as high as 50 mg kg^{-1}, rhuMAb-E25 did not induce systemic anaphylaxis (Shields et al. 1995). Together, these preclinical studies confirmed the safety and efficacy of the designed antibody rhuMAb-E25.

11.7.4 Clinical Studies

Early studies conducted in patients with mild allergic asthma showed that the anti-IgE concept translated into demonstrable clinical effects, as shown by inhibition of allergen-induced early and late bronchoconstrictor responses (Fig. 11.15) (Boulet et al. 1997; Fahy et al. 1997). A phase II study in patients with moderate-to-severe allergic asthma reported reduced asthma exacerbations and corticosteroid requirements with the use of an intravenous formulation of omalizumab

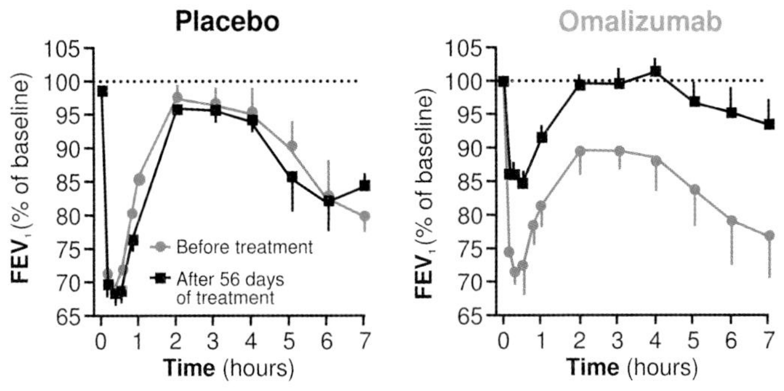

Fig. 11.15 Effect of omalizumab on the early- and late-phase asthmatic response. Right-hand panel: rhuMAb-E25 inhibits both the early asthmatic response and the late asthmatic response. Left-hand panel: placebo had no effect on either response, as would be expected. EAR = early asthmatic reaction; LAR = late asthmatic reaction. (Data from Fahy et al. 1997.)

(Milgrom et al. 1999). Subsequent phase III studies using the standard subcutaneous formulation of omalizumab were conducted in a total of 1405 children, adolescents and adults with moderate-to-severe allergic asthma and positive skin prick tests to one or more common environmental allergens (Busse et al. 2001; Milgrom et al. 2001; Soler et al. 2001). Subcutaneous injections of 150 to 375 mg omalizumab were given every 2 or 4 weeks, the dose being based on the patient's serum IgE level and bodyweight at baseline, in order to provide a dose of at least 0.016 $mg\,kg^{-1}$ omalizumab per $IU\,mL^{-1}$ IgE per 4 weeks (Hochhaus et al. 2003). Omalizumab was added on to inhaled corticosteroid (ICS) therapy for 16 weeks (steroid-stable phase) after which, during a 12-week steroid-reduction phase, the patients' doses of ICS were decreased to establish the optimal lowest dose required for an acceptable level of asthma control. The primary endpoint for the studies in adults was reduction in asthma exacerbations.

In all three studies, both the incidence and frequency of exacerbations (defined as a worsening of asthma requiring treatment with oral or intravenous corticosteroids or doubling of baseline ICS dose) were significantly reduced in the omalizumab treatment group compared with placebo, although there was also a reduction in asthma exacerbation incidence and frequency in this treatment group (Table 11.4). Patients significantly reduced their requirement for ICS, with a proportion totally withdrawing from this medication. Compared with the placebo group, a significantly greater reduction in ICS dose was achieved by the omalizumab-treated patients, and a substantially greater percentage of these were able to withdraw from ICS treatment completely. Improvements in asthma symptoms and rescue bronchodilator use were observed. Improvements in asthma-related quality of life also occurred with omalizumab treatment for both adults (Buhl et al. 2002; Finn et al. 2003) and children (Lemanske et al. 2002).

The prevention of potentially life-threatening episodes would not only provide significant benefits for patients but also reduce the overall cost of asthma care.

Table 11.4 Summary of the phase III, randomized, double-blind clinical trials comparing omalizumab with placebo in the treatment of allergic asthma.

Effect	*Omalizumab*	*Placebo*	*p-value*	*Reference*
Reduction of ICS dose (>50%) [%]	100	66.7	<0.001	Milgrom et al. (1999)
	72.4	54.9	<0.001	Busse et al. (2001)
	79	55	<0.001	Soler et al. (2001)
	74	51	0.001	Holgate et al. (2001)
Complete withdrawal of ICS [%]	55	39	0.004	Milgrom et al. (1999)
	39.6	19.1	<0.001	Busse et al. (2001)
	43	19	<0.001	Soler et al. (2001)
	21.4	15	0.198	Holgate et al. (2001)
Equivalent dose of inhaled BDP during extension phase [$mg\,day^{-1}$][a]	253	434	<0.001	Buhl et al. (2002)
	227	335	<0.001	Lanier et al. (2003)
Asthma exacerbations per patient (steroid-reduction phase)[a]	0.42	2.72	<0.001	Milgrom et al. (1999)
	0.39	0.66	0.003	Busse et al. (2001)
	0.36	0.75	<0.001	Soler et al. (2001)
Asthma exacerbations per patient (steroid-stable phase)[a]	0.28	0.54	0.006	Busse et al. (2001)
	0.28	0.66	<0.001	Soler et al. (2001)
	0.15	0.23	NS	Holgate et al. (2001)
Asthma exacerbations per patient (extension phase)[a]	0.48	1.14	<0.001	Buhl et al. (2002)
	0.60	0.83	0.023	Lanier et al. (2003)
Patients with ≥1 asthma exacerbation (steroid-stable phase) [%][a]	12.8	30.5	<0.001	Soler et al. (2001)
Subjects with ≥1 asthma exacerbation (steroid-reduction phase) [%][a]	15.7	29.8	<0.001	Soler et al. (2001)
Subjects with ≥1 asthma exacerbation (extension phase) [%][a]	24	40.6	<0.001	Buhl et al. (2002)
	31.8	42.8	0.015	Lanier et al. (2003)
Missed days of school[a]	0.65	1.21	0.04	Milgrom et al. (1999)

Values are median and ([a]) mean.
BDP = beclomethasone dipropionate; ICS = inhaled corticosteroids.

Therefore, lowering the rate of exacerbation is an important goal of asthma management (Masoli et al. 2004). The data from the three phase III studies were pooled and analyzed to determine the effect of omalizumab on serious exacerbations, which were measured on the basis of asthma-related emergency room visits and hospitalizations (Corren et al. 2003). Omalizumab-treated patients had significantly fewer unscheduled outpatient visits (21.3 versus 35.5, rate ratio 0.60, $p < 0.01$) and emergency room visits (1.8 versus 3.8, rate ratio 0.47, $p = 0.05$) per 100 patient-years compared with placebo-treated patients. Importantly, hospitalizations were markedly reduced from 3.42 events per 100 patient-years for placebo treatment to 0.26 for omalizumab treatment, a rate ratio of 0.08 ($p < 0.01$).

Although the majority of asthmatic conditions can be controlled by current treatment options, at least 5% of asthma patients have severe asthma that is often inadequately controlled by ICS and long-acting β_2-agonists (LABA). These patients are at high risk of severe exacerbations and death, and have the greatest medical need among the asthmatic population. A recent double-blind, parallel group, multicenter study (INNOVATE) was conducted in patients with severe persistent asthma that was inadequately controlled despite therapy with high-dose ICS and LABA (GINA step 4 treatment) (Humbert et al. 2005). Patients were randomized to receive omalizumab or placebo as add-on therapy for 28 weeks. Omalizumab significantly reduced ($p = 0.042$) the clinically significant asthma exacerbation rate compared with the placebo group (after adjustment for imbalance in baseline exacerbation history). Omalizumab treatment also halved the severe exacerbation rate (0.24 versus 0.48, $p = 0.002$), and the incidence of emergency visits was significantly lower for omalizumab patients (0.24 versus 0.43, $p = 0.038$). The results of the INNOVATE study indicate that omalizumab is an effective add-on therapy for difficult-to-treat patients with inadequately controlled severe persistent asthma.

The efficacy of omalizumab treatment in patients with severe persistent asthma has also been demonstrated in a pooled analysis of data from seven studies (Bousquet et al. 2005). Omalizumab treatment was added to current asthma therapy and compared with placebo (five double-blind studies) or with current asthma therapy alone (two open-label studies). The studies included 4308 patients, 93% of whom had severe persistent asthma according to the GINA 2002 classification. Omalizumab treatment significantly reduced the rate of asthma exacerbations by 38% and the rate of total emergency visits by 47% ($p < 0.001$ versus control), suggesting that omalizumab may fulfill an important need in this difficult-to-treat asthma population. Subgroup analysis revealed that omalizumab was effective irrespective of age, FEV_1, gender and IgE serum concentrations (Fig. 11.16). However, young patients (aged < 18 years) with an IgE level > 150 IU mL^{-1} and a severely impaired lung function ($FEV_1 < 60\%$ predicted) appeared to show a better response to treatment.

Evaluation of the treatment by the Cochran group involved an evidence-based survey of omalizumab's utility for asthma (Walker et al. 2004). This review found that omalizumab led to a significant reduction in inhaled steroid consumption compared with placebo: −114 mg per day (95% CI −150 to −78.13, two trials). The studies seemed to show increased health of patients and decreased steroid need, although the control groups had greater than expected responses. However, the Cochran reviewers stressed that the long-term clinical significance of omalizumab-induced IgE decrease still needs to be defined.

Altogether, the above clinical studies suggest that therapy with omalizumab can have a major effect on the treatment of patients with moderate-to-severe

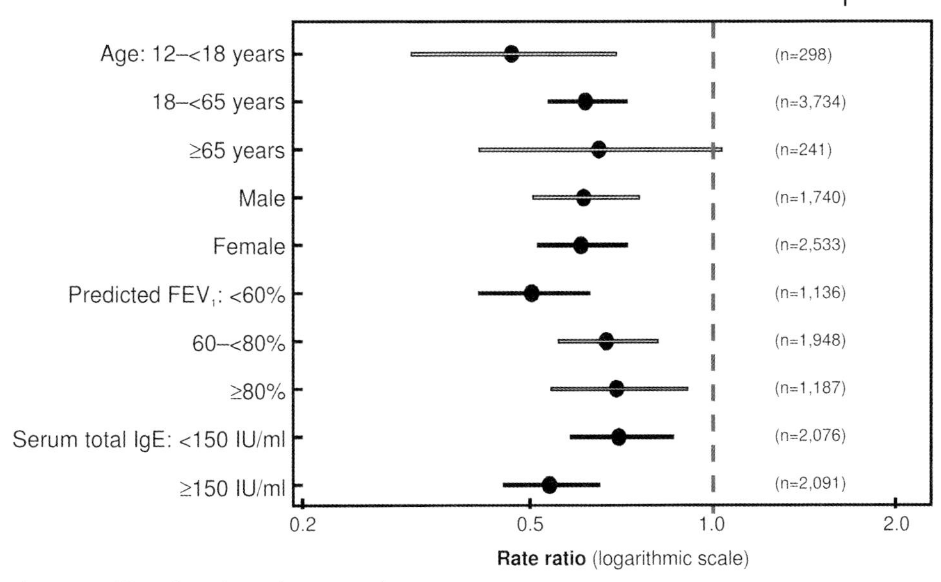

Fig. 11.16 Effect of omalizumab on exacerbation rate. A consistent reduction in asthma exacerbation rates can be observed, irrespective of baseline characteristics. (From Bousquet et al. 2005.)

Table 11.5 Summary of the beneficial effects of omalizumab in asthma.

- Reduces asthma exacerbations regardless of the type of allergic sensitization (seasonal or perennial)
- Improves asthma symptom scores
- Concomitantly improves upper airway symptoms in the case of co-existent allergic rhinitis
- Improves the quality of life of asthmatic subjects
- Steroid-sparing effect
- Reduces rescue medication

allergic asthma (Table 11.5). More importantly, omalizumab might be best used in the higher-risk patients who are at risk for serious asthma exacerbation requiring emergency department visitation, hospitalizations, or both.

11.8 Anti-Inflammatory Effects of Omalizumab

The biological effect of the anti-IgE antibody has been characterized in respect of its effect on humoral and cellular parameters, and shows actions at different levels of allergic inflammation (see Fig. 11.3).

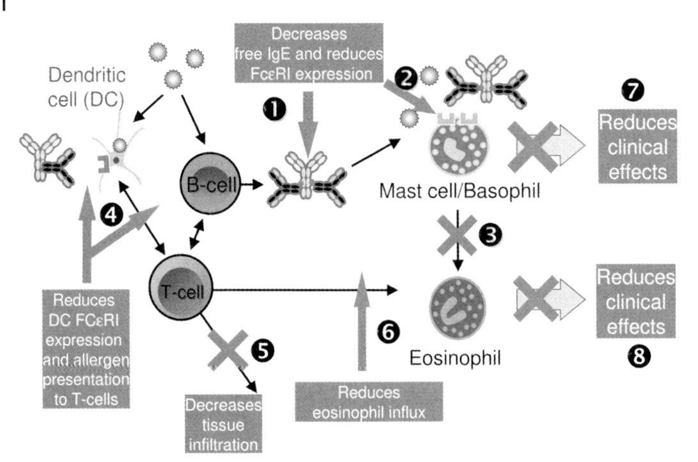

Fig. 11.17 Summary of anti-inflammatory actions of rhuMAb-E25 at different levels of allergic inflammation. ❶, decreases free IgE; ❷, reduces FcεRI expression on mast cells and basophils; ❸, blocks mast cell-mediated activation of eosinophils; ❹, reduces DC FCεRI expression on dendritic cells and allergen presentation to T cells; ❺, decreases T-cell tissue infiltration; and ❻ inhibits eosinophil tissue influx. As a consequence, the severity of both ❼ early and ❽ late asthmatic reactions are diminished.

11.8.1
Effects on Serum Free IgE Levels

The anti-IgEs under clinical evaluation and development have an association constant, K_a, for soluble IgE of approximately $10^{10}\,M^{-1}$ (Kolbinger et al. 1993), which is in the range of the affinity of the FcεRI receptor for IgE. Thus, if anti-IgE is maintained at concentrations in excess of IgE in the body, it should effectively compete with FcεRI for IgE (see Fig. 11.12). The concentrations of IgE in the blood vary widely among patients with allergy, ranging from 0.05 to $1\,mg\,mL^{-1}$ found in most patients. Taking this into account, anti-IgE given in excess to the basal concentration can bind most of the IgE, leaving a minimum of free IgE available for binding to IgE Fc-receptors.

11.8.2
Effect on Cytokines

Omalizumab induces complex changes in interleukin levels, and does not decrease all Th2-related interleukins. In a study of asthmatics, circulating levels of IL-5, IL-6, IL-8, IL-10, IL-13, and s-ICAM (Noga et al. 2003) were measured before and after 16 weeks of omalizumab treatment. Anti-IgE therapy also reduced cells staining for IL-4$^+$ cells in the airways of asthmatic subjects (Djukanovic et al. 2004). The results demonstrated a significant decreased IL-13 ($p < 0.01$) and reduced IL-5 and IL-8 levels in the omalizumab group compared to baseline, whereas the concentrations of other circulating mediators were unchanged. The reasons for this differential effect of omalizumab on circulating cytokine concentration are not known. However, cytokines in the peripheral blood represent a rather vague estimate of an inflammatory process which is taking place in the local inflammatory microenvironment of a defined tissue such as the airways.

11.8.3 Effects on FcεRI Cell Expression

The most readily appreciated pharmacological action of anti-IgE therapy is the indirect effect on the down-regulation of FcεRI on basophils. In earlier studies, the density of FcεRI on basophils was found to correlate strongly with the level of IgE in blood (see above). One of the clinical studies (MacGlashan et al. 1997b), in addition to *in-vitro* studies (MacGlashan et al. 1997a), show that the density of FcεRI on basophils decreased by between 95% and 99%, after anti-IgE had been administered to patients for up to 3 months. FcεRI expression on circulating basophils was significantly reduced by 99% in the presence of low serum levels of IgE during 90 days of treatment of ragweed-allergic individuals with the anti-IgE monoclonal antibody omalizumab (MacGlashan et al. 1997a). At the same time, the basophils isolated from patients after anti-IgE treatment were much less sensitive to allergen *ex-vivo* stimulation, as assessed by the degree of histamine release. In addition, dermal reagibility to allergen provocation (skin prick test) was reduced, indicating that mast cell function was also markedly decreased (MacGlashan et al. 1997a). Although the underlying mechanism for this is not completely understood, there is evidence that up-regulation is mediated by IgE interacting through FcεRI itself (MacGlashan et al. 1998).

In a phase I study, treatment with two intravenous doses of omalizumab decreased the number of FcεRI receptors from a median of 220000 to 8300 receptors per basophil at 3 months. Receptor density was decreased to 50% of baseline by day 3, and to 97% by 90 days, further supporting this hypothesis (MacGlashan et al. 1997a). In omalizumab studies evaluating its effects on early- and late-phase allergic reactions, the decrease in FEV_1 was reduced, and the allergen concentration required to elicit bronchoconstriction was increased (Boulet et al. 1997; Fahy et al. 1997). In addition, the maximum late-phase FEV_1 was increased 60% from baseline (Fahy et al. 1997).

11.8.4 Effect on Dendritic Cell APCs

With this background, clinical studies have shown evidence that a monoclonal anti-IgE antibody, omalizumab, affects DCs and therefore might be able to modulate T-cell responses. In addition to down-regulating FcεRI expression on basophils (MacGlashan et al. 1997a), therapy with omalizumab produced a rapid reduction in surface FcεRI expression of both precursor DCs type 1 and type 2 subsets (Prussin et al. 2003). A decrease or normalization in DC1s in patients treated with omalizumab during a grass pollen season, but not in the birch pollen season, has been shown (Feuchtinger et al. 2003). These findings might be explained by the fact that, in the absence of proinflammatory cytokines and mediators, DC1s promote the differentiation of Th2 cells. Furthermore, DC1s derived from monocytes and cultured with GM-CSF and IL-4 induce both Th1 and Th2 differentiation (Liu et al. 2001). Thus, decreased omalizumab-induced IL-4 production by mast cells during the pollen season might cause the apparent reduction in the number of myeloid DCs (Liu et al. 2001). Taken together, these findings imply that anti-IgE therapy might cause immunomodulation of T-cell responses and inhibit IgE-mediated type-I hypersensitivity.

Dendritic cells (DCs) are important in T-helper (Th) cell differentiation, with type I DCs (DC1) enhancing Th1 differentiation and type II DC (DC2) being an important factor for Th2 cell responses. DC2 cells are described as the "gatekeepers" of the immune response (Upham 2003), and IgE occupancy of DC2 FcεRI receptors is associated with greater allergen uptake and an increase in the resulting immune response (Maurer et al. 1995; Stingl and Maurer 1997). In ragweed-induced rhinitis, (Prussin et al. 2003), omalizumab treatment caused a significant decrease in basophil FcεRI expression at all time-points during the study (days 7, 14, 21, 28, 35, 42) compared with placebo ($p < 0.001$). In addition, the level of FcεRI expression on both DC1 and DC2 cells was reduced in omalizumab-treated individuals and correlated with serum levels of free IgE in patients receiving omalizumab. There also was a correlation between FcεRI expression on DC2 cells and that on DC1 cells and basophils. The maximum decrease in FcεRI expression was 73% on basophils, 52% on DC1 cells, and 83% on DC2 cells (Prussin et al. 2003). Interestingly, a significant reduction was observed as early as 7 days. These results show that the anti-IgE effects of omalizumab regulate FcεRI expression in these cell types in parallel with similar kinetics of inhibition.

11.8.5 Effect on Eosinophils

A recent study explored the effect of omalizumab on eosinophil survival induction of immunologic changes leading to eosinophil apoptosis, and also examined T-lymphocyte cytokine profiles in 19 patients with allergic asthma treated with omalizumab at a dose of at least $0.016\,mg\,kg^{-1}$ per IgE ($IU\,mL^{-1}$) every 4 weeks (Noga et al. 2006). Peripheral eosinophils and T-lymphocyte cytokine profiles were evaluated by fluorescence-activated cell sorting before treatment (baseline), at 12 weeks of treatment, and 12 weeks after discontinuation of treatment with omalizumab or placebo.

Markers of eosinophil apoptosis (Annexin V) were significantly increased in omalizumab recipients compared with placebo, whereas no changes in markers of necrosis (7-amino-actinomycin) or eosinophil activation CD69 or Fas receptor (CD95) were detected. GM-CSF^{+} lymphocytes were reduced in omalizumab recipients compared with placebo. Fewer IL-2^{+} and IL-13^{+} lymphocytes were also evident in omalizumab recipients than in the placebo group. There were no significant differences in IL-5, IFN-γ, or TNF-α between the omalizumab and placebo groups. In addition, anti-IgE therapy was associated with a decrease in sputum and tissue eosinophils accompanied by reduced $CD3^+$, $CD4^+$, and $CD8^+$ T lymphocytes (Djukanovic et al. 2004). These findings provide further evidence that omalizumab has additional anti-inflammatory activity, as demonstrated by the induction of eosinophil apoptosis and down-regulation of the inflammatory cytokines IL-2 and IL-13. Whilst further studies are needed to determine the underlying mechanisms, these findings support the critical role of IgE in the regulation of inflammation in allergic asthma, namely, influencing the inflammation is key to controlling the more severe type of asthma.

11.8.6
Effects on B Cells

Experimental evidence from both *in-vitro* and *in-vivo* studies supports the ability of anti-IgE in targeting IgE-expressing B cells and in inhibiting the continual production of IgE. The idea that anti-IgE can cause these effects is that anti-IgE binds to mIgE on IgE-expressing B cells. Because mIgE represents a part of the B-cell receptor, anti-IgE may interfere B-cell signaling or even cause their lysis, as has been shown for anti-IgM or anti-IgG (Eray et al. 1994; Warner and Scott, 1991; Warner et al. 1991). However, IgE-secreting plasma cells do not express mIgE, and presumably are not affected by anti-IgE. These cells reside in the bone marrow and probably have a life span of several weeks to several months. However, since new IgE-secreting plasma cells go through mIgE-expressing B-cell stages during differentiation, if their generation is abrogated by anti-IgE treatment, the existing plasma cells will die off over a period of several weeks to months, and thus, the production of IgE will also gradually abate. In addition, memory B cells may be affected by anti-IgE. The molecular mechanisms, leading to the depletion of these cells can be explained by apoptosis and reached by the immunologic process of tolerance and/or anergy induction. If this occurs, anti-IgE may have long-term effects on the fundamental disease process. Considering its effects on B cells and DCs, there is a chance that long-term treatment with omalizumab may eventually correct the pathogenic defect underlying allergic inflammation.

11.9
Pharmacological Properties of Omalizumab

11.9.1
Pharmacodynamics

Omalizumab forms complexes with IgE at the Cε3 domain (see Fig. 11.11), the principal site of effector cell binding. IgE is thereby prevented from binding to FcεRI on various effector cells, inhibiting activation and mediator release (Easthope and Jarvis 2001; Kelly and Sorkness 2002; see Package insert, Xolair 2003). In this manner, omalizumab decreases serum-unbound IgE levels, decreases the number of FcεRI receptors on basophils, and attenuates early- and late-phase allergic reactions (Boulet et al. 1997; Easthope et al. 2001; Fahy et al. 1997; Kelly et al. 2002; MacGlashan et al. 1997a; Package insert, Xolair 2003). Rapid, dose-dependent decreases (96–99%) in serum unbound IgE have been observed after subcutaneous administration (Package insert, Xolair 2003). A direct correlation between unbound IgE levels and the number of FcεRI receptors on circulating basophils has been shown (see above). Several large phase III clinical trials have demonstrated that omalizumab is more effective than placebo in controlling moderate to severe allergic asthma in patients who have poor disease control or exacerbations despite recommended therapy.

11.9.2
Pharmacokinetics

Omalizumab exhibits linear kinetics at doses > $0.5\,mg\,kg^{-1}$ (Package insert, Xolair 2003). After subcutaneous administration, omalizumab has an average bioavailability of 62%, reaching a maximum concentration of $30.9\,mg\,L^{-1}$ after 7 to 8 days. The volume of distribution is $78\,mL\,kg^{-1}$, suggesting limited distribution (i.e., plasma). In addition, an enhanced uptake into specific organs or tissues has not been observed (Easthope et al. 2001; Package insert, Xolair 2003). Omalizumab is eliminated primarily via hepatic degradation in the reticuloendothelial system and endothelial cells and, to a lesser degree, via biliary excretion (Fig. 11.13). After subcutaneous administration, the average elimination half-life of the drug is 26 days, and clearance appears to be body weight-related (i.e., doubling the body weight approximately doubles clearance). Although data regarding pharmacokinetic differences among age groups are lacking, the evaluation of small numbers of children aged between 12 and 18 years and elderly patients suggests no difference; however, the numbers of patients investigated were small (Package insert, Xolair 2003).

11.10
Adverse Effects

11.10.1
Systemic Side Effects

Overall adverse events reported for subjects treated with omalizumab compared with placebo have been similar and categorized as mild to moderate in severity. The most commonly reported effects are fatigue, arthralgias, rash, diarrhea, nausea, vomiting, dizziness, epistaxis, menorrhagia, itching, dermatitis, and hematoma [Food and Drug Administration (FDA) 2003]. Other systemic side effects included viral infection (23%), upper respiratory tract infection (20%), sinusitis (16%), headache (15%), and pharyngitis (11%). In clinical trials to date, there has been no evidence of anti-IgE–IgE immune complex disease, nor has there been any evidence of complement activation or fixation. The platelet abnormalities seen in a small segment of monkeys has not been observed in human trials (Lanier et al. 2003).

11.10.2 Local Reactions

Injection-site reactions, including bruising, redness, warmth, and burning, were observed in 45% of the treated patients and occurred more commonly with omalizumab treatment versus placebo (12% versus 9%). Generally, these reactions developed within 1 h of injection, but their incidence decreased with continued administration.

11.10.3 Serious Adverse Effects

Most adverse effects were mild to moderate, and incidence rates were similar to those observed with placebo. However, potential safety concerns identified by the FDA in reviewing trial data on omalizumab included risks of anaphylaxis and the development of cancer.

Anaphylaxis Treatment with omalizumab is meant to prevent any risk of anaphylaxis, as the agent cannot interact with IgE already bound to cell surfaces. However, in clinical trials, three patients (<0.01%) developed anaphylaxis, characterized by urticaria and angioedema (FDA 2003). Two of the reactions were temporally associated with omalizumab administration and occurred within 2 h after the first injection. One patient (<0.1%) developed antibodies to omalizumab, and one individual developed proteinuria deemed unrelated to immune-complex hypersensitivity (Package insert, Xolair 2003; Soler et al. 2001).

Malignancies Malignant neoplasms were observed in 20 (0.5%) of 4127 omalizumab-treated patients compared with five (0.2%) of 2236 control subjects. The observed malignancies included a wide variety of predominantly epithelial or solid-organ cancers, including breast, melanoma, non-melanoma, skin, prostate, and parotid; this suggested that there was no specific association with omalizumab treatment. Moreover, the difference in malignancy rate between omalizumab-treated patients and placebo was not statistically significant. In addition, as the rate of malignancy observed in the control group was lower than expected, no unequivocal association can be deduced from the data.

11.10.4 Immune Complex Diseases

In theory, the administration of omalizumab can induce antibodies against the murine components of the drug, but no immune complex-mediated pathologic conditions as a result of the formation of such antibodies have been observed. Although no clinical problems related to such an effect have been noted, this may represent a potential concern in specific populations. Again, further information is required on the safety profile after long-term use of omalizumab.

11.10.5 Long-Term Adverse Effects

As the majority of patients treated with omalizumab have been observed for only one year, the effect of longer use in patients who are at increased risk for cancer is not known. In particular, it is unknown whether the incidence of neoplasms is increased with long-term treatment or in high-risk patients. Therefore, omalizumab probably should not be used in patients with either a history of cancer or a strong family history of cancer until this risk relationship is better understood.

11.11 Indications

On the basis of clinical trials, the FDA approved omalizumab only for very specific indications (Table 11.6). Omalizumab is indicated for patients aged ≥12 years with moderate-to-severe persistent asthma who have a positive skin test or *in-vitro* reactivity to a perennial aeroallergen, and whose symptoms are inadequately controlled with inhaled corticosteroids. Omalizumab has not been approved for

Table 11.6 Indications for the use of omalizumab approved by the FDA (2003).

1. Second-line treatment (after first-line treatments have failed). In addition:
2. Patients must have moderate-persistent allergy-related asthma:
   - Daily symptoms
   - Daily use of inhaled short-acting beta-2-agonist
   - Exacerbations affect activity
   - Exacerbations at least twice per week, which may last for days
   - Night-time symptoms more frequently than once a week
   - Lung function of forced FEV_1 or peak expiratory flow
3. Patients must have severe-persistent allergy-related asthma:
   - Continual symptoms
   - Limited physical activity
   - Frequent exacerbations
   - Frequent night-time symptoms
   - Lung function of FEV_1 or PEF ≤60% predicted, and PEF variability >30%
4. Patients must be aged over 12 years.
5. Patients must have a positive skin test to a perennial aeroallergen (e.g., dust mite, cats, dogs, and fungi).
6. Patients must be symptomatic with inhaled corticosteroids.

children aged < 12 years, although this may change in future when relevant data are available. Omalizumab does not show any specific drug interactions, and can be used with corticosteroids and other anti-asthmatic drugs in most cases.

11.12 Contraindications

Absolute contraindications to the administration of omalizumab include a prior experience of severe hypersensitivity to omalizumab, although these reactions have only rarely been reported. Omalizumab treatment is also not indicated in patients with an IgE level $<30\,\text{IU}\,\text{mL}^{-1}$ or $>700\,\text{IU}\,\text{mL}^{-1}$, as studies proving the efficacy of the drug at these IgE concentrations are not available. Omalizumab does not alleviate asthma exacerbations acutely, and should not be used for the treatment of acute bronchospasm or status asthmaticus. The safety and efficacy of omalizumab treatment in patients with allergic conditions other than perennial aeroallergen has not been established. In addition, the clinical efficacy or safety of the use of omalizumab for the prevention of peanut or other food allergies, or for the prevention or treatment of allergic rhinitis, awaits further clinical studies (Table 11.7).

Table 11.7 Potential indications and contraindications (currently available) for the treatment of bronchial asthma and rhinitis with rhuMAb-E25 (omalizumab). Data are based on currently available information. (Modified after Kroegel et al. 2002).

Indications
- Allergic bronchial asthma
- Seasonal allergic asthma
- Perennial bronchial asthma due to unavoidable allergens (e.g., house-dust mite)
- Treatment of asthma in children/adolescent above the age of 12 years

Questionable indications
- Treatment of asthma in children under the age of 12 years
- Non-allergic, intrinsic asthma
- Food allergy
- Atopic dermatitis
- Idiopathic urticaria
- Churg–Strauss syndrome
- Specific immunotherapy

Contraindications
- Non-allergic, intrinsic bronchial asthma
- Acute asthma exacerbation
- Pregnancy
- Thrombocytopenia*

* Further studies warranted.

11.13
Preparation for Use

Omalizumab is available as a preservative-free, sterile, lyophilized powder for reconstitution with sterile water for injection. Prior to reconstitution, omalizumab should be refrigerated (2–8 °C). For single use, omalizumab is supplied in 5-mL vials designed to deliver either 150 or 75 mg on reconstitution with sterile water (not normal saline) for injection. The powder requires 15 to 30 min to dissolve; dissolution can be facilitated by constant stirring. After reconstitution, omalizumab is stable in the vial for 8 h under refrigeration, or for 4 h at room temperature. The solution should not be shaken and should be protected from direct sunlight.

The solution is viscous and must be carefully drawn up into the syringe before being administered. The injection itself may take 5 to 10 s to administer (Table 11.8). Fewer injection-site reactions are seen to occur when the solution injected is completely clear.

The baseline serum total IgE level and patient body weight are used to determine the dose (150–375 mg) and frequency (every 2 or 4 weeks) according to a standard nomogram (Table 11.9) (Package insert, Xolair 2003).

Because of the time-consuming requirements for drug preparation and the high cost of the drug, the following approach to handling should be considered when treating patients with omalizumab:

- scheduled appointments for injection of several patients; and
- preparation of the injection after the patient has arrived.

In total, a single administration of omalizumab may take more than 60 min, as a 30-min observation period is recommended following the injection.

11.14
Administration

The maximum dose per injection site is 150 mg; hence, doses exceeding 150 mg must be administered as multiple injections at different sites (Package insert,

Table 11.8 Points to be considered when administering omalizumab.

- Omalizumab 150 to 375 mg administered subcutaneously every 2 or 4 weeks.
- Dosing and frequency are determined by pretreatment serum total IgE level and body weight (Table 11.9), and should be in keeping with the FDA approved package insert.
- Doses should be adjusted for significant changes in body weight.
- Re-testing of IgE levels during treatment and up to 1 year after discontinuing treatment cannot be used as a guide for dose determination and is not medically necessary due to total serum IgE levels remaining elevated throughout this time.
- Injections are to be limited to not more than 150 mg per subcutaneous injection site, and to not more than 375 mg per day.

Table 11.9 Omalizumab dosage (in mg) and dosing schedule on baseline total serum IgE and body weight (kg) (Genentech Product Information).[a]

Baseline serum IgE [IU mL^{-1}]	*Body weight [kg]*			
	30–60	*>60–70*	*>70–90*	*>90–150*
≥30–100	150	150	150	300
>100–200	300	300	300	**225**
>200–300	300	**225**	**225**	**300**
>300–400	**225***	**225**	**300**	–
>400–500	**300**	**300**	**375**	–
>500–600	**300**	**375**	–	–
>600–700	**375**	–	–	–

* Bold values indicate dosing every 2 weeks; otherwise every 4 weeks.
a Patients whose pretreatment serum IgE levels or body weight are outside the limits of the dosing table (>30 or >700 IU mL^{-1} and <30 and >150 kg) should not be administered omalizumab.

Xolair 2003). Omalizumab is slowly absorbed after subcutaneous administration; thus, with a mean elimination half-life of 26 days, omalizumab (0.016 mg kg^{-1} IgE^{-1} [IU mL^{-1}] per 4 weeks) can be administered fortnightly or monthly. Following the injection of omalizumab, emergency medications should be available to treat severe hypersensitivity reactions, including anaphylaxis (Xolair Product Information 2004).

11.15 Clinical Dosing

The principal aim of anti-IgE mAb treatment with omalizumab is to eliminate serum free IgE levels. In early clinical trials a relationship between clinical efficacy variables and free IgE concentrations was demonstrated, with the best results obtained when serum free IgE levels were <25 ng mL^{-1} or 10.4 IU mL^{-1} (Casale et al. 1997). To achieve this goal, appropriate dosing depends on two factors (Bang and Plosker 2004):

- the subject's pretreatment serum IgE level (in IU mL^{-1}); and
- the patient's pretreatment body weight (in kg).

The minimal dosing is 0.016 mg kg^{-1} (IU mL^{-1}) IgE per 4 weeks in divided doses as necessary. On the basis of these data, the individual dosage required to reduce IgE levels to the threshold concentration mentioned above is calculated according to the nomogram given in Table 11.9. Pretreatment serum IgE levels or body weights outside the limits of the dosing table do not qualify for treatment because: (i) the volume of the drug would be too excessive; and (ii) the expense of the drug would be too high.

11.16 Dosing Adjustments

Dosing adjustments should not be made in response to subsequent total IgE levels (Kroegel et al. 2004); however, adjustments are required for patients experiencing significant body weight changes. Dosage adjustments for patients with renal or hepatic failure are not necessary.

11.17 Precautions and Contraindications

Omalizumab is indicated for the maintenance treatment of allergic asthma, but not indicated for the treatment of acute exacerbations (e.g., asthma attack, status asthmaticus). Although omalizumab may help reduce steroid requirements, systemic or inhaled steroids should be tapered gradually to lower doses rather than be abruptly discontinued (Package insert, Xolair 2003).

11.17.1 Drug Interactions

To date, no studies evaluating the potential for drug interactions involving omalizumab have been performed. Omalizumab is neither protein-bound nor metabolized by the cytochrome P450 isoenzyme system. This suggests a low probability of drug interactions (Package insert, Xolair 2003).

11.17.2 Pregnancy and Lactation

Human data on potential harmful effects during pregnancy and lactation are not available. In animal studies, omalizumab did not demonstrate maternal toxicity, embryotoxicity, or teratogenicity during organogenesis, or adverse fetal/neonatal growth effects during late gestation, delivery, and nursing. Conversely, omalizumab (similar to IgG) may cross the placental barrier and be secreted in breast milk. The risks of absorption and harm to the fetus or nursing infant are unknown. Omalizumab is a "category B" agent, and thus should be used only if clearly needed in pregnancy. In addition, until further data become available caution should be exercised during breast-feeding (Package insert, Xolair 2003).

11.18 Monitoring of Therapy

Serum-free IgE levels will decrease to approximately 1% of baseline values within 24 h of subcutaneous administration (Casale et al. 1997). Total serum IgE levels (free plus bound) increase with treatment due to the formation of circulating omalizumab–IgE complexes. Unfortunately, serum free IgE level is currently not available only for routine measurement; therefore, the monitoring of total serum IgE levels is not useful and should not be performed in a clinical setting once therapy has been initiated (Hamilton et al. 2005). A specified commercial assay may help to optimize dosing and maximize omalizumab therapy in future.

11.19 Cost

Omalizumab is considerably more expensive than conventional asthma therapy. Omalizumab has an average wholesale price of €424 ($541) per 150-mg vial, but the individual cost of treatment is variable. Based on current dosing recommendations, monthly treatment costs range from €424 ($541) to €2547 ($3247) (Package insert; Omalizumab, Xolair 2003). The related annual costs range from €3200 ($4000) to €15 700 ($20 000), depending on the dosage (Dacus 2003), with an average of approximately €9413 ($12 000) per year (Table 11.10). This compares with approximate costs per year of €1004 ($1280) for montelukast (Singulair, Merck), €1694 ($2160) for the combination of fluticasone dipropionate and salmeterol (Advair, GlaxoSmithKline), and €502 ($680) for extended-release theophylline (Uniphyl, Purdue). In addition, visit costs may further increase as omalizumab must be administered by a physician.

Table 11.10 Doses of omalizumab injection versus annual cost.

Omalizumab dose [mg]	*Treatment interval [weeks]*	*Annual cost*	
		US$	*Euro*
150	4	3879	3043
150	2	7759	6086
300	4	7759	6086
300	2	15518	12173
375	2	23277	18259

11.20 Response to Treatment

11.20.1 Onset of Action of Anti-Immunoglobulin E Effect

The onset of action is particularly important for seasonal allergic rhinitis, when patients need to respond quickly to changing pollen seasons. A recent study attempted to measure the onset of action of omalizumab by measuring the time taken to inhibit the nasal responses to ragweed allergen (Lin et al. 2004). Omalizumab had a substantial impact on ragweed-induced reductions in nasal volume within the first 7 days to up to 12% and a further reduction to 6% until day 42. These data suggest that omalizumab acts rapidly to significantly reduce serum free-IgE levels and substantially inhibits ragweed-induced reductions in the nasal volume. Interestingly, clinical improvement was paralleled by reductions in FcεRI expression on DC1 and DC2 cells as well as basophils. It might be proposed that administering omalizumab a week before the start of the pollen season, and 4 weeks later, would protect patients from AR symptoms during the pollen season.

Despite these observations, an apparent response to treatment can take several weeks (Chang 2000). Among patients in a clinical trial who had responded to omalizumab by 16 weeks, 87% had done so by 12 weeks (Bousquet et al. 2004). These data suggest that patients should be treated for at least 12 weeks before efficacy is assessed.

11.20.2 Duration of Treatment

The optimal duration of anti-IgE therapy is unknown, but it is likely that anti-IgE has to be administered continuously in a dose-dependent fashion. Even if the generation of new IgE-producing plasma cells is blocked by anti-IgE, new IgE-expressing B cells (and, thus, new plasma cells) will be regenerated in a few weeks to a few months in the absence of anti-IgE. Thus, repeated anti-IgE dosing appears to be necessary. While patients are under anti-IgE treatment, IgE-related immune mechanisms and their manifestations are abbreviated and remain exposed to the usual allergens. If these allergens drive the immune system toward non-IgE-related responses, the disease process may be gradually attenuated. If anti-IgE can inhibit IgE-expressing memory B cells that are responsible for recurrent IgE responses, then the continuous exposure of allergens may preferentially drive the immune system towards the production of antibodies of other subclasses. If anti-IgE can indeed influence a shift in the immune response, the combination of antigen immunotherapy with anti-IgE may present another new approach for individuals suffering from severe allergies (see below).

Given that serum IgE levels and the numbers of FcεRIs increase after therapy is discontinued (Saini et al. 1999), it seems that treatment needs to be continued for efficacy to persist, though no studies have yet been reported on the duration of effects after discontinuation. If administration is interrupted, then treatment should be resumed at the dose initially prescribed. Dosing may only need to be adjusted when substantial changes in body weight have occurred (see Table 11.9).

11.21 Non-Approved Diseases

11.21.1 Allergic Rhinitis

Omalizumab has been shown also to be useful in the treatment of allergic rhinitis, which affects a large proportion of the population (Adelroth et al. 2000; Casale et al. 1997, 1999; Plewako et al. 2002). Although often seasonal in nature, allergic rhinitis causes considerable distress to the sufferer and has a high social cost in terms of consumed healthcare resources and lost productivity. It is important to add that asthmatic patients frequently have associated rhinitis.

Several studies of omalizumab involving more than 1000 patients with severe seasonal allergic rhinitis induced by birch or ragweed pollen have been performed (Adelroth et al. 2000; Casale et al. 1997, 1999; Chervinsky et al. 2003), all of which showed a statistically significant improvement in nasal and ocular symptom severity scores, the use of rescue medication (antihistamines), and the quality of life (Table 11.11). Further studies have identified decreases in the number of inflammatory cells, including decreased eosinophil peroxidase-positive cells and decreased IgE^+-staining cells (Plewako et al. 2002). On the basis of these results it may be concluded that patients with concomitant allergic rhinitis and bronchial asthma may benefit more from treatment with omalizumab (Table 11.12) (D'Amato and Holgate 2002; D'Amato et al. 2002, 2004; Ayres et al. 2004; Bousquet et al. 2005; Holgate et al. 2005a,b).

11.21.2 Other Clinical Applications

In addition to moderate-to-severe persistent asthma (see above), a number of additional disorders relating to other organ manifestations of allergic diseases or nonallergic conditions associated with increased IgE levels may benefit from treatment with omalizumab (Table 11.13). There is clinical evidence that treatment with an anti-IgE mAb might diminish symptoms caused by peanut-induced allergic reactions in susceptible children (Leung et al. 2003), atopic dermatitis (Vigo et al. 2006), and idiopathic cold urticaria (Joshua and Boyce 2006). One study suggests that omalizumab can be used to treat healthcare workers with occupational latex allergy (Leynadier et al. 2004). In addition, there is evidence that a combination of anti-IgE with specific immunotherapy (SIT) might be

Table 11.11 Summary of the randomized, double-blind, clinical trials comparing omalizumab with placebo in the treatment of seasonal and perennial allergic rhinitis.

Parameter	*Omalizumab*	*Placebo*	*p-value*	*Reference*
Nasal symptom severity score	0.75	0.98	0.002	Casale et al. (1997)
	0.70	0.98	<0.001	Adelroth et al. (2000)
	1.0	1.4	0.001	Chervinsky et al. (2003)
Eye symptom severity score	0.41	0.67	0.001	Casale et al. (1997)
	0.43	0.54	0.031	Adelroth et al. (2000)
Use of rescue medication (mean number of tablets/day)	0.12	0.21	0.005	Casale et al. (1997)
	0.59	1.37	<0.001	Adelroth et al. (2000)
Rhinitis quality-of-life scores (% of patients with significant improvement)	44.1	30.6	<0.01	Adelroth et al. (2000)
	52	27	0.001	Chervinsky et al. (2003)
Total rhinitis quality-of-life score (five-point scale)	1.75	2.3	<0.001	Adelroth et al. (2000)
Global treatment effectiveness (%)	70.7	40.8	<0.001	Casale et al. (1997)
	59	35	0.001	Adelroth et al. (2000)
	53	34	0.001	Chervinsky et al. (2003)
Missed days of work and/or school	0.1	0.4	0.005	Casale et al. (1997)

Values shown are mean scores. Range of possible scores is 0 (no symptoms) to 3 (worst symptoms).

Table 11.12 Clinical efficacy of omalizumab in patients with allergic rhinitis and allergic asthma.

Allergic rhinitis

- Reduction in daily nasal severity score
- Reduction in the use of rescue antihistamines
- Improvement in the quality of life
- Blunting of the seasonal increase of nasal symptoms during the pollen season

Allergic asthma

- Reduction in early- and late-phase responses to allergen challenge
- Reduction in total asthma exacerbations
- Reduction in number of patients with ≥1 asthma exacerbations
- Reduction in dosage of ICS use
- More subjects achieve a complete discontinuation of corticosteroids
- Improvement in nocturnal asthma score
- Improvement in asthma quality-of-life score
- Reduction in hospitalizations and emergency department visits in high-risk asthmatic populations
- No significant change seen in FEV^1

ICS = Inhaled corticosteroid.

Table 11.13 Preliminary omalizumab therapy in non-asthmatic and non-rhinitic conditions, with beneficial effects.

Condition	*Reference*
Atopic dermatitis	Vigo et al. (2006)
Idiopathic cold urticaria	Joshua and Boyce (2006)
Protection from acute reactions after immunotherapy	Parks and Casale (2006)
Raised intensity (dose increase) of immunotherapy	Parks and Casale (2006)
Combination with allergen immunotherapy	Kuehr et al. (2002); Rolinck-Werninghaus et al. (2004)
Omalizumab in occupational allergy	Rambasek and Kavuru (2006)
Omalizumab plus immunotherapy	Parks and Casale (2006)
Food allergy (e.g., peanut)	Leung et al. (2003)
Occupational latex allergy	Leynadier et al. (2004)

beneficial. Combination therapy may permit a broader use of SIT by reducing the risk of anaphylactic side effects after SIT injections (Parks and Casale 2006). Furthermore, there is preliminary evidence that SIT in patients treated with omalizumab may improve the outcome of immunotherapy (Kuehr et al. 2002; Rolinck-Werninghaus et al. 2004). Taken together, these findings are encouraging and should prompt further studies in this regard.

11.22 Areas of Uncertainty

For the clinical trials of omalizumab, patients were enrolled with precisely defined characteristics of asthma, including sensitivity to specific perennial aeroallergens (i.e., dust mites, cockroaches, dog or cat dander). The role of omalizumab in patients with asthma who have allergies to other aeroallergens, such as molds or pollens, or who have negative allergy skin tests, has not been defined. It is also not clear to what extent omalizumab might be effective in patients with total serum IgE levels outside the trial ranges (30 to 700 $IU\,mL^{-1}$ for patients aged 12 to 75 years). In addition, in clinical practice, there is considerable variability of response to omalizumab therapy. The reasons for such variability have not been established; hence, studies are needed to determine whether specific characteristics of individual patients might help to predict response.

The clinical trials performed to date have evaluated omalizumab only as adjunctive therapy with inhaled corticosteroids as compared with placebo; they have not evaluated the relative benefit of omalizumab in comparison with other available therapies, such as leukotriene-modifiers or theophylline. Also needed are comparisons with asthma therapies that are available to patients for whom low-dose inhaled corticosteroids do not control the asthma and who require step-up management (i.e., an increased dose of the corticosteroid or the addition of another medication) (National Heart, Lung, and Blood Institute 1997, 2003).

A critical point should be raised regarding the actual clinical relevance of the moderate corticosteroid-sparing effects observed in the trials, even if these reductions were significant, as well as to the substantial improvements noted in placebo groups. Given that the cost of omalizumab is substantially greater than that of conventional asthma therapy, the potential cost-effectiveness of this form of treatment will be important to assess.

The efficacy and safety of omalizumab have not been established for durations of treatment that exceed 1 year, and it is not known for how long clinical effects might persist when therapy is discontinued. As asthma is a chronic disease, long-term studies – especially in children – are needed to evaluate the effect of serum IgE suppression throughout development. To date, it cannot be excluded that adverse effects may become apparent only with follow-up into adulthood after several years. To date, only one study is available that has been performed exclusively in the pediatric age group (Hochhaus et al. 2003). Similarly, efficacy and safety studies are also needed for geriatric and non-white patients.

On the other hand, studies have indicated that the expression of FceRI on basophils might be reduced by omalizumab therapy. Thus, there may be fewer targets for the anti-FceRI antibodies with clinical improvement, which may in turn permit a reduction of the omalizumab dose required to maintain treatment success.

11.23 Outlook

The introduction of omalizumab seems likely to have a major impact in the therapy of difficult-to-treat asthma. The fact that allergic diseases often occur at more than one anatomic locality – notably the upper respiratory tract and sinuses – means that patients receiving this therapy should experience improvements in the multiple manifestations of allergy. Some 80 years after the discovery of reagin (the first anti-IgE treatment), omalizumab now represents a promising new therapeutic option for patients with allergic diseases. These studies will also encourage further investigations which focus on the selective targeting of mIgE-bearing B cells, thus inhibiting IgE synthesis before IgE production starts. The action of omalizumab on proximal processes of the allergic immune response offers the potential that anti-IgE mAbs may modulate the immune regulation and lead to a profound and long-lasting clinical improvement. As allergic rhinitis, allergic gastroenteritis, anaphylaxy and atopic dermatitis are all diseases in which IgE participates – both in immediate-hypersensitivity response and in the induction of chronic allergic inflammation – the application of omalizumab is likely to be widened in future. As subgroups of the allergic manifestations in skin and airways occur even without increased total IgE levels and in the absence of specific IgE, the availability of anti-IgE treatment will help to establish whether there are true so-called “intrinsic” or “nonallergic” forms of the disease.

Acknowledgments

These studies were supported by the Deutsche Forschungsgemeinschaft, Bonn (Br 1949/1-1) and (Kr 956/2-1), the Bundesministerium für Bildung und Forschung (BMBF), Bonn, (01 ZZ9602, VKF Projekt 2.9), and the County of Thüringia, Germany (01KC8906/1)

References

Adelroth, E., Rak, S., Haahtela, T., Aasand, G., Rosenhall, L., Zetterstrom, O., Byrne, A., Champain, K., Thirwell, J., Della Gioppa, G., Sandstrom T. (2000), Recombinant humanized mAb E25, an anti-IgE mAb, in birch pollen-induced seasonal allergic rhinitis. *J Allergy Clin Immunol* 106: 253–259.

Allam, J.P., Niederhagen, B., Bücheler, M., Appel, T., Betten, H., Bieber, T., Bergé, S., Novak, N. (2006) Comparative analysis of nasal and oral mucosa dendritic cells. *Allergy* 61: 166–172.

Amiri, P., Haak-Frendscho, M., Robbins, K., Mckerrow, J.H., Stewart, T., Jardieu, P. (1994) Anti-immunoglobulin E treatment decreases worm burden and egg production in *Schistosoma mansoni*-infected normal and interferon gamma knockout mice. *J Exp Med* 180: 43–51.

Aubrey, J.P., Pochon, S., Graber, P., Jansen, K.U., Bonnefoy, J.Y. (1992) CD21 is a ligand for CD23 and regulates IgE production. *Nature* 358: 505–507.

Ayres, J.G., Higgins, B., Chilvers, E.R., Ayre, G., Blogg, M., Fox H. (2004) Efficacy and tolerability of anti-immunoglobulin E therapy with omalizumab in patients with poorly controlled (moderate-to-severe) allergic asthma. *Allergy* 59: 701–708.

Bang, L.M., Plosker, G.L. (2004) Omalizumab: a review of its use in the management of allergic asthma. *Treat Respir Med.* 3: 183–199.

Barnes, P.J. (2000) Anti-IgE therapy in asthma: rationale and therapeutic potential. *Int Arch Allergy Immunol* 123: 196–204.

Beasley, R., Crane, J., Lai, C.K., Pearce, N. (2000) Prevalence and etiology of asthma. *J Allergy Clin Immunol* 105(Suppl.): S466–S472.

Bieber, T., de la Salle, H., Wollenberg, A., Hakimi, J., Chizzonite, R., Ring, J., et al. (1992) Human epidermal Langerhans cells express the high affinity receptor for immunoglobulin E (FcεRI). *J Exp Med* 175: 1285–1290.

Bonnefoy, J.Y., Plater-Zyberk, C., Lecoanet-Henchoz, S., Gauchat, J.F., Aubry, J.P., Graber, P. (1996) A new role for CD23 in inflammation. *Immunol Today* 17: 418–420.

Borkowski, T.A., Jouvin, M.H., Lin, S.Y., Kinet, J.P. (2001) Minimal requirements for IgE-mediated regulation of surface FcεRI. *J Immunol* 167: 1290–1296.

Boulet, L.P., Chapman, K.R., Cote, J., Kalra, S., Bhagat, R., Swystun, V.A., Laviolette, M., Cleland, L.D., Deschesnes, F., Su, J.Q., DeVault, A., Fick, R.B., Cockcroft, D.W. (1997) Inhibitory effects of an anti-IgE antibody E25 on allergen-induced early asthmatic response. *Am J Respir Crit Care Med* 155: 1835–1840.

Boushey, H.A. (2001) Experiences with monoclonal antibody therapy for allergic asthma. J Allergy Clin Immunol 108: S77–S83.

Bousquet, J., Wenzel, S., Holgate, S., Lumry, W., Freeman, P., Fox, H. (2004) Predicting response to omalizumab, an anti-IgE antibody, in patients with allergic asthma. *Chest* 125: 1378–1386.

Bousquet, J., Cabrera, P., Berkman, N., et al. (2005) The effect of treatment with omalizumab, an anti-IgE antibody, on asthma exacerbations and emergency medical visits in patients with severe persistent asthma. *Allergy* 60: 302–308.

Burrows, B., Martinez, F.D., Halonen, M., Barbee, R.A., Cline M.G. (1989) Association of asthma with serum IgE levels and skin-test reactivity to allergens. *N Engl J Med* 320: 271–277.

Busse, W., Coren, J., Lanier, B.Q., McAlary, M., Fowler Taylor, A., Della Cioppa, G., van As, N., Gupta, N. (2001) Omalizumab, anti-IgE recombinant humanized monoclonal antibody, for the treatment of severe allergic asthma. *J Allergy Clin Immunol* 108: 184–190.

Buhl, R., Hanf, G., Soler, M., Bensch, G., Wolfe, J., Everhard, F., et al. (2002) The anti-IgE antibody omalizumab improves asthma-related quality of life in patients with allergic asthma. *Eur Respir J* 20: 1088–1094.

Casale, T.B., Bernstein, I.L., Busse, W.W., LaForce, C.R., Tinkelman, D.G., Stoltz, R.R., Dockhorn, R.J., Reimann, J., Su, J.Q., Fick, R.B., Adelman, D.C. (1997) Use of an anti-IgE humanized monoclonal antibody in ragweed-induced allergic rhinitis. *J Allergy Clin Immunol* 100: 110–121.

Casale, T., Condemi, J., Miller, S., Fickm, R.B., Jr., McAlarym, M., Fowler-Taylor, A., Gupta N. (1999) rhuMAb-E25 in the treatment of seasonal allergic rhinitis (SAR). *Ann Allergy Asthma Immunol* 82: 75–82.

Chang, T.W. (2000) The pharmacological basis of anti-IgE therapy. *Nat Biotechnol* 18: 157–163.

Chang, T.W., Davis, F.M., Sun, N.C., Sun, C.R., Macglashan, D.W., Jr., Hamilton, R.G. (1990) Monoclonal antibodies specific for human IgE-producing B cells: a potential therapeutic for IgE-mediated allergic diseases. *Biotechnology (NY)* 8: 122–126.

Chervinsky, P., Casale, T., Townley, R., et al. (2003) Omalizumab, an anti-IgE antibody, in the treatment of adults and adolescents with perennial allergic rhinitis. Ann Allergy Asthma Immunol 91: 1607.

Corren, J., Casale, T., Deniz, Y., Ashby, M. (2003) Omalizumab, a recombinant humanized antiIgE antibody, reduces asthma-related emergency room visits and hospitalizations in patients with allergic asthma. *J Allergy Clin Immunol* 111: 87–90.

Dacus, J.J. (2003) Disease of the month: asthma. *PEC Update* 3: 4–6.

D'Amato, G., Holgate, S. (2002) *The Impact of Air Pollution on the Respiratory Health. Monograph of the European Respiratory Society.* Sheffield, UK.

D'Amato, G., Oldani, V., Donner, C.F. (2002) Anti-IgE monoclonal antibody: a new approach to the treatment of allergic respiratory diseases. *Arch Chest Dis* 59: 29–33.

D'Amato, G., Liccardi, G., Noschese, P., Salzillo, A., D'Amato, M., Cazzola, M. (2004) Anti IgE Monoclonal antibody (Omalizumab) in the treatment of atopic asthma and allergic respiratory diseases. *Curr Drug Targets – Inflammation Allergy* 3: 227–231.

Davis, F.M., Gossett, L.A., Chang, T.W. (1991) An epitope on membrane-bound but not secreted IgE: implications in isotype-specific regulation. *Biotechnology (NY)* 9: 53–56.

Davis, F.M., Gossett, L.A., Pinkston, K.L., Liou, R.S., Sun, L.K., Kim, Y.W., Chang, N.T., Chang, T.W., Wagner, K., Bews, J., et al. (1993) Can anti-IgE be used to treat allergy? *Semin Immunopathol* 15: 51–73.

Djukanovic, R., Wilson, S.J., Kraft, M., Jarjour, N.N., Steel, M., Chung, K.F., Bao, W., Fowler-Taylor, A., Matthews, J., Busse, W.W., Holgate, S.T., Fahy, J.V. (2004) Effects of treatment with anti-immunoglobulin E antibody omalizumab on airway inflammation in allergic asthma. *Am J Respir Crit Care Med* 170: 583–593.

Easthope, S., Jarvis, B. (2001) Omalizumab. *Drugs* 61: 253–260.

Eray, M., Tuomikoski, T., Wu, H., Nordstrom, T., Andersson, L.C., Knuutila, S., et al. (1994) Cross-linking of surface IgG induces apoptosis in a bcl-2 expressing human follicular lymphoma line of mature B cell phenotype. *Int Immunol* 6: 1817–1827.

Fahy, J.V., Fleming, H.E., Wong, H.H., Liu, J.T., Su, J.Q., Reimann, J., Fick, R.B.,

Boushey, H.A. (1997) The effect of an anti-IgE monoclonal antibody on the early- and late-phase responses to allergen inhalation in asthmatic subjects. *Am J Respir Crit Care Med* 155: 1828–1834.

Feuchtinger, T., Bartz, H., von Berg, A., Riedinger, F., Brauburger, J., Stengle, S., et al. (2003) Treatment with omalizumab normalizes the number of myeloid dendritic cells during the grass pollen season. *J Allergy Clin Immunol* 111: 428–430.

Finn, A., Gross, G., van Bavel, J., et al. (2003) Omalizumab improves asthma-related quality of life in patients with severe allergic asthma. *J Allergy Clin Immunol* 111: 278–284.

Food and Drug Administration, Center for Biologics Evaluation and Research. (2003) BLASTN 103976/0, review of clinical safety data: original BLS submitted on June 2, 2000 and response to complete review letter submitted on December 18, 2002. Rockville, Md.: Department of Health and Human Services.

Fox, J.A., Hotaling, T.E., Struble, C., Ruppel, J., Bates, D.J., Schoenhoff, M.B. (1996) Tissue distribution and complex formation with IgE of an anti-IgE antibody after intravenous administration in cynomolgus monkeys. *J Pharmacol Exp Ther* 279: 1000–1008.

Garman, S.C., Kinet, J.P., Jardetzky, T.S. (1998) Crystal structure of the human high-affinity IgE receptor. *Cell* 95: 951–961.

Garman, S.C., Wurzburg, B.A., Tarchevskaya, S.S., Kinet, J.P., Jardetzky, T. S. (2000) Structure of the Fc fragment of human IgE bound to its high-affinity receptor FcepsilonRI alpha. *Nature* 406: 259–266.

Geha, R.S., Jabara, H.H., Brodeur, S.R. (2003) The regulation of immunoglobulin E class-switch recombination. *Nat Rev Immunol* 3: 721–732.

GINA – Global Initiative for Asthma. (1995, updated 2002, 2003). *Global strategy for asthma management and prevention.* NIH publication number 02-3659.

Gounni, A.S., Lamkhioued, B., Delaporte, E., Dubost, A., Kinet, J.P., Capron, A., et al. (1994) The high-affinity IgE receptor on eosinophils: from allergy to parasites or from parasites to allergy? *J Allergy Clin Immunol* 94: 1214–1216.

Gounni, A.S., Wellemans, V., Yang, J., Bellesort, F., Kassiri, K., Gangloff, S., et al. (2005) Human airway smooth muscle cells express the high affinity receptor for IgE (Fc epsilon RI): a critical role of Fc epsilon RI in human airway smooth muscle cell function. *J Immunol* 175: 2613–2621.

Hagan, P., Blumenthal, U.J., Dunn, D., Simpson, A.J., Wilkins, H.A. (1991) Human IgE, IgG4 and resistance to reinfection with Schistosoma haematobium. *Nature* 349: 243–245.

Hamilton, R.G., Marcotte, G.V., Saini, S.S. (2005) Immunological methods for quantifying free and total serum IgE levels in allergy patients receiving omalizumab (Xolair) therapy. *J Immunol Methods* 303: 81–91.

Herbelin, A., Elhadad, S., Ouaaz, F., de Groote, D., Descamps-Latscha, B. (1994) Soluble CD23 potentiates interleukin-1-induced secretion of interleukin-6 and interleukin-1 receptor antagonist by human monocytes. *Eur J Immunol* 24: 1869–1873.

Henchoz, S., Gauchat, J.F., Aubry, J.P., Graber, P., Pochon, S., Bonnefoy, J.Y. (1994) Stimulation of human IgE production by a subset of anti-CD21 monoclonal antibodies: requirement of a co-signal to modulate epsilon transcripts. *Immunology* 81: 285–290.

Heusser, C., Jardieu, P. (1997) Therapeutic potential of anti-IgE antibodies. *Curr Opin Immunol* 9: 805–813.

Hochhaus, G., Brookman, L., Fox, H., et al. (2003) Pharmacodynamics of omalizumab: implications for optimized dosing strategies and clinical efficacy in the treatment of allergic asthma. *Curr Med Res Opin* 19: 491–498.

Holgate, S.T. (1998) Asthma and allergy – disorders of civilization? *Q J Med* 91: 171–184.

Holgate, S.T., Bousquet, J., Wenzel, S., Fox, H., Liu, J., Castellsague, J. (2001) Efficacy of omalizumab, an anti-immunoglobulin E antibody, in patients with allergic asthma at high risk of serious asthma-related morbidity and mortality. *Curr Med Res Opin* 17: 233–240.

Holgate, S.T., Casale, T., Wenzel, S., Bousquet, J., Deniz, Y., Reisner, C. (2005a) The anti-inflammatory effects of omalizumab confirm the central role of IgE in allergic inflammation. *J Allergy Clin Immunol* 115: 459–465.

Holgate, S.T., Djukanovic, R., Casale, T., Bousquet, J. (2005b) Anti-immunoglobulin E treatment with omalizumab in allergic diseases: an update on anti-inflammatory activity and clinical efficacy. *Clin Exp Allergy* 35: 408–416.

Humbert, M., Beaskey, R., Ayres, J., Slavin, R., Hebert, J., Bousquet, J., Beeh, K.M., Ramos, S., Canonica, G.W., Hedgecock, S., Fox, H., Blogg, M., Surrey, K. (2005) Benefits of omalizumab as add-on therapy in patients with severe persistent asthma who are inadequately controlled despite best available therapy (GINA 2002 step 4 treatment): INNOVATE. *Allergy* 60: 309–316.

ISAAC-Steering Committee. (1998) Worldwide variation in prevalence of symptoms of asthma, allergic rhinoconjunctivitis, and atopic eczema: ISAAC. *Lancet* 351: 1225–1232.

Ishizaka, K., Ishizaka, T. (1970) Biological function of IgE antibodies and mechanisms of reaginic hypersensitivity. *Clin Exp Immunol* 6: 25–31.

Joshua, A., Boyce, J.A. (2006) Successful treatment of cold-induced urticaria/ anaphylaxis with anti-IgE. *J Allergy Clin Immunol* 117: 1415–1418.

Kalinski, P., Hilkens, C.M., Wierenga, E.A., Kapsenberg, M.L. (1999) T-cell priming by type-1 and type-2 polarized dendritic cells: the concept of a third signal. *Immunol Today* 20: 561–567.

Kehry, M.R., Yamashita, L.C. (1989) Low-affinity IgE receptor (CD23) function on mouse B cells: role in IgE-dependent antigen focusing. *Proc Natl Acad Sci USA* 86: 7556–7560.

Kelly, H.W., Sorkness, C.A. (2002) Asthma. In: Dipiro, J.T., Talbert, R.L., Yee, G.C., Matzke, G.R., Wells, B.G., Posey, L.M. (Eds.), *Pharmacotherapy: a pathophysiologic approach*. 5th edn. New York: McGraw-Hill, pp. 475–510.

Kijimoto-Ochiai, S. (2002) CD23 (the low-affinity IgE receptor) as a C-type lectin: a multidomain and multifunctional molecule. *Cell Mol Life Sci* 59: 648–664.

Kolbinger, F., Saldanha, J., Hardman, N., Bendig, M.M. (1993) Humanization of a mouse anti-human IgE antibody: a potential therapeutic for IgE-mediated allergies. *Protein Eng* 6: 971–980.

Kraft, S., Katoh, N., Novak, N., Koch, S., Bieber, T. (2001) Unexpected functions of FcepsilonRI on antigen-presenting cells. *Int Arch Allergy Immunol* 124: 35–37.

Kroegel, C. (2002) *Bronchial asthma. Pathogenic basis, diagnostic and therapy.* 2nd edn. Thieme, Stuttgart, New York.

Kroegel, C., Mock, B., Reißig, A., Hengst, U., Förster, M., Machnik, A., Henzgen, M. (2002) Antibodies in allergic diseases. Selective inhibition of IgE-mediated reactions by humanized IgE-anti-IgE-antibodies. *Arzneimitteltherapie* 20: 226–237.

Kroegel, C., Reissig, A., Henzgen, M. (2004) Therapeutic dosing of omalizumab = 20x[IgE]: is this equation correct or is enough too less? *Pneumologie* 58: 543–545.

Kuehr, J., Brauburger, J., Zielen, S., Schauer, U., Kamin, W., et al. (2002) Efficacy of combination treatment with anti-IgE plus specific immunotherapy in polysensitized children and adolescents with seasonal allergic rhinitis. *J Allergy Clin Immunol* 109: 274–280.

Lanier, B.Q., Corren, J., Lumry, W., Liu, J., Fowler-Taylor, A., Gupta, N. (2003) Omalizumab is effective in the long-term control of severe allergic asthma. *Ann Allergy Asthma Immunol* 91: 154–159.

Lemanske, R.F., Nayak, A., McAlary, M., Everhard, F., Fowler-Taylor, A., Gupta, N. (2002) Omalizumab improves asthma-related quality of life in children with allergic asthma. *Pediatrics* 110: E55.

Leung, D.Y., Sampson, H.A., Yunginger, J.W., Burks, A.W., Schneider, L.C., Wortel, C.H., et al. (2003) Effect of anti-IgE therapy in patients with peanut allergy. *N Engl J Med* 348: 986–993.

Leynadier, F., Doudou, O., Gaouar, H., Le Gros, V., Bourdeix, I., Guyomarch-Cocco, L., Trunet, P. (2004) Effect of omalizumab in health care workers with occupational latex allergy. *J Allergy Clin Immunol* 113: 360–361.

Lin, H., Boesel, K.M., Griffith, D.T., et al. (2004) Omalizumab rapidly decreases nasal allergic response and FceRI on basophils. *J Allergy Clin Immunol* 113: 297–302.

Liu, J., Lester, P., Builder, S., Shire, S.J. (1995) Characterization of complex formation by humanized anti-IgE antibody and monoclonal IgE *Biochemistry* 34: 10474–10482.

Liu, Y.J., Kanzler, H., Soumelis, V., Gilliet, M. (2001) Dendritic cell lineage, plasticity and cross-regulation. *Nat Immunol* 2: 585–589.

MacGlashan, D., Jr. (2005) IgE and Fc{epsilon}RI regulation. *Ann N Y Acad Sci* 1050: 73–88.

MacGlashan, D.W., Bochner, B.S., Adelman, D.C., Jardieu, P.M., Togias, A., Lichtenstein, L.M. (1997a) Serum IgE level drives basophil and mast cell IgE receptor display. *Int Arch Allergy Immunol* 113: 45–47.

MacGlashan, D.W., Jr., Bochner, B., Adelman, D., Jardieu, P., Togias, A., McKenzie-White, J., Sterbinsky, S., Hamilton, R., Lichtenstein, L.M. (1997b) Down-regulation of FceRI expression on human basophils during in vivo treatment of atopic patients with anti-IgE antibody. *J Immunol* 158: 1438–1445.

MacGlashan, D., McKenzie-White, J., Chichester, K., et al. (1998) In vitro regulation of FceRI expression on human basophils by IgE antibody. *Blood* 91: 1633–1643.

Maizels, N. (2003) Yin outwits yang at the IgE locus. *Nat Immunol* 4: 7–8.

Maizels, R.M., Yazdanbakhsh, M. (2003) Immune regulation by helminth parasites: cellular and molecular mechanisms. *Nat Rev Immunol* 3: 733–744.

Malveaux, F.J., Conroy, M.C., Adkinson, N.F.J., Lichtenstein, L.M. (1978) IgE receptors on human basophils: Relationship to serum IgE concentration. *J Clin Invest* 62: 176–181.

Masoli, M., Fabian, D., Holt, S., Beasley, R. (2004) Global Initiative for Asthma (GINA) Program. The global burden of asthma: executive summary of the GINA Dissemination Committee report. *Allergy* 59: 469–478.

Maurer, D., Ebner, C., Reininger, B., Fiebiger, F., Kraft, D., Kinet, J.P., et al. (1995) The high affinity IgE receptor (FcεRI) mediates IgE-dependent allergen presentation. *J Immunol* 154: 6285–6290.

Maurer, D., Ebner, C., Reininger, B., Petzelbauer, P., Fiebiger, E., Stingl, G. (1997) Mechanisms of Fc epsilon RI-IgE-facilitated allergen presentation by dendritic cells. *Adv Exp Med Biol* 417: 175–178.

Milgrom, H., Fick, R.B., Jr., Su, J.Q., Reimann, J.D., Bush, R.K., Watrous, M.L., et al. (1999) Treatment of allergic asthma with monoclonal anti-IgE antibody. *N Engl J Med* 341: 1966–1973.

Milgrom, H., Berger, W., Nayak, A., Gupta, N., Pollard, S., McAlary, M., Fowler-Taylor, A., Rohane, P. (2001) Treatment of childhood asthma with anti-immunoglobulin E antibody (omalizumab). *Pediatrics* 108: 36–45.

Nakamura, T., Kloetzer, W.S., Brams, P., Hariharan, K., Chamat, S., Cao, X., LaBarre, M.J., Chinn, P.C., Morena, R.A., Shestowsky, W.S., Li, Y.P., Chen, A., Reff, M.E. (2002) In vitro IgE inhibition in B cells by anti-CD23 monoclonal antibodies is functionally dependent on the immunoglobulin Fc domain. *Int J Immunopharmacol* 22: 131–141.

National Heart, Lung, and Blood Institute. (1997) National Asthma Education and Prevention Program Expert Panel Report 2: guidelines for the diagnosis and management of asthma. Bethesda, Md.: National Institutes of Health (Publication no. 97-4051).

National Heart, Lung, and Blood Institute. (2003) Education and Prevention Program Expert Panel Report: guidelines for the diagnosis and management of asthma – update on selected topics 2002. Bethesda, Md.: National Institutes of Health (Publication no. 02-5074).

Nissim, A., Eshhar, Z. (1992) The human mast cell receptor binding site maps to the third constant domain of immunoglobulin E. *Mol Immunol* 29: 1065–1072.

Nissim, A., Schwarzbaum, S., Siraganian, R., Eshhar, Z. (1993) Fine specificity of the IgE interaction with the low and high affinity Fce receptor. *J Immunol* 150: 1365–1374.

Noga, O., Hanf, G., Kunkel, G. (2003) Immunological changes in allergic asthmatics following treatment with omalizumab. *Int Arch Allergy Immunol* 131: 46–52.

Noga, O., Hanf, G., Brachmann, I., Klucken, A.C., Kleine-Tebbe, J., Rosseau, S., Kunkel, G., Suttorp, N., Seybold, J. (2006) Effect of omalizumab treatment on peripheral eosinophil and T-lymphocyte function in patients with allergic asthma. *J Allergy Clin Immunol* 117: 1493–1499.

Novak, N., Bieber T. (2003) Allergic and nonallergic forms of atopic diseases. *Allergy Clin Immunol* 112: 252–262.

Novak, N., Bieber, T., Katoh N. (2001) Engagement of FcεRI on human monocytes induces the production of IL-10 and prevents their differentiation in dendritic cells. *J Immunol* 167: 797–804.

Omalizumab (Xolair): an anti-IgE antibody for asthma. (2003) *Med Lett Drugs Ther* 45: 67–68.

Package insert. (2003) Xolair (omalizumab). East Hanover, NJ: Aventis, June 2003.

Parks, K.W., Casale, T.B. (2006) Anti–immunoglobulin E monoclonal antibody administered with immunotherapy. *Allergy Asthma Proc* 27(2 Suppl.1): S33–S36.

Platts-Mills, T.A. (2001) The role of immunoglobulin E in allergy and asthma. *Am J Respir Crit Care Med* 164(8 Pt 2): S1–S5.

Plewako, H., Arvidsson, M., Petruson, K., Oancca, I., Holmberg, K., Adelroth, E.A., Gustafsson, H., Sandstrom, T., Rak, S. (2002) The effect of omalizumab on nasal allergic inflammation, *J Allergy Clin Immunol* 110: 68–71.

Presta, L.G., Lahr, S.J., Shields, R.L., Porter, J.P., Gorman, C.M., Fendly, B.M., Jardieu, P.M. (1993) Humanization of an antibody directed against IgE. *J Immunol* 151: 2623–2632.

Presta, L.G., Shields, R.L., O'Connell, L., Lahr, S., Porter, J., Gorman, C., et al. (1994) The binding site on human immunoglobulin E for its high affinity receptor. *J Biol Chem* 269: 26368–26373.

Prussin, C., Griffith, D.T., Boesel, K.M., Lin, H., Foste, B., Casale, T.D. (2003) Omalizumab treatment downregulates dendritic cell FceRI expression, *J Allergy Clin Immunol* 112: 1147–1154.

Rambasek, T., Kavuru, M.S. (2006) Omalizumab dosing via the recommended card versus use of the published formula. *J Allergy Clin Immunol* 117: 708–709.

Rissoan, M.C., Soumelis, V., Kadowaki, N., Grouard, G., Briere, F., de Waal Malefyt, R., et al. (1999) Reciprocal control of T helper cell and dendritic cell differentiation. *Science* 283: 1183–1186.

Rolinck-Werninghaus, C., Hamelmann, E., Keil, T., Kulig, M., Koetz, K., Gerstner, B., et al. (2004) The co-seasonal application of anti-IgE after preseasonal specific immunotherapy decreases ocular and nasal symptom scores and rescue medication use in grass pollen allergic children. *Allergy* 59: 973–979.

Rosenwasser, L.J., Meng, J. (2005) Anti-CD23. *Clin Rev Allergy Immunol* 29: 61–72.

Saban, R., Haak-Frendscho, M., Zine, M., Ridgway, J., Gorman, C., Presta, L., Bjorling, D., Saban, M., Jardieu, P. (1994) Human FceRI-IgG and humanized anti-IgE monoclonal antibody MaE11 block passive sensitization of human and rhesus monkey lung. *J Allergy Clin Immunol* 94: 836–843.

Saini, S.S., MacGlashan, D. (2002) How IgE upregulates the allergic response. *Curr Opin Immunol* 14: 694–697.

Saini, S.S., MacGlashan, D.W., Jr., Sterbinsky, S.A., et al. (1999) Down-regulation of human basophil IgE and FcεRI surface densities and mediator release by anti-IgE-infusions is reversible in vitro and in vivo. *J Immunol* 162: 5624–5630.

Scharenberg, A.M., Kinet, J.P. (1995) Early events in mast cell signal transduction. *Chem Immunol* 61: 72–87.

Shields, R.L., Whether, W.R., Zioncheck, K., O'Connell, L., Fendly, B., Presta, L.G., Thomas, D., Saban, R., Jardieu, P. (1995) Inhibition of allergic reactions with antibodies to IgE. *Int Arch Allergy Immunol* 107: 308–312.

Soler, M., Matz, J., Townley, R., Buhl, R., O'Brien, J., Fox, H., Thirlwell, J., Gupta, N., Della Cioppa, G. (2001) The anti-IgE antibody omalizumab reduces exacerbations and steroid requirement in allergic asthmatics, *Eur Respir J* 18: 254–261.

Sutton, B.J., Gould, H.J. (1993) The human IgE network. *Nature* 366: 421–428.

Stingl, G., Maurer, D. (1997) IgE-mediated allergen presentation via FceRI on antigen-presenting cells. *Int Arch Allergy Immunol* 113: 24–29.

Turner, H., Kinet, J.P. (1999) Signalling through the high-affinity IgE receptor FcεRI. *Nature* 402(Suppl.): B24–B30.

Upham, J.W. (2003) The role of dendritic cells in immune regulation and allergic airway inflammation. *Respirology* 8: 1408–1418.

Vercelli, D., Helm, B., Marsh, P., Padlan, E., Geha, R.S., Gould, H. (1989) The B-cell binding site on human immunoglobulin E. *Nature* 338: 649–651.

Vigo, P.G., Girgis, K.R., Pfuetze, B.L., Critchlow, M.E., Fisher, J., Hussain, I. (2006) Efficacy of anti-IgE therapy in patients with atopic dermatitis. *J Am Acad Dermatol* 55: 168–170.

Walker, S., Monteil, M., Phelan, K., Lasserson, T.J., Walters, E.H. (2004) Anti-IgE for chronic asthma in adults and children. *Cochrane Database Syst Rev* 3: CD003559.

Wang, B., Rieger, A., Kilgus, O., Ochiai, K., Maurer, D., Fodinger, D., et al. (1992) Epidermal Langerhans cells from normal human skin bind monomeric IgE via FcεRI. *J Exp Med* 175: 1353–1365.

Warner, G.L., Scott, D.W. (1991) A polyclonal model for B cell tolerance. I. Fc-dependent and Fc-independent induction of nonresponsiveness by pretreatment of normal splenic B cells with anti-Ig. *J Immunol* 146: 2185–2191.

Warner, G.L., Gaur, A., Scott, D.W. (1991) A polyclonal model for B-cell tolerance. II. Linkage between signaling of B-cell egress from G0, class II upregulation and unresponsiveness. *Cell Immunol* 138: 404–412.

Warner, J.O., Kaliner, M.A., Crisci, C.D., Del Giacco, S., Frew, A.J., Liu, G.H., Maspero, J., Moon, H.B., Nakagawa, T., Potter, P.C., Rosenwasser, L.J., Singh, A.B., Valovirta, E., Van Cauwenberge, P. (2006) World Allergy Organization Specialty and Training Council. Allergy practice worldwide: a report by the World Allergy Organization Specialty and Training Council. *Int Arch Allergy Immunol* 139: 166–174.

Yazdanbakhsh, M., van den Biggelaar, A., Maizels, R.M. (2001) Th2 responses without atopy: immunoregulation in chronic helminth infections and reduced allergic disease. *Trends Immunol* 22: 372–377.

Websites

Xolair product labeling. Genentech, Inc. Accessed January, 2004, at: http://www.gene.com/gene/common/inc/pi/xolair.jsp.

Food and Drug Administration Web site. Patient safety news: first biologic for allergy-related asthma. Available at: www.accessdata.gov/scripts/cdrh/cfdocs. Accessed November 15, 2004

12
Palivizumab (Synagis)

Alexander C. Schmidt

12.1
Nature, Role in Disease, and Biology of the Target

12.1.1
Respiratory Syncytial Virus (RSV)-Induced Disease and RSV Epidemiology

Respiratory syncytial virus (RSV) infection is the single most common cause of hospitalization for acute respiratory disease among infants [39,81,82], and an important etiology of lower respiratory disease in young children and the elderly [28,69]. By the end of the second year of life, most children have experienced at least one RSV infection, and repeat infections with RSV are common. Approximately 1 to 3% of all healthy term infants are admitted to hospital for lower respiratory tract illness (LRTI) due to primary RSV infection [14,69,82], and this admission rate can rise to and even exceed 10% in high-risk populations. Risk factors for severe RSV disease include chronic lung disease of infancy/bronchopulmonary dysplasia (CLD/BPD), birth at less than 36 weeks of gestation, clinically significant congenital heart disease (CHD), and severe immunosuppression [2,29]. Of all children admitted to hospital for RSV acute LRTI, more than half do not have any of the above-mentioned risk factors, and most of these previously healthy term infants are aged less than 6 months [26,60]. Long-term sequelae associated with RSV LRTI include recurrent wheeze and an increased risk to develop an asthma phenotype [48,71].

Although RSV may be isolated from patients year-round, RSV epidemics of LRTI among infants and young children usually occur during late fall, winter, and early spring in countries with temperate climates. In tropical or semitropical climates, RSV epidemics appear equally as regular, but the pattern of seasonality is more complex and seems to depend on crowding, humidity, and temperature variables [76]. Primary RSV infection will usually lead to clinical disease that starts as an upper respiratory tract illness (URTI) which may subsequently progress to LRTI, commonly presenting as bronchiolitis and/or pneumonia. Less frequently, RSV LRTI will present as laryngotracheobronchitis (croup), bronchitis,

Handbook of Therapeutic Antibodies. Edited by Stefan Dübel
Copyright © 2007 WILEY-VCH Verlag GmbH & Co. KGaA, Weinheim
ISBN 978-3-527-31453-9

or with episodes of apnea (in very young and/or premature infants). Otitis media is another frequent sequela of RSV disease. The treatment of bronchiolitis is mostly supportive in nature, and includes adequate hydration, positioning, and respiratory support. Although widely used, pharmacological interventions offer limited benefit or none at all. There is limited evidence suggesting that nebulized epinephrine, as well as inhaled bronchodilators, may offer short-term benefits for bronchiolitis in outpatients, although evidence to support their use in inpatients is lacking [36,43]. No clinically significant improvement is associated with the systemic administration of glucocorticoids [57], and the use of ribavirin as an antiviral in RSV disease is controversial and limited to patients in need of mechanical ventilation or severely immunosuppressed patients [7,80].

12.1.2
The Target of the Antibody: The RSV Virion

RSV is an enveloped virus with a nonsegmented negative-sense RNA genome, a member of the Paramyxovirus family within the Mononegavirus order [21]. Its 15.2-kb genome consists of 10 genes that are transcribed sequentially from the 3′ end to the 5′ end by a virally encoded polymerase, and are translated into 11 protein species (Fig. 12.1).

The viral polymerase consists of the large polymerase protein L, the phosphoprotein P, and the transcription elongation factor M2-1. A second open reading frame (ORF) of the M2 gene encodes the M2-2 protein, a small protein that appears to mediate a shift from transcription to RNA replication during the infectious cycle [13]. The two nonstructural proteins NS-1 and NS-2 are not incorporated into the virion, but are important virulence factors that interfere with the host's interferon response [75]. The internal matrix protein M plays an important role during virus assembly, and the viral envelope is derived from the plasma membrane of the host cell as the virus buds from the apical surface of respiratory epithelial cells. The envelope contains three transmembrane surface glycoproteins – namely, the attachment protein G, the fusion protein F, and the small hydrophobic protein SH. The G and F glycoproteins are protective antigens that induce infectivity-neutralizing antibodies. Using monoclonal antibodies (mAbs), two antigenic subgroups (A and B) can be distinguished. While there is considerable divergence in G-protein sequence between the subgroup A and subgroup B viruses (44% amino acid identity with regard to the ectodomain [21]), the F protein sequence is more conserved (91% amino acid identity with regard to the ectodomain [51]), and antibodies directed against the F protein are generally cross-reactive between the two subgroups. The RSV F protein mediates fusion between the viral envelope and the host cell plasma membrane, thereby releasing the viral nucleocapsid into the cytoplasm. Later in the replicative cycle, F protein is expressed on the apical surface of epithelial cells and cell-to-cell fusion may occur, giving rise to multinucleated syncytia. The F protein is a type 1 glycoprotein that is synthesized as a precursor protein (F0) and activated through cleavage by a host protease into disulfide-linked subunits (F1 and F2). Cleavage releases

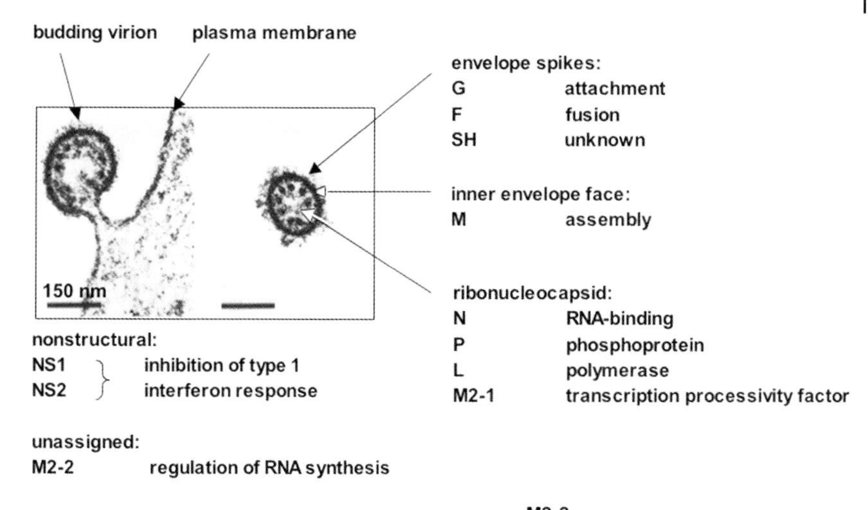

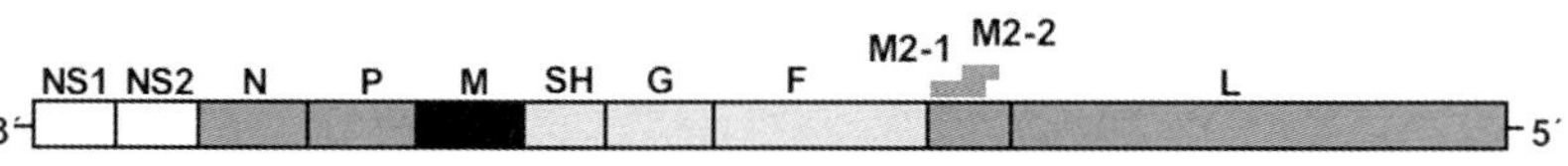

Fig. 12.1 Respiratory syncytial virus (RSV) virion, proteins and genome. Upper: Electron micrograph depicting a budding and a free RSV virion. The location of structural viral proteins is indicated by arrows. G = attachment glycoprotein; F = fusion glycoprotein; SH = small hydrophobic protein; M = matrix protein; N = nucleocapsid protein; P = phosphoprotein; L = large polymerase protein; M2-1 = transcription processivity factor; M2-2 = transcription/replication regulator protein. NS1 and NS2 = nonstructural proteins. Lower: Schematic drawing of the organization of the negative sense single-stranded RNA genome (not to scale). Graph courtesy of Peter Collins; electron micrograph by Tony Kalica, Laboratory of Infectious Diseases, NIAID, NIH.

a hydrophobic "fusion peptide" at the N-terminus of the F1 subunit that is thought to be directly involved in target membrane destabilization and membrane fusion. The protein forms homotrimers in the viral envelope that are stabilized by alpha-helical coiled coils [21].

12.1.3 Correlates of Protection from Disease

RSV infection does not induce sterilizing immunity – that is, re-infection, even with the same RSV strain, can and does occur. However, some protection is induced because subsequent RSV infection generally does not lead to severe disease in otherwise healthy subjects, and is only rarely a reason for hospitalization. RSV-neutralizing antibodies directed against the F or G glycoproteins are necessary and sufficient for protection from severe disease. Cell-mediated immune responses can also provide protection from disease, but this protection is not very durable [23]. The protective mechanism is not prevention of RSV infection but rather a restriction of virus replication. Maternally acquired IgG can protect the newborn infant from disease, and infants born to mothers with high RSV-neutralizing antibody titers are less likely to develop severe RSV disease during their first 6 months of life [54]. The quantitative aspects of passively acquired RSV-neutralizing antibody on RSV replication in the respiratory tract were initially determined in the cotton rat model of RSV pathogenesis [62]. In this animal model, a serum-neutralizing antibody titer of 1:380 or greater leads to complete or almost complete suppression of RSV replication in lung tissue (Fig. 12.2). While the effect of passively transferred IgG is impressive in the lungs, it is less obvious in the upper respiratory tract (URT), due to limited transudation of serum IgG to the URT mucosal surface. Antibodies can prevent infection if present in sufficiently high titer at the site of infection, but generally act to restrict virus replication.

Based on these data, a polyclonal human antibody preparation (RSV-IVIG; RespiGam, Medimmune) was prepared from multiple donors who were selected

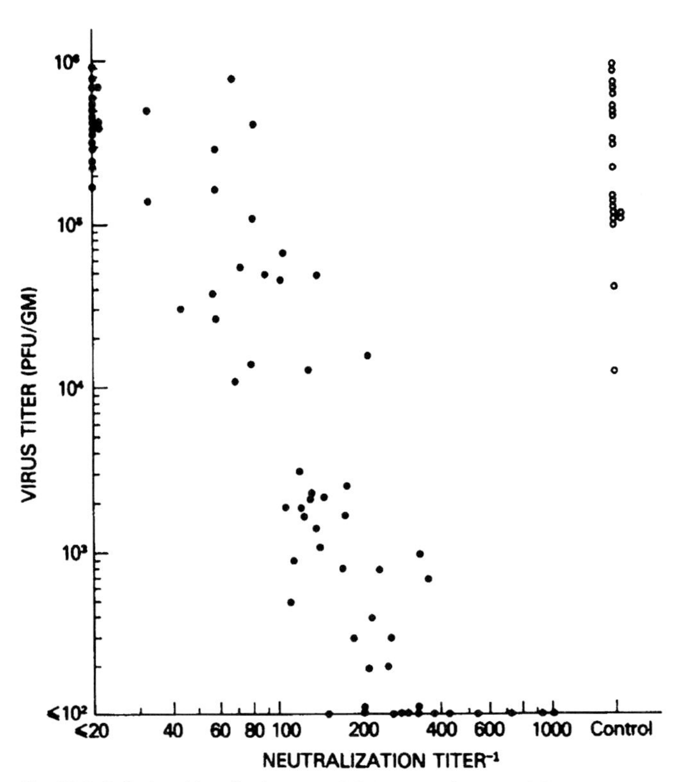

Fig. 12.2 Relationship of pulmonary RSV titer and serum RSV-neutralizing antibody titer in cotton rats receiving either RSV immune serum or control serum intraperitoneally 24 h before an intranasal challenge with 10^4 PFU of RSV. Lung titers were determined 4 days post infection. (Reprinted with permission from Ref. [62].)

for a high serum titer of RSV-neutralizing antibodies. When administered prophylactically to high-risk children at a dose of $15\,mL\,kg^{-1}$ body weight, RSV-IVIG was shown to decrease RSV LRTI, RSV disease severity and RSV-related hospitalization. RSV-IVIG did not, however, prevent RSV infection [4,35]. In order to be effective, the preparation had to be administered intravenously at monthly intervals during the RSV season. The most concerning serious adverse event (SAE) associated with RSV-IVIG administration was an increased mortality in children with CHD, possibly due to volume overload and/or increased viscosity associated with RSV-IVIG administration. RSV-IVIG was licensed by the FDA in 1996 for the prevention of serious lower respiratory tract infection caused by RSV in children under 24 months of age with bronchopulmonary dysplasia (BPD), or a history of premature birth (≤35 weeks' gestation). RSV-IVIG has been replaced by palivizumab (Synagis, Medimmune) and is no longer available for use.

12.2 Origin, Engineering, and Humanization of the Antibody

A detailed analysis of antigenic sites and neutralization epitopes of the RSV F glycoprotein – one of the two major envelope glycoproteins of RSV that induce a neutralizing antibody response – was performed in the Laboratory of Infectious Diseases (LID) at the National Institutes of Allergy and Infectious Diseases, National Institutes of Health. Beeler and Coelingh used a set of 18 RSV-neutralizing murine mAbs against the F protein of RSV subgroup A strain A2 for this analysis [11]. Eleven of these mAbs were generated at LID by immunizing BALB/c mice intranasally with the A2 strain of RSV, and boosting intraperitoneally 3 weeks later with a recombinant vaccinia virus expressing the RSV F protein. Four weeks later, mice were boosted once again with an intravenous injection of sucrose gradient-purified RSV. Splenic lymphocytes were then fused with NS-1 murine myeloma cells following the procedure of Kohler and Milstein [44].

Competitive binding assays identified three nonoverlapping antigenic sites (A, B, and C) and one bridging site (AB). Thirteen mAb-resistant mutants (MARMs) were selected, and the neutralization patterns of the mAbs with these MARMs and with clinical RSV isolates identified 16 epitopes on the F protein. Cross-neutralization studies using 23 clinical isolates (18 subgroup A and five subgroup B, isolated between 1956 and 1985) and 18 mAbs revealed that antigenic sites A and C were highly conserved, while antigenic site B was highly variable [11]. Each of nine virus-neutralizing mAbs directed against antigenic site A inhibited cell-to-cell fusion very efficiently, which suggested that the binding site might be close to the active site responsible for fusion-from-within. One of these murine antibodies directed against antigenic site A, mAb1129, was selected for humanization and clinical evaluation.

The development of the humanized version of mAb1129 was described in detail by Johnson et al. and revisited by Wu et al. [41,85]. The variable segments of the heavy chain (VH) and the light chain (VL) were assembled *de novo* using PCR and site-directed mutagenesis [45]. The human K102 sequence [12] with J-kappa-4 (NCBI gi:220083) was used as a template for the VL framework regions, and the human Cor and Cess sequences were used for VH FR1 and FR2 to FR4, respectively [61,78]. Complementarity-determining regions (CDRs) were defined using the coordinates of the known crystal structure of the MCPC603 antibody. The DNA sequences of the relevant CDRs were obtained from murine mAb1129 B cells and introduced into the DNA encoding the variable (Fab) and constant (Fc) regions of a human IgG1 molecule. With the exception of four amino acids in the VL CDR1, all CDRs were left as in mAb1129 (see Fig. 12.3). VL amino acid residues 24 through 27 (KCQL in the murine VL CDR1, underlined in Fig. 12.3) were substituted by the sequence SASS (neither murine nor human), due to a synthesis error during the humanization process [85].

Amino acids A105 in VH FR4 and L104 in VL FR4 (bold in Fig. 12.3) were not humanized, assuming (based on the coordinates of the MCPC603 antibody) that this would help maintain the structural integrity of the binding site [41]. Recently performed affinity maturation studies of palivizumab demonstrated that these two murine amino acid residues did not contribute significantly to the antibody's affinity, and they were humanized in a second-generation antibody to yield fully human germline sequences, thereby minimizing the potential immunogenicity of the antibody [85]. A model of palivizumab Fab indicating amino acid positions that affect the rate constant of association and the rate constant of dissociation (K_{on} and K_{off}) is shown in Fig. 12.4.

The prototype humanized mAb (MEDI-493, palivizumab) was initially expressed transiently in COS-1 cells (African green monkey kidney cells) using

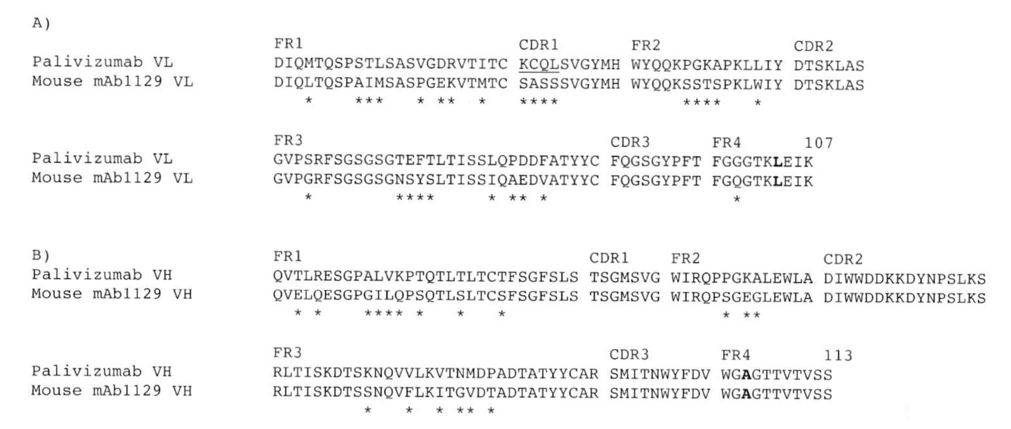

Fig. 12.3 Sequence comparison between palivizumab and mouse mAb1129 variable regions. Framework regions (FR) were humanized as described in the text; substitutions are indicated by asterisks. The human K102 sequence was used for VL (A), and the Cor and CESS sequences were used for VH FR1 and FR2-4 (B). Complementarity-determining regions (CDRs) were conserved as in mAb1129, with the exception of CDR1 which contains four random substitutions that are neither murine nor human (underlined). Amino acids L104 (VL FR4, in bold) and A105 (VH FR4, in bold) were not humanized to stabilize the binding site. (Modified from Ref. [41].)

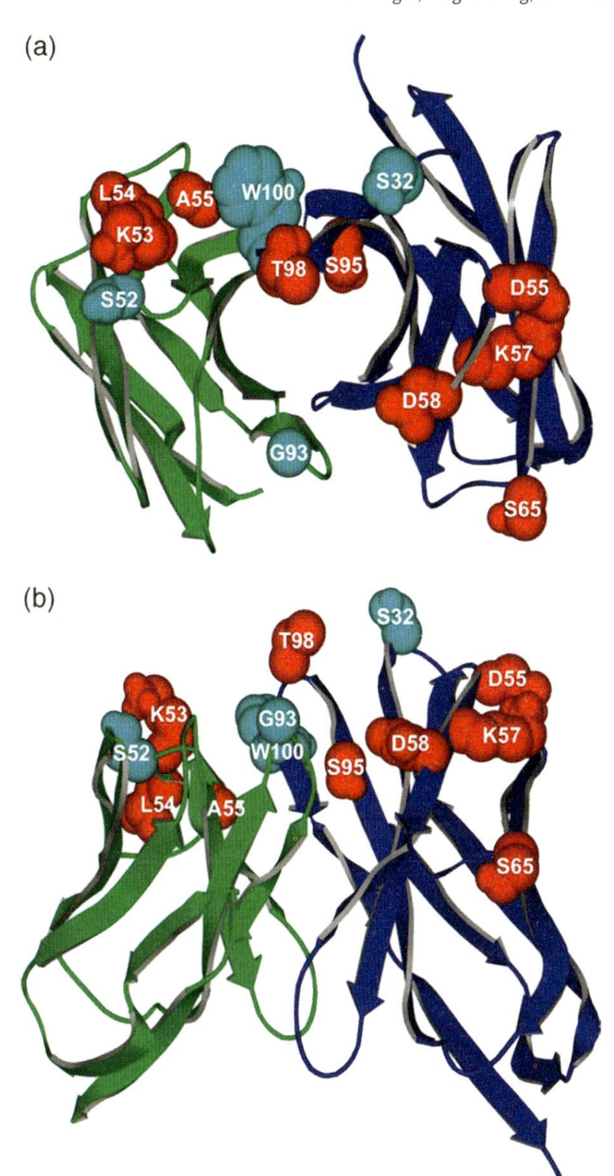

Fig. 12.4 Model of the palivizumab V region with amino acid positions that enhance antibody binding to (K_{on} positions, in red) or decrease antibody dissociation from (K_{off} positions, in blue) the RSV F protein. The heavy chain is depicted in blue, the light chain in green. It should be noted that all four positions that yielded improvements in K_{off} could also be mutated to improve K_{on}. (a) Top view (THE antigen-binding region is facing the reader). (b) Side view. The coordinates of the crystal structure of the palivizumab Fab were provided by Dr. B.C. Braden, Department of Natural Sciences, Bowie State University, Bowie, MD 20715, USA. (Reprinted with permission from Ref. [85].)

a cytomegalovirus (CMV) immediate-early promoter. Stable mouse myeloma cell lines (NSO) expressing MEDI-493 were then generated as described [10].

12.3 Mechanism of Action and Preclinical Results

Palivizumab neutralizes RSV infectivity by binding to the ectodomain of the fusion glycoprotein of RSV subgroups A and B. It also inhibits cell-to-cell fusion of RSV-infected cells. Preclinical studies of palivizumab were conducted to determine the kinetic parameters of antibody binding and to elucidate the mechanism of action. Binding properties were examined using the Biacore real-time biosensor system. Immunoaffinity-purified F protein from RSV strain A2 was used to coat the Biacore chip, and binding and dissociation were measured at antibody concentrations ranging from 12 to 400 nM. This analysis determined the rate constant of association (K_{on}) at $3 \times 10^5\,M^{-1}\,s^{-1}$ and the dissociation constant (K_{off}) at $4.3 \times 10^{-4}\,s^{-1}$ [41].

Three different biological assays were used to evaluate the *in-vitro* neutralizing activity of palivizumab [41]. First, in a classical plaque-reduction assay, serial dilutions of palivizumab were preincubated with 50 plaque-forming units of RSV A strain Long and RSV B strain 18537, and then placed onto HEp-2 monolayers for 1 h to allow infection of the cells. The virus–antibody suspension was then removed from the cells and cells were overlaid with methyl cellulose to prevent release of the virus into a liquid medium and spread via the liquid phase to other cells. After incubation for 5 to 6 days, the virus-induced plaques were enumerated. In this assay the effective concentration 50 (EC_{50}) – that is, 50% plaque reduction – for palivizumab was determined to be approximately $2\,\mu g\,mL^{-1}$.

In two additional *in-vitro* assays, RSV replication was quantified by measuring the amount of RSV F protein expression in 96-well plates using an RSV F protein-specific enzyme-linked immunosorbent assay (ELISA). RSV was either preincubated with antibody and HEp-2 cells were then added (microneutralization assay), or monolayers were infected first and antibody was added 4 h later, after removing the virus suspension (fusion-inhibition assay). The EC_{50} for palivizumab was calculated as $0.1\,\mu g\,mL^{-1}$ for the microneutralization assay, and at $0.2\,\mu g\,mL^{-1}$ for the fusion-inhibition assay, demonstrating that the monoclonal antibody preparation was approximately 20- to 30-fold more potent than the polyclonal preparation RSV-IVIG with regard to microneutralization and fusion-inhibition [41]. In this initial study, palivizumab was also shown to neutralize a wide spectrum of clinical isolates (34 subgroup A and 23 subgroup B isolates), although high antibody concentrations were used and EC_{50} values were not reported [41].

The *in-vivo* potency was determined in cotton rats – the same animal model that had been used previously for the evaluation of polyclonal antisera against RSV [62]. Palivizumab was administered intramuscularly (IM) or intravenously (IV) one day prior to intranasal inoculation of 10^5 plaque-forming units (PFU) of RSV subgroup A strain Long or subgroup B strain 18537, and the virus titer in

Table 12.1 Replication of RSV Long in the lungs of cotton rats following palivizumab prophylaxis (intravenous dosing).

Palivizumab dose[a] *[mg kg^{-1}]*	*N*[b]	*Serum concentration* *[μg mL^{-1}]*[c]	*Mean (±SE) RSV titer* *[log$_{10}$ PFU g^{-1} tissue]*[d]	*Mean reduction of RSV titer* *[log$_{10}$ PFU g^{-1} tissue]*
0	18	0	5.1 ± 0.1	0
0.312	7	2.7 ± 0.6	4.7 ± 0.1	0.4
0.625	17	5.3 ± 0.3	4.4 ± 0.1	0.7
1.25	18	10.1 ± 0.3	3.5 ± 0.2	1.6
2.5	17	28.6 ± 2.2	3.0 ± 0.2	2.1
5.0	15	55.6 ± 3.4	2.1 ± 0.1	3.0
10.0	18	117.6 ± 5.1	<2.0	>3.1

a Palivizumab was given an day –1 (i.e., 1 day before RSV challenge with 10^5 PFU intranasally).
b Data from three separate experiments were combined.
c Palivizumab serum concentration on Day 0 (i.e., the day of RSV challenge).
d Plaque-forming units of RSV g^{-1} homogenized lung tissue on Day 4 post challenge.
Modified from Ref. [41].

lung tissue was determined at 4 days after infection. For either virus, a 2-log$_{10}$ (i.e., 99%) reduction in virus titer was achieved using 2.5 mg kg^{-1} palivizumab IV or IM, corresponding to a serum concentration of approximately 30 μg mL^{-1} at the time of RSV challenge. Data for RSV Long are shown in Table 12.1 [41].

In addition to the above potency studies, a number of safety studies were conducted in mice to exclude the possibility that palivizumab might enhance viral replication or exacerbate RSV-induced lung pathology at noninhibitory antibody concentrations after primary or secondary RSV challenge. These safety concerns date back to the observation that children vaccinated with a formalin-inactivated experimental RSV vaccine during the 1960s experienced enhanced RSV disease with their first wildtype RSV infection after vaccination. One hypothesis at the time was that maternal antibodies might have been involved in enhanced disease, though it now appears that a lack of neutralizing antibodies and a Th2-biased T-cell response might have caused the increased severity of RSV disease [49]. With palivizumab, neither an increase in viral replication nor exacerbation of RSV-induced histopathology was observed after primary or secondary challenge with RSV [41].

12.4 Production, Downstream Processing, and Galenics of the Antibody

12.4.1 Production

Palivizumab is produced in a stable cell line that was developed from the NSO murine myeloma cell line. Twenty-one candidate production cell lines were screened for growth rate and rate of antibody secretion, and a suitable cell line was selected for biological cloning and generation of an accession cell bank and a master cell bank (MCB) [63]. The MCB was extensively characterized and shown to be sterile and free of adventitious agents. A working cell bank (WCB) was generated from the MCB and shown to be safe, sterile, free of adventitious agents, and comparable to the MCB with regard to DNA profile, etc.

A stirred-tank, fed-batch system is used for the production of palivizumab from the WCB. Production is initiated by thawing a frozen vial of the WCB and expanding cells in T-flasks and in bioreactors of increasing size. When the production capacity is reached, a production bioreactor (≥10 000 L capacity) is filled, and after 18 to 22 days of incubation the bioreactor content is harvested. The initial antibody yield was reported at approximately $1\,g\,L^{-1}$ of medium, but an enhanced yield process that was approved by the FDA in 2001 increased the palivizumab yield to over $3\,g\,L^{-1}$ of medium.

12.4.2 Downstream Processing

Downstream processing includes microfiltration to remove cells and debris, and an antibody purification process which involves three-stage chromatography, acid treatment, and nanofiltration. This purification process leads to clearance of potential (viral) contaminants, and in-process validation is performed to guarantee that manufacturing is safe and results in a product of consistent quality. The bulk material is sterile-filtered once again prior to filling of the final vials.

The initial FDA approval (June 19, 1998) was granted for palivizumab manufactured by Medimmune at their Gaithersburg, Maryland, pilot facility. Supplemental approval (September 8, 1999) was then granted by the FDA for large-scale manufacture at Boehringer Ingelheim (Biberach, Germany) under contract to Medimmune, and an additional supplemental approval for Medimmune's new facility in Frederick, Maryland was granted on December 17, 1999 [63].

12.4.3 Formulations

Palivizumab (Synagis) is available in two formulations: a lyophilized powder and a liquid solution. Both formulations are for IM injection only. The newer liquid formulation was approved by the FDA in July 2004, and currently both formulations are available. It is expected that the liquid formulation will replace the lyophilized formulation as it is easier to use (reconstitution of the lyophilized powder with sterile water for IM injection takes ca. 20 min). Each 100 mg single-use vial of the lyophilized powder is formulated in 67.5 mg of mannitol, 8.7 mg histidine and 0.3 mg of glycine, and is designed to deliver 100 mg of Synagis in 1.0 mL when reconstituted with 1.0 mL of sterile water for injection. Vials containing 50 mg are also available.

The liquid solution (100 mg mL^{-1}) is supplied as a sterile, preservative-free solution which should appear clear or slightly opalescent, with a pH of 6.0. Each 100-mg single-use vial of Synagis liquid solution is formulated in 4.7 mg of histidine and 0.1 mg glycine in a volume of 1.2 mL, and is designed to deliver 100 mg of Synagis in 1.0 mL. Vials containing 50 mg are also available. Notably, the liquid formulation contains no mannitol and has an osmolality of 52 mosm kg^{-1} compared to 462 mosm kg^{-1} for the reconstituted lyophilized powder.

12.4.4 Specifications

Following general FDA suggestions regarding the use of pre-approved comparability protocols to reduce reporting requirements for changes in the manufacturing process, Medimmune conducted a comparability assessment of palivizumab lots manufactured under varying conditions. The aim of these studies was to examine whether posttranslational modifications of the antibody (deamination, oxidation or oligosaccharide variation) occurred in the manufacture of palivizumab and could potentially lead to altered mAb binding or clearance [68].

Specification ranges to be used for comparabillity testing were generated using data from biochemical and functional analyses of over 25 preclinical, clinical, and manufacturing lots. Materials manufactured for phase I through phase III clinical trials in bioreactors of increasing size (20 L, 45 L, 100 L, and 200 L) and three manufacturing consistency lots made at the same facility in 500-L bioreactors were tested with regard to binding activity, purity and microheterogeneity using methods such as F-protein-specific ELISA, size-exclusion chromatography, monosaccharide composition, oligosaccharide profile and MALDI-TOF. SDS-PAGE, isoelectric focusing, capillary electrophoresis and Western blotting results of the individual lots were also performed, and yielded comparable results. The combined data from 20-L to 500-L lots was used for product registration with the FDA. Once manufacturing was initiated at a second site, lots manufactured during the scale-up process at that site were again tested in parallel with the reference standard – that is, the material made in a 200-L bioreactor for the phase III clinical trial that led to licensure. The results of these scale-up comparability tests are shown in Table 12.2.

Cell line stability studies were conducted using palivizumab purified from cells expanded from the WCB, cells similar to the extended cell bank, and from cells that were expanded well past the extended cell bank. With the exception of the *N*-acetylglucosamine composition in material from cells expanded past the extended cell bank, all tested parameters were within the specification range, and tryptic peptide maps were consistent for all lots [68]. Studies examining the effect of time of harvest (typically 18 to 22 days after inoculation of the production bioreactor) showed that material harvested on Day 11 was comparable to that harvested on Day 18, but an early harvest on Day five resulted in an altered

Table 12.2 Comparability test results for scale-up lots of palivizumab.

Test		*Specification*	*Lot 5 (500 L)*	*Test results for three lots (bioreactor size) Lot 7 (2,000 L)*	*Lot 10 (10,000 L)*
F protein binding activity (ELISA)		75–128% of R.S.[a]	96	89	93
Size Exclusion Chromatography		1 peak >99% AUC[b]	100.0	99.9	99.8
Monosaccharide composition (%)	GlcNAc[c]	40–60	55.6	50.8	44.8
	Mannose	20–45	30.5	33.3	37.2
	Fucose	5–15	7.8	10.1	10.6
	Galactose	2–12	6.3	5.8	7.6
Oligosaccharide profile	Peak 5 to 6	0.4–1.0	0.6	0.5	0.7
	Peak 5 to 7	1.5–4.0	2.3	1.5	2.7
Mass by MALDI-TOF[d] (Da)		147,700 ± 1,000	147,178	147,178	147,470

a R.S.: reference standard.
b AUC: area under the curve.
c GlcNAc: *N*-acetylglucosamine.
d MALDI-TOF: Matrix-assisted laser desorption/ionization-time of flight mass spectrometry.
Modified from Ref. [69].

oligosaccharide profile of the material [68]. Although media glucose concentrations below the process specification ($1.2\,g\,L^{-1}$) had an impact on cell viability and product yield, there was no discernable effect on product quality [68].

12.5 Summary of Results from Clinical Studies

12.5.1 Phase III Trials

Two pivotal phase III trials were conducted to examine the safety and efficacy of palivizumab in reducing the incidence of hospitalization due to RSV disease. The first trial enrolled premature children with or without BPD/CLD, and led to the approval of palivizumab by the FDA in 1998. The second phase III trial enrolled children aged under 2 years with significant CHD, and extended the indication for prophylaxis to this high-risk group in 2003. An overview of these two trials, which are described in more detail below, is provided in Table 12.3.

12.5.1.1 Palivizumab in Premature Infants and Children with BPD

The IMpact-RSV trial evaluated the safety and efficacy of prophylaxis with palivizumab in reducing the incidence of hospitalization due to RSV disease in high-risk infants and children [2]. The trial enrolled 1502 children with prematurity and/or BPD to receive five doses of palivizumab ($15\,mg\,kg^{-1}$) or placebo during

Table 12.3 Phase III trials of palivizumab.

Trial (Reference)	*Season*	*Study participants (n)*		*RSV hospitalization rate [%]*		*Reduction of RSV hospitalization [%]*	
		Placebo	*Palivizumab*	*Placebo*	*Palivizumab*	*Absolute*	*Relative*
IMpact-RSV [2]	1996–1997	500	1002	10.6	4.8	5.8	55
CHD Trial [29]	1998–2002	648	639	9.7	5.3	4.4	45

the 1996–1997 RSV season. The trial was designed as a double-blind multicenter trial with a 2:1 randomization. A total of 139 centers in the USA, the UK, and Canada participated, and enrolled children born at ≤35 weeks' gestation and aged less than 6 months, in addition to children aged ≤24 months with BPD that required treatment within 6 months prior to enrollment. Exclusion criteria included clinically significant CHD, active or recent RSV infection, hospitalization at the time of entry that was anticipated to last more than 30 days, mechanical ventilation at the time of entry, life expectancy less than 6 months, known hepatic or renal dysfunction, seizure disorder, immunodeficiency, allergy to IgG products, and receipt of RSV-IVIG within the past 3 months or previous receipt of palivizumab, other mAbs, RSV vaccines, or other investigational agents. Patients were followed until 30 days after the last scheduled injection, for a total of 150 days. All hospitalizations were identified and children with respiratory hospitalizations were tested for RSV antigen using commercially available tests.

The primary endpoint was defined as: (i) hospitalization for a respiratory illness and positive RSV antigen test of respiratory secretions; or (ii) hospitalization for any cause, a positive RSV test, and a minimum LRTI score of 3 that was at least one point higher than that recorded at the child's last pre-illness visit. LRTI was scored as follows: 0 = no respiratory illness; 1 = URTI; 2 = mild LRI; 3 = moderate LRI; 4 = severe LRI; and 5 = mechanical ventilation. Adverse events (AEs) were reported throughout the trial, and the potential relationship to the study drug was assessed. AE severity was assessed using a toxicity table modified from the pediatric AIDS vaccine trials group toxicity table. The two treatment groups were demographically similar, and compliance was excellent, with 99% of participants receiving all five doses and 93% of study participants completing the study. RSV-related hospitalization occurred in 10.6% of placebo recipients versus 4.8% of palivizumab recipients; that is, monthly prophylaxis with palivizumab was associated with a 5.8% absolute reduction or a 55% (95% CI 38–72%) relative reduction in the primary endpoint ($p = 0.00004$) (Table 12.3). Logistic regression analysis indicated that gestational age (GA) was not a significant predictor of RSV hospitalization, and that the palivizumab effect remained statistically significant in children born at 32 to 35 weeks of gestation [2].

Amongst the secondary efficacy endpoints, a significant reduction in total days in hospital per 100 children enrolled in the trial (63 versus 36 days), in total days with increased oxygen (51 versus 30 days), and in total days with an LRI score ≥3 (47 versus 30 days), were reported. These secondary endpoints are confounded, however, in that the total number of study participants in each arm was used as denominator, and therefore the reduction in RSV hospitalization rate resulted in a reduction in total days in hospital. The mean length of stay for those patients actually hospitalized was similar in the two study groups (7.6 days versus 5.9 days for palivizumab versus placebo). RSV disease was as severe among RSV hospitalized patients who received palivizumab as it was in those who received placebo. Mean days with increased oxygen and mean days with LRI score ≥3 were also similar in the two groups. The incidence of all hospital admissions (of any cause) and of all respiratory hospital admissions (but not of non-RSV hospital admissions) was significantly reduced in the palivizumab group, thus confirming the important role that RSV disease plays in this high-risk group. Otitis media occurred at a similar frequency in the two groups.

Adverse events that were judged to be related to the study drug occurred in 11% of the palivizumab recipients, and in 10% of those receiving placebo. In addition, 2.7% and 1.8% of the children in the palivizumab and placebo groups, respectively, experienced AEs related to the injection site. Only in 0.3% of palivizumab recipients were the injections discontinued due to local AEs. As with systemic AEs, the rate of aspartate aminotransferase (AST) elevation was slightly higher in the palivizumab group (3.6% versus 1.6%), but alanine aminotransferase (ALT) elevations were not associated, and overall hepatic AEs related to the study drug occurred at similar frequencies. Five children (1.0%) in the placebo group and four (0.4%) in the palivizumab group died during the trial; none of the deaths was considered to be related to palivizumab, although the study was not powered to detect inter-group differences in mortality.

12.5.1.2 Palivizumab in Children with Significant CHD

The second randomized, double-blind, placebo-controlled multicenter phase III trial was conducted to evaluate the safety, tolerance, and efficacy of palivizumab in children with hemodynamically significant CHD [29]. This trial enrolled 1287 children with CHD in four consecutive seasons (1998–2002), and randomly assigned them 1:1 to receive five monthly intramuscular injections of 15 $mg\,kg^{-1}$ palivizumab, or placebo. Among the study participants, 72% were recruited in the USA and Canada, and 28% in France, Germany, Poland, Sweden, and the UK.

Children were eligible if they were aged less than 24 months at the time of randomization, and had documented hemodynamically significant unoperated or only partially corrected CHD, as determined by the investigator. In general, patients with cyanotic CHD, with single ventricle physiology, and those with acyanotic CHD that required medical therapy, were considered to have significant CHD. Children were not eligible if they had unstable cardiac or respiratory status, including cardiac defects so severe that survival was not expected, or for which cardiac transplantation was planned or anticipated. Additional exclusion criteria were, amongst others, noncardiac anomalies or end-organ dysfunction with anticipated survival less than 6 months, or current RSV, HIV or other acute infection [29]. Children were followed for 150 days – that is, until 30 days after the last dose of palivizumab.

All hospitalizations were identified, and RSV antigen testing of respiratory secretions was performed in all children with an acute cardiorespiratory hospitalization. The primary endpoint of the trial was the incidence of RSV hospitalization, including primary RSV hospitalizations and nosocomial RSV hospitalizations. Primary RSV hospitalization was defined as hospitalization for an acute cardiorespiratory illness in which the RSV antigen test was positive within 48h before or after admission.

The types of cardiac lesions were similar in the two treatment groups. Within the cyanotic stratum (53% of all patients), single ventricle (hypoplastic left or right heart) and tetralogy of Fallot accounted for 21.9% and 11.4%, respectively. In the acyanotic stratum, ventricular septal defect and atrioventricular septal defect accounted for 18.0% and 7.2%, respectively, of all patients enrolled. This distribution of cardiac lesions in the study population indicates that patients with single ventricle hemodynamics – that is, those with the most severe CHD – were over-represented in this study population.

Compliance was excellent in this trial, and more than 95% of all patients completed the study. In total, 63 of 648 placebo recipients (9.7%) and 34 of 639 palivizumab recipients (5.3%) met the primary endpoint RSV hospitalization – that is, monthly palivizumab administration was associated with an absolute reduction in the RSV hospitalization rate of 4.4%, or a relative reduction by 45%. With regard to secondary endpoints, a significant reduction in the total days in hospital with RSV hospitalization (from 129 to 57 days per 100 children; 56% relative reduction) and a significant reduction in the RSV hospitalization days with increased supplemental oxygen (from 102 to 28 days per 100 children; 73% relative reduction) were reported. Again, these secondary endpoints are confounded by the reduction in RSV hospitalization rate and the use of all study children in each arm as denominator. The mean duration of RSV hospitalization was 13.3 versus 10.8 days, and the mean duration of supplemental oxygen requirement 10.4 versus 5.2 days in the placebo and palivizumab groups, respectively. Intensive care unit (ICU) admission rate, days of ICU stay, mechanical ventilation rate, and days of mechanical ventilation were all lower in the palivizumab group, but did not reach statistical significance. Heart surgery with cardiopulmonary bypass led to a 58% decrease in the serum palivizumab concentration (through dilution and bleeding), indicating that re-dosing is necessary after cardiopulmonary bypass [29].

Both phase III trials showed unequivocally that palivizumab is efficacious in reducing hospitalization for RSV disease by approximately 50% in children at high risk for severe RSV disease, including infants born at ≤35 weeks' gestation, children aged under 2 years with BPD, and children with significant CHD. Efficacy was demonstrated in infants born at <32 weeks' gestation and in infants born at 32 to 35 weeks' gestation. Although the CHD study was not powered to permit subgroup analyses, a reduction in RSV hospitalization was observed in both the cyanotic and noncyanotic populations.

12.6 Indications and Usage

When the results of the palivizumab CHD trial had been reviewed by the FDA, the package insert for palivizumab (Syangis) was modified to include the following indications and usage instructions: "Synagis is indicated for the prevention of serious lower respiratory tract disease caused by respiratory syncytial virus (RSV) in pediatric patients at high risk of RSV disease. Safety and efficacy were established in infants with bronchopulmonary dysplasia (BPD), infants with a history of premature birth (≤35 weeks gestational age), and children with hemodynamically significant congenital heart disease (CHD)."

The recommended dose of Synagis is 15 $mg\,kg^{-1}$ body weight. Patients, including those who develop an RSV infection, should continue to receive monthly doses throughout the RSV season. The first dose should be administered prior to commencement of the RSV season, and then at monthly intervals throughout the RSV season. Synagis should be administered intramuscularly using aseptic technique, preferably in the anterolateral aspect of the thigh. The gluteal muscle should not be used routinely as an injection site because of the risk of damage to the sciatic nerve. Injection volumes in excess of 1 mL should be given as a divided dose.

12.7 Clinical Reports after Approval

Palivizumab received initial approved for the prevention of RSV disease in high-risk infants in 1998 in the US, followed by approval in the European Union and in Japan in 1999 and 2002, respectively. It is now approved in over 50 countries. Following FDA approval, a considerable number of clinical studies and surveys examined the efficacy of palivizumab prophylaxis in study settings, as well as its effectiveness in field use, mostly in terms of hospitalization rates in infants receiving palivizumab. The majority of these studies were observational in nature and without concurrent controls instead, the majority of these studies compared the hospitalization rate in their palivizumab-treated population to that in a previous study, most commonly to that observed in the phase III IMpact-RSV study. As Heikkinen et al. pointed out [37], such comparisons between studies are often problematic because of differences in demographics and study design. Of particular interest in this regard is the proportion of children with BPD/CLD, as these children are at much higher risk for severe disease and hospitalization than premature infants without CLD or children with CHD [14]. In several of the studies mentioned below, subgroup analyses were performed to address this problem (see Table 12.4); moreover, meta-analyses of the hospitalization data by subgroup have been published elsewhere [59,72].

The definition of BPD/CLD (the terms are used synonymously in this chapter) and the type of disease described by the term BPD has changed considerably since its original description by Northway et al. in 1967 [52]. Antenatal steroid therapy, postnatal surfactant therapy and advances in ventilation techniques have helped to reduce the frequency of lung injury in less immature babies, and the disease is now infrequent in infants born at ≥30 weeks' gestation or 1200 g birth weight.

Table 12.4 RSV hospitalization rate in palivizumab recipients: phase III trial and post-approval observations.

Study (Reference)	*Design*	*Season*	*N*	*CLD [%]*	*Percent RSV hospitalization rate*			*Fraction of respiratory admissions tested for RSV (% positive)*
					Overal	*CLD only*	*No CLD only*	
Impact-RSV (phase III trial)	Prospective randomized controlled trial	1996/1997	1002	50	4.8	7.9	1.8	0.95 (30)
19	Telephone survey (USA)	1998/1999	7013	n.a.	1.5	n.a.	n.a.	n.a.
74	Retrospective chart review (USA)	1998/1999	1839	22	2.3	4.0	2.1	n.a.
33	Prospective, observational, single arm (Canada, Europe)	1998/1999	565	n.a.	2.1	n.a.	n.a.	0.57 (24)
46	Prospective, observational, single arm (France)	1999/2000	516	81	7.6	9.0	3.0	0.90 (46)
55	Prospective, observational, single arm (Canada)	1999/2000	444	23	2.4	6.0	1.6	0.84 (43)
56	Prospective, observational, single arm (USA)	2000/2001	2116	24	2.9	5.8	2.1	n.a.
58	Prospective, observational, single arm (Spain)	2000/2001 2001/2002	1919	11	4.0	5.6	3.2	n.a.
50	Prospective, observational, population-based (Sweden)	2000/2001 2001/2002	390	52	4.1	n.a.	n.a.	n.a.

n.a. = not available.

A BPD workshop in 1978 defined BPD as an oxygen requirement for the first 28 days of life with radiographic changes (of the lungs), and a decade later oxygen requirement at 36 weeks' gestation (postmenstrual age) was suggested as a better predictor of long-term respiratory outcomes and became a widely used definition of BPD [70]. It is clear that for very premature infants (e.g., ≤28 weeks' gestation) the definition of BPD using 36 weeks of gestation describes a much smaller population of infants with more severe lung disease than the 28 days of age definition. A BPD workshop in 2000 suggested new definitions for mild, moderate and severe BPD [40], but this classification is not yet universally used. Many centers still use oxygen requirement at 36 weeks' gestation or at 28 days of postnatal age as BPD/CLD definitions, although often the definition is not specified in palivizumab publications.

In addition to difficulties with study population demographics and varying BPD definitions, the estimation of RSV hospitalization rates in the absence of palivizumab prophylaxis can be difficult. A concurrent control group is usually unavailable as a majority of high-risk infants in high-income countries receive palivizumab, and local pretreatment surveillance data for the baseline incidence of RSV hospitalization often do not exist. This makes calculation of the attributable risk reduction for RSV-related hospitalization impossible, and as a result cost-effectiveness analyses often rely on vague assumptions. Baseline RSV hospitalization rates in the absence of palivizumab prophylaxis can differ greatly between different locations, and for many regions they are not known [25,37]. A Finnish retrospective cohort study of children born between 1991 and 2000 determined that the RSV hospitalization rate in their population of nonprophylaxis children with CLD was 12%, while for all premature infants it was 4.9%. Of 586 children who would have met the criteria for palivizumab administration, only 4.6% were hospitalized for RSV disease during the RSV season [37], compared to 10.6% in the IMpact-RSV study. As almost all hospitalized children in this study were tested for RSV on admission, this difference in hospitalization rates seems to reflect a real difference between countries, and is not likely due to under-reporting. Similarly, hospitalization rates for children with clinically significant CHD are reported to differ greatly in different settings, and a lively debate continues as to the attributable risk reduction of palivizumab prophylaxis [30].

On the other hand, the incidence of RSV LRTI and severe RSV LRTI can be similar in very different climates and different countries. A recent population-based study, for example, found almost identical RSV-attributable incidence rates for severe LRI in children aged under 1 year (15–16 per 1000 child-years) in Mozambique, South Africa and Indonesia, although the incidence of all LRTI varied more than 10-fold between these countries [64].

As palivizumab is an expensive drug and administration is invasive, there is an ongoing debate with regard to who should and who should not receive palivizumab prophylaxis. Approximately 10% of all newborns in developed countries are born prematurely (i.e., at <37 weeks' gestational age), and approximately 1% of all newborn babies are born with very low birth weight (VLBW; i.e. <1500 g, and usually <32 weeks' gestation). The prevalence of CHD in infants is approximately 0.7 to 1%, and no more than half would qualify as significant CHD. Although the efficacy of palivizumab in reducing RSV hospitalization was found to be similar in infants born at a GA <32 weeks and in those born at a GA of 32 to 35 weeks [2], the inclusion of the more mature group would multiply the cost of palivizumab prophylaxis. On the other hand, several studies have suggested that infants born at a GA of 32 to 35 weeks experience considerable morbidity with, and also subsequent to, RSV hospitalization [17,67,83].

The American Academy of Pediatrics recommends palivizumab prophylaxis for the following high-risk groups [9]:

1. Infants and children aged less than 2 years with CLD who have required medical therapy (supplemental oxygen, bronchodilator, diuretic or corticosteroid therapy) for CLD within 6 months before the anticipated start of the RSV season.
2. Infants born at 28 weeks' gestation or earlier may benefit from prophylaxis during their first RSV season, whenever that occurs during the first 12 months of life.
3. Infants born at 29 to 32 weeks' gestation may benefit most from prophylaxis up to 6 months of age.
4. Infants born between 32 and 35 weeks' of gestation, only if two additional risk factors (child care attendance, school-aged siblings, exposure to environmental air pollutants, congenital abnormalities of the airways, or severe neuromuscular disease) are present.
5. Children who are aged 24 months or younger with hemodynamically significant CHD.

Professional societies in a number of countries have published their own guidelines [1,3,5,6,18,31,32,79], but many countries include the above-mentioned groups 1, 2 and 5 which, taken together, represent less than 2% of a country's birth cohort [15]. A considerable number of reports describing the local, regional or national experience with palivizumab use after the initial FDA approval have been published. Not all can be discussed in this overview, but a selection with preference for larger and/or prospective studies are summarized here.

In the USA, an RSV Education and Compliance Helpline (REACH) program was implemented as part of a pharmacovigilance program to enhance parent education and compliance (see Table 12.4) [19,34]. Adverse event reports were collected via bimonthly telephone contacts with the parents of palivizumab recipients. The hospitalization rate for palivizumab recipients in REACH was reported as 1.5%, compared to 4.8% in the IMpact-RSV trial. A total of 2.8% of the parents reported SAEs [34]. An additional report on the REACH program determined the SAE rate among 19958 children receiving palivizumab to be 2.5%, and the death report rate to be 3.4 per 1000, compared to a mortality rate of 4 per 1000 in the IMpact-RSV trial [22]. Except for very rare (<1:100000) reports of anaphylaxis, no new SAEs were reported in the REACH and Outcome Registry pharmacovigilance programs [22].

Nine centers in the US retrospectively reviewed the RSV hospitalization rate of their population of 1839 patients that had received palivizumab in the first season after licensure (i.e., 1998–1999) [74]. Patients included in the study had a gestational age of ≤35 weeks, were aged less than 2 years at their first injection, and had received at least one dose of palivizumab. The RSV hospitalization rate for the total population of infants receiving palivizumab was 2.3% (42 of 1839) compared to 4.8% in the IMpact-RSV trial (see Table 12.4; [74]). It should be noted, however, that this study population had less chronic lung disease than the cohort studied in the IMpact-RSV population (i.e., 22% versus 50%, respectively). The hospitalization rate for infants with CLD was 4.0% (compared to 7.9% in the IMpact-RSV) and 2.1% for infants born at ≤35 weeks' gestation without chronic lung disease (compared to 1.8% in the IMpact-RSV).

For the same season (1998–1999), a phase III/IV multicenter, single-arm, open-label study was conducted in Canada and in a number of European countries [33]. A total of 565 preterm children aged <6 months at enrollment and born at ≤35 weeks of gestation, and children with BPD aged ≤24 months at enrollment who required medical intervention during the 6-month period prior to recruitment, were enrolled and followed for 150 days for AEs and hospitalization. RSV testing was not required in this study. In total, 94% of the study participants completed the study, 14 (2.5%) discontinued for personal reasons, 1% was lost to follow-up, and 2% (11 reports) discontinued because of AEs. Among these latter 11 reports of AEs, three were considered to be possibly or probably related to palivizumab: one patient developed oxygen desaturation immediately after the third injection; another patient developed gastroenteritis; and a third patient developed abdominal and peripheral edema. Likewise, 39 patients (6.9%) reported 40 AEs that were considered related to drug administration. No SAEs were considered related to palivizumab. Two patients died from causes not related to palivizumab or RSV. A total of 51 respiratory hospitalizations was recorded, but only 29 (57%) were tested for RSV, of which seven (24%) tested positive. If it were to be assumed that 24% of all respiratory hospitalizations were due to RSV, this would translate to a 2.1% RSV hospitalization rate (see Table 12.4) [33], but this figure should be treated with caution as only half the patients admitted were tested.

A similar observational field survey (not blinded, no control group) was conducted in France in 1998/1999, prior to the approval of palivizumab in Europe [46]. A total of 516 preterm infants with a median gestational age of 28 weeks (88% born at ≤32 weeks' gestation) and a very high BPD rate of 81% (at 28 days of age) were followed (see Table 12.4) [46]. Thirty-nine infants (7.6%) were hospitalized for RSV disease, 10 required intensive care treatment, and four mechanical ventilation [46].

In the following RSV season (1999/2000), a prospective observational multicenter study (COMPOSS study) was conducted in Canada to evaluate compliance and outcomes in 444 infants born at ≤32 weeks of gestation (see Table 12.4) [55]. The study participants were aged <6 months at the onset of the RSV season or aged <2 years and had CLD requiring oxygen therapy within the past 6 months before the onset of the season. This study cohort represented 16% of the total population who received palivizumab in Canada during that season. Compliance was good, and 77% of all doses administered were given within 30 ± 5 days. Forty hospitalizations for respiratory events were observed, including 28 admissions for LRTI, two for URTI, seven for respiratory distress due to underlying lung disease, and three for croup. Twenty-five children were admitted 28 times for LRTI, and nine of the 21 tested for RSV were positive, resulting in a 2.4% RSV hospitalization rate. This hospital admission rate was only half that reported in the IMpact-RSV study, but again it should be considered that in the IMpact-RSV study 50% of the study participants had BPD, compared to 23% in the present study. The hospitalization rate for children with BPD receiving palivizumab was similar in the two studies (6% in COMPOSS, 7.9% in IMpact-RSV). Premature infants without BPD had a hospitalization rate of 1.6%, compared to 1.8% in IMpact-RSV. However, as this was an observational study, not all RTIs were recorded and not all not RTI hospital admissions were tested for RSV.

Data on RSV hospitalization rates during the 2000/2001 RSV season in the US were reported by the Palivizumab Outcomes Registry Study Group (see Table 12.4) [56]. A total of 2116 children who received palivizumab were enrolled at 63 sites, mostly in pediatric office settings. There were no exclusion criteria; 47% of the study participants were born at 32 weeks' gestation, 45% at 32 to 35 weeks, and 8% at >35 weeks. Some 24% of the study population had CLD, and 5% had CHD. Of these patients, 14% received palivizumab for a second season (54% of these had CLD and 9% CHD). The RSV hospitalization rate was 5.8% for children with CLD, 4.3% for those with CHD, and 2.1% for premature infants without CLD (Table 12.4). The overall RSV hospitalization rate was 2.9% for the 2116 children in the registry. Almost half of all hospitalizations (and 75% of all hospitalizations in subjects receiving all injections within 35-day intervals) occurred within the first and second injection intervals, suggesting that trough serum palivizumab concentration after the first and second doses might not have been protective in some of the infants.

A prospective multicenter study enrolling infants born at ≤32 weeks' gestation and aged <6 months at the start of the RSV season was conducted in Spain during the 2000/2001 and 2001/2002 RSV seasons to assess the RSV hospitalization rate in infants receiving palivizumab (see Table 12.4) [58]. In this cohort of 1919 infants (median gestational age at birth 29 weeks, mean birth weight 1261 g, CLD in 11%), the RSV hospitalization rate was determined as 4%. A historical cohort of infants who did not receive palivizumab was derived by combining the data from two previous studies that followed infants born at ≤32 weeks of gestation and aged <6 months during the 1998/1999 and 1999/2000 RSV seasons [16]. In this control cohort (median gestational age at birth 31 weeks, mean birth weight 1426 g, 5% CLD), the RSV hospitalization rate was determined as 13%. Although it is difficult in this type of historical comparison to control for differences in RSV season severity, induced parental behavior changes that might influence exposure to RSV, and changes in a country's hospital admission policies, the 70% reduction in RSV hospitalization suggests good efficacy of palivizumab in preventing RSV hospitalization. One important difference between the two cohorts that remained unexplained was that the number of deaths in the control cohort was four times as high (13/1000) as in the palivizumab cohort (3/1000) [58].

In Sweden, a population-based, nationwide prospective study of RSV hospitalization of preterm children was conducted in the 2000/2001 and 2001/2002 RSV seasons [50]. This study was unique in covering 85% of all preterm children aged <2 years in a single country. The study aim was to assess the appropriateness of Sweden's restrictive use of palivizumab. The Swedish recommendations limited the use of palivizumab to children aged <2 years with CLD (36 weeks' GA definition) requiring active treatment during the 6 months prior to the RSV season, and to infants with a gestational age of ≤26 weeks at the start of the RSV season. Over two seasons, the RSV hospitalization rate for all 5800 children born at ≤35 weeks GA (5410 [93%] without palivizumab prophylaxis, 390 receiving palivizumab) was 3.8% (n = 218) and 5.4% (n = 97) for children born at ≤32 weeks GA. In total, 390 children were treated with palivizumab according to the Swedish recommendations, and 16 (4.1%) of those were hospitalized for RSV (see Table 12.4) [50].

In Japan, Saji et al. conducted a retrospective survey of palivizumab use in children with CHD [66]. Surveys were mailed to 476 centers using palivizumab, and 61 centers reported 108 CHD patients that were given palivizumab. Complex CHD accounted for 40% of the population, ventricular septal defect for 21%, coarctation for 11%, atrial septal defect for 9%, and hypoplastic left heart syndrome for 4%. Five of 108 (4.6%) children were hospitalized for RSV disease.

In the USA, the 2003–2004 Outcomes Registry reported for 664 children with CHD receiving palivizumab that persistent ductus arteriosus (20%), ventricular septal defect (16%) and atrial septal defect (10%) were the most common CHD indications for palivizumab prophylaxis [20].

Singleton et al. [73] compared the rate of first RSV hospitalization in eligible premature Alaska Native infants born in the Yukon Kuskokwim region at ≤35 weeks' gestation before and after the introduction of palivizumab prophylaxis. In this region of southwest Alaska, acute respiratory tract infections (RTI) are a major cause of morbidity [42]. Typically, two-thirds of all hospitalizations in children aged <3 years are due to RTI, and RSV hospitalization rates are as high as 156 per 1000 infants aged <1 year (which is much higher than rates reported in other region of North America or Europe). The observed hospitalization rate for first-time RSV disease in children born at ≤35 weeks' gestation decreased from 18/41 (44%) premature infants born between 1993 and 1996 to 9/60 (15%) premature infants born between 1998 and 2001 – that is, after the introduction of palivizumab. Although it is difficult to ascertain the palivizumab eligibility of the cohort born in 1993 to 1996, the fact that RSV hospitalization remained unchanged in term infants indicated that the decline observed in high-risk children was likely due to prophylaxis and not to decreased RSV circulation or decreased RSV testing.

12.8 Is Protective Efficacy a Function of Palivizumab Serum Concentration?

Based on the assumption that severe RSV disease is a consequence of active high-titer RSV replication in the LRT, immunoprophylaxis that restricts viral replication 100- to 1000-fold should lead to abrogation, or least amelioration, of RSV-induced LRTI. Although this assumption is not universally accepted, there is good evidence to support it, not least from studies with live-attenuated vaccines that do not cause disease in seronegative infants if their level of replication is at least 100-fold attenuated compared to wildtype RSV [84]. In cotton rats, a serum palivizumab concentration of $30\,\mu g\,mL^{-1}$ reduced the RSV titer in the LRTI approximately 100-fold (by 10^2 $TCID_{50}\,g^{-1}$ lung tissue), while a concentration of $56\,\mu g\,mL^{-1}$ led to a 1000-fold reduction in RSV titer (see Table 12.1) [41].

For the preclinical and clinical development of palivizumab it seemed reasonable, therefore, to design an IM palivizumab dosing regimen that maintained trough concentrations of at least $30\,\mu g\,mL^{-1}$ and, as a margin of safety for person-to-person variability in palivizumab pharmacokinetics, ideally a trough concentration of greater than $40\,\mu g\,mL^{-1}$ [65].

A phase I/II trial in premature infants and children found the pharmacokinetics of palivizumab to be similar to those of other human IgG1 antibodies [65]. Mean serum concentrations were $91\,\mu g\,mL^{-1}$ (range: 52 to $174\,\mu g\,mL^{-1}$) at 2 days after the first IM dose of $15\,mg\,kg^{-1}$, and $49\,\mu g\,mL^{-1}$ (range: 14 to $132\,\mu g\,mL^{-1}$) at 30 days after IM injection (Fig. 12.5). The serum half-life was calculated to be approximately 24 days. The percentages of children with trough concentrations $>40\,\mu g\,mL^{-1}$ after each of five monthly IM injections were 66, 86, 91, 96, and 95% [65].

The IMpact-RSV trial determined mean (±SE) trough concentrations in the serum of 1396 infants with a mean age of 5.7 months and a mean body weight of 4.8 kg as to be $37 \pm 1.2\,\mu g\,mL^{-1}$ prior to dose 2, with concentrations increasing to 57 ± 2.3, 68 ± 2.9, and $72 \pm 1.7\,\mu g\,mL^{-1}$ at 30 days after doses 2, 3, and 4, respectively [2].

Wu et al. reported that, in their patient population of 24 infants born at ≤30 weeks of gestation with a mean (±SD) age of 31 ± 4 days and a mean body weight of 1293 ± 236 g at study entry, only 23% of those tested had palivizumab trough concentrations $>40\,\mu g\,mL^{-1}$ at 28 days after the first dose. The mean palivizumab concentration at that time point was $32 \pm 11\,\mu g\,mL^{-1}$, with a range of 16 to $56\,\mu g\,mL^{-1}$, indicating that a significant proportion of infants might not be optimally protected. Following the report of Wu et al., data at two teaching hospitals in Italy were reviewed retrospectively [47]. Of 389 infants at risk treated with palivizumab, nine experienced breakthrough RSV infections that led to hospitalization. All nine RSV cases occurred at least 19 days after the previous palivizumab dose – five after dose 2, and four after dose 3, coinciding with the local RSV epidemic peak [47].

Singleton et al. also reported the first-time RSV hospitalization rate for protected versus unprotected days in a cohort of 335 Alaska Native palivizumab

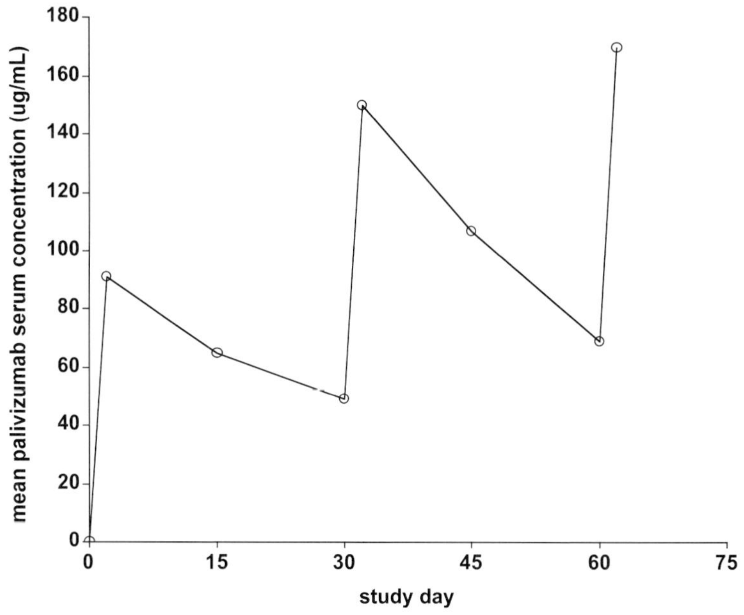

Fig. 12.5 Mean serum palivizumab concentrations in infants born at ≤35 weeks' gestation. Palivizumab (15 mg kg^{-1} body weight) was administered IM on Days 0 and 30, and then at 30-day intervals until Day 120. Data for Days 0, 2, 15, 30, 32, 45, and 60 are shown to visualize the accumulation effect that occurs with monthly palivizumab administrations. (Data from Ref. [65].)

recipients (1998–2001). Protected days were defined as days within 32 days of receiving palivizumab, while all other days within the RSV season were defined (arbitrarily) as unprotected days, based on the assumption that serum concentrations were too low to prevent RSV disease. Protection was provided for approximately two-thirds of all days in the RSV seasons. Among 69 RSV hospitalizations, 54 were first-time RSV hospitalizations, and 42 of these occurred during the RSV seasons. Twenty-one of 42 (50%) first-time RSV hospitalizations occurred on unprotected days, for a rate of 1.07 admissions per 1000 unprotected days and 0.55 per 1000 protected days – a relative risk reduction similar to that observed in the IMpact-RSV study.

Afghani et al. [8] reviewed the RSV hospitalization rate of 250 infants born at ≤35 weeks' gestation who received palivizumab in the 2003/2004 and 2004/2005 seasons at a single center. The six infants hospitalized for RSV disease were matched (1 : 2) with control subjects of the same gestational age and born within 2 months of the case patient. All cases and all controls had CLD. Five of the six cases were born at ≤28 weeks' gestation. In contrast to the control patients, all case patients had additional complications, such as tracheostomy, congenital anomalies, or severe intraventricular hemorrhage. There was no indication that nonprotective palivizumab trough levels contributed to hospitalization. The mean number of palivizumab injections prior to hospitalization was 2.8, and hospitalization occurred at a mean of 15.5 days after the last injection.

Zaaijer et al. [86] used a simple pharmacokinetic model to predict palivizumab trough concentrations, and suggested an alternative dosage regimen with a shortened first interval and a dose reduction from dose 2 (15 mg kg^{-1} on day 0, followed by 10 mg kg^{-1} on days 23, 53, 83, and 113). The authors suggested that, by using this alternative regimen, mean trough levels might be maintained above 45 μg mL^{-1}, while palivizumab expenses would be reduced considerably. No clinical results using this dosing regimen have been reported to date, however.

12.9 Post-Marketing Experience with Regard to Adverse Events

During the 8-year period since FDA approval of palivizumab, over four million doses of the antibody have been administered, and today the global safety databank is of considerable size. The most common possibly related AEs reported since the approval of palivizumab are similar in frequency to those reported in the IMpact-RSV phase III trial, namely injection site reactions (2–3%), fever (1–3%), diarrhea (up to 1%), and nervousness/irritability (up to 2.5%). Ongoing pharmacovigilance programs (REACH, Outcomes Registry) are in place to detect infrequent or unexpected side effects [38]. Acute hypersensitivity reactions such as urticaria (16/100 000 exposures) and anaphylaxis (<1 : 100 000 patients) were detected as rare AEs, and the package insert was modified to include these AEs.

As palivizumab is a humanized antibody containing both human (95%) and mouse (5%) amino acid sequence, it is possible that palivizumab is immunogenic – that is, anti-palivizumab antibodies might be induced. As a follow-up to the IMpact-RSV study, the safety and immunogenicity of palivizumab administered over two consecutive seasons was evaluated in 87 children. Thirty-two placebo-recipients and 56 palivizumab recipients received five monthly doses of 15 mg kg^{-1} palivizumab in the season following the IMpact-RSV trial. Only one child (receiving palivizumab for a second season) developed an anti-palivizumab titer of greater than 1/40 (1/160 on study Day 30, falling to 1/10 on Day 120, i.e., 30 days after the fourth injection). This child did not show accelerated palivizumab clearance and did not experience any SAEs. Two of the children in this study had developed transient anti-palivizumab titers during the IMpact-RSV trial, but neither child showed reactivity upon entry or during the second season [53]. In previously conducted studies in adults, low titer anti-palivizumab antibodies could be detected in approximately one-third of adult volunteers after the first dose, but not after repeat administrations of palivizumab. These antibodies were thought to be anti-idiotypic antibodies, but they had no clinical effect and no effect on palivizumab clearance [53]. In children, low anti-palivizumab titers were observed infrequently but, in contrast to adults, they were nonspecific and directed against the Fc portion of palivizumab and other IgGs. Again, clinical effects were not associated with an identifiable anti-palivizumab response [2,65,77].

As indicated by the manufacturer of palivizumab (see Synagis package insert), the frequency of adverse reactions that have been identified and reported during post-approval use of palivizumab is difficult to estimate as reports of these reactions are voluntary and the population is of uncertain size. Based on experience in over 400 000 patients who have received Synagis (>2 million doses), however, very rare severe acute hypersensitivity reactions were the only new SAEs detected in pharmacovigilance programs.

Severe acute hypersensitivity reactions have been reported on initial or subsequent exposure. Very rare cases of anaphylaxis (<1 case per 100 000 patients) have also been reported following re-exposure. None of the reported hypersensitivity reactions was fatal. The hypersensitivity reactions mentioned in the package insert include dyspnea, cyanosis, respiratory failure, urticaria, pruritus, angioedema, hypotonia, and unresponsiveness. The relationship between these reactions and the development of antibodies to Synagis is unknown. Limited data from post-marketing reports suggest that giving more than five doses in a single season does not increase the risk of hypersensitivity reactions.

The development of palivizumab-resistant RSV mutants is an important theoretical risk to consider. *In vitro*, palivizumab-resistant RSV can be selected in the presence of palivizumab, and these resistant mutants can replicate in cotton rats at palivizumab serum concentrations that would normally restrict RSV replication [87]. However, resistance does not seem to be a relevant issue in breakthrough RSV disease leading to RSV hospitalization. A study at eight centers in the US evaluated palivizumab binding to 371 RSV isolates – including 25 isolates from children receiving palivizumab concurrently – and found that all of them were recognized by palivizumab [24]. A total of 124 of these RSV isolates was evaluated in an RSV microneutralization test, and all of them were neutralized by palivizumab [24]. The risk of clinically significant palivizumab resistance seems therefore very low in the immunocompetent host, where RSV is eliminated within 2 weeks of infection. Even in immunosuppressed patients in whom RSV can replicate for months and cause severe disease with high mortality, palivizumab escape mutants do not seem to develop [27]. It should be noted, however, that URT secretions are commonly tested for RSV while selection-pressure for palivizumab escape mutants is higher in the LRT. Most of the RSV replication in severe disease occurs in the LRT, and palivizumab is present at higher concentration in the lungs.

12.10 Ongoing Clinical Studies and Outlook

As indicated above, Medimmune has developed a second-generation humanized RSV mAb based on palivizumab by designing and screening palivizumab mutants [85]. This second-generation mAb, motavizumab (MEDI-524, NuMax), has an improved association rate (K_{on}) and neutralizes RSV more effectively. Motavizumab is currently being studied in comparison with palivizumab in two large clinical trials in premature infants and in infants with CHD. A third study is evaluating motavizumab versus placebo in term infants at high risk. Motavizumab is projected to enter the market in 2008/2009 if shown to be superior to palivizumab.

A large motavizumab versus palivizumab phase III study began enrollment in 2004, and is designed to compare the safety and efficacy of motavizumab with that of palivizumab for the reduction of RSV hospitalization in high-risk infants. Medically attended LRTI and otitis media are used, amongst others, as secondary endpoints. This randomized, double-blind trial enrolled approximately 6600 high-risk infants at 300 sites in 24 countries in both the Northern and the Southern hemispheres. As with the IMpact-RSV trial, the study population consists of premature infants born at ≤35 weeks GA and aged ≤6 months at randomization, as well as children with CLD aged ≤24 months who required CLD treatment during the 6-month period prior to enrollment. Five monthly IM injections of 15 $mg\,kg^{-1}$ of either motavizumab or palivizumab during the RSV season are compared for their safety and efficacy in reducing RSV hospitalization, medically attended outpatient visits for LRTI, and otitis media.

The second motavizumab versus palivizumab trial is a randomized, double-blind trial that evaluates the safety, tolerability, immunogenicity and pharmacokinetics of the two mAbs in children with CHD. Recruitment took place at approximately 160 sites in the Northern Hemisphere during the 2005–2006 RSV season. Inclusion criteria are similar to those used in the palivizumab CHD phase III trial – that is, infants with hemodynamically significant CHD aged ≤24 months at randomization. Approximately 620 patients received five monthly doses of either 15 $mg\,kg^{-1}$ motavizumab or palivizumab, and were followed until day 150 – that is, 30 days after the last injection. Hospitalization due to RSV is assessed as a secondary endpoint.

Motavizumab versus placebo is being studied in a randomized, double-blind, placebo-controlled phase III trial designed to enroll approximately 3000 healthy full-term Navajo and White Mountain Apache infants over four RSV seasons. Study participants will receive up to five monthly doses (15 $mg\,kg^{-1}$) of motavizumab or placebo during the RSV season for two consecutive seasons. During the first year, the plans are to enroll approximately 450 infants. Similar to the data reported for Alaskan infants, full-term Native American infants are at high risk of developing severe RSV disease. As Native American children also have a high rate of wheezing and asthma, the study participants will be followed until they are aged 5 years, in order to collect data on whether reducing RSV disease during the first 2 years of life can reduce the incidence of wheezing and asthma in full-term children.

12.11 Summary

Palivizumab was the first mAb product to be approved for an infectious disease indication, and quickly developed into a very successful biopharmaceutical product. The safety and efficacy of palivizumab in reducing severe RSV disease in high-risk infants and children was established in large and well-designed phase III trials, and its safety and effectiveness in premature children and children with CHD seems to have been confirmed in a large number of clinical reports after approval. One major problem in using palivizumab relates to its cost; RSV prophylaxis using palivizumab may be recommended in more than 5% of all infants, but this places a high financial burden on healthcare systems. In order to deal with the enormous burden of disease caused by RSV, an affordable live-attenuated vaccine would be most desirable from a public health point of view. However such a vaccine it not available at present.

Acknowledgments

The author thanks Drs. Peter Collins, Brian Murphy, Sheila Nolan, Ruth Karron, Niki Oquist, and Robert Fuentes for reading the manuscript and providing valuable comments and suggestions. The author received financial support as part of the NIAID Intramural Program.

References

1 Swedish Consensus Group. (2001) Management of infections caused by respiratory syncytial virus. Scand J Infect Dis 33: 323–328.
2 The IMpact-RSV Study Group. (1998) Palivizumab, a humanized respiratory syncytial virus monoclonal antibody, reduces hospitalization from respiratory syncytial virus infection in high-risk infants. Pediatrics 102: 531–537.
3 [Recommendations for immunologic prevention of respiratory syncytial virus (RSV) infections]. (1999) Z Geburtshilfe Neonatol 203(Suppl.2): 16.
4 The PREVENT Study Group. (1997) Reduction of respiratory syncytial virus hospitalization among premature infants and infants with bronchopulmonary dysplasia using respiratory syncytial virus immune globulin prophylaxis. Pediatrics 99: 93–99.
5 American Academy of Pediatrics Committee on Infectious Diseases and Committee on Fetus and Newborn. (2003) Revised indications for the use of palivizumab and respiratory syncytial virus immune globulin intravenous for the prevention of respiratory syncytial virus infections. Pediatrics 112: 1442–1445.
6 National Advisory Committee on Immunization. (2003) Statement on the recommended use of monoclonal anti-RSV antibody (palivizumab). Can Commun Dis Rep 29: 1–15.
7 Adams, R., Christenson, J., Petersen, F., Beatty, P. (1999) Pre-emptive use of aerosolized ribavirin in the treatment of asymptomatic pediatric marrow transplant patients testing positive for RSV. Bone Marrow Transplant 24: 661–664.
8 Afghani B., Ngo, T., Zeitany R., Amin A. (2006) Premature infants who get hospitalized for respiratory syncytial virus infection despite receiving palivizumab. J Invest Med 54: S133.
9 American Academy of Pediatrics. (2006). Respiratory syncytial virus. In: Pickering, L.K., Baker, C.J., Long, S.S., McMillan, J.A. (Eds.), Red Book: 2006 Report of the Committee on Infectious

Diseases. 27th edn. American Academy of Pediatrics, Elk Grove Village, IL, USA, pp. 560–566.

10 Bebbington, C.R., Renner, G., Thomson, S., King, D., Abrams, D., Yarranton, G.T. (1992) High-level expression of a recombinant antibody from myeloma cells using a glutamine synthetase gene as an amplifiable selectable marker. Biotechnology (New York) 10: 169–175.

11 Beeler, J.A., van Wyke Coelingh, K. (1989) Neutralization epitopes of the F glycoprotein of respiratory syncytial virus: effect of mutation upon fusion function. J Virol 63: 2941–2950.

12 Bentley, D.L., Rabbits, T.H. (1980) Human immunoglobulin variable region genes – DNA sequences of two V kappa genes and a pseudogene. Nature 288: 730–733.

13 Bermingham, A., Collins, P.L. (1999) The M2-2 protein of human respiratory syncytial virus is a regulatory factor involved in the balance between RNA replication and transcription. Proc Natl Acad Sci USA 96: 11259–11264.

14 Boyce, T.G., Mellen, B.G., Mitchel, E.F., Jr., Wright, P.F., Griffin, M.R. (2000) Rates of hospitalization for respiratory syncytial virus infection among children in medicaid. J Pediatr 137: 865–870.

15 Carbonell Estrany, X., Quero Jimenez, J. (2000) [Recommendations for the prevention of respiratory syncytial virus infections. Standards Committee of the Spanish Society of Neonatology. Board of Directors of the Spanish Society of Neonatology]. Ann Esp Pediatr 52: 372–374.

16 Carbonell-Estrany, X., Quero, J. (2001) Hospitalization rates for respiratory syncytial virus infection in premature infants born during two consecutive seasons. Pediatr Infect Dis J 20: 874–879.

17 Carbonell-Estrany, X., Quero, J., Bustos, G., Cotero, A., Domenech, E., Figueras-Aloy, J., Fraga, J.M., Garcia, L.G., Garcia-Alix, A., Del Rio, M.G., Krauel X., Sastre, J.B., Narbona, E., Roques, V., Hernandez, S.S., Zapatero, M. (2000) Rehospitalization because of respiratory syncytial virus infection in premature infants younger than 33 weeks of gestation: a prospective study. IRIS Study Group. Pediatr Infect Dis J 19: 592–597.

18 Chantepie, A. (2004) [Use of palivizumab for the prevention of respiratory syncytial virus infections in children with congenital heart disease. Recommendations from the French Paediatric Cardiac Society.]. Arch Pediatr 11: 1402–1405.

19 Cohen, A., Hirsch, R.L., Sorrentino, M., Top, J.R., Carlin, D., McClain, B. (1999) First year experience using palivizumab humanized monoclonal antibody for protection from RSV lower respiratory tract infection. Second World Congress of Pediatric Infectious Diseases Abstract Book. Manila, Philippines. Abstract A16.

20 Cohen, S., Boron, M., Cohen M., Rankin A., M. and the Palivizumab Outcomes Registry Study Group. (2004) Patterns of RSV prophylaxis in children with congenital heart disease (CHD): preliminary results from the 2003 to 2004 palivizumab outcomes registry. J Perinatol 24: 605.

21 Collins, P.L., Chanock, R.M., Murphy, B.R. (2001) Respiratory syncytial virus. In: Knipe, D.M., Howley, P.M., Griffin, D.E., Lamb, R.A., Martin, M.A., Roizman, B., Strauss, S.E. (Eds.), Fields Virology. 4th edn. Lippincott Williams & Wilkins, Philadelphia. Vol. 1. pp. 1443–1486.

22 Connor, E.M., McClain, J.B., Sorrentino, M., Hirsch, R.L., Top, F.H. (2002) Safety surveillance in children treated with palivizumab (PLV). J Perinatol 22: 619.

23 Crowe, J.E., Jr., Firestone, C.Y., Murphy, B.R. (2001) Passively acquired antibodies suppress humoral but not cell-mediated immunity in mice immunized with live attenuated respiratory syncytial virus vaccines. J Immunol 167: 3910–3918.

24 DeVincenzo, J.P., Hall, C.B., Kimberlin, D.W., Sanchez, P.J., Rodriguez, W.J., Jantausch, B.A., Corey, L., Kahn, J.S., Englund, J.A., Suzich, J.A., Palmer-Hill, F.J., Branco, L., Johnson, S., Patel, N.K., Piazza, F.M. (2004) Surveillance of clinical isolates of respiratory syncytial virus for palivizumab (Synagis)-resistant mutants. J Infect Dis 190: 975–978.

25 Duppenthaler, A., Ammann, R.A., Gorgievski-Hrisoho, M., Pfammatter, J. P., Aebi, C. (2004) Low incidence of respiratory syncytial virus hospitalisations in haemodynamically significant congenital heart disease. Arch Dis Child 89: 961.

26 Duppenthaler, A., Gorgievski-Hrisoho, M., Aebi, C. (2001) Regional impact of prophylaxis with the monoclonal antibody palivizumab on hospitalisations for respiratory syncytial virus in infants. Swiss Med Wkly 131: 146–151.

27 El Saleeby, C.M., Suzich, J., Conley, M.E., DeVincenzo, J.P. (2004) Quantitative effects of palivizumab and donor-derived T cells on chronic respiratory syncytial virus infection, lung disease, and fusion glycoprotein amino acid sequences in a patient before and after bone marrow transplantation. Clin Infect Dis 39: e17–e20.

28 Falsey, A.R., Hennessey, P.A., Formica, M.A., Cox, C., Walsh, E.E. (2005) Respiratory syncytial virus infection in elderly and high-risk adults. N Engl J Med 352: 1749–1759.

29 Feltes, T.F., Cabalka, A.K., Meissner, H.C., Piazza, F.M., Carlin, D.A., Top, F. H., Jr., Connor, E.M., Sondheimer, H.M. (2003) Palivizumab prophylaxis reduces hospitalization due to respiratory syncytial virus in young children with hemodynamically significant congenital heart disease. J Pediatr 143: 532–540.

30 Feltes, T.F., Simoes, E. (2005) Palivizumab prophylaxis in haemodynamically significant congenital heart disease. Arch Dis Child 90: 875–876.

31 Figueras Aloy, J., Quero, J., Domenech, E., Lopez Herrera, M.C., Izquierdo, I., Losada, A., Perapch, J., Sanchez-Luna, M. (2005) [Recommendations for the prevention of respiratory syncytial virus infection]. An Pediatr (Barc) 63: 357–362.

32 Forster, J. (1999) [Practice guideline by the German Society for Pediatric Infectious Diseases with respect to prevention of RSV infections through immunoglobulin administration]. Klin Padiatr 211: 476.

33 Groothuis, J.R. (2001) Safety and tolerance of palivizumab administration in a large Northern Hemisphere trial. Northern Hemisphere Expanded Access Study Group. Pediatr Infect Dis J 20: 628–630.

34 Groothuis, J.R., Nishida, H. (2002) Prevention of respiratory syncytial virus infections in high-risk infants by monoclonal antibody (palivizumab). Pediatr Int 44: 235–241.

35 Groothuis, J.R., Simoes, E.A., Levin, M.J., et al. (1993) Prophylactic administration of respiratory syncytial virus immune globulin to high-risk infants and young children. The Respiratory Syncytial Virus Immune Globulin Study Group. N Engl J Med 329: 1524–1530.

36 Hartling, L., Wiebe, N., Russell, K., Patel, H., Klassen, T.P. (2004) Epinephrine for bronchiolitis. The Cochrane Database of Systematic Reviews 2004: 1–35.

37 Heikkinen, T., Valkonen, H., Lehtonen, L., Vainionpaa, R., Ruuskanen, O. (2005) Hospital admission of high risk infants for respiratory syncytial virus infection: implications for palivizumab prophylaxis. Arch Dis Child Fetal Neonatal Ed 90: F64–F68.

38 Hudak, M. (2002) Palivizumab prophylaxis of RSV disease – results of 5097 children – the 2001 to 2002 outcomes registry. J Perinatol 22: 619.

39 Iwane, M.K., Edwards, K.M., Szilagyi, P.G., Walker, F.J., Griffin, M.R., Weinberg, G.A., Coulen, C., Poehling, K.A., Shone, L.P., Balter, S., Hall, C.B., Erdman, D.D., Wooten, K., Schwartz, B. (2004) Population-based surveillance for hospitalizations associated with respiratory syncytial virus, influenza virus, and parainfluenza viruses among young children. Pediatrics 113: 1758–1764.

40 Jobe, A.H., Bancalari, E. (2001) Bronchopulmonary dysplasia. Am J Respir Crit Care Med 163: 1723–1729.

41 Johnson, S., Oliver, C., Prince, G.A., Hemming, V.G., Pfarr, D.S., Wang, S.C.,

Dormitzer, M., O'Grady, J., Koenig, S., Tamura, J.K., Woods, R., Bansal, G., Couchenour, D., Tsao, E., Hall, W.C., Young, J.F. (1997) Development of a humanized monoclonal antibody (MEDI-493) with potent in vitro and in vivo activity against respiratory syncytial virus. J Infect Dis 176: 1215–1224.

42 Karron, R.A., Singleton, R.J., Bulkow, L., Parkinson, A., Kruse, D., DeSmet, I., Indorf, C., Petersen, K.M., Leombruno, D., Hurlburt, D., Santosham, M., Harrison, L.H. (1999) Severe respiratory syncytial virus disease in Alaska native children. RSV Alaska Study Group. J Infect Dis 180: 41–49.

43 Kellner, J.D., Ohlsson, A., Gadomski, A.M., Wang, E.E.L. (1999) Bronchodilators for bronchiolitis. The Cochrane Database of Systematic Reviews 1999: 1–18.

44 Kohler, G., Milstein, C. (1976) Derivation of specific antibody-producing tissue culture and tumor lines by cell fusion. Eur J Immunol 6: 511–519.

45 Kunkel, T.A. (1985) Rapid and efficient site-specific mutagenesis without phenotypic selection. Proc Natl Acad Sci USA 82: 488–492.

46 Lacaze-Masmonteil, T., Roze, J.C., Fauroux, B. (2002) Incidence of respiratory syncytial virus-related hospitalizations in high-risk children: follow-up of a national cohort of infants treated with Palivizumab as RSV prophylaxis. Pediatr Pulmonol 34: 181–188.

47 Manzoni, P., Sala, U., Gomirato, G., Coscia, A., Fabris, C. (2005) Optimal timing and dosing intervals of palivizumab in premature neonates: still some work to do. Pediatrics 115: 1439–1440.

48 Martinez, F.D. (2003) Respiratory syncytial virus bronchiolitis and the pathogenesis of childhood asthma. Pediatr Infect Dis J 22: S76–S82.

49 Murphy, B.R., Prince, G.A., Walsh, E.E., Kim, H.W., Parrott, R.H., Hemming, V. G., Rodriguez, W.J., Chanock, R.M. (1986) Dissociation between serum neutralizing and glycoprotein antibody responses of infants and children who received inactivated respiratory syncytial virus vaccine. J Clin Microbiol 24: 197–202.

50 Naver, L., Eriksson, M., Ewald, U., Linde, A., Lindroth, M., Schollin, J. (2004) Appropriate prophylaxis with restrictive palivizumab regimen in preterm children in Sweden. Acta Paediatr 93: 1470–1474.

51 Naylor, C.J., Britton, P., Cavanagh, D. (1998) The ectodomains but not the transmembrane domains of the fusion proteins of subtypes A and B avian pneumovirus are conserved to a similar extent as those of human respiratory syncytial virus. J Gen Virol 79 (Pt 6): 1393–1398.

52 Northway, W.H., Jr., Rosan, R.C., Porter, D.Y. (1967) Pulmonary disease following respirator therapy of hyaline-membrane disease. Bronchopulmonary dysplasia. N Engl J Med 276: 357–368.

53 Null, D., Jr., Pollara, B., Dennehy, P.H., Steichen, J., Sanchez, P.J., Givner, L.B., Carlin, D., Landry, B., Top, F.H., Jr., Connor, E. (2005) Safety and immunogenicity of palivizumab (Synagis) administered for two seasons. Pediatr Infect Dis J 24: 1021–1024.

54 Ogilvie, M.M., Vathenen, A.S., Radford, M., Codd, J., Key, S. (1981) Maternal antibody and respiratory syncytial virus infection in infancy. J Med Virol 7: 263–271.

55 Oh, P.I., Lanctjt, K.L., Yoon, A., Lee, D.S., Paes, B.A., Simmons, B.S., Parison, D., Manzi, P. (2002) Palivizumab prophylaxis for respiratory syncytial virus in Canada: utilization and outcomes. Pediatr Infect Dis J 21: 512–518.

56 Parnes, C., Guillermin, J., Habersang, R., et al. (2003) Palivizumab prophylaxis of respiratory syncytial virus disease in 2000–2001: results from The Palivizumab Outcomes Registry. Pediatr Pulmonol 35: 484–489.

57 Patel, H., Platt, R., Lozano, J.M., Wang, E.E.L. (2004) Glucocorticoids for acute viral bronchiolitis in infants and young children. The Cochrane Database of Systematic Reviews 2004: 1–46.

58 Pedraz, C., Carbonell-Estrany, X., Figueras-Aloy, J., Quero, J. (2003) Effect of palivizumab prophylaxis in decreasing respiratory syncytial virus hospitalizations in premature infants. Pediatr Infect Dis J 22: 823–827.
59 Pollack, P., Groothuis, J.R. (2002) Development and use of palivizumab (Synagis): a passive immunoprophylactic agent for RSV. J Infect Chemother 8: 201–206.
60 Prais, D., Danino, D., Schonfeld, T., Amir, J. (2005) Impact of palivizumab on admission to the ICU for respiratory syncytial virus bronchiolitis: a national survey. Chest 128: 2765–2771.
61 Press, E.M., Hogg, N.M. (1970) The amino acid sequences of the Fd fragments of two human gamma-1 heavy chains. Biochem J 117: 641–660.
62 Prince, G.A., Horswood, R.L., Chanock, R.M. (1985) Quantitative aspects of passive immunity to respiratory syncytial virus infection in infant cotton rats. J Virol 55: 517–520.
63 Rader, R. (2005) Biopharmaceutical Products in the U.S. and European Markets. BioPlan Associates, Inc., Rockville, MD, USA.
64 Robertson, S.E., Roca, A., Alonso, P., Simoes, E.A., Kartasasmita, C.B., Olaleye, D.O., Odaibo, G.N., Collinson, M., Venter, M., Zhu, Y., Wright, P.F. (2004) Respiratory syncytial virus infection: denominator-based studies in Indonesia, Mozambique, Nigeria and South Africa. Bull World Health Org 82: 914–922.
65 Saez-Llorens, X., Castano, E., Null, D., Steichen, J., Sanchez, P.J., Ramilo, O., Top, F.H., Jr., Connor, E. (1998) Safety and pharmacokinetics of an intramuscular humanized monoclonal antibody to respiratory syncytial virus in premature infants and infants with bronchopulmonary dysplasia. The MEDI-493 Study Group. Pediatr Infect Dis J 17: 787–791.
66 Saji, T., Nakazawa, M., Harada, K. (2005) Safety and efficacy of palivizumab prophylaxis in children with congenital heart disease. Pediatr Int 47: 397–403.
67 Sampalis, J.S. (2003) Morbidity and mortality after RSV-associated hospitalizations among premature Canadian infants. J Pediatr 143: S150–S156.
68 Schenerman, M.A., Hope, J.N., Kletke, C., Singh, J.K., Kimura, R., Tsao, E.I., Folena-Wasserman, G. (1999) Comparability testing of a humanized monoclonal antibody (Synagis) to support cell line stability, process validation, and scale-up for manufacturing. Biologicals 27: 203–215.
69 Shay, D.K., Holman, R.C., Newman, R.D., Liu, L.L., Stout, J.W., Anderson, L. J. (1999) Bronchiolitis-associated hospitalizations among US children, 1980–1996. JAMA 282: 1440–1446.
70 Shennan, A.T., Dunn, M.S., Ohlsson, A., Lennox, K., Hoskins, E.M. (1988) Abnormal pulmonary outcomes in premature infants: prediction from oxygen requirement in the neonatal period. Pediatrics 82: 527–532.
71 Sigurs, N., Gustafsson, P.M., Bjarnason, R., Lundberg, F., Schmidt, S., Sigurbergsson, F., Kjellman, B. (2005) Severe respiratory syncytial virus bronchiolitis in infancy and asthma and allergy at age 13. Am J Respir Crit Care Med 171: 137–141.
72 Simoes, E.A. (2002) Immunoprophylaxis of respiratory syncytial virus: global experience. Respir Res. 3(Suppl.1): S26–S33.
73 Singleton, R., Dooley, L., Bruden, D., Raelson, S., Butler, J.C. (2003) Impact of palivizumab prophylaxis on respiratory syncytial virus hospitalizations in high risk Alaska Native infants. Pediatr Infect Dis J 22: 540–545.
74 Sorrentino, M., Powers, T. (2000) Effectiveness of palivizumab: evaluation of outcomes from the 1998 to 1999 respiratory syncytial virus season. The Palivizumab Outcomes Study Group. Pediatr Infect Dis J 19: 1068–1071.
75 Spann, K.M., Tran, K.C., Collins, P.L. (2005) Effects of nonstructural proteins NS1 and NS2 of human respiratory syncytial virus on interferon regulatory factor 3, NF-kappaB, and

proinflammatory cytokines. J Virol 79: 5353–5362.

76 Stensballe, L.G., Devasundaram, J.K., Simoes, E.A. (2003) Respiratory syncytial virus epidemics: the ups and downs of a seasonal virus. Pediatr Infect Dis J 22: S21–S32.

77 Subramanian, K.N., Weisman, L.E., Rhodes, T., Ariagno, R., Sanchez, P.J., Steichen, J., Givner, L.B., Jennings, T.L., Top, F.H., Jr., Carlin, D., Connor, E. (1998) Safety, tolerance and pharmacokinetics of a humanized monoclonal antibody to respiratory syncytial virus in premature infants and infants with bronchopulmonary dysplasia. MEDI-493 Study Group. Pediatr Infect Dis J 17: 110–115.

78 Takahashi, N., Noma, T., Honjo, T. (1984) Rearranged immunoglobulin heavy chain variable region (VH) pseudogene that deletes the second complementarity-determining region. Proc Natl Acad Sci USA 81: 5194–5198.

79 Tulloh, R., Marsh, M., Blackburn, M., Casey, F., Lenney, W., Weller, P., Keeton, B.R. (2003) Recommendations for the use of palivizumab as prophylaxis against respiratory syncytial virus in infants with congenital cardiac disease. Cardiol Young 13: 420.

80 Ventre, K., Randolph, A.G. (2004) Ribavirin for respiratory syncytial virus infection of the lower respiratory tract in infants and young children. The Cochrane Database of Systematic Reviews 2004: 1–20.

81 Weigl, J.A., Puppe, W., Belke, O., Neususs, J., Bagci, F., Schmitt, H.J. (2005) Population-based incidence of severe pneumonia in children in Kiel, Germany. Klin Padiatr 217: 211–219.

82 Weigl, J.A., Puppe, W., Schmitt, H.J. (2001) Incidence of respiratory syncytial virus-positive hospitalizations in Germany. Eur J Clin Microbiol Infect Dis 20: 452–459.

83 Willson, D.F., Landrigan, C.P., Horn, S.D., Smout, R.J. (2003) Complications in infants hospitalized for bronchiolitis or respiratory syncytial virus pneumonia. J Pediatr 143: S142–S149.

84 Wright, P.F., Karron, R.A., Madhi, S.A., Treanor, J.J., King, J.C., O'Shea, A., Ikizler, M.R., Zhu, Y., Collins, P.L., Cutland, C., Randolph, V.B., Deatly, A. M., Hackell, J.G., Gruber, W.C., Murphy, B.R. (2006) The interferon antagonist NS2 protein of respiratory syncytial virus is an important virulence determinant for humans. J Infect Dis 193: 573–581.

85 Wu, H., Pfarr, D.S., Tang, Y., An, L.L., Patel, N.K., Watkins, J.D., Huse, W.D., Kiener, P.A., Young, J.F. (2005) Ultra-potent antibodies against respiratory syncytial virus: effects of binding kinetics and binding valence on viral neutralization. J Mol Biol 350: 126–144.

86 Zaaijer, H.L., Vandenbroucke-Grauls, C.M., Franssen, E.J. (2002) Optimum dosage regimen of palivizumab? Ther Drug Monit 24: 444–445.

87 Zhao, X., Chen, F.P., Megaw, A.G., Sullender, W.M. (2004) Variable resistance to palivizumab in cotton rats by respiratory syncytial virus mutants. J Infect Dis 190: 1941–1946.

13
Rituximab (Rituxan)

Michael Wenger

13.1
Introduction

Rituximab (MabThera, F. Hoffmann La-Roche; Rituxan, Genentech and Biogen Idec) – a chimeric monoclonal antibody (mAb) produced by recombinant technology – entered clinical development during the early 1990s, and subsequently received approval by American and European regulatory authorities in 1997 and 1998, respectively. In this chapter we will first discuss rituximab's production and mode of action, before focusing on its efficacy and safety in non-Hodgkin's lymphoma (NHL) and chronic lymphocytic leukemia (CLL). Finally, the evaluation of rituximab in new therapeutic areas including rheumatoid arthritis (RA) will be briefly described.

13.1.1
Production, Design, and Structure of Rituximab

Rituximab is a genetically engineered human/mouse chimeric mAb that is specific for the CD20 antigen on the surface of B cells (Reff et al. 1994). It is a fusion of the light- and heavy-chain variable (antigen-binding) domains of 2B8 (a murine monoclonal anti-CD20 antibody) and human kappa light-chain and gamma 1 heavy-chain constant regions (Fig. 13.1) (Boye et al. 2003; Reff et al. 1994).

Rituximab consists of two heavy chains of 451 amino acids and two light chains of 213 amino acids, with an approximate molecular weight of 145 kDa. It has a binding affinity for the CD20 antigen of approximately 8.0 nM (Reff et al. 1994).

Because the majority of the antibody is of human origin, rituximab has a low potential for immunogenicity. Indeed, no human anti-mouse immunoglobulin antibodies (HAMA) have been detected in patients treated with rituximab to date. The development of human anti-chimeric antibodies (HACA) occurs rarely (documented in <1% of rituximab-treated patients), and it is not clear whether these antibodies have any neutralizing effect (MabThera SmPC; see link on EMEA

Handbook of Therapeutic Antibodies. Edited by Stefan Dübel
Copyright © 2007 WILEY-VCH Verlag GmbH & Co. KGaA, Weinheim
ISBN 978-3-527-31453-9

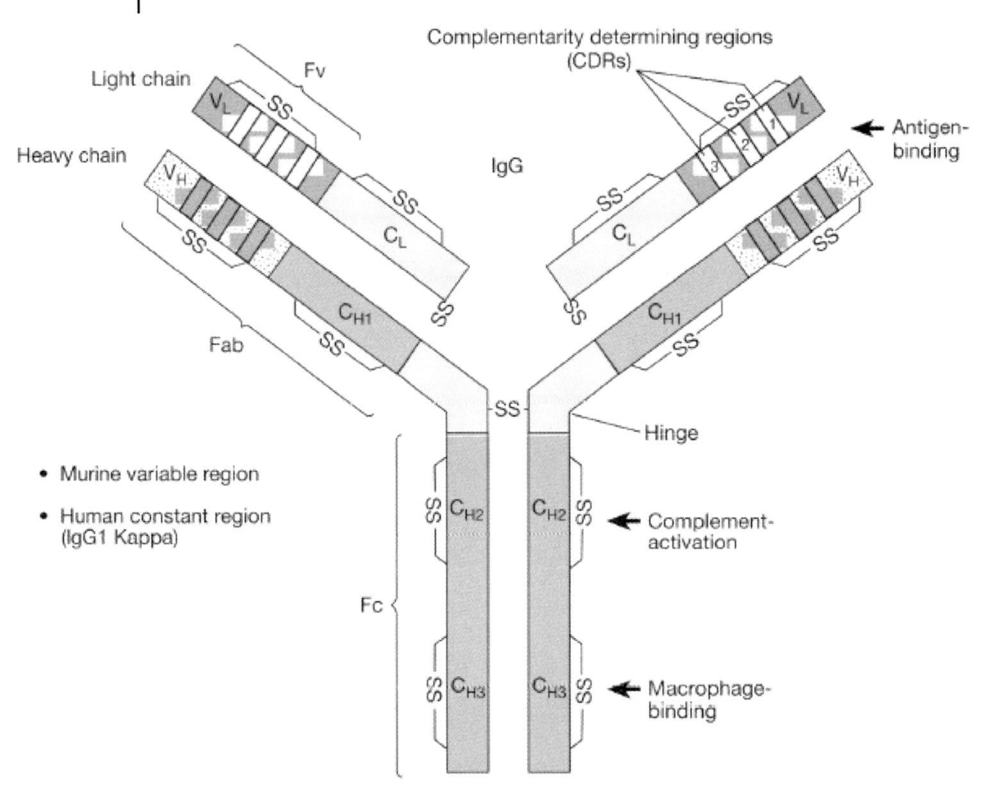

Fig. 13.1 The structure of rituximab. (Reproduced with permission from Boye et al. 2003.)

website, http://www.emea.eu.int; Rituxan datasheet, see link on Genentech website, http://www.gene.com).

Rituximab is produced by Chinese hamster ovary cells in suspension, and purified by affinity chromatography and ion-exchange chromatography. The purification process also incorporates specific viral inactivation and removal procedures. Rituximab is supplied as a sterile, clear, colorless, preservative-free liquid concentrate for intravenous (IV) infusion.

Thus, rituximab binds to the target CD20 antigen via the variable murine regions, while the remainder of the antibody interacts with human effector mechanisms to kill the target cells.

13.1.2
CD20 as a Therapeutic Target

CD20 is a transmembrane surface antigen, which is expressed only by B-cell precursors and mature B cells. This antigen appears to be involved in the regulation of B lymphocyte growth and differentiation; indeed, recent data suggest that CD20 could play an important role in the influx of calcium across cell membranes, sustaining intracellular calcium concentrations and allowing the activation of B cells (Li et al. 2003).

CD20 makes an attractive target for monoclonal therapy for a number of reasons:

- It is reliably expressed on a large proportion of B cells in a number of diseases (e.g., NHL), while it is not expressed on stem cells, normal mature plasma cells or other normal tissues. Its expression is lost when normal B cells differentiate into antibody-secreting plasma cells (Nadler et al. 1981; Anderson et al. 1984).
- CD20 is present on malignant plasma cells in 20% of patients with multiple myeloma, up to 50% of patients with plasma cell leukemia, 75–100% of patients with Waldenström's macroglobulinemia, and >95% of patients with B-cell lymphoma and leukemia (Treon et al. 2003; Olszewski and Grossbard 2004).
- $CD20^+$ cells can be completely eradicated from the body without causing excessive toxicity because normal B cells will re-emerge following differentiation from stem cells, while serum immunoglobulin levels can be maintained by persisting plasma cells.
- It is not internalized after binding by antibody, and its expression is stable (Press et al. 1987). Antibodies targeting CD20 therefore remain bound to the antigen on the cell surface at a constant density determined by the level of antigen expression. The bound antibody then initiates immune processes and induces apoptosis.
- CD20 is not normally shed from the cell surface, and serum levels of the antigen are undetectable in most patients (Einfeld et al. 1988); however, soluble CD20 has been observed in some patients with CLL (Manshouri et al. 2003).

13.1.3 Mode of Action

Evidence suggests that several mechanisms may be involved in providing rituximab's therapeutic efficacy via the CD20 antigen (Fig. 13.2; Olszewski and Grossbard 2004). Because CD20 is neither shed from the cell surface nor internalized upon antibody binding, sustained binding of rituximab to the CD20 antigen

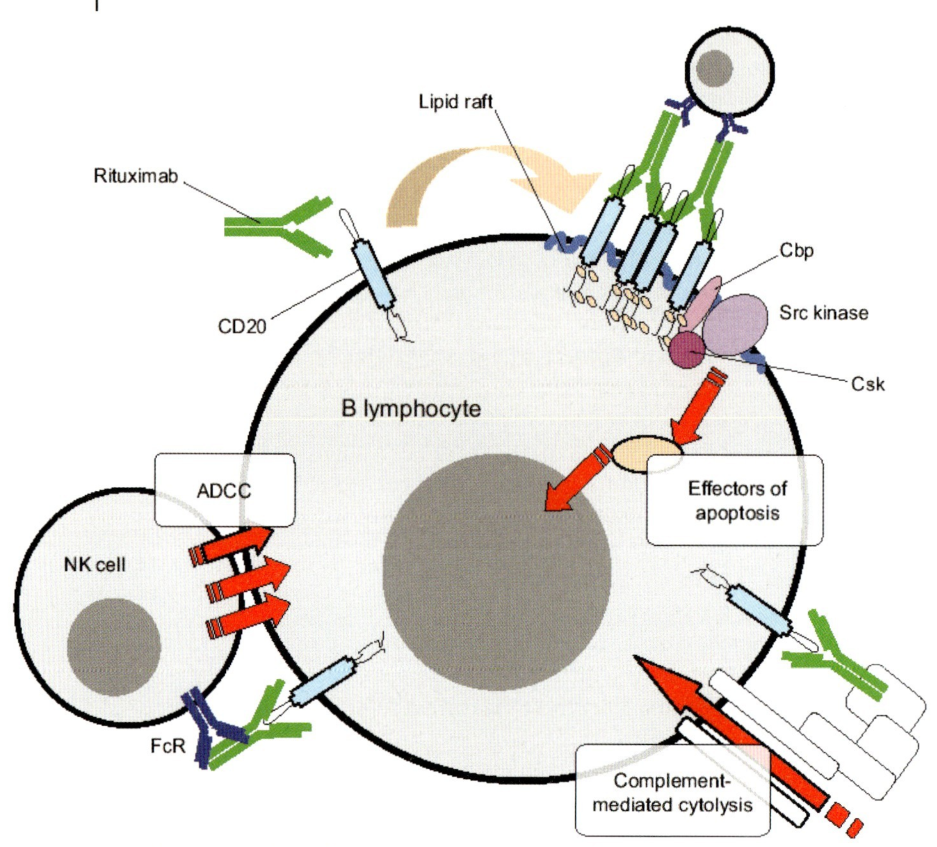

Fig. 13.2 Rituximab mechanisms of action. (Reproduced with permission from Olszewski and Grossbard 2004.)

occurs. The mechanism of cytotoxicity induced by rituximab is not fully understood, and is thought to include several mechanisms. The dominant mechanism may vary according to cellular compartment, presence or absence of concurrent chemotherapy, and type of tumor treated (Friedberg 2005).

Antibody-dependent cellular cytotoxicity (ADCC) appears to be a major *in-vivo* mechanism of rituximab. ADCC is mediated through ligation of the Fc portion of the antibody to Fc receptors (FcR) expressed by accessory cells. Complement-dependent cell lysis (CDC) occurs when rituximab binds to the complement factor C1q; it appears that translocation of CD20 into lipid rafts precedes the activation of lytic complement. The development of lipid rafts also appears to play a key role in rituximab-induced apoptosis. Increased calcium conductivity and induction of a close proximity with Src kinases leads to caspase activation and then to apoptosis. The cellular microenvironment also appears to contribute to rituximab-induced cytotoxicity, with differential responses to rituximab occurring in lymphomas of various histologies. Finally, rituximab therapy of lymphoma cells may promote uptake and cross-presentation of lymphoma-derived peptides by antigen-presenting dendritic cells (DC), induce the maturation of DC, and allow the generation of specific cytotoxic T-cell responses – a so-called "vaccinal" effect (Friedberg 2005).

In-vitro data confirm that rituximab may potentiate the cytotoxic activity of other drugs used in the oncology setting, suggesting a synergistic effect when rituximab is used in combination with other chemotherapy agents.

Preclinical data suggesting synergy between rituximab and cytotoxic or biologic therapy includes the following:

- DHL-4 – a drug-resistant B-cell lymphoma cell line – becomes susceptible to doxorubicin and cisplatin following exposure to rituximab (Demidem et al. 1995, 1997).
- Rituximab inhibits the production of interleukin (IL)-10, which in turn down-regulates the expression of the anti-apoptotic protein, bcl-2, sensitizing B cells to the apoptotic effects of chemotherapy (Alas and Bonavida 2001).
- Fludarabine down-modulates the expression of the complement inhibitor, CD55, and increases the susceptibility of malignant B cells to lysis by rituximab (Di Gaetano et al. 2001).
- Rituximab reduces the IC_{30} and IC_{50} of cladribine, doxorubicin, mitoxantrone and bendamustine necessary for induction of apoptosis in neoplastic lymphocytes (Chow et al. 2002).

13.1.4 Preclinical Studies

The pharmacology and toxicology profiles of rituximab have been evaluated in preclinical studies in macaque cynomolgus monkeys (Reff et al. 1994). Rituximab was administered by IV injection, resulting in potent but reversible B-cell depletion, with no notable toxic effects. The maximum single dose administered was 100 $mg\,kg^{-1}$, and the highest repeat dose was 20 $mg\,kg^{-1}$ weekly for 8 weeks. There were no significant toxicological effects at any of the doses or schedules studied, and no animal died as a result of treatment with rituximab. Depletion of B cells – the intended pharmacological effect of rituximab – was observed in all animals studied, with >95% depletion after the first or second dose. B-cell recovery occurred over a variable period of time; partial recovery occurred most commonly after 4 to 8 weeks, while recovery to baseline levels required a minimum of approximately 3 months.

13.1.5 Pharmacokinetic Studies

Extensive pharmacokinetic (PK) studies of rituximab have been carried out in patients with NHL (Onrust et al. 1999). After administration of rituximab by IV infusion at a dose of 375 $mg\,m^{-2}$ once weekly for 4 weeks, peak plasma concentration (C_{max}) was 465 $mg\,L^{-1}$ (Berinstein et al. 1998). Clearance of rituximab decreased following multiple infusions, from 38.2 $mL\,h^{-1}$ after a single infusion to 9.2 $mL\,h^{-1}$ after the fourth.

Accumulation of rituximab occurs, with the area under the concentration–time curve (AUC) increasing from 16320 $mg\,L^{-1}\,h$ after the first infusion to 86125 $mg\,L^{-1}\,h$ after the fourth (Berinstein et al. 1998). Importantly, the accumulation of rituximab is not accompanied by any increase in toxicity.

The terminal elimination half-life ($t_{1/2}$) of rituximab increased from 3.2 days after a single infusion to 8.6 days following a fourth infusion, thus confirming accumulation of the antibody (Berinstein et al. 1998). At 3 months after completion of therapy, rituximab was still detectable at a median concentration of 20.3 $mg\,L^{-1}$ (range: 0.0 to 96.8 $mg\,L^{-1}$) in 63% of 116 evaluable patients, and remained detectable in 11.2% of these patients after 6 months (Berinstein et al. 1998). Ongoing studies suggest a terminal half-life of approximately 3 weeks – somewhat longer than had been previously assumed (Data on file, Roche).

Serum levels of rituximab have been shown to be dose-dependent over the dose range of 100 to 500 $mg\,m^{-2}$, given as an IV infusion (Maloney et al. 1994). An analysis of B-cell depletion showed the peripheral B lymphocyte count to be reduced by approximately 90% within 3 days of a single infusion of rituximab at a dose of 250 or 500 $mg\,m^{-2}$ (Maloney et al. 1994). The clinical response to rituximab has been shown to correlate with serum concentrations of the antibody (Berinstein et al. 1998; Maloney et al. 1997a), with responders having significantly higher median serum concentrations than nonresponders throughout the 4-week treatment period ($p < 0.01$).

A study of the feasibility and efficacy of PK-based maintenance dosing of rituximab showed that serum drug levels increased appropriately after a single bolus for most patients, and were maintained for several months afterwards (Gordan et al. 2005). The majority of patients studied (28 of 29) required no more than three additional single-dose rituximab infusions in 12 months of follow-up, with monthly serum level monitoring to maintain a level at or above an arbitrarily defined threshold of 25 $mg\,L^{-1}$. A trend was noted for higher rituximab levels in patients considered to be responders (Gordan et al. 2005).

13.2
Rituximab Clinical Data in NHL and CLL

13.2.1
Overview of NHL and CLL

NHL is a heterogeneous group of malignancies originating in lymphoid tissue, with 90% being of B-cell origin. Data from Surveillance Epidemiology and End Results (SEER) reveal an age-adjusted incidence of 19.1 per 100000 men and women per year. Notably, the median age at diagnosis is 66 years, meaning that many NHL patients may be frail and/or suffering from comorbid disease (see link on SEER website, National Cancer Institute, http://seer.cancer.gov). The incidence of NHL has doubled over the past two decades in most westernized countries. Whilst improved cancer reporting and changes in disease classification may have contributed to this dramatic increase, other etiological factors are clearly important. These are thought to include congenital or acquired immunosuppression, exposure to infectious organisms (including human immunodeficiency virus, HIV), *Helicobacter pylori* and Epstein–Barr virus (EBV), as well as genetic susceptibility and exogenous factors including ultraviolet radiation, pesticides and hair dyes (Fisher and Fisher 2004). Recently, stable incidences in Western developed countries have been observed, suggesting a decline of the 1 to 2% increase in incidence observed during the 1960s to 1990s.

The cellular classification [using the World Health Organization (WHO) modification of the Revised European American Lymphoma (REAL) Classification] divides the disease into more than 20 clinicopathological entities (Pileri et al. 1998). A simplified version of the WHO Classification, together with the incidence of these types of NHL reported by the Non-Hodgkin's Lymphoma Classification Project (1998) is provided in Table 13.1. The most frequent 11 diseases occur in more than 2% of patients, and all the remaining categories occur with a frequency of less than 1%.

A clinically useful division of the cellular classification into prognostic groups is frequently used. Indolent lymphomas comprise 25 to 40% of NHLs, with follicular lymphoma (FL) being the most common type. The marginal zone lymphomas (MZL), a heterogeneous group of disorders consisting of mucosa-associated lymphoid tissue (MALT) lymphoma, nodal MZL and splenic MZL, account for approximately 8% of NHLs. Small lymphocytic lymphoma (SLL) is considered to have the same morphologic and immunophenotypic features as CLL (see below), and to constitute the same disease entity. Lymphoplasmacytic lymphoma (Waldenström's macroglobulinemia, WM) is a rare indolent lymphoma characterized primarily by the infiltration of lymphophoplasmacytic cells into bone marrow and the demonstration of an IgM monoclonal gammopathy.

Indolent NHL is typically characterized by a remitting and relapsing course, and median survival after first relapse is 4 to 5 years (Johnson et al. 1995). Transformation to an aggressive histological subtype may occur at any stage of the disease, and frequently has a fatal outcome (Yuen et al. 1995). The majority of

Table 13.1 REAL Classification of non-Hodgkin's lymphoma: subtypes occurring in at least 2% of patients in non-Hodgkin's Lymphoma Classification Project (Armitage and Weisenburger 1998).

Category	*Frequency [%]*
1 Diffuse large B-cell lymphoma (DLBCL)	31
2 Follicular lymphoma (FL)	22
3 Peripheral T-cell	6
4 Small lymphocytic lymphoma (SLL); Mantle cell lymphoma (MCL)	6
5 Marginal zone, mucosal-associated lymphoid tissue (MALT)	5
6 Primary mediastinal large B-cell (PMBCL)	2
7 Anaplastic large-T/null cell	2
8 High-grade B-cell, Burkitt-like	2
9 Lymphoblastic lymphoma	2

patients present with advanced disease, and the decision to treat is based on a number of factors including symptoms, rate of disease progression, presence of bulky disease, cytopenia or impaired end-organ function as a consequence of lymphoma, as well as patient and physician preference. Advanced-stage FL, MALT lymphoma and nodal MZL are generally managed in a similar way; treatment options include single-agent or combination chemotherapy, single-agent rituximab, rituximab in combination with chemotherapy, or high-dose chemotherapy with autologous or allogeneic stem cell support. Therapy should be individualized, taking into account the patient's age, extent of disease, comorbid conditions and the goals of therapy. Splenectomy is often the treatment of choice for splenic MZL, and the sole initial treatment for gastric MALT lymphoma confined to the gastric wall should be eradication of *H. pylori* with antibiotics (Bertoni and Zucca 2005). Treatment options for WM include single-agent or combination chemotherapy, with or without rituximab (Treon et al. 2006). SLL is managed in the same way as CLL (see below).

The majority (60–75%) of NHLs are aggressive, with the most common type being diffuse large B-cell lymphoma (DLBCL). DLBCL may be cured in a significant percentage of patients, depending on baseline tumor and patient characteristics; however, therapeutic challenges persist, especially for patients with high-risk disease (Coiffier 2005). Primary mediastinal B-cell lymphoma (PMBCL) is a DLBCL that arises in the thymus and mainly affects young adults. Burkitt's and Burkitt-like lymphoma (BL/BLL) are aggressive B-cell malignancies characterized by a rapid proliferative rate and a propensity for extranodal sites of involvement such as the gastrointestinal tract and central nervous system (CNS). Lymphoblastic lymphoma (LBL) is a rare subtype of aggressive NHL with biological features similar to those of acute lymphoblastic leukemia (ALL). In the majority of cases, LBL shows a T-cell phenotype, and mediastinal tumors are the most frequent manifestation. Peripheral T-cell lymphomas – including precursor T-lymphoblastic lymphoma and anaplastic large-cell lymphoma T/null-cell types – represent a heterogeneous group of diseases, most of which have disappointing cure rates.

Mantle cell lymphoma (MCL) accounts for approximately 6–8% of all lymphoma diagnoses; patients usually respond well to induction therapy but have a poor median survival of only 3 to 4 years (Witzig 2005). There is no established standard of care in MCL, and patients with this disease should be treated in clinical trials wherever possible.

Chronic lymphocytic leukemia (CLL) is the most common form of leukemia in Western countries, with most cases occurring in elderly patients. The age-adjusted incidence rate according to SEER data is 3.6 cases per 100000 men and women per year, and the median age at diagnosis is 72 years (see link on SEER website, National Cancer Institute, http://seer.cancer.gov). CLL follows a variable course, with survival ranging from months to decades. Recently, considerable progress in the identification of molecular and cellular prognostic markers has been made; in particular, the mutational profile of Ig genes and some cytogenetic abnormalities have been shown to be strongly predictive of clinical outcomes (Dighiero 2005).

13.2.2 Rituximab plus Chemotherapy Induction Therapy in Indolent NHL

For many years, patients with advanced symptomatic indolent NHL were treated either with single-agent chemotherapy (e.g., chlorambucil or cyclophosphamide) or combination chemotherapy [e.g., cyclophosphamide, vincristine, prednisone (CVP) or cyclophosphamide, doxorubicin, vincristine, prednisone (CHOP)]. Despite the exploration of many different regimens, chemotherapy has had no major impact on survival in indolent NHL. The use of autologous stem cell transplant (ASCT) in relapsed/refractory indolent NHL may prolong disease-free survival (DFS), but its value in this setting remains uncertain due to the lack of conclusive phase III data. The use of ASCT as consolidation therapy in the first remission of advanced-stage FL indicates that improvement in time to disease progression (TTP) and even overall survival (OS) are possible, although an increase in secondary hematologic neoplasms can occur (Hiddemann 2005a). However, age and comorbidity may preclude the use of ASCT in many patients with indolent NHL.

13.2.2.1 Rituximab plus Chemotherapy in Previously Untreated Indolent NHL

Based on their different modes of action, promising *in-vitro* data and nonoverlapping toxicity profiles, rituximab and chemotherapy combinations were predicted to have additive and even synergistic efficacy in indolent NHL. This hypothesis was first examined in phase II trials and subsequently in phase III trials

(Table 13.2). The majority of investigators have examined the concurrent administration of chemotherapy and rituximab, although several groups have assessed sequential rituximab following chemotherapy. Most of the earliest trials employed four infusions of rituximab, either alone, concurrently with or sequentially after chemotherapy. As it became clear that an increase in the number of infusions of rituximab led to an increase in efficacy with a similar toxicity profile, studies using up to eight infusions of rituximab in combination with chemotherapy were initiated.

13.2.2.1.1 Concurrent Rituximab and Chemotherapy: Phase II Studies

The earliest phase II trial was initiated in 1996, and examined the safety and efficacy of six doses of rituximab together with six cycles of CHOP chemotherapy in 40 patients with indolent NHL, 31 of whom were previously untreated (Czuczman et al. 1999). All 38 evaluable patients responded, with 55% achieving a complete response (CR) using protocol-defined response criteria (Czuczman et al. 1999). Updated response rates based on the standard criteria used to assess response in lymphoma (Cheson et al. 1999) were as follows: ORR 100%; CR/unconfirmed complete response (CRu) 87%; partial response (PR) 13% (Czuczman et al. 2004). Long-term follow-up has revealed these responses to be highly durable, with the median TTP being almost 7 years. Sixteen patients remained in remission at 6 to 9 years after treatment (Czuczman et al. 2004).

A number of other phase II studies subsequently examined rituximab in combination with a variety of chemotherapeutic regimens (Table 13.2). Combining rituximab with other cyclophosphamide-based regimens [e.g., mitoxantrone and cyclophosphamide (CM) and cyclophosphamide, mitoxantrone, vincristine and prednisone (CNOP) (Economopoulos et al. 2003; Emmanouilides et al. 2003)] has resulted in high ORR and CR rates (Table 13.2), and these regimens have been well tolerated. A recently updated phase II trial conducted in Mexico randomized patients to receive either six infusions of rituximab, six cycles of CNOP or six cycles of R-CNOP (Rivas-Vera et al. 2005). There were no significant differences in the ORR and CR rates between the various arms, and median TTP and OS were not reached in all three treatment groups after a median of 24 months' follow-up. Infections were more common with R-CNOP (15%) than CNOP (6%) or rituximab alone (5%). The authors noted that their study included fewer patients with Stage IV disease than many other trials of rituximab and chemotherapy, which might explain the apparent lack of benefit of R-CNOP over CNOP alone (Rivas-Vera et al. 2005).

A study of seven infusions of rituximab plus six cycles of the purine analog fludarabine in patients with low-grade or FL enrolled 40 subjects, including 27 chemotherapy-naïve patients. An ORR of 90% (80% CR) was seen in the intention-to-treat (ITT) population, with similar outcomes observed both in chemotherapy-naïve and in relapsed patients (Czuczman et al. 2005). The median duration of response (DR), TTP and OS were not reached after a median follow-up of 44 months. Following unexpected hematological toxicities in the first 10 patients studied, a change in study design was implemented: prophylactic

Table 13.2 Trials of rituximab (R) in induction therapy regimens in previously untreated indolent non-Hodgkin's lymphoma (NHL).

Reference	*Regimen of R + other agents × no. of cycles*	*Evaluable patients*	*ORR*	*CR*	*PR*	*Median TTP*[a]	*Median OS*
Rituximab with concurrent chemotherapy phase II trials							
Czuczman et al. (1999, 2004)	R × 6 plus CHOP × 6	29 UT + 9 PT	100	87	13	82.3 months	
Czuczman et al. (2005)	R × 7 plus F × 6	27 UT + 13 PT	90	80	10	44+ months	44+ months
Hainsworth et al. (2005a)	R × 4 then R × 3 plus CHOP or CVP × 3 then R × 2	86 UT	93	55	38	4-year PFS: 62%	3-year OS: 95%
Martinelli et al. (2003)	R × 4 plus Chl × 6 wk, then R × 4 plus Chl × 4	15 UT + 12 PT	89	63	26		
McLaughlin et al. (2005)	R × 6 plus FND × 6	161 UT	100	88	12	4-year FFS: 70%	5-year OS: 89%
	FND × 6 then R (randomization)		96	85	11	4-year FFS: 59%	5-year OS: 86%
Economopoulos et al. (2003)	R × 6 plus CNOP × 6	42 UT	90	71	19	19.5+ months	19.5+ months
Rivas-Vera et al. (2005)	R × 6 plus CNOP × 6	144 UT	90	66	24	24+ months	24+ months
	CNOP × 6		83	63	22	24+ months	24+ months
	R × 6 (randomization)		85	49	36	24+ months	24+ months
Drapkin et al. (2003)	R × 1 then R plus P × 4 then R × 1 then R plus P × 4	43 UT + 16 PT	72	26	46	15 months	25+ months
Di Bella et al. (2005)	R plus P × 2 plus M × 1 then R × 2 plus P × 2 plus M (for total max 10 cycles)	24 UT	83	45	38	14 months	81% at 18 months

Table 13.2 *Continued*

Reference	*Regimen of R + other agents × no. of cycles*	*Evaluable patients*	*ORR*	*CR*	*PR*	*Median TTP*[a]	*Median OS*
Phase III trials							
Marcus et al. (2005); Solal-Celigny et al. (2005)	R × 8 plus CVP × 8	162 UT	81***	41***	40	33.6 months***	3-year OS: 89%
	CVP × 8 (randomization)	159 UT	57	10	47	14.5 months	3-year OS: 81%
Herold et al. (2004, 2005)	R × 8 plus MCP × 8	105 UT	92***	50***	42	2.5 year PFS:82%***	2.5 year OS: 89%**
	MCP × 8 (randomization)	96 UT	75	25	50	2.5 year PFS: 51%	2.5 year OS: 76%
Hiddemann et al. (2005b)	R × 6–8 plus CHOP × 6–8	222 UT	96*	20	76	Not reached***	2-year OS: 95%*
	CHOP × 6–8 (randomization)	205 UT	90	17	73	31 months (TTF)	2-year OS: 90%
Salles et al. (2004)	R × 6 plus CHVP-IFN × 6	184 UT	94***	76	18	2.5-year EFS: 78%**	
	CHVP-IFN × 6 (randomization)	175 UT	85	49	36	2.5-year EFS: 62%	
Chemotherapy followed by sequential rituximab phase II trials							
Cohen et al. (2002)	F plus C × 4–6, then R × 4	33 UT	88	85	3		
Emmanouilides et al. (2003)	C plus M × 2, then R plus M × 4	22 UT, 10 PT	90	72	19	30 (actuarial) months	
Gregory et al. (2002)	F plus M × 4–6, then R × 4	31 UT	97	45	52		
Jaeger et al. (2002)	CHOP × 3–8, then R × 4	41 UT	100	88	12		
Rambaldi et al. (2002, 2005)	CHOP × 6, then R × 4	77 UT	82	69	13		
Vitolo et al. (2004)	FND × 4, then R × 4	70 UT	90	83	7	3-year FFS: 50%	
Zinzani et al. (2004)	FM × 6, then R × 4[b]	72 UT	96	90	6	19+ months	19+ months
	CHOP × 6, then R × 4[b] (randomization)	68 UT	96	80	16	19+ months (PFS)	19+ months

Rituximab and immune system modulators						
McLaughlin et al. (2005)	R × 4 plus GM-CSF	14 UT + 25 PT	79	36	43	
Kimby et al. (2002)	R × 4 (induction) then R × 4	81 UT +	56	11	44	
	or R × 4 plus IFN-α-2a	45 PT	78	22	56	
	(randomization)	36	94	48	45	
		33				
Rituximab monotherapy						
Colombat et al. (2001); Solal-Celigny et al. (2004)	R × 4	49 UT	73	27	47	18 months (PFS)
Witzig et al. (2005)	R × 4	36 UT	72	36	36	2.2 years
Hainsworth et al. (2002)	R × 4, then further R × 4 at 6-month intervals for a total of 4 courses	60 UT	47	7	40	
Conconi et al. (2003)	R × 4	23 UT + 11 PT[c]	73	44	29	14.2 months (TTF)

All doses of R 375 $mg\,m^{-2}$.

a Or other parameter as indicated.

b R only given to patients with a bcl-2-positive CR or a PR after chemotherapy.

c Patients with extranodal marginal zone B-cell lymphoma of MALT type.

* $p < 0.05$; **$p < 0.01$; ***$p < 0.001$ versus comparator.

C = cyclophosphamide; Chl = chlorambucil; CR = complete response; CHOP = cyclophosphamide, doxorubicin, vincristine, prednisolone; CHVP-IFN = cyclophosphamide, doxorubicin, etoposide, prednisone, interferon-α-2a; CNOP = cyclophosphamide, mitoxantrone, vincristine, prednisolone; CLL = chronic lymphocytic leukemia; CVP = cyclophosphamide, vincristine, prednisolone; EFS = event-free survival; F = fludarabine; FFS = failure-free survival; FM = fludarabine, mitoxantrone; FND = fludarabine, mitoxantrone, dexamethasone; GM-CSF = granulocyte-macrophage colony-stimulating factor; HDMP = high-dose methyl prednisolone; M = mitoxantrone; MCP = mitoxantrone, chlorambucil, prednisolone; ORR = overall response rate; P = pentostatin; PFS = progression-free survival; PT = previously treated; PR = partial response; OS = overall survival; TTF = time to treatment failure; TTP = time to progression; UT = previously untreated.

trimethoprim/sulfamethoxazole was discontinued, the fludarabine dose was decreased by 40% in cases of prolonged cytopenia, and growth factor support was not used as prophylaxis. The hematological toxicity profile was improved in subsequently accrued patients, although grade 3/4 neutropenia was common (71%) and 24% of patients required transient growth factor support.

Pentostatin is another nucleoside analog which is less myelosuppressive than fludarabine. In a phase II trial of rituximab and pentostatin, responses were seen in 83% of previously untreated patients and 63% of pretreated patients. The median TTP was 15 months, and neutropenia was the only adverse event seen in ≥10% of patients (Drapkin et al. 2003). A further trial examined the use of these two products together with mitoxantrone (PMR) in the first-line setting. The ORR was 83% and the median response duration 10.0 months. Toxicities were increased with the PMR regimen compared with the PR regimen, with Grade 3/4 neutropenia occurring in 67% of patients. Sepsis and febrile neutropenia occurred in 8% and 17% of patients, respectively (Di Bella et al. 2005).

The safety and efficacy of rituximab in combination with chlorambucil was examined in newly diagnosed (n = 15) and relapsed/refractory (n = 14) low-grade and FL. Only one patient was withdrawn from the study due to progressive disease. No major neutropenic-related infections were observed, and no transfusion or growth factor support was required (Martinelli et al. 2003). This combination is currently being evaluated in a phase III trial by the International Extranodal Lymphoma Study Group (IELSG) in MALT lymphoma.

A phase II trial conducted between 1997 and 2002 randomized patients with previously untreated Stage IV indolent lymphoma and demonstrable bcl-2 rearrangement to either concurrent rituximab (six doses) and eight cycles of fludarabine, mitoxantrone and dexamethasone (FND) chemotherapy, or FND followed sequentially by rituximab (McLaughlin et al. 2005a). Maintenance interferon (IFN) was subsequently given in both treatment arms. There were no significant differences in ORR or CR rates between the two arms, but the concurrent regimen was associated with significantly higher molecular responses both at 6 and 12 months post therapy (89 versus 60%, $p < 0.01$ and 89 versus 68%, $p = 0.01$, respectively). Both regimens were well tolerated (McLaughlin et al. 2005a).

13.2.2.1.2 Chemotherapy and Sequential Rituximab: Phase II Studies

The sequential administration of fludarabine, mitoxantrone and dexamethasone (FND) and rituximab was evaluated in the US trial described above (McLaughlin et al. 2005a) and in an Italian study in elderly patients (aged >60 years) with advanced-stage FL (Vitolo et al. 2004). In this study, which assessed four cycles of FND followed by four infusions of rituximab, the ORR was 90%, with 83% CR. Patients with adverse prognostic features, such as bone marrow involvement, a poor International Prognostic Index (IPI) score and bulky disease responded as well to therapy as patients with a more favorable prognosis. The regimen was well tolerated, with only three patients experiencing Grade 3/4 infections (Vitolo et al. 2004).

Several groups have examined the sequential administration of CHOP chemotherapy and rituximab (Table 13.2) (Jaeger et al. 2002; Zinzani et al. 2004; Rambaldi et al. 2005). Sequential rituximab improved response status, both in clinical and molecular terms in all studies. The trial conducted by Zinzani and colleagues randomized patients to either CHOP or fludarabine/mitoxantrone (FM) induction, followed in both cases by four doses of rituximab for patients who failed to achieve a bcl-2/IgH negative CR (CR–). The final CR– rate was higher in the FM arm than the CHOP arm, although no statistically significant differences in progression-free survival (PFS) or OS between treatment groups were observed (Zinzani et al. 2004). Gregory et al. (2002) have also shown that FM followed by rituximab is an effective treatment option, while the use of fludarabine and cyclophosphamide (FC) with sequential rituximab is another feasible therapeutic approach (Cohen et al. 2002).

Hainsworth and colleagues evaluated 4-weekly infusions of rituximab monotherapy followed by a short (three-cycle) course of CHOP or CVP and then two further infusions of rituximab in stage II–IV follicular lymphoma (Hainsworth et al. 2005a). Actuarial PFS at 4 years was 62%, and rates of neutropenia or fever were low (7%). A longer follow-up is required to determine whether the combination of rituximab with a shorter course of chemotherapy is as effective as longer-duration chemotherapy regimen combined with rituximab. This therapeutic approach could be particularly attractive for elderly or frail patients who tolerate chemotherapy poorly.

13.2.2.1.3 Chemotherapy plus Rituximab: Phase III Studies

Following the promise demonstrated by rituximab–chemotherapy combinations in phase II trials in indolent NHL, a number of prospective randomized phase III studies were initiated (Table 13.2). These studies were designed to evaluate the effect of the addition of rituximab to standard first-line chemotherapeutic regimens: CHOP (Hiddemann et al. 2005b), CVP (Marcus et al. 2005), mitoxantrone, chlorambucil, prednisolone (MCP) (Herold et al. 2004, 2005) or cyclophosphamide, doxorubicin, etoposide, prednisone plus interferon (CHVP-IFN) (Salles et al. 2004).

In the international trial evaluating the addition of eight cycles of rituximab to CVP chemotherapy, overall and complete response rates were highly significantly improved in the immunochemotherapy arm compared with the chemotherapy alone arm (81 versus 57%, $p < 0.0001$ and 41 versus 10%, $p < 0.0001$, respectively) (Marcus et al. 2005). A recently presented update after a median of 42 months' follow-up reveals that the increased response seen with R-CVP translates into highly durable and significant improvements in median TTP (33.6 versus 14.5 months, $p < 0.0001$) (Fig. 13.3), median time to next lymphoma treatment (TNLT) (46.3 versus 12.3 months, $p < 0.0001$), median DR (37.7 versus 13.5 months, $p < 0.0001$) and median DFS (44.8 versus 20.5 months, $p = 0.0005$) (Solal-Celigny et al. 2005). Median OS was not reached in both arms, but showed a strong trend

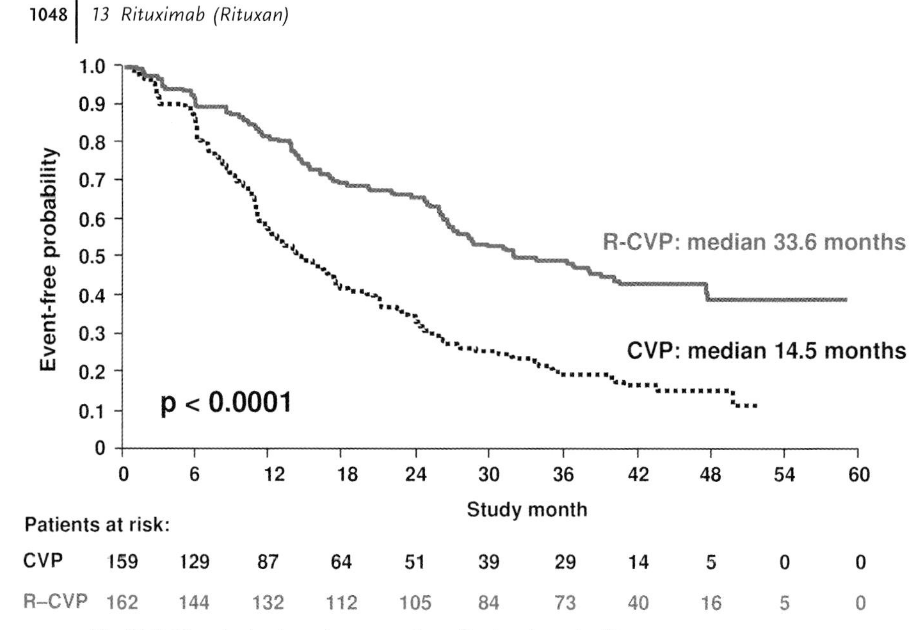

Fig. 13.3 Time to treatment progression after treatment with eight cycles of cyclophosphamide, vincristine, prednisone (CVP) chemotherapy or eight cycles of rituximab plus CVP chemotherapy in previously untreated follicular lymphoma. (Reproduced with permission from Solal-Celigny et al. 2005.)

towards an overall survival benefit for the rituximab-containing arm ($p = 0.0553$), while 3-year OS rates were estimated to be 81% in the CVP arm and 89% in the R-CVP arm. Notably, significantly more patients died due to lymphoma progression after CVP (25 deaths) than after R-CVP (12 deaths, $p = 0.02$) (Solal-Celigny et al. 2005).

A multivariate Cox regression analysis of outcomes according to baseline prognostic factors revealed that a significant improvement in outcomes was achieved with the addition of rituximab to CVP therapy, regardless of baseline patient or disease characteristics (Imrie et al. 2005). The incidence of Grade 3/4 neutropenia was higher during treatment with R-CVP (24%) than CVP (14%), but the incidence of infections and neutropenic sepsis was similar in the two groups (Marcus et al. 2005). The results of this trial led to EU regulatory approval of the R-CVP combination for first-line treatment of FL in 2004.

The German Low-Grade Lymphoma Study Group (GLSG) has conducted a randomized comparison of six or eight cycles of CHOP and R-CHOP in advanced-stage indolent FL (Hiddemann et al. 2005b). R-CHOP was superior with regard to ORR (96 versus 90%, $p = 0.011$) and TTF ($p < 0.001$). Importantly, OS was also significantly prolonged after immunochemotherapy compared with chemotherapy ($p = 0.016$). There was a 10% increase in granulocytopenia with R-CHOP compared with CHOP, but rates of infections and other therapy-associated adverse events were similar (Hiddemann et al. 2005b).

A further study conducted by Herold and colleagues in the Ostdeutsche Studiengruppe für Hämatologie und Onkologie (OSHO) evaluated the addition of eight cycles of rituximab to mitoxantrone, chlorambucil and prednisolone (MCP) chemotherapy in the first-line treatment of advanced FL (Table 13.2) (Herold et al. 2004, 2005). Once again, the addition of rituximab to chemotherapy significantly improved ORR (92 versus 75%, $p < 0.0001$) and CR rates (50 versus 25%, $p < 0.0003$) (Herold et al. 2004, 2005). After a median of 30 months' follow-up, this translated into significantly improved rates of event-free survival (EFS) (79 versus 44%, $p < 0.0001$), PFS (82 versus 51%, $p < 0.0001$) and, most importantly, OS (89 versus 76%, $p = 0.007$) (Fig. 13.4). Both R-MCP and MCP were well tolerated (Herold et al. 2004, 2005).

In the fourth randomized phase III trial, adding rituximab to the CHVP-IFN regimen led to a significant increase in ORR and CR/CRu rates (94 versus 85% and 76 versus 49%, $p < 0.0001$) (Salles et al. 2004). After a median follow-up of 30 months, the estimated 2.5-year EFS was 62% with CHVP-IFN and 78% with R-CHVP-IFN ($p = 0.003$) (Salles et al. 2004).

Taken together, these results show that rituximab has a significantly beneficial effect when added to initial chemotherapy in patients with advanced FL. Subanalyses indicate that the benefits are experienced irrespective of baseline characteristics, suggesting that an immunochemotherapy approach may overcome adverse prognostic factors in FL. Further studies are needed to establish optimal regimens for different types of patients with indolent NHL.

13.2.2.2 Rituximab plus Chemotherapy in Relapsed/Refractory Indolent NHL

Patients with relapsing/refractory indolent NHL have been traditionally treated with further courses of chemotherapy, which may include the use of non-cross-

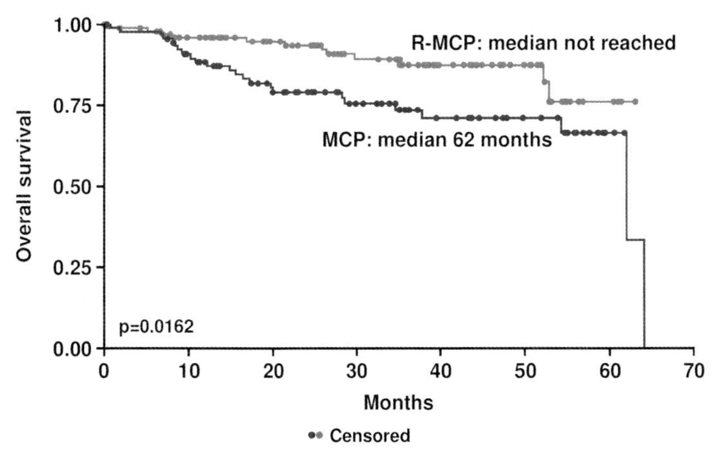

Fig. 13.4 Overall survival after treatment with eight cycles of mitoxantrone, chlorambucil, prednisolone (MCP) chemotherapy or eight cycles of rituximab plus MCP chemotherapy in previously untreated follicular lymphoma. (Reproduced with permission from Herold et al. 2004.)

resistant drugs not typically used during previous treatment. Younger, fitter patients may be eligible for ASCT and, in a small number of cases, for an allogeneic stem cell transplant (SCT). The combination of rituximab and chemotherapy is another therapeutic strategy in this setting.

13.2.2.2.1 Rituximab and Chemotherapy: Phase II Trials

As discussed above, phase II trials including patients with relapsed/refractory indolent NHL have assessed the safety and tolerability of a variety of regimens including R-CHOP (Czuczman et al. 1999, 2004), R-CNOP (Economopoulos et al. 2003), R-F (Czuczman et al. 2005), R-CM (Emmanouilides et al. 2003), R-pentostatin (Drapkin et al. 2003), and R-chlorambucil (Martinelli et al. 2003) (Table 13.2). Other investigators have also conducted phase II trial of rituximab together with chemotherapy in the relapsed setting (Table 13.3). Sacchi et al. (2003) assessed four cycles of FCR in 39 previously treated patients: ORR and CR rate according to ITT analysis were 94% and 75%, respectively and median duration of response after a median follow-up of 20 months was 26 months. There were 20 cases of Grade 3/4 neutropenia and 12 patients required treatment delays, indicating that this regimen requires careful management (Sacchi et al. 2003).

Rituximab has also been administered in combination with bendamustine (BR) (Rummel et al. 2005) and bendamustine and mitoxantrone (BMR) (Weide et al. 2004) in patients with relapsed/refractory indolent NHL (Table 13.3). These regimens are associated with high ORR and CR rates and, as expected, myelosuppression appears to be the major toxicity (Rummel et al. 2005; Weide et al. 2004). Other rituximab–chemotherapy regimens which have shown promise in phase II trials in this setting are R-CVP (Garcia-Conde et al. 2000), R-CHOP (Domingo-Domenech et al. 2002), and rituximab with vincristine and 5-day cyclophosphamide (Lazzarino et al. 2005) (Table 13.3).

13.2.2.2.2 Rituximab and Chemotherapy: Phase III Trials

In 1999, a phase III intergroup trial coordinated by the European Organization for Research and Treatment of Cancer (EORTC) Lymphoma Group was initiated by a number of groups in Europe, Canada, and Australasia. The EORTC 20891 study was designed to investigate first the effect of the addition of six doses of rituximab to six cycles of CHOP chemotherapy on the rate and quality of response in patients with relapsed FL, and second the effect of rituximab maintenance therapy (a single infusion every 3 months for 2 years) on progression-free survival (van Oers et al. 2006) (Table 13.3).

Patients who received R-CHOP induction experienced significantly improved ORR and CR rates compared with those who received CHOP induction (85 versus 72%, 29.5 versus 16%, $p < 0.0001$). This translated into significantly lengthened median PFS with R-CHOP compared with CHOP (33.1 versus 20.2 months, $p = 0.0003$, $HR = 0.65$) (Table 13.3). Rates of Grade 3/4 neutropenia were similar with R-CHOP and CHOP (54.7% and 48.2%, respectively). More patients on R-CHOP experienced grade 3 to 4 allergy (8 versus 0 individuals) or skin reactions

Table 13.3 Trials of rituximab (R) in induction therapy regimens in relapsed/refractory indolent NHL.

Reference	*Regimen of R + other agents × no. of cycles*	*Evaluable patients*	*ORR*	*CR*	*PR*	*Median TTP*[a]	*Median OS*
Rituximab with concurrent chemotherapy phase II trials							
Domingo-Domenech et al. (2002)	R × 6 plus CHOP × 6	16 PT	88	75	13		
Garcia-Conde et al. (2000)	R × 6 plus CVP × 6	32 PT	81	50	31		
Sacchi et al. (2003)	R × 4 plus F and C × 4	39 PT	94	75	19		
Rummel et al. (2005)	R × 1 then R plus B × 4 then R × 1	24 PT	96	71	25		
Weide et al. (2004)	R × 4 plus B and M × 4	19 PT	100	59	41	9+ months (PFS)	
Lazzarino et al. (2005)	P × 4 plus C and V × 4	29 PT	65	55	10	16.1 months (EFS)	
phase III trials							
van Oers et al. (2006)	R × 6 plus CHOP × 6	231 UT	85	29.5***	56	33.1 months***	3-year OS: 82.5%
	CHOP × 6 (randomization)	230 UT	72	16	57	20.2 months (PFS)	3-year OS: 71.9%
Forstpointner et al. (2004);	R × 4 plus FCM × 4	67 UT	96*	39*	57	3.9 years*	4-year OS: 74%*
Dreyling et al. (2005a)	FCM × 4 (randomization)		71	23	48	1.7 years (PFS)	median 3.8 years
Rituximab and immune system, modulators: phase I/II trials							
Davis et al. (2000a)	R × 4 plus IFN-α-2a	38 PT	45	11	34	8.95 months	
Sacchi et al. (2001)	R × 4 plus IFN-α	64 PT	70	33	37	19 months (DR)	3-year OS: 80%
Friedberg et al. (2002)	R × 4 plus IL-2	20 PT	55	5	50	13+ months (PFS: responders)	
Ansell et al. (2003)	R × 2 then R plus IL-2 × 2	43 PT + UT	67	25	42		
van der Kolk et al. (2003)	R × 4 plus G-CSF	26 PT	42	16	26	24 months	
Gluck et al. (2004)	R × 4 plus IL-2 daily	17 PT	29	6	24	14.9 months	
	R × 4 plus IL-2 thrice weekly	13 PT	46	38	23	16.1 months (responders)	

Table 13.3 *Continued*

Reference	*Regimen of R + other agents × no. of cycles*	*Evaluable patients*	*ORR*	*CR*	*PR*	*Median TTP*[a]	*Median OS*
Rituximab monotherapy							
Maloney et al. (1997b)	R × 4	37 PT	46	8	38	10.2 months	
McLaughlin et al. (1998)[b]	R × 4	151 PT	50	6	44	12.5 months	
Davis et al. (1999)	R × 4	28 PT[c]	43	4	39	8.1 months (responders)	
Davis et al. (2000b)	R × 4	57 PT	40	11	30	16.3 months	
Cortes-Funes et al. (2000)	R × 4	31 PT	62	23	39		
Feuring-Buske et al. (2000)	R × 4	30 PT	47	17	30	201 days	
Foran et al. (2000a)	R × 4	70 PT	46	3	43		
Igarashi et al. (2002)	R × 4	61 PT	61	23	38	245–376 days	
Martinelli et al. (2005)	R × 4	26 PT + UT[d]	77	46	31		28+ months
Piro et al. (1999)	R × 8	35 PT	60	14	46	19.4+ months (responders)	

All doses of R 375 $mg\,m^{-2}$.

a Or other parameter as indicated.

b Reanalysis according to Cheson criteria revealed ORR 56% and CR 32% (Grillo-Lopez et al. 2000).

c All patients had bulky disease (lesions >10 cm in diameter).

d All patients had histologically-proven gastric MALT lymphoma.

* $p < 0.05$; **$p < 0.01$; ***$p < 0.001$ versus comparator.

B = bendamustine; DR = duration of response; FCM = fludarabine, cyclophosphamide, mitoxantrone; G-CSF = granulocyte colony-stimulating factor; IFN = interferon; IL-12 = interleukin-12; V = vincristine. Other abbreviations as Table 13.2.

(31 versus 17), but rates of withdrawd from treatment because of taxicity were similar between groups (van Oers et al. 2006).

In a phase III trial conducted by the GLSG, a significant improvement in ORR and CR rate (96 versus 71%, p = 0.011 and 39 versus 23%, p = 0.013, respectively) occurred when rituximab was added to a regimen of fludarabine, cyclophosphamide and mitoxantrone (FCM) in patients with recurrent FL (Table 13.3) (Dreyling et al. 2005a; Forstpointner et al. 2004). Progression-free and overall survival were also significantly extended after R-FCM therapy compared with FCM therapy (median: 3.9 versus 1.7 years, p = 0.029 and 74% at 4 years versus a median of 3.8 years, p = 0.033) (Dreyling et al. 2005a).

13.2.2.3 Meta-Analysis of Rituximab and Chemotherapy in Indolent NHL

A meta-analysis to determine the effectiveness of combination chemotherapy plus rituximab versus combination chemotherapy alone, with respect to overall survival in indolent lymphoma, has been performed by the Cochrane Group (Schulz et al. 2005a). Medical databases were searched to identify randomized controlled trials of immunochemotherapy versus chemotherapy in this setting. Six eligible trials were identified as follows: rituximab plus CVP versus CVP alone in FL (Marcus et al. 2005; Solal-Celigny et al. 2005); rituximab plus MCP versus MCP alone in FL and MCL (Herold et al. 2004, 2005); rituximab plus CHOP versus CHOP alone in FL (Hiddemann et al. 2005b); rituximab plus CHOP versus CHOP alone in MCL (Lenz et al. 2005); rituximab plus FCM versus FCM alone in FL and MCL (Forstpointner et al. 2004); and rituximab plus CNOP versus CNOP alone versus rituximab alone (Rivas-Vera et al. 2005). The studies were combined and hazard ratios for overall survival were determined. The hazard ratio for all trials was 0.62 (95% CI: 0.49–0.77). For FL, the hazard ratio was 0.57 (0.43–0.77), and for MCL it was 0.60 (0.37–0.98). The authors stated that this preliminary meta-analysis demonstrated evidence for improved survival among patients with FL and MCL treated with rituximab plus chemotherapy compared with chemotherapy alone. Longer follow-up periods in existing trials and further randomized controlled trials are needed to improve the robustness of these data (Schulz et al. 2005a).

13.2.3 Induction Therapy with Rituximab plus Immune System Modulators in Indolent NHL

Not all patients are suitable for induction treatment with chemotherapy ± rituximab, particularly those who are elderly, have comorbid disease, or are unwilling to receive chemotherapy. An alternative therapeutic approach is the administration of rituximab in combination with immune system modulators, including IFN-α, interleukin (IL)-2, IL-12, and colony stimulating factors (CSFs). Immune modulators may increase the potency of rituximab by potentiation of CD20 expression, increasing Fc receptor density on effector cells or increasing numbers of effector cells (Dillman, 2003).

13.2.3.1 Rituximab plus Immune System Modulators

A regimen of four cycles of rituximab followed by an 8-week course of granulocyte-macrophage CSF (GM-CSF) was assessed in 39 evaluable patients with indolent NHL, 14 of whom were previously untreated (McLaughlin et al. 2005b). Overall and complete response rates were 79% and 36%, respectively. Tolerance of the rituximab plus GM-CSF regimen was comparable to that seen with rituximab alone. The investigators postulate that the addition of GM-CSF to rituximab enhances ADCC activity, which leads to an improved clinical response (McLaughlin et al. 2005b).

In a phase II study conducted in Denmark, Finland, Norway and Sweden, patients with symptomatic, advanced-stage CD20+ low-grade lymphoma (untreated or after first relapse) received rituximab once weekly for 4 weeks (Jurlander et al. 2004; Kimby et al. 2002). Patients with a PR or minor response (MR) were randomized to another four infusions of rituximab or to IFN-α-2a (3–4.5 MIU day^{-1}) before and during four infusions of rituximab. The ORR and CR rate in patients receiving two cycles of rituximab were 78% and 22%, respectively, and higher rates were observed in patients who received IFN-α-2a with the second cycle (94 and 48%, respectively) (Kimby et al. 2002). The frequency of MRD-negativity was 44% in patients who received a single cycle of rituximab, 66% in patients who received two cycles, and 77% in those who received two cycles of rituximab with IFN-α-2a. This trend towards a dose–response relationship was not significant due to the small number of patients in each group (Jurlander et al. 2004). The finding that one cycle of rituximab was less effective than two cycles, which in turn was less effective than two cycles plus IFN, suggests that the results with the combination regimen can be improved by dose intensification. A randomized phase III study is currently ongoing to further investigate this possibility.

13.2.3.2 Rituximab plus Immune Modulators in Relapsed/Refractory Indolent NHL

Two studies have examined the combination of rituximab and IFN-α-2a in relapsed/refractory indolent NHL (Davis et al. 2000a; Sacchi et al. 2001). Both demonstrated promising overall response rates (45% and 70%, respectively) and good tolerability (Table 13.3).

In a study of rituximab and IL-2, the ORR in 20 patients was 55%. The infusional toxicity associated with rituximab was not exacerbated by IL-2 (Friedberg et al. 2002). Gluck and colleagues (2004) found that a thrice-weekly IL-2 schedule in combination with rituximab was safe and effective. The combination of rituximab and IL-12 has also shown promising efficacy (ORR 67%, CR rate 25%), but was less well tolerated, with liver enzyme abnormalities and cytopenias being the most commonly observed toxicities (Ansell et al. 2003).

A trial of G-CSF with rituximab led to ORR and CR rates of 42% and 26%, respectively, and the median TTP was an encouraging 24 months (van der Kolk et al. 2003). The toxicity profile for the combination appeared similar to that reported for rituximab monotherapy (McLaughlin et al. 1998).

13.2.4
Induction with Rituximab Monotherapy in Indolent NHL

The earliest clinical studies of rituximab explored its use as monotherapy in relapsed/refractory indolent NHL. Following the encouraging results obtained in the relapsed setting, its efficacy and safety was subsequently examined in previously untreated patients.

13.2.4.1 Rituximab Monotherapy in Previously Untreated Indolent NHL

An early study of rituximab monotherapy as first-line treatment for indolent lymphoma enrolled patients with a low tumor burden FL (lesions ≤7 cm) (Colombat et al. 2001). Patients received four once-weekly doses of rituximab, whereby 36 of 49 evaluable patients (73%) responded, including 27% CR. In some patients, response improved over time, so that within the year following treatment the objective response rate rose to 80% with 40% CR. After 1 year, 16 of 26 patients (62%) were PCR-negative in the peripheral blood. Only two Grade 3 adverse events were reported – one case each of hypotension and hypertension. Long-term follow-up results from this trial were subsequently reported (Solal-Celigny et al. 2004). After a median follow-up of 60 months, the median PFS was 18 months and overall survival 94%.

Witzig et al. (2005) studied four once-weekly doses of rituximab in patients with previously untreated advanced Grade 1 FL. The ORR was 72% (36% CR) and the median TTP 2.2 years. Grade 3/4 adverse events occurred in 14% of patients, with 6% being considered possibly related to rituximab (one case each of neutropenia and urticaria/rash).

A further phase II study conducted by Hainsworth et al. (2002) examined the same induction schedule of rituximab in previously untreated patients with FL (61%) or SLL (39%). The ORR at 6 weeks was 47%, and only two patients had Grade 3/4 adverse events (one with chills/rigors, one with flushing/dyspnea/chest pain). Patients in this study received further maintenance courses of rituximab, and outcomes following further rituximab treatment are detailed below.

13.2.4.2 Rituximab Monotherapy in Relapsed Indolent NHL

The largest phase II study of rituximab in this setting led to the approval of rituximab for patients with relapsed/refractory indolent NHL in the US in 1997 and worldwide in 1998 (McLaughlin et al. 1998). In the analysis of 151 evaluable patients who received four once-weekly doses of rituximab, ORR was 50% with 6% CR. After a median follow-up period of 11.8 months, the projected median TTP for responders was 12.5 months. After a median follow-up period of more than 3 years, the median duration of remission was reported to be 11.2 months (McLaughlin et al. 1999). Re-evaluation of the study results following development of the Cheson criteria (Cheson et al. 1999) showed an ORR of 56% and CR of 32% (Grillo-Lopez et al. 2000). The toxicity profile demonstrated in this study has come to be recognized as characteristic for the product, and has been replicated in subsequent investigations (Kimby 2005). The majority of patients (71%)

experienced adverse events during the first infusion (usually Grade 1/2), but 55% remained free of any adverse events during infusions 2 to 4 (McLaughlin et al. 1998). The most common symptoms were transient fever, chills, nausea, asthenia, and headache. Thrombocytopenia and neutropenia occurred in less than 9% and 14% of patients, respectively, and infectious adverse events occurred in 30% (4% Grade 3).

A number of other smaller trials examined the safety and efficacy of four once-weekly doses of rituximab in relapsed/refractory NHL and are summarized in Table 13.3 (Conconi et al. 2003; Cortes-Funes et al. 2000; Davis et al. 1999, 2000b; Feuring-Buske et al. 2000; Foran et al. 2000a; Igarashi et al. 2002; Maloney et al. 1997b). Piro and colleagues (1999) examined an extended dosing schedule of eight once-weekly infusions and obtained an ORR of 60% and CR rate of 14%. Median TTP and response duration were not reached after 19.4+ and 13.4+ months, respectively. The authors reported that extending rituximab dosing neither delayed B-cell recovery, stimulated an increased incidence of HACA, nor resulted in an increased incidence of infections compared with a shorter rituximab treatment regimen.

13.2.5 Rituximab in Other Subtypes of Indolent Lymphoma

13.2.5.1 Rituximab in Marginal Zone Lymphoma

Conconi et al. (2003) investigated the clinical activity of rituximab in patients with previously untreated or relapsed MALT lymphoma. Rituximab monotherapy was effective and well tolerated, but the relapse rate (36%) was relatively high. Martinelli et al. (2005) studied rituximab in 27 patients with gastric MALT lymphoma, resistant to – or not eligible for – anti-*H. pylori* therapy. Twenty of 26 evaluable patients (77%) achieved an objective response, including 46% CR. After a median follow-up of 33 months, only two patients had relapsed, one of whom subsequently achieved a CR after treatment with rituximab and chlorambucil.

As discussed above, Martinelli et al. (2003) have also found the combination of rituximab and chlorambucil to be active and well tolerated in newly diagnosed and relapsed/refractory low-grade and FL. The study population included six patients with previously extranodal MZL. The results of the IELSG phase III trial evaluating this combination should be available in the next few years.

Bennett and co-workers (2005) performed a retrospective analysis of 11 patients with splenic MZL who received four cycles of rituximab therapy. Rituximab therapy was chosen because the patients were considered poor candidates for surgery, or had refused surgery. The instigation of rituximab resulted in prompt reduction in splenomegaly in nine of 10 patients, while a further patient with a pleural effusion demonstrated complete clearance of the effusion. Eight of the 10 responding patients had complete resolution of cytopenias. The median response duration was 21 months, and two patients who relapsed at 21 and 23 months responded to rituximab retreatment. These observations suggest that rituximab has considerable activity in splenic MZL, and could be an appropriate therapeutic option in this disease, especially for elderly patients, or those with concomitant disease.

13.2.5.2 Rituximab in Small Lymphocytic Lymphoma

As discussed above, SLL and CLL represent different manifestations of the same disease. Treatment guidelines recommend that SLL and CLL be managed in the same way (National Comprehensive Cancer Network physician prescribing guidelines; see link on NCCN website, http://www,nccn.org). The pivotal phase II trial of rituximab in relapsed/refractory indolent NHL included 30 evaluable patients with SLL, who achieved an ORR of only 13% – markedly lower than the ORR of 60% seen in 118 evaluable patients with FL (McLaughlin et al. 1998). Another early study of four cycles of rituximab monotherapy in newly diagnosed and relapsed B-cell malignancies included 29 patients with relapsed/refractory SLL (Foran et al. 2000b). The ORR was only 14%, and the authors speculated that this might be explained by the low density of cell-surface CD20 in SLL. The use of a thrice-weekly dosing schedule of rituximab led to a somewhat higher objective response in three of seven patients with SLL (43%) (Byrd et al. 2001).

A study of rituximab in combination with cyclophosphamide and mitoxantrone included five patients with SLL/CLL (Emmanouilides et al. 2003). One patient achieved a CR and two a PR, giving an ORR of 60%. Hainsworth and colleagues assessed the use of rituximab monotherapy as maintenance therapy (four once-weekly infusions at 6-month intervals for 2 years) in patients with previously untreated CLL/SLL (Hainsworth et al. 2003a). The outcome in five patients with SLL was not reported separately. The ORR and CR rates in 44 patients were 58% and 9%, respectively, and the estimated median progression-free time was 19 months.

13.2.5.3 Rituximab in Waldenström's Macroglobulinemia

A number of investigators have studied the efficacy and safety of rituximab, either alone or in combination with other agents, in WM. In the largest study of single-agent rituximab, four once-weekly doses in 69 patients led to an ORR of 27% (Gertz et al. 2004). Extended rituximab dosing (eight infusions) has also been evaluated and ORRs of 27 to 44% were seen (Dimopoulos et al. 2002; Treon et al. 2001, 2005a). The use of rituximab in combination with chemotherapy appears to be a more effective treatment strategy in WM. For example, the use of PCR was associated with an ORR of 77% (Hensel et al. 2005), 85% of patients responded to R-CHOP therapy (Treon et al. 2005b), and the combination of rituximab, cladribine and cyclophosphamide was associated with an ORR of 94% (Weber et al. 2003). Further studies are currently under way to more fully assess the role of rituximab in this disorder.

The Third International Workshop on WM recently reported updated treatment recommendations for this disease (Treon et al. 2006). For patients with previously untreated or relapsed WM, appropriate therapeutic options include rituximab monotherapy or rituximab together with nucleoside analogs, alkylating agents or combination chemotherapeutic regimens (Treon et al. 2006).

13.2.6 Rituximab Maintenance Therapy

As discussed above, the typical course of indolent NHL is one of repeated relapses and remissions. While recent advances in NHL treatment have improved the outlook for many patients, there remain considerable unmet needs. Strategies for extending remission duration in indolent NHL without significantly increasing toxicity are eagerly sought. One approach that has the potential to do this is the use of maintenance therapy in patients who have responded to initial induction therapy.

Maintenance therapy with intermittent chemotherapy (Ezdinli et al. 1987; Stewart et al. 1988) and interferon (Rohatiner et al. 2005) has not been shown to alter the natural history of the disease. Single-agent rituximab may highly suitable for use as a maintenance therapy, as it is effective and very well tolerated. A number of investigators have evaluated different schedules of maintenance therapy in recent years (Table 13.4). Further investigation of rituximab maintenance, including its safety and efficacy and optimal duration of therapy, is currently under way in prospective randomized trials. These include the Primary RItuximab and MAintenance (PRIMA) and OSHO/GLSG trials comparing rituximab maintenance with observation after immunochemotherapy induction in previously untreated indolent NHL.

13.2.6.1 Rituximab Maintenance Therapy Following Monotherapy Induction

In a study conducted by Ghielmini and colleagues (2004), patients with previously untreated or relapsed/refractory FL who did not progress after the standard (4-week) rituximab regimen either received maintenance rituximab (single infusion at 3, 5, 7, and 9 months) or were observed without further rituximab infusions. After a median follow-up of 35 months, EFS was significantly prolonged in patients who had received rituximab maintenance therapy compared with nontreated patients (23 versus 12 months; $p = 0.024$); this effect was more pronounced in chemotherapy-naïve patients (36 versus 19 months, $p = 0.009$) (Ghielmini et al. 2004). The median duration of response was also longer after maintenance therapy than after no further treatment, both in previously untreated patients and in those with previously treated disease (NR versus 20 months, $p = 0.079$ and 25 versus 13 months, $p = 0.065$, respectively) (Table 13.4). Rituximab was well tolerated, with the incidence of late toxicity in patients evaluable beyond 1 year being only 7% in both treatment arms (Ghielmini et al. 2004).

In a phase II study of 62 patients with previously untreated but symptomatic indolent lymphoma, Hainsworth et al. (2002) investigated the impact of rituximab maintenance therapy in patients who had not progressed following four once-weekly doses of rituximab. Maintenance therapy was administered according to a schedule of four once-weekly doses at 6-monthly intervals for a maximum of four courses, or until disease progression. Following maintenance therapy, 16 of the 27 patients who had SD after induction therapy achieved a response, with ORR increasing from 47% to 73% (the CR increased from 4% to 37%)

Table 13.4 Trials of rituximab (R) maintenance therapy in patients with indolent NHL.

Reference	*Regimen of induction (I) and maintenance (M) therapy*	*Evaluable patients*	*Median follow-up*	*Key outcomes of maintenance arms R-M vs. Obs (unless otherwise stated)*		
				Median PFS[a]	*Median DR*	*Median OS*
Rituximab maintenance after rituximab monotherapy						
Ghielmini et al. (2004)	I: R × 4 M: R × 1 every 2 months for 4 doses vs. Obs	51 UT + 100 PT	35 months	23* vs. 12 months (EFS) 36** vs. 19 months (EFS for UT pts)	35+ vs. 20 months (UT pts) 25 vs. 13 months (PT pts)	
Hainsworth et al. (2002, 2003b)	I: R × 4 M: R × 4, every 6 months for 2 years	60 UT	55 months	37 months		5-year actuarial OS: 70%
Hainsworth et al. (2005b)	I: R × 4 M: R × 4, every 6 months for 2 years vs. retreatment at progression with R × 4	90 PT	41 months	31** vs. 7 months		3-year OS: 72 vs. 68%
Rituximab maintenance after chemotherapy						
Hochster et al. (2004)	I: CVP × 6–8 M: R × 4 every 6 months for 2 yrs vs. Obs	120 UT (FL) 117 UT (FL)	42 months	61*** vs. 15 months		3.5 year OS: 91* vs. 75%

Table 13.4 *Continued*

Reference	*Regimen of induction (I) and maintenance (M) therapy*	*Evaluable patients*	*Median follow-up*	*Key outcomes of maintenance arms R-M vs. Obs (unless otherwise stated)*		
				Median PFS[a]	*Median DR*	*Median OS*
Rituximab maintenance after (immuno)chemotherapy						
van Oers et al. (2006)	I: CHOP × 6 vs. R × 6 + CHOP × 6 M: R × 1 every 3 months for 2 years vs. Obs	159 PT 160 PT	36 months	52*** vs. 15 months; 42*** vs. 12 months (post-CHOP induction); 52** vs. 23 months (post-R-CHOP induction)		3-year OS: 85* vs. 77%;
Forstpointner et al. (2006)	I: FCM × 4 vs. R × 4 plus FCM × 4 M: R × 4, 3 and 9 months after induction vs. Obs	176 PT (FL and MCL)	26 months		26+*** vs. 17 months; (post R-FCM induction)	

All doses of R 375 $mg\,m^{-2}$.
a Or other parameter as indicated.
* $p < 0.05$; ** $p < 0.01$; *** $p < 0.001$ versus comparator.
FCM = fludarabine, cyclophosphamide, mitoxantrone; FL = follicular lymphoma; I = induction; M = maintenance; MCL = mantle cell lymphoma; MCP = mitoxantrone, chlorambucil, prednisolone; NR = not reached; Obs = observation; pts = patients. Other abbreviations as Table 13.2.

(Hainsworth et al. 2002). After a median of 55 months' follow-up, the median PFS was 37 months and the 5-year PFS and OS rates were 24% and 70%, respectively (Hainsworth et al. 2003b). Rituximab therapy was well tolerated, and rituximab maintenance was not associated with any Grade 3/4 toxicity. No cumulative or late toxicity was observed and there were no opportunistic infections (Hainsworth et al. 2002, 2003b).

In a further phase II trial, patients with relapsed or refractory indolent NHL responding to, or with stable disease after four weekly doses of rituximab therapy, were randomized to rituximab maintenance (four weekly doses every 6 months for a total of four occasions) or to retreatment with four weekly doses of rituximab at disease progression (Hainsworth et al. 2005b). The median PFS was significantly prolonged in the rituximab maintenance group compared with those receiving rituximab retreatment (31 versus 7 months, $p = 0.007$) (Table 13.4; Fig. 13.5). However, the median duration of rituximab benefit and the 3-year survival rates were similar for patients receiving rituximab maintenance and rituximab retreatment (31 versus 27 months, $p = 0.94$ and 72 versus 68%, respectively) (Hainsworth et al. 2005b). Both treatment arms were well tolerated, and there were no treatment-related hospitalizations or patient discontinuations because of therapy-related adverse effects.

13.2.6.2 **Rituximab Maintenance Therapy Following Chemotherapy Induction**

Extremely promising results have been obtained in an ongoing Eastern Cooperative Oncology Group (ECOG) phase III study in which patients with advanced indolent NHL responding to CVP chemotherapy were randomized to receive rituximab maintenance or observation only (Hochster et al. 2004, 2005). The majority of

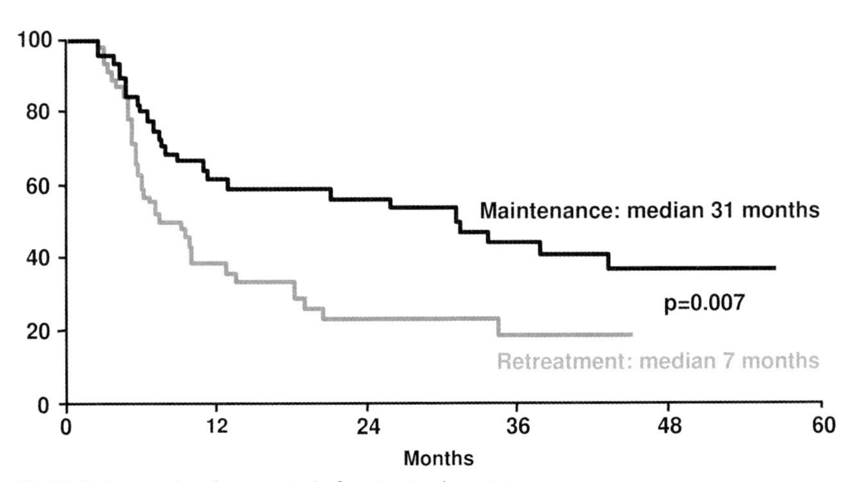

Fig. 13.5 Progression-free survival after rituximab maintenance therapy or retreatment with rituximab at progression following rituximab induction therapy for relapsed/refractory indolent NHL: the Minnie Pearl Cancer Research Network trial. (Reproduced with permission from Hainsworth et al. 2005b.)

patients studied (237/305) had FL. Patients randomized to maintenance had a significantly extended PFS (61 versus 15 months, p = 0.0000003, HR = 0.4) and OS (91 versus 75%, p = 0.03, HR = 0.5) compared to those randomized to observation. The improvements in PFS were evident regardless of FLIPI score, tumor burden or extent of residual disease (Hochster et al. 2005). Rituximab maintenance therapy was well tolerated, and did not lead to significantly higher rates of neutropenia, thrombocytopenia or infection compared with observation (Hochster et al. 2004).

13.2.6.3 Rituximab Maintenance Therapy Following Rituximab Chemotherapy Induction

As discussed above, the second primary objective of the EORTC 20891 study was to evaluate the effect of rituximab maintenance therapy (a single infusion every 3 months for 2 years) on progression-free survival (van Oers et al. 2006). Median PFS was significantly improved after maintenance therapy compared with observation (51.6 versus 15 months, p < 0.0001, HR = 0.4) and importantly, these improvements were seen both in patients treated with CHOP induction (42.2 versus 11.6 months, p < 0.0001, HR = 0.3) and R-CHOP induction (51.9 versus 23.1 months, p = 0.0043, HR = 0.54). Median OS was also significantly extended after maintenance compared with observation: 3-year OS rates 85.1 versus 77.1%, p = 0.0111, HR = 0.52. Grade 3/4 neutropenia rates were slightly higher in the maintenance arm (10.8 versus 5.4%), as were Grade 3/4 infection rates (9 versus 2.4%) (van Oers et al. 2006). This study formed the basis of the submission for approval of rituximab for maintenance therapy made to the EMEA in 2006.

The phase III GLSG trial of FCM versus R-FCM in relapsed FL discussed above also included a second randomization to rituximab maintenance (four weekly doses at 3 and 9 months post-induction) or observation only (Table 13.4). The median response duration after rituximab maintenance therapy was significantly longer than after observation only in 176 evaluable patients with FL and MCL (NR at 26 months versus 17 months, p < 0.001) (Table 13.4). The significant improvement in response duration for patients receiving rituximab maintenance therapy was also seen in the subgroup of FL patients who had received R-FCM induction therapy (n = 81) (NR at 26 months versus 26 months, p = 0.035) and those with MCL receiving R-FCM induction therapy (n = 47) (p = 0.049) (Table 13.4). This suggests that rituximab maintenance can benefit all patients with FL and MCL, regardless of whether or not they have received rituximab as part of their induction therapy (Forstpointner et al. 2006).

13.2.7 Rituximab Retreatment

Retreatment with rituximab – either alone or in combination with chemotherapy – appears to be an efficient treatment strategy. An early study conducted by Davis et al. (2000b) in patients with relapsed indolent NHL observed an ORR of 38% after retreatment of 58 patients with rituximab, and TTP for responders at retreatment was longer than that observed after initial therapy (17.4 versus 12.4 months). Retreatment was not associated with induction of HACA or cumulative myelosuppression. Igarashi and colleagues (2001) also observed an ORR of 38% in 13 patients with relapsed indolent NHL receiving a second course of rituximab, which was well tolerated.

Lemieux et al. (2004), at the CHU-Lyon-Sud, subsequently reported the effect of a second treatment of rituximab either alone or in combination with chemotherapy in patients with a variety of lymphoma subtypes, the majority (82%) of which were indolent. The ORR was 73%, and median TTP was longer for the second treatment than the first (15.2 versus 11.3 months). The second treatment was well tolerated.

In the study of BMR in relapsed/refractory indolent NHL discussed above (Weide et al. 2004), ORR and CR rates were as good in 16 patients previously treated with rituximab as in the full study population (88 versus 94% and 56 versus 44%, respectively). Recently, Bairey and associates (2005) retrospectively evaluated rituximab retreatment in 66 patients with indolent NHL, 18 with aggressive NHL and six with MCL. The ORR to the second rituximab treatment (with or without chemotherapy) was 71%, including 39% CR. Median TTP after the second treatment was similar to that seen after the first (14 versus 12 months). In a Spanish retrospective analysis of patients with DLBCL relapsing after rituximab, the ORR to a second course of rituximab ± chemotherapy in 25 assessable patients was 92% (56% CR) (Canales et al. 2005). Nine patients received a third course of rituximab ± chemotherapy and yielded an ORR of 78% (33% CR). Adverse events following retreatment were described as not being different to those usually observed with these regimens.

As discussed above, in a phase II trial comparing rituximab maintenance (four weekly doses every 6 months for a total of four occasions) and retreatment (four weekly doses of rituximab at disease progression), median PFS was significantly prolonged in the rituximab maintenance group compared with the rituximab retreatment group (31 versus 7 months, $p = 0.007$) (Table 13.4; Fig. 13.5) (Hainsworth et al. 2005b). However, the median duration of rituximab benefit and the 3-year survival rates were similar for patients receiving rituximab maintenance and rituximab retreatment (31 versus 27 months, $p = 0.94$ and 72 versus 68%, respectively) (Hainsworth et al. 2005b).

Taken together, these observations suggest that rituximab retreatment is safe and feasible in patients with relapsed B-cell NHL. Prospective studies of rituximab retreatment are currently in progress.

13.2.8 Rituximab in Aggressive NHL

For decades, aggressive NHL was considered an incurable disease, but during the 1960s the development of anthracycline-containing combination chemotherapies meant that cure finally became an achievable goal for some patients (DeVita et al. 1975). CHOP (cyclophosphamide, doxorubicin, vincristine and prednisolone) has been extensively studied in randomized clinical trials, but is associated with a cure rate of only 30 to 40% in nonlocalized aggressive NHL (Fisher 1994), leaving much room for improvement.

Despite early optimism, second- and third-generation chemotherapy regimens were unable to improve on overall survival compared with CHOP; moreover, the toxicity levels and costs of these newer regimens were higher (Fisher et al. 1994). A reduction of the CHOP treatment interval from 21 to 14 days is also an option. In both elderly and young patients with good-prognosis disease, an interval reduction has been shown to result in a modest improvement in overall survival, but at the cost of increased toxicity in elderly patients (Pfreundschuh et al. 2004a,b).

High-dose therapy (HDT) followed by ASCT has been shown to improve survival compared with chemotherapy when used as front-line therapy (Milpied et al. 2004), and to consolidate complete remission after front-line chemotherapy in younger patients with poor-risk aggressive NHL (Haioun et al. 2000). While intensive chemotherapy regimens with or without ASCT may improve outcomes for younger healthier patients, the morbidity/mortality associated with this strategy makes it generally unsuitable for patients aged over 60 years (who comprise more than half of all patients with the most common form of aggressive NHL, DLBCL) (The Non-Hodgkin's Lymphoma Classification Project, 1997a,b).

The successful use of rituximab in patients with indolent NHL prompted its further evaluation in aggressive NHL. Early success in a phase II trial of R-CHOP led to the initiation of multicenter phase III trials assessing the impact of adding rituximab to CHOP/CHOP-like chemotherapy in patients with DLBCL. Rituximab has subsequently been evaluated with a number of different treatment regimens both in previously untreated (Table 13.5) and relapsed/refractory aggressive NHL (Table 13.6).

13.2.8.1 Rituximab plus Chemotherapy in Previously Untreated Aggressive NHL: Phase II Studies

An early study of R-CHOP in aggressive NHL produced extremely encouraging results (Vose et al. 2001, 2002). Patients with newly diagnosed, aggressive lymphoma were treated with six cycles of R-CHOP every 21 days. The ORR was 94%, with 61% patients achieving a CR. Even poor-prognosis patients (IPI score ≥2) had an ORR of 89% (Vose et al. 2001). These excellent results were achieved without adding significantly to the toxicity of CHOP chemotherapy. Even more encouragingly, the presence or absence of bcl-2, an anti-apoptotic protein which is often overexpressed in NHL cells and has been associated with poor prognosis in chemotherapy-treated aggressive NHL patients (Hermine et al. 1996), did not significantly affect the CR rate, and nor did patient age (Vose et al. 2001). Recently published long-term follow-up data reveal 5-year PFS and OS rates of 82% and 88%, respectively and no long-term adverse events directly related to rituximab were observed (Vose et al. 2005).

The combination of rituximab and VNCOP-B (etoposide, mitoxantrone, cyclophosphamide, vincristine, prednisolone, bleomycin) – a chemotherapy regimen developed to treat elderly patients for whom CHOP chemotherapy was not feasible

Table 13.5 Trials of rituximab (R) in previously untreated aggressive NHL.

Reference	*Regimen of R plus other agents × no. of cycles*	*Patients*	*ORR*	*CR*	*PR*	*Median PFS*[a]	*Median OS*
phase II trials							
Vose et al. (2002, 2005)	R × 6 plus CHOP × 6	33 UT[b]	94	61	33	5-year PFS: 82%	5-year OS: 88%
Rodriguez et al. (2005)	R × 6–8 plus CHOP[c] × 6–8	68 UT[d]	93	91	2		OS: 94%
Canales et al. (2005)	R × 6–8 plus CHOP/CHOP-like chemo × 6–8	80 UT	86	72	14	2-year EFS: 60%	2-year OS: 73%
Hainsworth et al. (2003c)	R × 7 plus VNCOP-B × 8	27 UT	93	37	56	21 months+	21 months+
Wilson et al. (2005)	R-DA-EPOCH + filgrastim × 6–8	60 UT 11 PT[e]	100	68	32	1.5-year PFS: 80%	1.5-year OS: 88%
Lavilla et al. (2005)	R × 6 plus CHOP-14 × 6	28 UT[g]	96	89	7		
Rigacci et al. (2006)	R × 6 plus CHOP-14 × 6	26 UT	100	77	23	DFS: 70% after 17 months	OS: 79% after 23 months
Glass et al. (2005)	R × 6 plus MegaCHOEP × 4	72 UT				3-year FFTF: 70%*	3-year OS: 75%
	MegaCHOEP × 4 (historical controls)	35 UT				50%	57%
Intragumtornchai et al. (2006)	R × 7 + CHOP × 3 plus ESHAP × 4	84 UT		67	14	5-year FFS:	5-year OS:
	CHOP × 3 plus ESHAP × 4 plus HDT			44	22	61%***	61%
	CHOP × 8 (randomization)			36	24	34% 16%	43% 24%
phase III trials							
Coiffier et al. (2005);	R × 8 plus CHOP × 8	197 UT	83	76**	8	5+ years***	5+ years**
Feugier et al. (2005)	CHOP × 8 (randomization)	202 UT	69	63	6	1 year	3.1 years

Table 13.5 *Continued*

Reference	*Regimen of R plus other agents × no. of cycles*	*Patients*	*ORR*	*CR*	*PR*	*Median PFS*[a]	*Median OS*
Pfreundschuh et al. (2006)	R × 6 plus CHOP/CHOP-like chemo × 6	413 UT		86***		3-year EFS:	3-year OS:
	CHOP/CHOP-like chemo × 6 (randomization)	410 UT		68		79%*** 59%	93%*** 84%
Habermann et al. (2006)	R × 4–5 plus CHOP × 6–8	318 UT	77	32	42	3-year FFS:	3-year OS:
	CHOP × 6–8 (randomization)	314 UT	76	31	41	53%* 46%	67%* 58%
Sonneveld et al. (2005)	R × 6 plus CHOP-14 × 8	99 UT[h]		34		2-year FFS:	2-year OS:
	CHOP-14 × 8 (randomization)	98 UT[i]		30		51%** 23%	63%* 46%
Pfreundschuh et al. (2005b)	R × 8 plus CHOP-14 × 8	203 UT				70%	74% for 8 × R plus 6–8 × CHOP-14
	R × 8 plus CHOP-14 × 6	211 UT				70%	
	CHOP-14 × 8	210 UT				58%	
	CHOP-14 × 6 (randomization)	203 UT				53% (TTF)	78% for 6–8 × CHOP-14

All doses of R 375 $mg\,m^{-2}$.

a Or other parameter as indicated.

b Included 7 patients with follicular large cell lymphoma, 2 with immunoblastic lymphoma, and 2 with other aggressive lymphomas.

c Regimen contains sphingosomal vincristine in place of free vincristine.

d 56 patients had DLBCL, 4 Grade, 3 FL, 4 PTCL, 4 anaplastic large cell lymphoma, and 2 transformed indolent lymphoma.

e Included a few patients with PMBCL and aggressive B-cell lymphoma, unspecified.

f Included 7 patients with Grade 3 FL.

g Included 3 patient with Grade 3 FL.

h Included 15 patients with MCL and 6 with Grade 3 FL.

i Included 14 patients with MCL and 3 with Grade 3 FL.

* $p < 0.05$; ** $p < 0.01$; *** $p < 0.001$ versus comparator.

CHOP-14 = 2-week cycles of CHOP; DFS = disease-free survival; EPOCH = etoposide, prednisone, vincristine, cyclophosphamide, doxorubicin; ESHAP = etoposide, solumedrol, cytosine arabinoside, cisplatin; FFS = failure-free survival; FFTF = freedom from treatment failure; HDT = high-dose therapy; MegaCHOEP = cyclophosphamide, adriamycin, vincristine, etoposide, prednisone; PFS = progression-free survival; VNCOP = etoposide, mitoxantrone, cyclophosphamide, vincristine, prednisolone, bleomycin. Other abbreviations as Table 13.2.

Table 13.6 Trials of rituximab (R) in relapsed/refractory aggressive NHL.

Reference	*Regimen of R plus other agents × no. of cycles*	*Patients*	*ORR*	*CR*	*PR*	*Median PFS*[a]	*Median OS*
Trials of rituximab plus chemotherapy							
Kewalramani et al. (2004)	R × 4 plus ICE × 3	36 PT	78	53	25		
Mey et al. (2006)	R × 4 plus DHAP × 4	53 PT	62	32	30	6.7 months	8.5 months
El Gnaoui et al. (2005)	R × 4 plus GEMOX × 4	40 PT[b]	85	53	32	20 months (TTP)	
Jermann et al. (2004)	R × 4–6 plus EPOCH × 4–6	50 PT[c]	68	28	40	11.8 months (EFS)	17.9 months
Sirohi et al. (2005)	R × 4 plus GEM-P × 4	24 PT	67			1-year PFS: 51%	3-year OS: 65%
Hicks et al. (2005)	R × 8 plus ESHAP × 2–4	20 PT[d]	89			9+ months	9+ months
Venugopal et al. (2004)	R × 6 plus ESHAP × 6	14 PT[e]	64	50	14		
Trials of rituximab monotherapy							
Coiffier et al. (1998)	R × 8	45 PT +	31	9	22	105+ days (TTP)	
	R × 1 then R 500 × 7	9 UT[f]				121+ days (TTP)	
Tobinai et al. (2004)	R × 8	57 PT	37	26	11	52 days	
Rothe et al. (2004)	R × 4	21 PT	38	5	33	3.8 months (EFS)	8.6 months

All doses of R 375 g m^{-2} unless otherwise indicated.

a Or other parameter as indicated.

b 16 patients with DLBCL, 6 with FL and 2 with MCL.

c 25 patients with DLBCL, 18 with transformed large B-cell lymphoma and 7 with MCL.

d 9 patients with relapsed aggressive lymphoma, 2 with refractory aggressive lymphoma and 9 with transformed indolent lymphoma.

e Predominantly DLBCL patients.

f Includes a few patients with MCL, FL or unclassified histology.

DHAP = dexamethasone, cytarabine, cisplatin; DOC = docetaxel; EPOCH = etoposide, prednisone, vincristine, cyclophosphamide, doxorubicin; ESHAP = etoposide, solumedrol, cytosine arabinoside, cisplatin; GEMOX = gemcitabine, oxaliplatin; GEM-P = gemcitabine, cisplatin, methylprednisolone; ICE = ifosfamide, carboplatin, etoposide; PAC = paclitaxel; T = topotecan. Other abbreviations as Table 13.2.

(Zinzani et al. 1999) – is also effective. The treatment of 41 elderly patients, the majority of whom had an age-adjusted IPI score of 2 or 3, with rituximab plus VNCOP-B resulted in an ORR of 66% and 2-year PFS and OS rates of 59% and 57%, respectively (Hainsworth et al. 2003c).

Another effective regimen for patients with aggressive NHL is dose-adjusted EPOCH (etoposide, vincristine, doxorubicin, cyclophosphamide, prednisolone) chemotherapy (DA-EPOCH). In this regimen the doses of doxorubicin, etoposide and cyclophosphamide are adjusted to achieve neutrophil nadirs of $<500\,mm^{-3}$. However, treatment with DA-EPOCH produces a significantly worse outcome for patients overexpressing bcl-2 compared with those expressing normal levels of this protein (Wilson et al. 2002, 2003).

The addition of rituximab to DA-EPOCH may overcome the adverse outcome both in DLBCL patients overexpressing bcl-2 and in those with high levels of p53 (another anti-apoptotic protein; Wilson et al. 2003). Furthermore, rituximab plus DA-EPOCH has recently been shown to be effective in all subtypes of *de-novo* DLBCL (Wilson et al. 2005), reinforcing the value of rituximab in combination with DA-EPOCH in this clinical setting.

13.2.8.2 Rituximab plus Chemotherapy in Previously Untreated DLBCL: Phase III Studies and Population Analysis

13.2.8.2.1 The GELA LNH98-5 Trial

The phase III randomized controlled study LNH98-5 conducted by the Groupe d'Etude des Lymphomes de l'Adulte (GELA), randomized 399 elderly patients (aged 60–80 years) with DLCBL to treatment either with eight cycles of CHOP or eight cycles of R-CHOP. A significant number of patients had high-risk disease; 60% of patients had IPI scores ≥2 and 65% had elevated lactate dehydrogenase (LDH) levels.

After eight cycles of treatment, the ORRs for patients treated with R-CHOP and CHOP were 83% and 69%, respectively, with significantly more R-CHOP-treated patients achieving a CR compared with those receiving CHOP alone (76 versus 63%; $p = 0.005$) (Coiffier et al. 2002). This higher CR rate translated into substantially improved long-term outcomes for patients treated with R-CHOP compared with those receiving CHOP alone (Coiffier et al. 2002; Feugier et al. 2005). With a median follow-up of 5 years, DFS, PFS (Fig. 13.6) and OS were all significantly prolonged in patients receiving R-CHOP compared with those receiving CHOP alone (Feugier et al. 2005).

Subgroup analysis demonstrates that R-CHOP confers significant benefits compared with CHOP alone regardless of age, co-morbidity or IPI score (Coiffier et al. 2002, 2003; Feugier et al. 2005). Furthermore, adding rituximab to CHOP chemotherapy appears to overcome chemotherapy failure associated with bcl-2 overexpression (Mounier et al. 2003). There was no significant difference in clinically relevant toxicity, late toxicities or the incidence of secondary tumors between patients treated with R-CHOP and those receiving CHOP (Coiffier et al. 2002; Feugier et al. 2005).

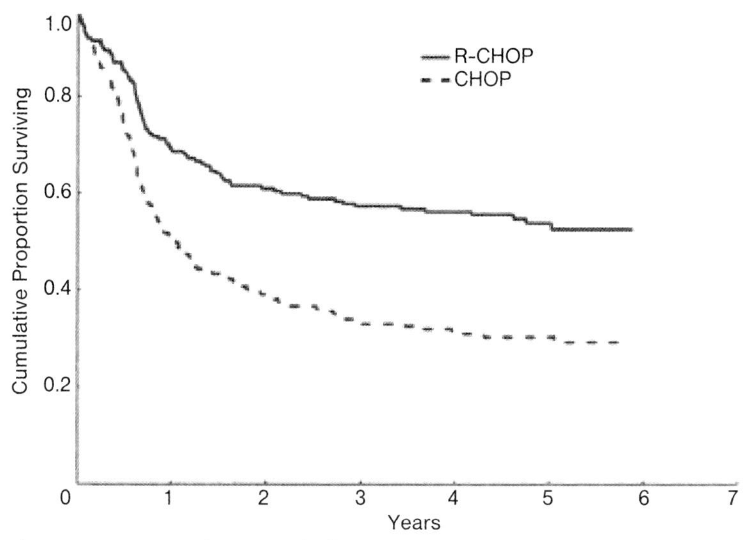

Fig. 13.6 Progression-free survival after treatment with eight cycles of CHOP chemotherapy or eight cycles of rituximab plus CHOP chemotherapy in elderly patients with previously untreated aggressive lymphoma: Groupe d'Etudes des Lymphomes de l'Adultes (GELA) LNH 98-5 trial. (Reproduced with permission from Feugier et al. 2005.)

13.2.8.2.2 **The Intergroup E4494 Trial**

Further support for the use of R-CHOP as first-line therapy in elderly patients with aggressive lymphoma in elderly patients comes from an Intergroup study performed in the US (Habermann et al. 2006). Patients aged at least 60 years were randomized to receive either R-CHOP or CHOP and were then randomized to rituximab maintenance therapy or observation only. The study was not designed to compare R-CHOP and CHOP induction therapy directly, and the ORR was similar with the two induction regimens (77 versus 76%).

It was recognized that the second randomization may have confounded interpretation of the effect of induction therapy, and secondary ("weighted") analyses were therefore performed.

Failure-free survival (FFS) at 3 years was significantly better in the R-CHOP arm (53 versus 46% for CHOP; $p = 0.04$). Moreover, when the effect of maintenance therapy was removed from analysis of the data, OS rates were also significantly improved in the R-CHOP induction group compared with the CHOP group (estimated 3-year OS, 67 versus 58%, respectively; $p = 0.05$). Furthermore, maintenance therapy with rituximab appears to provide additional benefits, with prolonged FFS rates in this arm of the study compared with the observation arm ($p = 0.009$) (Habermann et al. 2006). FFS and OS were analyzed according to the four treatment strategies assessed in the trial. The addition of rituximab to CHOP improved the outcome of CHOP, irrespective of whether it was combined with CHOP in the induction phase or given as maintenance therapy after CHOP induction (Habermann et al. 2006).

Consistent with other studies of the addition of rituximab to chemotherapy, no significant difference in toxicity was observed between the two induction arms, with fatal infection rates of 3% and 2.5% for R-CHOP and CHOP patients, respectively (Habermann et al. 2006). Rituximab maintenance therapy was also well tolerated, although granulocytopenia was significantly higher in patients receiving maintenance rituximab compared with those being observed (p = 0.008).

13.2.8.2.3 The MInT Trial

The phase III, multicenter, open-label MabThera International Trial (MInT) randomized patients aged 18 to 60 years with previously untreated low-risk (IPI score ≤1) DLBCL to receive six cycles of CHOP-like chemotherapy (CHEMO) or CHOP-like chemotherapy plus rituximab (R-CHEMO) every 21 days. The chemotherapy regimens used were CHOP, CHOEP (CHOP plus etoposide), MACOP-B (methotrexate, doxorubicin, cyclophosphamide, vincristine, prednisolone and bleomycin) and PMitCEBO (prednisolone, mitoxantrone, cyclophosphamide, etoposide, bleomycin, vincristine).

The first interim analysis of this trial took place in November 2003, and included 326 evaluable patients. At this point, the log-rank p-value for the primary endpoint of TTF was 0.0000041 after 97 events had been observed over 24 months. As this was considerably less than the critical threshold for alpha-spending for 97 events (0.00192), the Data Safety and Monitoring Committee stopped recruitment to the trial (Pfreundschuh et al. 2004c).

Data from the full-set analysis demonstrated that the 413 patients receiving R-CHEMO achieved a significantly higher CR than the 410 patients treated with CHEMO alone (Table 13.5) (Pfreundschuh et al. 2006). Moreover, significantly more CHEMO patients had progressive disease (PD) during treatment than R-CHEMO patients. As with the GELA trial, the excellent responses seen in patients receiving chemotherapy plus rituximab translated into significantly improved long-term outcomes, with a significantly longer TTF and OS for R-CHEMO patients compared with patients receiving CHEMO alone (Table 13.5; Fig. 13.7). The rate of lymphoma-associated deaths was reduced by two-thirds with R-CHEMO.

Patients benefited from the addition of rituximab to chemotherapy regardless of their IPI or bulky disease status. R-CHEMO was particularly effective in patients with an IPI score of 0 and no bulk, with a 3-year FFS of 89%, compared with 76% for patients with an IPI score of 1 and/or bulky disease (p = 0.0162). However, the OS rates were not significantly different between these two groups of patients, with a 3-year OS of 98% for patients with an IPI score of 0 and no bulk and 91% for patients with an IPI score of 1 and/or bulky disease (p = 0.8) (Pfreundschuh et al. 2006).

A secondary analysis focusing on outcomes according to chemotherapy regimen in different patient subgroups has also been presented (Pfreundschuh et al.

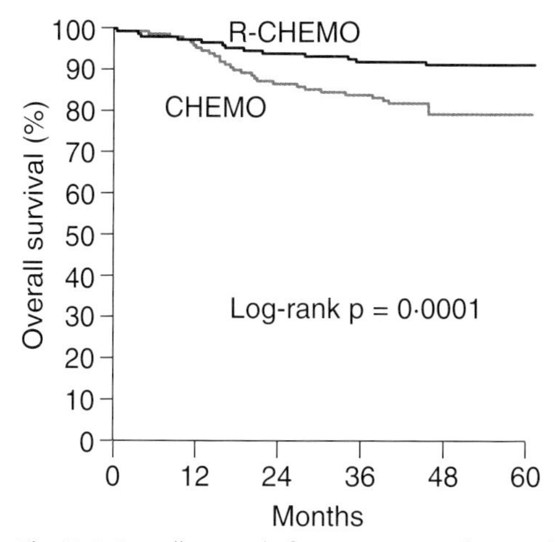

Fig. 13.7 Overall survival after treatment with six cycles of chemotherapy or six cycles of rituximab plus chemotherapy in young patients with previously untreated aggressive lymphoma: MabThera International Trial (MInT). (Reproduced with permission from Pfreundschuh et al. 2006.)

2005a). Too few patients received (R)-PMitCEBO and (R)-MACOP-B for any meaningful conclusions to be drawn. The 2-year TTF rates were significantly better after CHOEP than CHOP (65 versus 55%, p = 0.04), but not significantly different after R-CHOEP and R-CHOP (80 versus 82%, p = 0.67). This suggests that the addition of rituximab neutralizes the superiority of CHOEP over CHOP (Pfreundschuh et al. 2005a).

The benefits of adding rituximab to chemotherapy were gained without any increase in toxicity, with grade 3/4 hematological toxicity being observed in 6% to 7% of patients in each group (Pfreundschuh et al. 2006).

13.2.8.2.4 **BCCA Population Analysis**

In March 2001, the British Columbia Cancer Agency (BCCA) implemented a new provincial policy recommending that all patients with newly diagnosed advanced-stage DLBCL be treated with R-CHOP. A subsequent retrospective study demonstrated a significantly improved PFS and OS in the 152 patients treated during the 18 months after this policy change compared with the 140 patients treated during the 18 months beforehand (p = 0.002 and p < 0.0001, respectively) (Sehn et al. 2005). Multivariate analysis demonstrated that the era of treatment remained a strong predictor of both PFS (p = 0.005) and OS (p < 0.001) after controlling for age and IPI score (Sehn et al. 2005). A retrospective review of young patients with DLBCL in the Czech Lymphoma Group Registry included 90 intermediate-high/high risk (IH/H) patients treated with anthracyline-based chemotherapy (CT) and 55 IH/H patients who received rituximab plus CT (Trneny et al. 2005). The 2-year PFS and OS rates were both significantly better after R-CT compared with CT in these patients: 75.8 versus 41.8%, p = 0.0003 and 83.4 versus 57.0%, p = 0.0007, respectively (Trneny et al. 2005).

13.2.8.3 Rituximab plus Dose-Densified Chemotherapy in Previously Untreated Aggressive NHL: Phase III Studies

The multicenter prospective randomized controlled RICOVER-60 trial was designed to investigate whether an increase in the number of CHOP-14 cycles of chemotherapy administered (from six to eight) or the additional administration of eight cycles of rituximab resulted in improvements in treatment outcome in elderly patients with previously untreated DLBCL (Pfreundschuh et al. 2005b). After a median follow-up of 26 months, there was a trend towards better TTF after eight cycles of CHOP-14 compared with six cycles of CHOP-14 (58 versus 53%, $p = 0.13$). This trend was neutralized after the addition of rituximab: TTF after both six and eight cycles of R-CHOP-14 was 70%. There were no statistically significant differences in OS in the four different treatment arms. Adverse events were generally more common after eight cycles than after six cycles of therapy; the incidence of peripheral neuropathy was 14% and 8% after eight and six cycles, respectively.

The HOVON-46 study was instigated by the Nordic Lymphoma Group and the Dutch HOVON group in an effort to improve clinical outcomes in elderly patients with intermediate or high-risk NHL (Sonneveld et al. 2005). At the time of study initiation (in 2001), preliminary results from the GELA LNH98-5 trial had shown that the addition of rituximab to CHOP improved outcomes in elderly patients with aggressive NHL. It had also been shown that the use of G-CSF allowed the interval between CHOP cycles to be reduced from 3 to 2 weeks, without a significant increase in toxicity. The HOVON-46 study randomized elderly patients with previously untreated intermediate or high-risk DLBCL, MCL or Grade 3 FL to receive either six cycles of CHOP-14 chemotherapy or six cycles of rituximab in addition to six cycles of CHOP-14 (R-CHOP-14). An interim analysis after a median follow-up of 15 months showed that the R-CHOP-14-treated patients achieved significantly better FFS and OS than the CHOP-14-treated patients (2-year FFS: 51 versus 28%, $p = 0.005$; 2-year OS: 63 versus 46%, $p = 0.03$). There was no significant difference in toxicity between the two treatment arms. However, these results are immature and a final analysis will be performed in 2006.

Taken together, the results of these two trials suggest that the addition of rituximab to dose-densified CHOP chemotherapy may improve outcomes compared with dose-densified CHOP chemotherapy alone. As the follow-up period for both trials remains relatively short, a longer follow-up is needed to determine the impact of these regimens on clinical end-points, most notably survival. It is also important to note the importance of using G-CSF in order to minimize treatment-related toxicity. Further studies are currently under way to more fully assess the combination of rituximab with dose-densified CHOP chemotherapy.

13.2.8.4 Rituximab plus Chemotherapy in Relapsed Aggressive NHL

Rituximab has been combined with a number of chemotherapy regimens in the treatment of patients with relapsed or refractory aggressive lymphoma (Table 13.6). Rituximab in combination with ifosfamide, carboplatin and etoposide (R-ICE) induced very high CR rates in 34 patients with DLBCL who received this regimen prior to ASCT (Kewalramani et al. 2004). The investigators compared the CR rates of R-ICE with those of 147 similar historical control patients who received ICE alone, and found significantly better responses with the immunochemotherapy combination (CR rates: 53% for R-ICE versus 27% for ICE; $p = 0.01$). Further follow-up is required to elucidate the effect of this combination on long-term outcomes.

Rituximab has been successfully combined with platinum-based combination chemotherapy regimens, including dexamethasone, high-dose cytarabine and cisplatin (DHAP) (Mey et al. 2006) and etoposide, solumedrol, cytarabine and cisplatin (ESHAP) (Hicks et al. 2005; Venugopal et al. 2004) in patients with relapsed or refractory aggressive lymphoma. Results suggest that the addition of rituximab to these established regimens improves efficacy compared with the chemotherapy regimen alone, without adding significantly to the toxicity of chemotherapy. Prospective phase III studies are required to fully assess the safety and efficacy of these and other immunochemotherapy combinations in relapsed/refractory aggressive NHL.

Gemcitabine-based regimens have shown promise in NHL, and the addition of rituximab to such regimens has been investigated. Rituximab in combination with gemcitabine and oxaliplatin (R-GEMOX) resulted in an ORR of 85% in a study of 40 patients with refractory or relapsed B-cell lymphoma who were ineligible for high-dose chemotherapy with ASCT. As in studies of other rituximab–combination chemotherapy regimens, the toxicity was acceptable in these patients (El Gnaoui et al. 2005).

13.2.8.5 Rituximab Monotherapy in Aggressive NHL

Although rituximab in combination with chemotherapy is widely accepted as standard therapy for the majority of patients with aggressive NHL, some patients may be unsuitable for chemotherapy because of frailty or comorbid disease, or they may be unwilling to receive such therapy. For these patients rituximab monotherapy may be an appropriate treatment option, as several studies have demonstrated anti-tumor activity and a good tolerability profile (Coiffier et al. 1998; Rothe et al. 2004; Tobinai et al. 2004) (Table 13.6).

13.2.8.6 Rituximab in Other Subtypes of Aggressive B-Cell NHL

13.2.8.6.1 Rituximab in PMBCL

The optimal management of primary mediastinal large B-cell lymphoma (PMBCL) has been widely debated. CHOP chemotherapy and more aggressive regimens such as MACOP-B and VACOP-B are widely used. At the BCCA, the addition of rituximab to CHOP chemotherapy for previously untreated PMBCL was recommended in March 2001 (Savage et al. 2006). A retrospective analysis of patients treated between 1980 and 2003 revealed 5-year OS rates of 87%, 81%, and 71% for patients treated with MACOP-B/VACOP-B ($n = 47$), R-CHOP ($n = 19$) and CHOP-like regimens ($n = 67$), respectively. Longer-term follow-up of larger numbers of patients treated with R-CHOP are required in order to assess the impact of this regimen on outcomes in PMBCL.

13.2.8.6.2 Rituximab in Burkitt's and Burkitt-Like Lymphoma or Lymphoblastic Lymphoma/Leukemia

Burkitt and Burkitt-like lymphomas represent clonal proliferations of poorly differentiated B-lineage lymphocytes. Lymphoblastic lymphoma is a rare subtype of NHL with biological features similar to those of ALL. These diseases are all highly aggressive and require intensive therapy. Multi-agent chemotherapy has typically been given in a dose-dense fashion (Kasamon and Swinnen 2004). CD20 is present on one-third of B-precursor blast cells and the majority of mature blasts, providing a rationale to explore rituximab in Burkitt's lymphoma and B-cell lymphoblastic leukemias and lymphomas (Gökbudget and Hoelzer 2004).

Thomas and colleagues (2006) have studied the addition of rituximab to hyper-CVAD in adults with newly diagnosed HIV-negative Burkitt/Burkitt-like lymphoma (n = 15) or ALL (n = 16). The 3-year OS, EFS and DFS rates were 89%, 80%, and 88%, respectively. The R-hyper-CVAD regimen was associated with a significantly lower rate of relapse than hyper-CVAD alone (7% versus 34% for all patients, p = 0.008; 0% versus 50% for elderly patients, p = 0.02). Toxicity profiles for hyper-CVAD and R-hyper-CVAD were similar (Thomas et al. 2006). Hoelzer and associates (2003), in the German Multicenter ALL Study Group (GMALL), have reported an interim analysis from a study evaluating the combination of rituximab with an established GMALL regimen in high-grade CD20-positive lymphomas. The mid-treatment ORR in 26 patients with Burkitt's lymphoma was 96%, including 60% CR, and no excess toxicity was apparent from the addition of rituximab. Further studies are currently under way to define the role of rituximab in these highly aggressive hematological malignancies.

13.2.9 Rituximab in MCL

There is currently no standard therapy for newly diagnosed or relapsed MCL. Many chemotherapy regimens including CHOP, FCM, and fractionated cyclophosphamide, vincristine, doxorubicin and dexamethasone hyper-CVAD have been shown to be highly active in producing tumor responses, but relapses typically occur. The use of HDT with ASCT and allogeneic SCT has also been extensively studied in MCL, and a randomized controlled trial has shown that for patients achieving a CR or PR after induction chemotherapy, ASCT results in a significantly better PFS compared with interferon maintenance (Dreyling et al. 2005b).

Rituximab was first studied as monotherapy in MCL, and showed some activity in this setting (Table 13.7). However, it is the combination of rituximab with chemotherapeutic regimens which has shown the greatest promise, both in previously untreated and relapsed MCL (Table 13.7).

13.2.9.1 Rituximab plus Chemotherapy in Previously Untreated MCL

A study by the GLSG compared the efficacy of CHOP alone with that of a combination of rituximab with CHOP in 122 patients with previously untreated MCL

Table 13.7 Trials of rituximab (R) in mantle cell lymphoma (MCL).

Reference	Regimen of R plus other agents × no. of cycles	Evaluable patients	ORR	CR	PR	Median PFS[a]	Median OS
Trials of rituximab plus chemotherapy							
Lenz et al. (2005)	R × 6 plus CHOP × 6	62 UT	94**	34***	60	21 months*	NR
	CHOP × 6 (randomization)	60 UT	75	7	68	14 months (TTF)	NR
Forstpointner et al. (2004);	R × 4 plus FCM × 4	24 PT	58	29**	29		2.5 years*
Dreyling et al. (2005a)	FCM × 4 (randomization)	24 PT	46	0	46		11 months
Howard et al. (2002)	R × 6 plus CHOP × 6	40 UT	96	48	48	16.6 months	
Romaguera et al. (2005)	R × 6–8 plus hyper-CVAD/M-A × 6–8	97 UT	97	87	10	3-year FFS = 64%	3-year OS = 82%
Romaguera et al. (2005)	R × 6–8 plus hyper-CVAD/M-A × 6–8	21 PT	95	43	52	18 months (FFS)	
Herold et al. (2004)	R × 8 plus MCP × 8	201 UT	86	42	38	19 months	
	MCP × 8 (randomization)		66	20	48	19 months	
Kahl et al. (2006)	R × 6 plus modified hyper-CVAD × 6 followed by R × 4 every 6/12 for 2 years	22 UT	77	64	14	37 months	Not reached at 37 months
Drach et al. (2005)	R × 4 plus thalidomide	18 PT	83	28	60	20.6 months	44.1 months
Trials of rituximab monotherapy							
Ghielmini et al. (2005)	Induction phase: R × 4	34 UT	27	3	24		
	Maintenance phase: R × 4	54 PT	28	2	26		
	observation (randomization)	34	41[b]	11[c]		6 months	
		27	55[b]	12[c]		12 months (EFS)	
Foran et al. (2000b)	R × 4	34 UT	38	16	22	1.2 years (median DR for all MCL patients)	
		40 PT	37	14	23		

All doses of R 375 $mg\,m^{-2}$.
a Or other parameter as indicated.
b Overall best response.
c % experiencing a CR at some time during the study.
* $p < 0.05$; ** $p < 0.01$; *** $p < 0.001$ versus comparator.
DR = duration of response; EPOCH = etoposide, prednisone, vincristine, cyclophosphamide, doxorubicin; FCM = fludarabine, cyclophosphamide, mitoxantrone; hyperCVAD = fractionated cyclophosphamide, vincristine, doxorubicin, dexamethasone; M-A = methotrexate, high-dose cytarabine; MCP = mitoxantrone, chlorambucil, prednisolone. Other abbreviations as Table 13.2.

(Lenz et al. 2005). Significant improvements were seen in overall response (94 versus 75%; p = 0.0054), CR (34 versus 7%; p = 0.00024) and median TTF (21 versus 14 months; p = 0.0131) in patients who received R-CHOP compared with patients who received CHOP alone. The authors did not notice a major difference in toxicities between the two treatment arms, and suggested that the R-CHOP combination may serve as a new baseline regimen for advanced-stage MCL (Lenz et al. 2005).

Similarly high rates of response were demonstrated in a prospective study conducted at the MD Anderson Cancer Center, which combined rituximab with a fractionated regimen of cyclophosphamide, vincristine, doxorubicin and dexamethasone (hyper-CVAD) alternating with rituximab plus high-dose methotrexate and cytarabine (Romaguera et al. 2005a). In the 97 assessable patients, a response rate of 97% was achieved, with a CR/CRu rate of 87%. Patients aged ≤65 years (n = 65) achieved significantly better outcomes than those aged >65 years (n = 32): 3-year FFS rates: 73 versus 50% (p = 0.02); 3-year OS rates: 86 versus 74% (p = 0.047). As expected, hematological toxicity was prominent and neutropenic fever occurred after 15% of treatment courses, with three deaths due to neutropenic sepsis. In addition, three patients developed and died from MDS while in remission from MCL and one patient remains alive with acute myeloid leukemia (AML). The authors concluded that the alternating R-hyper-CVAD/R-high-dose methotrexate and cytarabine regimen is an effective therapy for younger patients with previously untreated aggressive MCL. The significantly lower FFS rate seen in patients aged >65 years, and the significant hematological toxicity associated with this intense regimen, mean that it cannot be recommended as standard therapy for older patients.

Kahl and colleagues (2006) have assessed the safety and efficacy of modified hyper-CVAD followed by rituximab maintenance in previously untreated MCL. After a median follow-up of 37 months, the median PFS was 37 months, and the median OS had not been reached. As would be anticipated, the major toxicity was hematological and included 59 Grade 3/4 neutropenic events.

The phase III study of R-MCP versus MCP discussed above also included 90 patients with previously untreated MCL (Herold et al. 2004). The ORR and CR rates were higher in the R-MCP group than the MCP group (70 versus 63%, p = not significant; 32 versus 15%, p = 0.06, respectively), but the differences were not statistically significant. The median PFS was 20.5 months after R-MCP and 18.9 months after MCP (p = not significant), while the median EFS with R-MCP and MCP was 20 months and 13.9 months, respectively (p = 0.20) (Herold et al. 2004). The addition of rituximab to MCP does not appear therefore to improve outcomes as markedly in MCL patients as in FL patients.

13.2.9.2 Rituximab plus Chemotherapy in Relapsed MCL

There has also been much interest in studying the combination of rituximab with chemotherapy in the relapsed MCL setting. The GLSG study of R-FCM versus FCM in the relapsed setting included 50 patients with MCL. Overall and complete response rates were higher after R-FCM than FCM (58 versus 46%, p = 0.282 and

29 versus 0%, $p = 0.004$, respectively) (Dreyling et al. 2005a). Median overall survival was significantly extended after R-MCP compared with MCP (2.5 versus 0.9 years, $p = 0.031$).

The fractionated regimen of hyper-CVAD alternating with rituximab plus high-dose methotrexate and cytarabine described above has also been studied in the relapsed setting (Romaguera et al. 2005b). Preliminary results show an ORR of 95%, including 43% CR/CRu. Grade 4 neutropenia and thrombocytopenia were observed in 58% and 53% of patients, receiving rituximab plus intensive chemotherapy versus intensive chemotherapy alone, respectively.

A small study of patients with relapsed or refractory MCL using a combination of rituximab with thalidomide as salvage therapy demonstrated marked antitumor activity for this regimen (Kaufmann et al. 2004). In a recently updated analysis, the ORR was 83%, including 28% CR (Drach et al. 2005). The estimated 4-year survival was 48%. The promising results of this study have prompted the investigators to initiate a trial of rituximab in combination with CHOP and thalidomide in previously untreated patients with MCL (Drach et al. 2005).

13.2.9.3 Rituximab Monotherapy in MCL

An early phase II study conducted by Foran et al. (2000b) showed that rituximab has single-agent activity in previously untreated and relapsed MCL (Table 13.7). This was subsequently confirmed by Ghielmini and colleagues (2005). Rituximab monotherapy can thus be considered an appropriate treatment for patients unsuitable for more aggressive interventions.

13.2.10 Rituximab in CLL

CLL has traditionally been treated with alkylating agents such as chlorambucil and combination chemotherapeutic regimens such as CVP. The development of purine analogs has led to their integration into CLL therapy regimens, with fludarabine being the most widely used. However, despite improving response rates and response durations, these newer regimens are not curative and the search for more effective therapeutic options has continued.

13.2.10.1 Rituximab plus Chemotherapy in Previously Untreated CLL

A phase II study performed by the Cancer and Leukemia Group B (CALGB) examined the concurrent and sequential administration of fludarabine and rituximab in previously untreated CLL (Byrd et al. 2003). Concurrent FR was associated with higher ORR and CR rates than sequential $F \rightarrow R$ (90 versus 77% and 43 versus 28%, respectively). Infusion-related and hematological toxicity rates were higher with concurrent than sequential therapy, but infectious toxicities occurred at similar rates in the two treatment arms (Byrd et al. 2003). A retrospective comparison of outcomes with fludarabine–rituximab and fludarabine alone showed that both 2-year PFS (67 versus 45%, respectively; $p < 0.0001$) and 2-year OS (93 versus 81%, respectively; $p = 0.0006$) were significantly better in fludarabine–rituximab recipients than fludarabine monotherapy recipients (Byrd et al. 2005). The rate of infection-related toxicities was similar in both groups. The investigators recently reported the impact of genetic prognostic factors on outcomes in trial patients for whom cryopreserved cells were available (Byrd et al. 2006). Median PFS and OS were shorter in patients with unmutated IgV_H status and high-risk interphase cytogenetics. These observations suggest that, in the future, genetic features may be used to stratify therapies, with low-risk patients being assigned to well-tolerated therapies such as fludarabine-rituximab and more intensive therapies applied to patients with a poor prognosis.

The addition of cyclophosphamide to the fludarabine–rituximab combination (FCR) induced very high response rates in previously untreated CLL patients (Keating et al. 2005a). A total of 224 patients (median age 58 years) with progressive or advanced CLL was included in the study, which demonstrated an ORR of 95% (CR 70%, nodular PR (nPR) 10% and PR 15%), the highest ever observed in a clinical trial in this setting. Two-thirds of patients showed elimination of minimal residual disease analyzed by flow cytometry. The projected 4-year TTF was estimated to be 69%. The FCR regimen was generally well tolerated, with myelosuppression and infections – as anticipated – being the most common adverse events. Minor infections were reported in 10% of treatment courses and major infections (including pneumonia and septicemia) in 2.6% of courses (Keating et al. 2005a). Extended follow-up reveals a 4-year OS rate of 83% (Keating et al. 2005b). There were three cases of AML, three of myelodysplastic syndrome (MDS), and a total of 53 second cancers occurred. There were also 25 cases of autoimmune hemolytic leukemia (AIHA) and six of red cell aplasia (RCA), indicating that patients must be carefully monitored for complications after FCR therapy (Keating et al. 2005b).

The combination of rituximab with pentostatin and cyclophosphamide (PCR) is also showing promise in CLL (Kay et al. 2006). In 64 previously untreated patients, the ORR was 91% (41% CR, 22% nPR and 28% PR). Grade 3/4 neutropenia occurred in 41% of patients, and the most common Grade 3/4 nonhematological toxicities were nausea ($n = 6$), vomiting ($n = 4$), infection ($n = 6$) and fever without neutropenia ($n = 4$). Other studies of rituximab in combination with chemotherapy in previously untreated CLL are summarized in Table 13.8.

13.2.10.2 **Rituximab plus Chemotherapy in Relapsed CLL**

Rituximab in combination with fludarabine-based regimens has also been investigated in the relapsed CLL setting. The ORR in 177 patients who received FCR was 73% (25% CR, 16% nPR and 32% PR). Almost one-third of patients achieved molecular remission in bone marrow. As in the first-line setting, myelosuppression was the most common toxicity, with Grade 3/4 neutropenia occurring in 62% of assessable treatment courses and major infections noted in 16% of treated patients (Wierda et al. 2005a). A retrospective comparison with other fludarabine-based regimens showed that statistically significantly longer estimated median overall survival was noted in patients treated with FCR compared

Table 13.8 Trials of rituximab (R) in chronic lymphocytic leukemia (CLL).

Reference	*Regimen of R plus other agents × no. of cycles*	*Evaluable patients*	*ORR*	*CR*	*PR*	*Median PFS*[a]	*Median OS*
Trials of rituximab plus chemotherapy/immunotherapy in previously untreated CLL							
Keating et al. (2005a)	R × 1 then R 500 × 5 plus FC × 6	224 UT	95	70	25	48+ months	48+ months
Byrd et al. (2003)	R × 7 plus F × 6 then R × 4	51 UT	90	47	43	2-year PFS: 70%	
	F × 6 then R × 4 (randomization)	53 UT	77	28	49	2-year PFS: 70% (estimated)	
Schulz et al. (2002)	F × 2 then R × 2 plus F × 2 then R × 2	20 UT	85	20	65		
		11 PT	90	27	63		
Savage et al. (2003)	R 125–775 × 8 plus F × 8	6 UT	100	50	50		
		4 PT	75	25	50		
Kay et al. (2006)	R × 6 plus PC × 6	64 UT	91	41	50	32.6 months	
Yunus et al. (2003)	R × 7 plus P × 6	12 UT + 22 PT	50	33	17		
Mena et al. (2005)	R × 7 plus dose-intensive P × 6	133 UT + PT[b]	42[c]	18	24	25.6 months[c]	
			34[d]			13.9 months[d]	
Ferrajoli et al. (2005)	R × 4 plus GM-CSF × 8	29 UT + 26 PT	69	9	60		
Castro et al. (2005)	R × 4 plus HDMP × 3	16 UT	88	6	81	12+ months (TTP)	
Trials of rituximab monotherapy in previously untreated CLL							
Hainsworth et al. (2003a)	R × 4, then further R every 6/12 for total of 4 courses	44 UT	58	9	49	18.6 months	
Thomas et al. (2001)	R × 8	21 UT	86	19	67		

Table 13.8 *Continued*

Reference	*Regimen of R plus other agents × no. of cycles*	*Evaluable patients*	*ORR*	*CR*	*PR*	*Median PFS*[a]	*Median OS*
Trials of rituximab plus chemotherapy/immunotherapy in relapsed/refractory CLL							
Wierda et al. (2005a)	R × 1 then R 500 × 5 plus FC × 6	177 PT	73	25	48	28 months (TTP)	42 months
Lamanna et al. (2005)	R × 5 plus PC × 6	32 PT	75	25	50	40 months (TTF)	44 months
Hensel et al. (2004)	PC × 6 then R × 9	22 PT[e]	64	5	59		
Eichhorst et al. (2005)	R × 5–7 plus CHOP × 6–8	17 PT	70				
Mavromatis et al. (2005)	R × 6 plus FG × 6	5 UT	100	20	80		
		19 PT	53	5	47		
Wierda et al. (2005b)	R 375–500 × 6 plus FCA × 6	44 PT	65	27	38	19+ months all (estimated TTP; responders)	16 months (estimated)
Faderl et al. (2005)	R × 1 then R 500 × 3 plus A	20 PT	55	30	25		
Trials of rituximab monotherapy in relapsed/refractory CLL							
Byrd et al. (2001)	R 100 × 1 then R × 11	27 PT + 6 UT	45	3	42	6 months (TTP)	
Huhn et al. (2001)	R × 4	28 PT	25	0	25	16 weeks	
Itala et al. (2002)	R × 4	24 PT	35	0	35		
O'Brien et al. (2001)	R × 1 then R 500–2250 × 3	40 PT	36	0	36	8 months (TTP)	1-yr OS: 80% (estimated)

All doses of R 375 $mg m^{-2}$ unless otherwise indicated.
a Or other parameter as indicated.
b Mixture of NHL and CLL patients.
c ORR in CLL-UT.
d ORR in CLL-PT.
e Includes several patients with Waldenström's macroglobulinemia.
A = alemtuzumab; C = cyclophosphamide; F = fludarabine; FC = fludarabine, cyclophosphamide; FCA = fludarabine, cyclophosphamide, alemtuzumab; FG = fludarabine, genansense; HDMP = high-dose methyl prednisolone; PC = pentostatin, cyclophosphamide. Other abbreviations as Table 13.2.

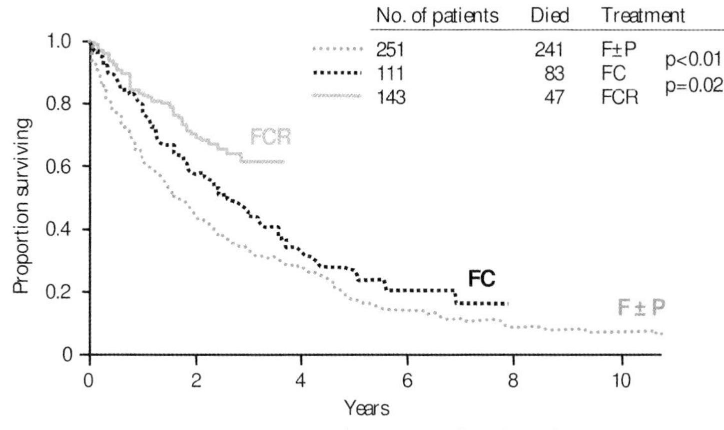

Fig. 13.8 Retrospective comparison of treatments for relapsed/refractory CLL: overall survival. (Reproduced with permission from Wierda et al. 2006.)

with patients treated with FC or fludarabine ± prednisolone (Fig. 13.8) (Wierda et al. 2006).

Wierda and colleagues (2005b) have also studied the addition of alemtuzumab to the FCR regimen in heavily pretreated patients with CLL. The ORR in 44 patients was 65%, with 27% CR and elimination of minimal residual disease by flow cytometry occurred in 92% of patients. This regimen was highly myelosuppressive, however, with nearly all patients experiencing at least one episode of Grade 3/4 neutropenia and infections occurring in 26% of patients. Faderl and associates (2005) administered rituximab together with alemtuzumab to patients with relapsed CLL. The ORR was 55%, with 30% CR. Infections occurred in 50% of patients, including cytomegalovirus (CMV) reactivation in 21%.

The PCR regimen is also under investigation in CLL patients who have previously received treatment. Lamanna et al. (2005) have reported very promising results for 32 patients with CLL and for 14 patients with other low-grade lymphoid neoplasms. The ORR was 75%, with 25% of patients achieving CR. The regimen was described as being well-tolerated, with the principal toxicity being myelosuppression (Lamanna et al. 2005).

The German CLL Study Group (GCLLSG) gave six cycles of R-CHOP to patients with fludarabine-refractory CLL or CLL with AIHA or Richter's transformation (Eichhorst et al. 2005). The ORR in 17 evaluable patients was 70%. The main toxicity was myelosuppression (59% of all documented courses), and 22% of patients developed infections. Other studies using rituximab-based regimens in relapsed/refractory CLL are summarized in Table 13.8.

13.2.10.3 Rituximab with Immune System Modulators in CLL

Patients with previously untreated or relapsed/refractory CLL were given four once-weekly infusions of rituximab plus thrice-weekly GM-CSF for 8 weeks

(Ferrajoli et al. 2005). The ORR in 55 patients was 69% with 9% CR, 9% nPR and 51% PR. There was a reduction in self-reported fatigue symptoms in two-thirds of the patients studied. The regimen was very well tolerated, with Grade 3/4 neutropenia and thrombocytopenia occurring in only 4% and 2% of patients, respectively. These results are very encouraging, and accrual to the study is continuing.

13.2.10.4 Rituximab Monotherapy in CLL

Rituximab monotherapy appears to have only modest efficacy in the CLL setting, possibly due to relatively low levels of CD20 expression on B cells in this disease compared with that of normal B cells or B cells in other neoplasms. Another possible reason is the presence of soluble CD20 in the plasma of CLL patients, which may inhibit the capacity of rituximab to bind to CLL B cells *in vivo* (Wierda et al. 2000c). Single-agent rituximab appears to have only modest activity in CLL (Table 13.8), but dose-densification (Byrd et al. 2001) or dose-intensification (O'Brien et al. 2001) can significantly improve response rates. The use of rituximab monotherapy as maintenance therapy (four once-weekly infusions at 6-month intervals for 2 years) has been shown to be a successful strategy in previously untreated CLL/SLL (Hainsworth et al. 2003a). At completion of therapy, ORR and CR rates were 58% and 9%, respectively and the estimated median progression-free time was 19 months.

13.2.11 Rituximab in the Transplant Setting

High-dose chemotherapy (HDT) followed by ASCT is employed in a variety of lymphomas. In order to reduce risk of relapse, *in-vitro* or *in-vivo* graft purging processes are typically employed. Rituximab has been used as an *in-vivo* purging process, being administered to patients before and/or during stem cell mobilization. Gisselbrecht and Mounier (2003) have reviewed early studies of rituximab as an *in-vivo* purging agent, which showed that rituximab effectively depletes the stem cell harvest without adversely impacting stem cell yield or engraftment or causing adverse clinical reactions.

A recently reported trial randomized patients with poor-risk FL to receive either HDT with rituximab *in-vivo* purging followed by ASCT or rituximab in combination with CHOP chemotherapy (Ladetto et al. 2005). The rate of progression or non-response was 35% in the R-CHOP arm compared with only 13% in the R-HDT + ASCT arm ($p < 0.05$). After a median of 24 months' follow-up, EFS was significantly extended in the transplant arm (66% versus 41%, $p < 0.001$). However, a longer-term follow-up is needed to determine impact on survival.

Recently, much interest has been expressed in the role of rituximab as a consolidation or maintenance therapy post-ASCT, in indolent, aggressive or MCL. Woods et al. (2005) have studied the use of two, 4-week courses of rituximab at 2 and 6 months post-ASCT in patients with recurrent FL. After a median follow-up of 4.8 years, median PFS and OS have not been reached, and 6/12 patients achieving PCR-negativity after rituximab remained in clinical remission at a median of 4.6 years post-ASCT.

Mangel and colleagues (2004) compared outcomes in patients who received HDT with a rituximab *in-vivo* purge and ASCT followed by posttransplant rituximab maintenance with historical controls treated with conventional chemotherapy alone. Rates of PFS and OS after 3 years were significantly extended in the rituximab-ASCT arm compared with the chemotherapy arm (89 versus 29%, $p < 0.00001$ and 88 versus 65%, $p = 0.052$, respectively). The Nordic Lymphoma Group has examined the use of pre-emptive rituximab for patients with MCL relapsing after HDT with rituximab purge and ASCT (Geisler et al. 2005). Eight of 10 patients who received pre-emptive rituximab for PCR-positivity became PCR-negative, and six remained in remission for between 200 and 600 days post-rituximab. Brugger and associates (2005) have also evaluated the impact of posttransplant rituximab in patients with FL and MCL. Recently updated results with a median follow-up of 61 months showed that a median PFS had not been reached, although 5-year PFS and OS rates were higher in FL patients (90% and 100%) than MCL patients (46 and 79%).

The use of posttransplant maintenance in aggressive NHL has also yielded promising results. Patients received two courses of four once-weekly rituximab infusions after HDT and ASCT: one course started at 6 weeks post-ASCT and the second at 6 months post-ASCT (Horwitz et al. 2004). With 30 months' median follow-up, the 2-year EFS and OS rates were 83% and 88%, respectively. Neutropenia was the most common toxicity, but all cases resolved spontaneously within 7 days, or within 2 to 4 days of G-CSF administration. The LNH98-B3 GELA study has examined the impact of four once-weekly rituximab infusions given 2 months posttransplant on relapse rate in patients with poor-risk DLBCL treated with HDT and ASCT (Haioun et al. 2005). With a median follow-up of 3 years, patients randomized to receive posttransplant rituximab ($n = 139$) had a tendency towards a better EFS than patients randomized to observation only ($n = 130$) (80 versus 72%, $p = 0.10$). Rituximab was well tolerated in this setting; the only clinically relevant infections observed were two cases of varicella zoster virus infection, both of which resolved.

The use of rituximab in the transplant setting thus appears to have a beneficial effect upon outcomes, and remains under active investigation. A large trial in relapsed FL coordinated by the European Blood and Bone Marrow Transplant (EBMT) Group is assessing the impact of rituximab *in-vivo* purging and post-transplant maintenance on outcomes. The results are eagerly awaited.

13.2.12 Rituximab in Other Malignancies of B-Cell Origin

The efficacy of rituximab has been investigated on other malignancies of B-cell origin, including posttransplant lymphoproliferative disorder (PTLD), HIV-associated lymphoma, primary CNS lymphoma (PCNSL), hairy cell leukemia (HCL), and Hodgkin's disease (HD). Although the evidence available for the efficacy and safety of rituximab in these settings is limited, the preliminary results are promising.

13.2.12.1 Rituximab in PTLD

PTLD is a life-threatening complication that occurs in patients receiving immunosuppressive therapy following organ transplantation. Following reports of the successful use of rituximab in this disease, prospective trials were initiated. Choquet et al. (2006) performed a multicenter phase II trial of four once-weekly infusions of rituximab in patients with untreated PTLD, not responding to tapering of immunosuppression. The ORR in 43 patents was 44%, and 12 patients achieved CR or CRu. The overall survival rate at 1 year was 67%. Oertel and colleagues (2005) also examined this regimen in 17 patients with PTLD. Nine patients (53%) achieved a complete remission with a mean duration of 17.8 months, and no severe adverse events were reported. A third trial in 11 patients found an ORR of 64% with six CRs, and rituximab was well tolerated (Blaes et al. 2005). These promising results have led to the instigation of new trials examining the efficacy and safety of rituximab in combination with other therapies in PTLD. The early results from a trial of sequential rituximab, CHOP and G-CSF have shown an ORR of 83%, including 62% CR, in 24 evaluable patients (Trappe et al. 2005). Grade 3/4 leucopenia occurred in 29 of 84 chemotherapy cycles applied.

13.2.12.2 Rituximab in HIV-Associated Lymphoma

HIV-associated lymphomas are a major course of morbidity and mortality in patients with HIV infection. Rituximab has been administered in combination with CHOP chemotherapy (Boue et al. 2006; Kaplan et al. 2005), cyclophosphamide, doxorubicin and etoposide (CDE) chemotherapy (Spina et al. 2005) and DA-EPOCH chemotherapy (Dunleavy et al. 2004) in this setting. The largest study published to date examined outcomes with CHOP or R-CHOP in 150 HIV-infected patients (Kaplan et al. 2005). The R-CHOP regimen was associated with longer median TTP, PFS and OS than CHOP, but these differences did not reach statistical significance because the study was underpowered to detect clinical benefit. Fewer patients died from NHL progression with R-CHOP than CHOP (14 versus 29%), but significantly more died due to treatment-related infection in the R-CHOP arm than the CHOP arm (14 versus 2%, $p = 0.035$). Most of these infectious deaths occurred in severely immunocompromised patients with a CD4+ lymphocyte count of $<50\,mm^{-3}$. A low CD4+ count is a well-recognized risk factor for febrile neutropenia and early death with cytotoxic chemotherapy, and rituximab might increase this risk. Ongoing studies will determine whether the sequential administration of chemotherapy and rituximab or the use of prophylactic antibiotics can reduce the incidence of infectious complications.

13.2.12.3 Rituximab in PCNSL

The majority of PCNSLs in immunocompetent patients are DLBCLs, and rituximab is therefore a rational treatment option. However, its potential efficacy is limited by its high molecular weight, which prevents penetration into the CNS through an intact blood–brain barrier (Hoang-Xuan et al. 2004). Preliminary investigations of the intraventricular/intrathecal administration of rituximab – which allowed a higher concentration to be achieved in the cerebrospinal fluid (CSF) – have yielded encouraging results (Rubenstein et al. 2004; Schulz et al. 2004). For instance, Schulz and co-workers (2004) observed total clearance of malignant cells in the CSF following intraventricular rituximab, in two patients with PCNSL and in one patient with Burkitt's lymphoma with meningeal involvement. The administration of intravenous rituximab in combination with the alkylating agent temozolomide is another promising strategy (Enting et al. 2004; Wong et al. 2004). Further studies of rituximab in PCNSL are ongoing.

13.2.12.4 **Rituximab in HCL**

Hairy cell leukemia is a chronic CD20-positive B-cell malignancy which traditionally is treated with nucleoside analogs or interferon. However, many patients are refractory to or intolerant of standard therapies. Rituximab is showing promise in the treatment of this disease. Thomas et al. (2003) administered eight doses of rituximab (and a further four doses to responders who failed to achieve CR) and obtained an ORR of 80%, including 53% CR (2004). Toxicity was minimal, and no infectious events were observed. The use of chlorodeoxyadenosine followed by eight once-weekly infusions of rituximab was recently shown to eradicate minimal residual disease in all 12 patients studied, and no patients had relapsed after a median of 11 months follow-up (Ravandi-Kashani et al. 2005).

13.2.12.5 **Rituximab in HD**

Classical HD expresses CD20 on only a minority of cells, but the lymphocyte-predominant subtype of the disease (LPHD) has high CD20 expression. A phase II study conducted by the German Hodgkin Lymphoma Study Group (GHSG) assessed four once-weekly infusions of rituximab in 21 patients with relapsed LPHD (Schulz et al. 2005b). The ORR was 90% and the median TTP 31 months. The combination of rituximab and chemotherapy has been examined in classical HD. It is suggested that in these patients, rituximab may eliminate B cells from the tumor and deprive malignant lymphoid cells of important growth signals. Younes and colleagues (2005a) gave six weekly doses of rituximab and doxorubicin, bleomycin, vinblastine, dacarbazine (ABVD) chemotherapy to 72 newly diagnosed patients with classical HD. With a median follow-up of 32 months, estimated EFS and OS were 82% and 100%, respectively, In another study, the combination of gemcitabine and rituximab was evaluated in relapsed or refractory classical HD (Younes et al. 2005b); here, 13 patients (50%) responded to treatment. Studies are ongoing to further elucidate the safety and efficacy of rituximab in HD.

13.3 Rituximab in Autoimmune Disorders

The issue of the relative contributions made by T cells, B cells, cytokines and other elements to the pathogenesis of autoimmunity has been debated for many decades. Over the past 10 years, there has been an upsurge of interest in the role of B cells and the notion that blocking them may be beneficial. Edwards and Cambridge (2001) were the first to show that B-cell depletion is a successful treatment strategy in rheumatoid arthritis (RA), and subsequently a number of groups have examined B-cell depletion in other autoimmune diseases including systemic lupus erythematosus (SLE) and autoimmune cytopenias (AC).

13.3.1
Rituximab in RA

The first evidence that rituximab might be effective against RA came from early case reports describing patients with NHL and coexisting RA. These data, coupled with evidence that B cells play a key role in RA pathogenesis, led initially to small open-label studies and later to a phase II clinical trial program. Two large phase IIb studies investigating the safety and efficacy of rituximab in RA have recently reported extremely encouraging results.

Preliminary results from the Dose-ranging Assessment iNternational Clinical Evaluation of Rituximab in rheumatoid arthritis (DANCER) study have revealed that rituximab in combination with methotrexate induced significantly better responses than methotrexate plus placebo in patients with active disease, despite previous therapy with traditional disease-modifying anti-rheumatic drugs (DMARDS) (other than methotrexate) or an anti-tumor necrosis factor (TNF) agent. (Emery et al. 2005). Efficacy was examined based on achievement of American College of Rheumatology (ACR) 20, 50, and 70 responses at 24 weeks compared with equivalent responses after methotrexate treatment (Fig. 13.9).

As its name suggests, the DANCER study was designed to investigate different dosing regimens of rituximab; 2 × 500 mg and 2 × 1000 mg doses were used (Emery et al. 2005). Although no dose–response relationship was evident with ACR20 and ACR50 responses, there was a trend towards a greater number of patients achieving an ACR70 response with the higher rituximab dose. The pattern of more patients achieving other high-hurdle endpoints (European League Against Rheumatism (EULAR) "good" response, Disease Activity Score (DAS) "remission" and "low disease") was also evident for the higher rituximab dose.

Another important finding from the DANCER study was the significant positive impact that rituximab had on patient-reported outcomes, including health-related quality of life assessments (Mease et al. 2005). This is especially significant considering the chronic, painful nature of RA.

The higher (2 × 1000 mg) dose of rituximab was studied in the Randomised Evaluation oF Long term Efficacy or rituXimab in rheumatoid arthritis (REFLEX) study, performed in patients who had active disease despite treatment with one or more anti-TNF therapies plus methotrexate (Cohen et al. 2006). The primary ITT efficacy population comprised 499 patients, including both rheumatoid factor

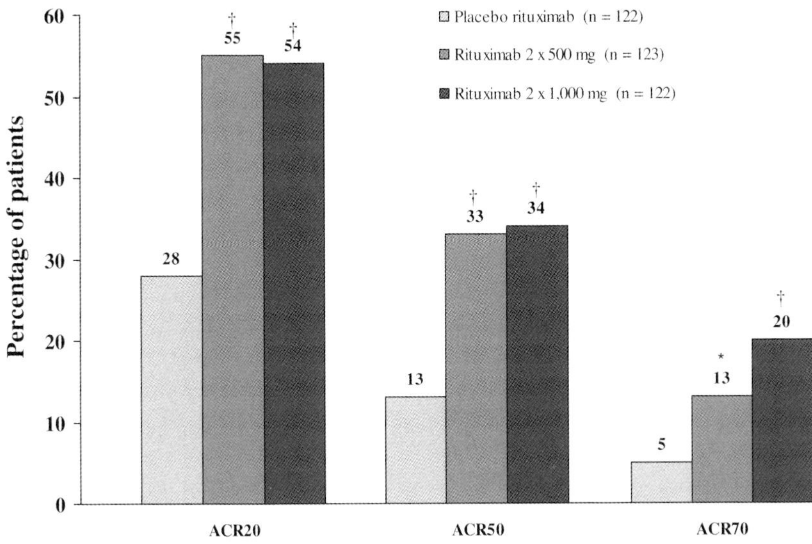

Fig. 13.9 Response rates following administration of two different doses of rituximab or placebo together with methotrexate in patients refractory to prior therapy: Dose-ranging Assessment iNternational Clinical Evaluation of Rituximab in rheumatoid arthritis (DANCER) trial.

(RF)-positive and -negative patients. Again, rituximab demonstrated a favorable efficacy profile, with significantly more rituximab-treated patients achieving ACR20, ACR50 and ACR70 responses at 24 weeks compared with those who received placebo. Symptomatic improvement in disease was again mirrored by improvements in patient-related outcomes (Keystone et al. 2005).

Preliminary data suggest that a second course of rituximab is safe and efficacious in RA (Fleischmann et al. 2005). Further studies are under way to establish the optimal dosing regimen for rituximab in RA, including time to retreatment, and to assess efficacy in earlier lines of therapy.

13.3.2
Rituximab in SLE

Recent research in SLE has led to the discovery that B cells play a key role in the pathogenesis of the disease (Anolik and Aringer 2005). As a consequence, rituximab has been used to directly target the immune cells that are abnormal in SLE, with promising results. Early studies of rituximab in this setting revealed a good tolerability profile and significant improvements in disease activity scores at 2 and 3 months, which persisted for 12 months (Looney et al. 2004). Randomized control trials of rituximab are under way in patients with SLE, and the results are eagerly awaited (Eisenberg and Looney 2005).

13.3.3
Rituximab in Autoimmune Cytopenias and Hemophilia

The autoimmune cytopenias (AC) include AIHA, pure red cell aplasia (PRCA), immune thrombocytopenia (ITP), autoimmune neutropenia (AIN), and various combinations of these disorders such as Evans' syndrome. B cells play a crucial role in the pathogenesis of these disorders, providing a rationale for the use of rituximab in their treatment. The majority of patients with AC respond to standard immunosuppressive therapy (e.g., steroids), but a proportion require second-line treatment, including cytotoxic therapy, G-CSF, or splenectomy. Some patients remain resistant even after these interventions, and the use of monoclonal antibodies has become a new approach in treating patients with severe, refractory AC. Robak (2004) has reviewed the use of rituximab in AC, and concluded that it has significant efficacy in the treatment of refractory or relapsed ITP, AIHA, and PRCA.

The treatment of acquired hemophilia typically includes steroids, chemotherapy, or intravenous immunoglobulin. Recently, interest has been shown in the use of rituximab for selective B-cell depletion in acquired hemophilia, and the results of preliminary studies have been promising (Kessler 2005). Prospective controlled studies have not yet been performed, however.

13.3.4
Rituximab in Chronic Graft-versus-Host Disease (GvHD)

Chronic GvHD is an important cause of morbidity and mortality in patients receiving allogeneic transplants. A coordinated B-cell and T-cell response occurs in at least some patients with chronic GvHD (Cutler and Antin 2006). Corticosteroids are the mainstay of therapy, but many patients are refractory to this approach. Ratanatharathorn et al. (2003) observed a sustained response in four of eight patients with chronic GvHD who received four cycles of rituximab. In a further case series, patients were permitted to receive up to eight cycles of rituximab, and five of six patients had objective responses (Canninga-van Dijk et al. 2004). Several larger series have been reported recently: Cutler and Antin (2006) obtained an ORR of 82% in 18 patients, predominantly in those with cutaneous and rheumatologic involvement, while Zaja et al. (2005) found that 64% of 25 patients with chronic GvHD with cutaneous involvement responded. Further clinical trials of rituximab in chronic GvHD are currently in progress.

13.3.5
Rituximab in Other Autoimmune Disorders

Altered development and function of B cells may be important in the pathogenesis of a range of other autoimmune disorders, including Sjögren's syndrome, antineutrophil cytoplasmic antibody (ANCA)-associated vasculitis, and dermatomyositis. The strategy of B-cell depletion with rituximab is being investigated as a therapeutic option in these diseases. In a phase II study in Sjögren's syndrome patients, significant improvements in subjective symptoms were achieved in 15 patients, but four developed HACAs and three had associated serum sickness (Pijpe et al. 2005). In a study of 11 patients with ANCA-associated vasculitis, four infusions of rituximab + prednisolone produced complete remission in 3 months and were well tolerated (Keogh et al. 2006). In a small pilot study in dermatomyositis, all six evaluable patients who received four once-weekly infusions of rituximab exhibited major clinical improvement, with muscle strength increasing over baseline by between 36 and 113% (Levine 2005). Maximal improvements in muscle strength occurred as early as 12 weeks after the initial infusion of rituximab. CD20+ B cells were effectively depleted in all patients by 12 weeks. Rituximab was well tolerated, with no treatment-related severe or serious adverse events during the observation period of this study. Further studies are under way in these settings.

Case reports describing the use of rituximab in a number of other autoimmune disorders have also appeared in the literature. Opsoclonus-myoclonus syndrome (OMS) is a rare serious neurological disorder with significant unmet needs. The administration of rituximab to a child with OMS led to the elimination of CSF B cells and dramatic clinical improvement (Pranzatelli et al. 2005). Myasthenia gravis, a B-cell-mediated autoimmune neuromuscular disorder, is characterized by weakness and fatigability of skeletal muscles. Rituximab has been shown to have a favorable impact on this disorder (Gajra et al. 2004). B cells are also thought to play a key role in the pathogenesis of pemphigus, and successful reports of rituximab's use have been published (Arin and Hunzelmann 2005). It should be emphasized that these data are extremely limited and do not support the routine use of rituximab in these settings. However, for refractory patients, enrolment in a clinical trial assessing the benefits of rituximab may be appropriate. Many such trials are registered on www.clinicaltrials.gov.

13.4 Summary and Conclusions

As rituximab targets the CD20 antigen on B cells, producing selective and sustained B-cell depletion, it is an attractive therapeutic option not only for B-cell malignancies but also for a range of systemic autoimmune diseases in which altered development and/or function of B cells appear to play a prominent role.

A very large body of evidence now exists to support the use of rituximab in NHL. When used in combination with a variety of chemotherapeutic regimens in indolent and aggressive NHL, rituximab increases the efficacy of these treatments. An increase in the incidence of neutropenia has been observed when rituximab is combined with some chemotherapies, but this does not appear to translate into an increased risk of infections. The combination of rituximab with CHOP chemotherapy was the first regimen in over 20 years to improve survival over CHOP chemotherapy alone, and R-CHOP is now accepted worldwide as standard therapy for previously untreated aggressive NHL. For patients with advanced symptomatic indolent NHL, the addition of rituximab to various chemotherapies – including CHOP, CVP and MCP – also appears to improve clinical outcomes. Very encouragingly, significant improvements in overall survival have been reported with immunochemotherapy regimens in this setting.

Recently there has been much interest in the use of rituximab maintenance therapy to prolong remission in patients with previously untreated or relapsed/refractory indolent NHL who respond to induction therapy, either with chemotherapy, rituximab alone or rituximab in combination with chemotherapy. For patients with relapsed/refractory FL responding to (immuno)chemotherapy induction, the use of rituximab maintenance therapy has been shown to confer a survival advantage compared with observation only. Patients with previously untreated FL who responded to CVP induction have also been shown to have improved overall survival after maintenance therapy compared with observation.

Rituximab has also been investigated in CLL. Phase II trials of rituximab in combination with fludarabine and cyclophosphamide (FCR) have demonstrated extremely encouraging results, and the FCR regimen is currently being investigated in phase III trials. For patients with MCL – an NHL with a notoriously poor outcome – the use of rituximab in combination with chemotherapy or transplantation strategies has also been shown to improve outcomes.

Combining rituximab with immune system modulators such as cytokines provides alternative, effective options to standard chemotherapy combinations. Rituximab/cytokine combinations or rituximab monotherapy may be appropriate treatment options for patients who are unable or unwilling to receive standard chemotherapy regimens. Rituximab monotherapy may also be a suitable alternative to "watch and wait" for patients with asymptomatic indolent NHL who seek a therapeutic intervention for their disease.

The positive results obtained in hematological malignancies have prompted the study of rituximab in a number of other B-cell-associated diseases, including RA, SLE, and autoimmune cytopenias. In RA, rituximab appears to be a useful addition to the therapeutic armamentarium, inducing high rates of response in patients who are refractory to prior anti-TNF therapies.

The extensive clinical experience with rituximab has yielded a consistent safety and tolerability profile, as recently reviewed by Kimby (2005). The majority of patients receiving their first infusion of rituximab experience flu-like symptoms, while other common symptoms include headache, nausea, rash, and fatigue. Approximately 10% of patients with hematological malignancies experience more severe symptoms with their first infusion, including bronchospasm, hypoxia, and hypotension. These necessitate the interruption of the infusion and the instigation of supportive therapies as required, including oxygen, steroids, and bronchodilators. The vast majority of patients are able to receive further infusions of rituximab without experiencing severe adverse events.

Rituximab is associated with a low incidence of severe hematological adverse effects. There have been some reports of late-onset neutropenia after rituximab, with most cases occurring in experimental studies of rituximab in the setting of transplantation. Further investigation of the mechanism of this adverse effect is required. Rarely, mucocutaneous reactions and pulmonary adverse events may occur in association with rituximab treatment. The safety and tolerability profile of rituximab in RA appears similar to that observed in oncology, although the incidence and severity of adverse events is lower. This may reflect the fact that RA patients do not experience the cytokine release or tumor lysis syndromes observed in patients with B-cell malignancies.

Rituximab has become an integral part of treatment for many patients with NHL and CLL. It is also showing great promise in a variety of autoimmune diseases, including RA and autoimmune cytopenias. A major clinical trial program continues to assess the optimal use of this therapy in patients with hematological malignancies and autoimmune disorders.

References

Alas, S., Bonavida, B. (2001) Rituximab inactivates signal transducer and activation of transcription 3 (STAT3) activity in B-non-Hodgkin's lymphoma through inhibition of the interleukin 10 autocrine/paracrine loop and results in down-regulation of Bcl-2 and sensitization to cytotoxic drugs. *Cancer Res* 61: 5137–5144.

Anderson, K.C., Bates, M.P., Slaughenhoupt, B.L., et al. (1984) Expression of human B-cell associated antigens on leukemias and lymphomas: a model of human B cell differentiation. *Blood* 63: 1424–1433.

Anolik, J.H., Aringer, M. (2005) New treatments for SLE: cell-depleting and anti-cytokine therapies. *Best Pract Res Clin Rheumatol* 19: 859–878.

Ansell, S.M. (2003) Adding cytokines to monoclonal antibody therapy: does the concurrent administration of interleukin-12 add to the efficacy of rituximab in B-cell non-Hodgkin's lymphoma? *Leuk Lymphoma* 44: 1309–1315.

Arin, M.J., Hunzelmann, N. (2005) Anti-B-cell-directed immunotherapy (rituximab) in the treatment of refractory pemphigus – an update. *Eur J Dermatol* 15: 224–230.

Armitage, J.O., Weisenburger, D.D. (1998) New approach to classifying non-Hodgkin's lymphomas: clinical features of the major histologic subtypes. Non Hodgkin's Lymphoma Classification Project. *J Clin Oncol* 16: 2780–2795.

Bairey, O., Dann, E.J., Herishanu, Y., et al. (2005) Rituximab Retreatment in B-Cell Non-Hodgkins Lymphoma Patients. Abstract 2455, updated in a poster presentation at the 47th Annual Meeting of the American Society of Hematology (ASH), 10–13 December 2005, Atlanta, USA.

Bennett, M., Sharma, K., Yegena, S., Gavish, I., Dave, H.P., Schechter, G.P. (2005) Rituximab monotherapy for splenic marginal zone lymphoma. *Haematologica* 90: 856–858.

Berinstein, N.L., Grillo-López, A.J., White, C.A., et al. (1998) Association of serum rituximab (IDEC-C2B8) concentration and anti-tumor response in the treatment of recurrent low-grade or follicular non-Hodgkin's lymphoma. *Ann Oncol* 9: 995–1001.

Bertoni, F., Zucca, E. (2005) State-of-the-art therapeutics: marginal zone lymphoma. *J Clin Oncol* 23: 6415–6420.

Blaes, A.H., Peterson, B.A., Bartlett, N., et al. (2005) Rituximab therapy is effective for posttransplant lymphoproliferative disorders after solid organ transplantation: results of a phase II trial. *Cancer* 104: 1661–1667.

Boue, F., Gabarre, J., Gisselbrecht, C., et al. (2006) Phase II trial of CHOP plus rituximab in patients with HIV-associated non-Hodgkin's lymphoma. *J Clin Oncol* 24: 4123–4128.

Boye, J., Elter, T., Engert, A. (2003) An overview of the current clinical use of the anti-CD20 monoclonal antibody rituximab. *Ann Oncol* 14: 520–535.

Brugger, W., Hirsch, J., Repp, R., et al. (2005) Long Term Remission in Follicular Lymphoma (FL) and Mantle Cell Lymphoma (MCL) with Rituximab Consolidation after High-Dose Chemotherapy and Autologous Stem Cell Transplantation: Extended Follow-Up of a Multicenter phase II Study. Abstract 2078, updated in a poster presentation at the 47th Annual Meeting of the American Society of Hematology (ASH), 10–13 December 2005, Atlanta, USA.

Byrd, J.C., Murphy, T, Howard, R.S., et al (2001) Rituximab using a thrice-weekly dosing schedule in B-cell chronic lymphocytic leukemia and small lymphocytic lymphoma demonstrates clinical activity and acceptable toxicity. *J Clin Oncol* 19: 2153–2164.

Byrd, J.C., Peterson, B.L., Morrison, V.A., et al. (2003) Randomized phase 2 study of fludarabine with concurrent versus sequential treatment with rituximab in symptomatic untreated patients with B-cell chronic lymphocytic leukemia: results from Cancer and Leukemia Group B 9712 (CALGB 9712). *Blood* 101: 6–14.

Byrd, J.C., Rai, K., Peterson, B.L., et al. (2005) Addition of rituximab to fludarabine may prolong progression-free survival and overall survival in patients with previously untreated chronic lymphocytic leukemia: an updated retrospective comparative analysis of CALGB 9712 and CALGB 9011. *Blood* 105: 49–53.

Byrd, J.C., Gribben, J., Peterson, B.L., et al. (2006) Select high-risk genetic features predict earlier progression following chemoimmunotherapy with fludarabine and rituximab in chronic lymphocytic leukemia: justification for risk-adapted therapy. *J Clin Oncol* 24: 437–443.

Canales, M.A., Palacios, A., Martinez-Chamorro, C., et al. (2005) Re-Treatment with Rituximab Plus Chemotherapy in Patients with Diffuse Large-B Cell Lymphoma (DLBCL): A Spanish Multicenter Study. Abstract 2457, updated in a poster presentation at the 47th Annual Meeting of the American Society of Hematology (ASH), 10–13 December 2005, Atlanta, USA.

Canninga-van Dijk, M.R., van der Straaten, H.M., Fijnheer, R., Sanders, C.J., van den Tweel, J.G., Verdonck, L.F. (2004) Anti-CD20 monoclonal antibody treatment in 6 patients with therapy-refractory chronic graft-versus-host disease. *Blood* 104: 2603–2606.

Castro, J.E., Bole, J.E., Prado, C.E., et al. (2005) Rituximab and high dose Methylprednisolone (HDMP) as a first line treatment for patients with chronic lymphocytic leukemia. Abstract 2969, updated in a poster presentation at the 47th Annual Meeting of the American Society of Hematology (ASH), 10–13 December 2005, Atlanta, USA.

Cheson, B.D., Horning, S.J., Coiffier, B., et al. (1999) Report of an international workshop to standardize response criteria for non-Hodgkin's lymphoma. *J Clin Oncol* 17: 1244–1253.

Choquet, S., Leblond, V., Herbrecht, R., et al. (2006) Efficacy and safety of rituximab in B-cell post-transplant lymphoproliferative disorders: results of a prospective multicentre phase II study. *Blood* 107: 3053–3057.

Chow, K.U., Sommerlad, W.D., Boehrer, S., et al. (2002) Anti-CD20 antibody (IDEC-C2B8, rituximab) enhances efficacy of cytotoxic drugs on neoplastic lymphocytes in vitro: role of cytokines, complement, and caspases. *Haematologica* 87: 33–43.

Cohen, A., Polliack, A., Ben-Bassat, I., et al. (2002) Results of a phase II study employing a combination of fludarabine, cyclophosphamide, plus rituximab (FC + R) in relapsed follicular lymphoma patients. Abstract, updated in a poster presentation at the 44th Annual Meeting of the American Society of Hematology (ASH), 7–10 December 2002, Philadelphia, USA.

Cohen, S.B., Emery, P., Greenwald, M.W., et al. (2006) Rituximab for rheumatoid arthritis refractory to anti-tumor necrosis factor therapy: Results of a multicenter, randomized, double-blind, placebo-controlled, phase III trial evaluating primary efficacy and safety at twenty-four weeks. *Arthritis Rheum* 54: 2793–2806.

Coiffier, B., Haioun, C., Ketterer, N., et al. (1998) Rituximab (anti-CD20 monoclonal antibody) for the treatment of patients with relapsing or refractory aggressive lymphoma: a multicenter phase II study. *Blood* 92: 1927–1932.

Coiffier, B., Lepage, E., Brière, J., et al. (2002) CHOP chemotherapy plus rituximab compared with CHOP alone in elderly patients with diffuse large-B-cell lymphoma. *N Engl J Med* 346: 235–242.

Coiffier, B., Herbrecht, R., Tilly, H., et al. (2003) GELA study comparing CHOP and R-CHOP in elderly patients with DLCL: 3-year median follow-up with an analysis according to co-morbidity factors. Abstract 2395, updated in a poster presentation at the 39th Annual Meeting of the American Society of Clinical Oncology, 29 May–2 June 2003, Chicago, USA.

Coiffier, B. (2005) State-of-the-art therapeutics: diffuse large B-cell lymphoma. *J Clin Oncol* 23: 6387–6393.

Colombat, P., Salles, G., Brousse, N., et al. (2001) Rituximab (anti-CD20 monoclonal antibody) as single first-line therapy for patients with follicular lymphoma with a low tumor burden: clinical and molecular evaluation. *Blood* 97: 101–106.

Conconi, A, Martinelli, G, Thieblemont, C, et al. (2003) Clinical activity of rituximab in extranodal marginal zone B-cell lymphoma of MALT type. *Blood* 102: 2741–2745.

Cortes-Funes, H., de la Serna, J., Flores, E., et al. (2000) Rituximab in monotherapy in patients with relapsed follicular or low grade B-cell non-Hodgkin's lymphoma: results after a 6 month follow-up cut-off. Abstract 114, updated in a poster presentation at the 36th Annual Meeting of the American Society of Clinical Oncology (ASCO), 20–23 May, 2000, New Orleans, USA.

Cutler, C., Antin, J.H. (2006) Chronic graft-versus-host disease. *Curr Opin Oncol* 18: 126–131.

Czuczman, M.S., Grillo-López, A.J., White, C.A., et al. (1999) Treatment of patients with low-grade B-cell lymphoma with the combination of chimeric anti-CD20 monoclonal antibody and CHOP chemotherapy. *J Clin Oncol* 17: 268–276.

Czuczman, M., Weaver, R., Alkuzweny, B., et al. (2004) Prolonged clinical and molecular remission in patients with low-grade or follicular non-Hodgkin's lymphoma treated with rituximab plus CHOP chemotherapy: 9-year follow-up. *J Clin Oncol* 22: 4659–4664.

Czuczman, M.S., Koryzna, A., Mohr, A., et al. (2005) Rituximab in combination with fludarabine chemotherapy in low-grade or follicular lymphoma. *J Clin Oncol* 23: 694–704.

Davis, T.A., White, C.A., Grillo-López, A.J. (1999) Single-agent monoclonal antibody efficacy in bulky non-Hodgkin's lymphoma: results of a phase II trial of rituximab. *J Clin Oncol* 17: 1851–1857.

Davis, T.A., Maloney, D.G., Grillo-López, A.J., et al. (2000a) Combination immunotherapy of relapsed or refractory low-grade or follicular non-Hodgkin's lymphoma with rituximab and interferon-α-2a. *Clin Cancer Res* 6: 2644–2652.

Davis, T.A., Grillo-López, A.J., White, C.A., et al. (2000b) Rituximab anti-CD20 monoclonal antibody therapy in non-Hodgkin's lymphoma: safety and efficacy of re-treatment. *J Clin Oncol* 18: 3135–3143.

Demidem, A., Hanna, N., Hariharan, H., et al. (1995) Chimeric anti-CD20 antibody (IDEC C2B8) is apoptotic and sensitises drug-resistant human B cell lymphomas to the cytotoxic effect of CDDP, VP-16 and toxins. *FASEB J* 9: A206.

Demidem, A., Lam, T., Alas, S., et al. (1997) Chimeric anti-CD20 (IDEC-C2B8) monoclonal antibody sensitises a B cell lymphoma cell line to cell killing by cytotoxic drugs. *Cancer Biother Radiopharm* 12: 177–186.

DeVita, V.T., Jr., Canellos, G.P., Chabner, B., et al. (1975) Advanced diffuse histiocytic lymphoma, a potentially curable disease. *Lancet* 1: 248–250.

Di Bella, N., Reynolds, C., Faragher, D., et al. (2005) An open-label pilot study of pentostatin, mitoxantrone and rituximab in patients with previously untreated, Stage III or IV, low-grade non-Hodgkin lymphoma. *Cancer* 103: 978–984.

Di Gaetano, N., Ziao, Y., Erba, E., et al. (2001) Synergism between fludarabine and rituximab revealed in a follicular lymphoma cell line resistant to the cytotoxic activity of either drug alone. *Br J Haematol* 114: 800–809.

Dighiero, G. (2005) CLL Biology and Prognosis. *Hematology (Am Soc Hematol Educ Program)*: 278–284.

Dillman, R.O. (2003) Treatment of low-grade B-cell lymphoma with the monoclonal antibody rituximab. *Semin Oncol* 30: 434–437.

Dimopoulos, M.A., Zervas, C., Zomas, A., et al. (2002) Treatment of Waldenström's macroglobulinemia with rituximab. *J Clin Oncol* 20: 2327–2333.

Domingo-Domenech, E., Gonzalez-Barca, E., Estany, C., et al. (2002) Combined treatment with anti-CD20 (rituximab) and CHOP in relapsed advanced-stage follicular lymphomas. *Haematologica* 87: 1229–1230.

Drach, J., Kaufmann, H., Seidl, L., et al. (2005) Updated analysis of thalidomide plus rituximab in relapsed mantle cell lymphoma. Abstract 340, updated in a poster presentation at the 9th International Congress on Malignant Lymphoma (ICML), 8–11 June 2005, Lugano, Switzerland.

Drapkin, R., Di Bella, N., Faragher, D.C., et al. (2003) Results of a phase II multicenter trial of pentostatin and rituximab in patients with low-grade B-cell non-Hodgkin's lymphoma; an effective and minimally toxic regimen. *Clin Lymphoma* 4: 169–175.

Dreyling, M.H., Forstpointner, R., Ludwig, W., et al. (2005a) Combined immuno-chemotherapy (R-FCM) results in superior remission rates and overall survival in recurrent follicular lymphoma and mantle cell lymphoma – follow-up of a prospective randomized trial of the German Low Grade Lymphoma Study Group. Abstract 6528, updated in an oral presentation at the 41st Annual Meeting of the American Society of Clinical Oncology (ASCO), May 13–17 2005, New Orleans, USA.

Dreyling, M., Lenz, G., Hoster, E., et al (2005b) Early consolidation by myeloablative radiochemotherapy followed by autologous stem cell transplantation in first remission significantly prolongs progression-free survival in mantle-cell lymphoma: results of a prospective randomized trial of the European MCL Network. *Blood* 105: 2677–2684.

Dunleavy, K., Little, R., Gea-Banacloche, J., et al. (2004) Abbreviated Treatment with Short-Course Dose-Adjusted EPOCH and Rituximab (DA-EPOCH-R) Is Highly Effective in AIDS-Related Lymphoma (ARL). Abstract 3111, updated in a poster presentation at the 45th Annual Meeting of the American Society of Hematology, 6–9 December 2003, San Diego, USA.

Economopoulos, T., Fountzilas, G., Pavlidis, N., et al. (2003) Rituximab in combination with CNOP chemotherapy in patients with previously untreated indolent non-Hodgkin's lymphoma. *Hematol J* 4: 110–115.

Edwards, J.C., Cambridge, G. (2001) Sustained improvement in rheumatoid arthritis following a protocol designed to deplete B lymphocytes. *Rheumatology (Oxford)* 40: 205–211.

Eichhorst, B.F., Busch, R., Duehrsen, U., et al. (2005) CHOP Plus Rituximab (CHOP-R) in Fludarabine (F) Refractory Chronic Lymphocytic Leukemia (CLL) or CLL with Autoimmune Hemolytic Anemia (AIHA) or Richter's Transformation (RT): First Interim Analysis of a phase II Trial of the German CLL Study. Abstract 2125, updated in a poster presentation at the 47th Annual Meeting of the American Society of Hematology (ASH), 10–13 December 2005, Atlanta, USA.

Einfeld, D.A., Brown, J.P., Valentine, M.A., et al. (1988) Molecular cloning of the human B cell CD20 receptor predicts a hydrophobic protein with multiple transmembrane domains. *EMBO J* 7: 711–717.

Eisenberg, R. and Looney, R.J. (2005) The therapeutic potential of anti-CD20 "What do B-cells do?" *Clin Immunol* 117: 207–213.

El Gnaoui, T., Dupuis, J., Joly, B., et al. (2005) Rituximab, gemcitabine and oxaliplatin (R-GEMOX): a promising regimen for refractory/relapsed B-cell lymphoma, a single institutional pilot study. Abstract 496, 9th International Congress on Malignant Lymphoma (ICML), 8–11 June 2005, Lugano, Switzerland.

Emery, P., Filipowicz-Sosnowska, A., Szczepański, L., et al. (2005) Primary analysis of a double-blind, placebo-controlled, dose-ranging trial of rituximab, an anti-CD20 monoclonal antibody, in patients with rheumatoid arthritis receiving methotrexate (DANCER trial). *Ann Rheum Dis* 64(Suppl.3): 58.

Emmanouilides, C., Territo, M., Menco, R., et al. (2003) Mitoxantrone–cyclophosphamide–rituximab: an effective and safe combination for indolent NHL. *Hematol Oncol* 21: 99–108.

Enting, R.H., Demopoulos, A., DeAngelis, L.M., Abrey, L.E. (2004) Salvage therapy for primary CNS lymphoma with a combination of rituximab and temozolomide. *Neurology* 63: 901–903.

Ezdinli, E.Z., Harrington, D.P., Kucuk, O., et al. (1987). The effect of intensive intermittent maintenance therapy in advanced low-grade non-Hodgkin's lymphoma. *Cancer* 60: 156–160.

Faderl, S., Ferrajoli, A., Wierda, W., et al. (2005) Continuous Infusion/Subcutaneous Alemtuzumab (Campath-1H) Plus Rituximab Is Active for Patients with Relapsed/Refractory Chronic Lymphocytic Leukemia (CLL). Abstract 2963, updated in a poster presentation at the 47th Annual Meeting of the American Society of Hematology (ASH), 10–13 December 2005, Atlanta, USA.

Ferrajoli, A., O'Brien, S.M., Faderl, S.H., et al. (2005) Rituximab Plus GM-CSF for Patients with Chronic Lymphocytic Leukemia. Abstract 721, updated in an oral presentation at the 47th Annual Meeting of the American Society of Hematology (ASH), 10–13 December 2005, Atlanta, USA.

Feugier, P., Van Hoof, A., Sebban, C., et al. (2005) Long-term results of the R-CHOP study in the treatment of elderly patients with diffuse large B-cell lymphoma: a study by the Groupe d'Étude des Lymphomes de l'Adulte. *J Clin Oncol* 23, 4117–4126

Feuring-Buske, M., Kneba, M., Unterhalt, M., et al. (2000) IDEC-C2B8 (Rituximab) anti-CD20 antibody treatment in relapsed advanced-stage follicular lymphomas: results of a phase-II study of the German Low-Grade Lymphoma Study Group. *Ann Hematol* 79: 493–500.

Fisher, R.I. (1994) Treatment of aggressive non-Hodgkin's lymphomas. Lessons from the past 10 years. *Cancer* 74: 2657–2681.

Fisher, R.I., Gaynor, E.R., Dahlberg, S., et al. (1994) A phase III comparison of CHOP versus m-BACOD versus ProMACE-CytaBOM versus MACOP-B in patients with intermediate- or high-grade non-Hodgkin's lymphoma: results of SWOG-8516 (Intergroup 0067), the National High-Priority Lymphoma Study. *Ann Oncol* 5(Suppl.2): 91–95.

Fisher, S.G., Fisher, R.I. (2004) The epidemiology of non-Hodgkin's lymphoma. *Oncogene* 23: 6524–6534.

Fleischmann, R.M., Pavelka, K., Baldassare, A., et al. (2005) Preliminary efficacy results of rituximab retreatment in patients with active rheumatoid arthritis. *Arthritis Rheum* 52(Suppl.9): S131.

Foran, J.M., Gupta, R.K., Cunningham, D., et al. (2000a) A UK multicentre phase II study of rituximab (chimaeric anti-CD20 monoclonal antibody) in patients with follicular lymphoma, with PCR monitoring of molecular response. *Br J Haematol* 109: 81–88.

Foran, J.M., Rohatiner, A.Z., Cunningham, D., et al. (2000b) European phase II study of rituximab (chimeric anti-CD20 monoclonal antibody) for patients with newly diagnosed mantle-cell lymphoma and previously treated mantle-cell lymphoma, immunocytoma, and small B-cell lymphocytic lymphoma. *J Clin Oncol* 18: 317–324.

Forstpointner, R., Dreyling, M., Repp, R., et al. (2004) The addition of rituximab to a combination of fludarabine, cyclophosphamide, mitoxantrone (FCM) significantly increases the response rate and prolongs survival as compared with FCM alone in patients with relapsed and refractory follicular and mantle cell lymphomas: results of a prospective randomized study of the German Low-Grade Lymphoma Study Group. *Blood* 104: 3064–3071.

Forstpointner, R., Unterhalt, M., Dreyling, M., et al. (2006) Maintenance therapy with rituximab leads to a significant prolongation of response duration after salvage therapy with a combination of rituximab, fludarabine, cyclophosphamide and mitoxantrone (R-FCM) in patients with relapsed and refractory follicular and mantle cell lymphomas—results of a prospective randomized study of the German low grade lymphoma study group (GLSG). *Blood* Aug 31; [Epub ahead of print].

Friedberg, J.W. (2005) Unique toxicities and resistance mechanisms associated with monoclonal antibody therapy. *Hematology (Am Soc Hematol Educ Program)* 2005: 329–334.

Friedberg, J.W., Neyberg, D., Gribben, J.G., et al. (2002) Combination immunotherapy with rituximab and interleukin 2 in patients with relapsed or refractory follicular non-Hodgkin's lymphoma. *Br J Haematol* 117: 828–834.

Gajra, A., Vajpayee, N., Grethlein, S.J. (2004) Response of myasthenia gravis to rituximab in a patient with non-Hodgkin lymphoma. *Am J Hematol* 77: 196–197.

Garcia-Conde, J.M. Conde, E., Sierra, J., et al. (2000) Rituximab (IDEC-C2B8) and CVP chemotherapy in follicular or low grade B-cell lymphoma after relapse: results after 6 months of follow-up. Abstract 96, updated in a poster presentation at the 36th Annual Meeting of the American Society of Clinical Oncology (ASCO), 20–23 May, 2000, New Orleans, USA.

Geisler, C.H., Elonen, E., Kolstad, A., et al. (2005) Nordic Mantle Cell Lymphoma (MCL) Project: Preemptive Rituximab Treatment of Molecular Relapse Following Autotransplant Can Reinduce Molecular Remission and Prolonged Disease-Free Survival. Abstract 2429, updated in a poster presentation at the 47th Annual Meeting of the American Society of Hematology (ASH), 10–13 December 2005, Atlanta, USA.

Gertz, M.A., Rue, M., Blood, E., et al. (2004) Multicenter phase 2 trial of rituximab for Waldenström macroglobulinemia (WM): an Eastern Cooperative Oncology Group Study (E3A98). *Leuk Lymphoma* 45: 2047–2055.

Ghielmini, M., Schmitz, S.F., Cogliatti, S.B., et al. (2004) Prolonged treatment with rituximab in patients with follicular lymphoma significantly increases event-free survival and response duration compared with the standard weekly x 4 schedule. *Blood* 103: 4416–4423.

Ghielmini, M., Schmitz, S.F., Cogliatti, S.B., et al. (2005). Effect of single-agent rituximab given at the standard schedule or as prolonged treatment in patients with mantle cell lymphoma; a study of the Swiss Group for Clinical Cancer Research (SAKK). *J Clin Oncol* 23: 705–711.

Gisselbrecht, C., Mounier, N. (2003) Rituximab: enhancing outcome of autologous stem cell transplantation in non-Hodgkin's lymphoma. *Semin Oncol* 30(1 Suppl.2): 28–33.

Glass, B., Kloess, M., Reiser, M., et al. (2005) Dose Escalated CHOP + Etoposide Followed by Repetitive Autologous Stem Cell Transplantation (MegaCHOEP) with or without Rituximab for Primary Treatment of Aggressive NHL. Abstract 1492, updated in a poster presentation at the 47th Annual Meeting of the American Society of Hematology (ASH), 10–13 December 2005, Atlanta, USA.

Gluck, W.L., Hurst, D., Yuen, A., et al. (2004) Phase I studies of interleukin (IL)-2 and rituximab in B-cell non-Hodgkin's lymphoma: IL-2 mediated natural killer cell expansion correlations with clinical response. *Clin Cancer Res* 10: 2253–2264.

Gokbuget, N., Hoelzer, D. (2004) Treatment with monoclonal antibodies in acute lymphoblastic leukemia: current knowledge and future prospects. *Ann Hematol* 83: 201–205.

Gordan, L.N., Grow, W.B., Pusateri, A., et al. (2005). Phase II trial of individualized rituximab dosing for patients with CD20-positive lymphoproliferative disorders. *J Clin Oncol* 23: 1096–1102.

Gregory, S.A., Venugopal, P., Adler, S., et al. (2002) Combined fludarabine, mitoxantrone and rituximab achieves a high initial response as initial treatment for advanced low grade non-Hodgkin's lymphoma. Abstract, updated in a poster presentation at the 44th Annual Meeting of the American Society of Hematology (ASH), 7–10 December 2002, Philadelphia, USA.

Grillo-López, A.J., Cheson, B.D., Horning, S.J., et al. (2000) Response criteria for NHL: importance of 'normal' lymph node size and correlations with response rates. *Ann Oncol* 11: 399–408.

Habermann, T.M., Weller, E., Morrison, V.A., et al. (2006) Rituximab-CHOP versus CHOP alone or with maintenance rituximab in older patients with diffuse large B-cell lymphoma. *J Clin Oncol* 24: 3121–3217.

Habermann, T.M., Weller, E., Morrison, V.A., et al. (2004) Rituximab-CHOP versus CHOP with or without maintenance rituximab in patients 60 years of age or older with diffuse large B-cell lymphoma (DLBCL): an update. Abstract 127, updated in an oral presentation at the 46th Annual Meeting of the American Society of Hematology, 3–7 December 2004, San Diego, USA.

Hainsworth, J.D., Litchy, S., Burris, H.A., et al. (2002) Rituximab as first-line and maintenance therapy for patients with indolent non-Hodgkin's lymphoma. *J Clin Oncol* 20: 4261–4267.

Hainsworth, J.D., Litchy, S., Barton, J.H., et al. (2003a) Single agent rituximab as first-line and maintenance therapy for patients with chronic lymphocytic leukemia or small lymphocytic lymphoma. a phase II trial of the Minnie Pearl Cancer Research Network. *J Clin Oncol* 21: 1746–1751.

Hainsworth, J.D., Litchy, S., Morrissey, L., et al. (2003b) Rituximab as first-line and maintenance therapy for indolent non-Hodgkin's lymphoma (NHL): long-term follow-up of a Minnie Pearl Cancer Research Network phase II trial. Abstract 1496, updated in a poster presentation at the 45th Annual Meeting of the American Society of Hematology (ASH), 6–9 December 2003, San Diego, USA.

Hainsworth, J.D., Litchy, S., Lamb, M.R., et al. (2003c) First-line treatment with brief-duration chemotherapy plus rituximab in elderly patients with intermediate-grade non-Hodgkin's lymphoma: phase II trial. *Clin Lymphoma* 4: 36–42.

Hainsworth J.D., Litchy S., Morrissey L.H., et al. (2005a) Rituximab plus short-duration chemotherapy as first-line treatment for follicular non-Hodgkin's lymphoma: a phase II trial of the Minnie Pearl Cancer Research Network. *J Clin Oncol* 23: 1500–1506.

Hainsworth, J.D., Litchy, S., Shaffer, D.W., et al. (2005b) Maximizing therapeutic benefit of rituximab: maintenance therapy versus re-treatment at progression in patients with indolent non-Hodgkin's lymphoma – a randomized phase II trial of the Minnie Pearl Cancer Research Network. *J Clin Oncol* 23: 1088–1095.

Haioun, C., Lepage, E., Gisselbrecht, C., et al. (2000) Survival benefit of high-dose therapy in poor-risk aggressive non-Hodgkin's lymphoma: final analysis of the prospective LNH87-2 protocol – a Groupe d'Étude des Lymphomes de l'Adulte study. *J Clin Oncol* 18: 3025–3030.

Haioun, C., Mounier, N., Emile, J.F., et al. (2005) Rituximab vs. Observation after High-Dose Consolidative First-Line Chemotherapy (HDC) with Autologous Stem Cell Transplantation in Poor Risk Diffuse Large B-Cell Lymphoma. Final Analysis of the LNH98-B3 GELA Study. Abstract 677, updated in an oral presentation at the 47th Annual Meeting of the American Society of Hematology (ASH), 10–13 December 2005, Atlanta, USA.

Hensel, M., Krasniqi, F., Villalobos, M., et al. (2004) Pentostatin, cyclophosphamide and rituximab is an active regimen with low toxicity for previously treated patients with B-cell chronic lymphocytic leukemia and Waldenström's macroglobulinemia. Abstract 6557, updated in a poster presentation at the 40th Annual Meeting of the American Society of Clinical Oncology (ASCO), 5–8 June 2004, New Orleans, USA.

Hensel, M., Villalobos, M., Kornacker, M., et al. (2005) Pentostatin/cyclophosphamide with or without rituximab: an effective regimen for patients with Waldenström's macroglobulinemia/lymphoplasmacytic lymphoma. *Clin Lymphoma Myeloma* 6: 131–135.

Hermine, O., Haioun, C., Lepage, E., et al. (1996) Prognostic significance of bcl-2 protein expression in aggressive non Hodgkin's lymphoma. Groupe d'Étude des Lymphomes de l'Adulte (GELA). *Blood* 87: 265–272.

Herold, M., Pasold, R., Srock, S., et al. (2004) Results of a prospective randomised open label phase II study comparing rituximab plus mitoxantrone, chlorambucil and prednisolone chemotherapy (R-MCP) versus MCP alone in untreated advanced indolent non-Hodgkin's lymphoma (NHL) and mantle cell lymphoma (MCL). Abstract 584, updated in an oral presentation at the 46th Annual Meeting of the American Society of Hematology, (ASH), 3–7 December 2004, San Diego, USA.

Herold, M. Pasold, R. Srock, S., et al. (2005) Rituximab plus mitoxantrone, chlorambucil, prednisolone (R-MCP) is superior to MCP alone in advanced indolent and follicular lymphoma – results of a phase III study (OSHO39). Abstract 60, updated in an oral presentation at the 9th International Conference on Malignant Lymphoma (ICML), June 8–11, 2005, Lugano, Switzerland.

Hicks, L., Buckstein, R., Mangel, J., et al. (2005) Rituximab increases response rate to ESHAP in relapsed, refractory, or transformed aggressive B-cell lymphoma. Abstract 490, 9th International Conference on Malignant Lymphoma (ICML), June 8–11, 2005, Lugano, Switzerland.

Hiddemann, W., Buske, C., Dreyling, M., et al. (2005a) Treatment strategies in follicular lymphomas: current status and future perspectives. *J Clin Oncol* 23: 6394–6399.

Hiddemann, W., Kneba, M., Dreyling, M., et al. (2005b) Front-line therapy with rituximab added to the combination of cyclophosphamide, doxorubicin, vincristine and prednisone (CHOP) significantly improves the outcome of patients with advanced stage follicular lymphomas as compared to CHOP alone – results of a prospective randomized study of the German Low Grade Lymphoma Study Group (GLSG). *Blood* 106: 3725–3732.

Hoang-Xuan, K., Camilleri-Broet, S., Soussain, C. (2004) Recent advances in primary CNS lymphoma. *Curr Opin Oncol* 16: 601–606.

Hochster, H.S., Weller, E., Ryan, T., et al. (2004) A phase III trial of CVP with or without maintenance rituximab in advanced indolent lymphoma (NHL). Abstract 6502, updated in an oral presentation at the 40th Annual Meeting of the American Society of Clinical Oncology (ASCO), 5–8 June 2004, New Orleans, USA.

Hochster, H.S, Weller, E., Gascoyne, R., et al. (2005) Maintenance rituximab after CVP results in superior clinical outcome in advanced follicular lymphoma (FL): results of the E1496 trial from the Eastern Cooperative Oncology Group and the Cancer and Leukemia Group B. Abstract 349, updated in an oral presentation at the 47th Annual Meeting of the American Society of Hematology (ASH), 10–13 December 2005, Atlanta, USA.

Hoelzer, D., Bauer, K-H., Giagounidis, A., et al. (2003) Short intensive chemotherapy seems successful in Burkitt NHL, Mature B-ALL and Other High-Grade B-NHL. Abstract 236, updated in an oral presentation at the 45th Annual Meeting of the American Society of Hematology, 6–9 December 2003, San Diego, USA.

Horwitz, S.M., Negrin, R.S., Blume, K.G., et al. (2004) Rituximab as adjuvant to high-dose therapy and autologous hematopoietic cell transplantation for aggressive non-Hodgkin lymphoma. *Blood* 103: 777–783.

Howard, O.M., Gribben, J.G., Neuberg, D.S., et al. (2002) Rituximab and CHOP induction therapy for newly diagnosed mantle-cell lymphoma: molecular complete responses are not predictive of progression-free survival. *J Clin Oncol* 20: 1288–1294.

Huhn, D., von Schilling, C., Wilhelm, M., et al. (2001) Rituximab therapy of patients with B-cell chronic lymphocytic leukemia. *Blood* 98: 1326–1331.

Igarashi, T., Ohtsu, T., Fujii, H., et al. (2001) Re-treatment of relapsed indolent B-cell lymphoma with rituximab. *Int J Hematol* 73: 213–221.

Igarashi, T., Kobayashi, Y., Ogura, M., et al. (2002) Factors affecting toxicity, response and progression-free survival in relapsed patients with indolent B-cell lymphoma and mantle cell lymphoma treated with rituximab: a Japanese phase II study. *Ann Oncol* 13: 928–943.

Imrie, K., Belch, A., Pettengell, R., et al. (2005) Rituximab plus CVP chemotherapy versus CVP alone as first-line treatment for follicular lymphoma: treatment effect according to baseline prognostic factors. Abstract 6525, updated in an oral presentation at the 41st Annual Meeting of the American Society of Oncology (ASCO), 13–17 May 2005, New Orleans, USA.

Intragumtornchai, T., Bunworasate, U., Nakorn, T.N., et al. (2006) Rituximab-CHOP-ESHAP vs CHOP-ESHAP-mgh-dose therapy vs conventional CHOP chemotherapy in high-intermediate and high-risk aggressive non-Hodgkin's lymphoma. *Leuk Lymphoma* 47: 1306–1314..

Itala, M., Geisler, C.H., Kimby, E., et al. (2002) Standard-dose anti-CD20 antibody rituximab has efficacy in chronic lymphocytic leukaemia: results from a Nordic multicentre study. *Eur J Haematol* 69: 129–134.

Jaeger, G., Neumeister, P., Brezinschek, R., et al. (2002) Rituximab (anti-CD20 monoclonal antibody) as consolidation of first-line CHOP chemotherapy in patients with follicular lymphoma: a phase II study. *Eur J Haematol* 69: 21–26.

Jermann, M., Jost, L.M., Taverna, Ch., et al. (2004) Rituximab-EPOCH, an effective salvage therapy for relapsed, refractory or transformed B-cell lymphomas: results of a phase II study. *Ann Oncol* 15: 511–516.

Johnson, P.W., Rohatiner, A.Z., Whelan, J.S., et al. (1995) Patterns of survival in patients with recurrent follicular lymphoma: a 20-year study from a single center. *J Clin Oncol* 13: 140–147.

Jurlander, J., Geisler, C., Hagberg, H., et al. (2004) Long-term complete molecular remissions in untreated symptomatic follicular lymphoma treated with rituximab as single agent and in combination with interferon-alpha-2a: analysis of minimal residual disease in the randomized phase II study M39035. Abstract 1393, updated in a poster presentation at the 46th Annual Meeting of the American Society of Hematology, (ASH), 3–7 December 2004, San Diego, USA.

Kahl, B., Longo, W., Eickhoff, J., et al (2006) Maintenance rituximab following induction chemoimmunotherapy may prolong progression-free survival in mantle cell lymphoma: a pilot study from the Wisconsis Oncology Network. *Ann Oncol* 17: 1418–1423.

Kaplan, L.D., Lee, J.Y., Ambinder, R.F., et al. (2005) Rituximab does not improve clinical outcome in a randomized phase 3 trial of CHOP with or without rituximab in patients with HIV-associated non-Hodgkin lymphoma: AIDS-Malignancies Consortium Trial 010. *Blood* 106: 1538–1543.

Kasamon, Y.L., Swinnen, L.J. (2004) Treatment advances in adult Burkitt lymphoma and leukemia. *Curr Opin Oncol* 16: 429–435.

Kaufmann, H., Raderer, M., Wöhrer, S., et al. (2004) Antitumor activity of rituximab plus thalidomide in patients with relapsed/refractory mantle cell lymphoma. *Blood* 104: 2269–2271.

Kay, N.E., Geyer, S.M., Call, T.G., et al. (2006) Combination chemoimmunotherapy with pertostatin, cyclophosphomide and rituximab shows significant clinical activity with low accomponying toxicity in previously untreated B-chronic lymphocytic leukemia. *Blood* Sep 28; [Epub ahead of print].

Keating, M.J., O'Brien, S., Albitar, M., et al. (2005a) Early results of a chemoimmunotherapy regimen of fludarabine, cyclophosphamide, and rituximab as initial therapy for chronic lymphocytic leukemia. *J Clin Oncol* 23: 4079–4088.

Keating, M.J., O'Brien, S., Albitar, M., et al. (2005b) Extended Follow-Up of a Chemo-Immunotherapy Regimen FCR (Fludarabine, F; Cyclophosphamide, C; and Rituximab, R) as Initial Therapy for Chronic Lymphocytic Leukemia (CLL).

Abstract 2118, updated in a poster presentation at the 47th Annual Meeting of the American Society of Hematology (ASH), 10–13 December 2005, Atlanta, USA.

Keogh, K.A., Ytterberg, S.R., Fervenza, F.C., et al. (2006) Rituximab for Refractory Wegener's Granulomatosis: Report of A Prospective, Open-Label Pilot Trial. *Am J Respir Crit Care Med* 173: 180–187.

Kessler, C.M. (2005) New perspectives in hemophilia treatment. *Hematology (Am Soc Hematol Educ Program)*, 429–435.

Kewalramani, T., Zelenetz, A.D., Nimer, S. D., et al. (2004) Rituximab and ICE as second-line therapy before autologous stem cell transplantation for relapsed or primary refractory diffuse large B cell lymphoma. *Blood* 103: 3684–3688.

Keystone, E.C., Burmester, G.R., Furie, R., et al. (2005) Improved quality of life with rituximab plus methotrexate in patients with active rheumatoid arthritis who experienced inadequate response to one or more anti-TNF-α therapies. *Arthritis Rheum* 52(Suppl.9): S141.

Kimby, E. (2005) The safety and tolerability of rituximab (MabThera). *Cancer* 31: 456–473.

Kimby, E., Geisler, C., Hagberg, H., et al. (2002) Rituximab as a single agent and in combination with interferon-alpha-2a as treatment of untreated and first relapse follicular or other low-grade lymphomas. A randomized phase II study. Abstract 289, 8th International Conference on Malignant Lymphoma (ICML), 12–15 June 2002, Lugano, Switzerland.

Ladetto, M., Ricca, I., Benedetti, F., et al. (2005) Rituximab-Supplemented High-Dose Sequential Chemotherapy (HDS) Has Superior Response Rate and Event-Free Survival (EFS) Compared to R-CHOP in Poor Risk Follicular Lymphoma (FL) at Diagnosis: Results from a Multicenter Randomized GITMO Trial. Abstract 675, updated in an oral presentation at the 47th Annual Meeting of the American Society of Hematology (ASH), 10–13 December 2005, Atlanta, USA.

Lamanna, N., Kalaycio, M., Maslak, P., et al. (2005) Pentostatin, Cyclophosphamide, and Rituximab (PCR) Has Comparable Activity but Appears To Be Better Tolerated Than Fludarabine, Cyclophosphamide, and Rituximab (FCR) in Patients with Previously Treated Chronic Lymphocytic Leukemia. Abstract 2127, updated in a poster presentation at the 47th Annual Meeting of the American Society of Hematology (ASH), 10–13 December 2005, Atlanta, USA.

Lavilla, E., Perez Encinas, M., Romero, E., et al. (2005) Combination of CHOP-14 with rituximab for newly-diagnosed aggressive B cell lymphoma. Preliminary data of a prospective multicentric regimen. Abstract 480, 9th International Conference on Malignant Lymphoma, 8–11 Jun 2005, Lugano, Switzerland.

Lazzarino, M., Arcaini, L., Orlandi, E., et al. (2005) Immunochemotherapy with rituximab, vincristine and 5-day cyclophosphamide for heavily pretreated follicular lymphoma. *Oncology* 68: 146–153.

Lemieux, B., Bouafia, F., Thieblemont, C., et al. (2004) Second treatment with rituximab in B-cell non-Hodgkin's lymphoma: efficacy and toxicity on 41 patients treated at CHU-Lyon Sud. *Hematol J* 5: 467–471.

Lenz, G., Dreyling, M., Hoster, E., et al. (2005) Immunochemotherapy with rituximab and cyclophosphamide, doxorubicin, vincristine, and prednisone significantly improves response and time to treatment failure, but not long-term outcome in patients with previously untreated mantle cell lymphoma: results of a prospective randomized trial of the German Low Grade Lymphoma Study Group (GLSG). *J Clin Oncol* 23: 1984–1992.

Levine, T.D. (2005) Rituximab in the treatment of dermatomyositis: an open-label pilot study. *Arthritis Rheum* 52: 601–607.

Li, H., Ayer, L.M., Lytton, J., Deans, J.P. (2003) Store-operated cation entry mediated by CD20 in membrane rafts. *J Biol Chem* 278: 42427–42434.

Looney, R.J., Anolik, J.H., Campbell, D., et al. (2004) B cell depletion as a novel treatment for systemic lupus erythematosus: a phase I/II dose-escalation trial of rituximab. *Arthritis Rheum* 50: 2580–2589.

Maloney, D.G., Lies, T.M., Czerwinski, D.K., et al. (1994) phase I clinical trial using escalating single-dose infusion of chimeric anti-CD20 monoclonal antibody (IDEC-C2B8) in patients with recurrent B-cell lymphoma. *Blood* 84: 2457–2466.

Maloney, D.G., Grillo-López, A.J., Bodkin, D.J., et al. (1997a) IDEC-C2B8: results of a phase I multiple-dose trial in patients with relapsed non-Hodgkin's lymphoma. *J Clin Oncol* 15: 3266–3274.

Maloney, D.G., Grillo-López, A.J., White, C.A., et al. (1997b) IDEC-C2B8 (rituximab) anti-CD20 monoclonal antibody in patients with relapsed low-grade non-Hodgkin's lymphoma. *Blood* 90: 2188–2195.

Mangel, J., Leitch, H.A., Connors, J.M., et al. (2004) Intensive chemotherapy and autologous stem-cell transplantation plus rituximab is superior to conventional chemotherapy for newly diagnosed advanced stage mantle-cell lymphoma: a matched pair analysis. *Ann Oncol* 15: 283–290.

Manshouri, T., Do, K.A., Wang, X., et al. (2003) Circulating CD20 is detectable in the plasma of patients with chronic lymphocytic leukaemia and is of prognostic significance. *Blood* 101: 2507–2513.

Marcus, R., Imrie, K., Belch, A., et al. (2005) CVP chemotherapy plus rituximab compared with CVP as first-line treatment for advanced follicular lymphoma. Blood 105, 1417–1423.

Martinelli, G., Laszlo, D., Bertolini, F., et al. (2003) Chlorambucil in combination with induction and maintenance rituximab is feasible and active in indolent non-Hodgkin's lymphoma. *Br J Haematol* 123: 271–277.

Martinelli, G., Laszlo, D., Ferreri, A.J., et al. (2005) Clinical activity of rituximab in gastric marginal zone non-Hodgkin's lymphoma resistant to or not eligible for anti-*Helicobacter pylori* therapy. *J Clin Oncol* 23: 1979–1983.

Mavromatis, B., Rai, K.R., Wallace, P.K., et al. (2005) Efficacy and Safety of the Combination of Genasense (Oblimersen Sodium, Bcl-2 Antisense Oligonucleotide), Fludarabine and Rituximab in Previously Treated and Untreated Subjects with Chronic Lymphocytic Leukemia. Abstract 2129, updated in a poster presentation at the 47th Annual Meeting of the American Society of Hematology (ASH), 10–13 December 2005, Atlanta, USA.

McLaughlin, P., Grillo-López, A.J., Link, B.K., et al. (1998) Rituximab chimeric anti-CD20 monoclonal antibody therapy for relapsed indolent lymphoma: half of patients respond to a four-dose treatment program. *J Clin Oncol* 16: 2825–2833.

McLaughlin, P., Hagemeister, F.B., Grillo-López, A.J. (1999) Rituximab in indolent lymphoma: the single-agent pivotal trial. *Semin Oncol* 26 (Suppl.14): 79–87.

McLaughlin, P., Rodriguez, M., Hagemeister, F., et al. (2005a) Stage IV indolent lymphoma: a randomized trial of concurrent versus sequential FND (fludarabine, mitoxantrone, dexamethasone) and rituximab, with interferon maintenance. Abstract 247, updated in a poster presentation at the 9th International Conference on Malignant Lymphoma (ICML), June 8–11 2005, Lugano, Switzerland.

McLaughlin, P., Liu, N., Poindexter, N., et al. (2005b) Rituximab plus GM-CSF (Leukine) for indolent lymphoma. Abstract 104, updated in an oral presentation at the 9th International Conference on Malignant Lymphoma (ICML), June 8–11 2005, Lugano, Switzerland.

Mease, P., Szechiński, J., Greenwald, M., et al. (2005) Improvements in patient reported outcomes over 24 weeks for rituximab with methotrexate in rheumatoid arthritis patients in phase IIb trial (DANCER). *Arthritis Rheum* 52(Suppl.9): S138.

Mena, R.M., Smith, J., Geils, G.F., et al. (2005) Enhanced Safety and Tolerability of Dose Intensive Pentostatin and Rituximab in Patients with CLL or Low-Grade B-Cell NHL. Abstract 937, updated in a poster presentation at the 47th Annual Meeting of the American Society of Hematology (ASH), 10–13 December 2005, Atlanta, USA.

Mey, U.J., Orlopp, K.S., Flieger, D., et al. (2006) Dexamethasone, high-dose cytarabine, and cisplatin in combination with rituximab as salvage treatment for patients with relapsed or refractory aggressive non-Hodgkin's lymphoma. *Cancer Invest* 24: 593–600.

Milpied, N., Deconinck, E., Gaillard, F., et al. (2004) Initial treatment of aggressive lymphoma with high-dose chemotherapy and autologous stem-cell support. *N Engl J Med* 350: 1287–1295.

Mounier, N., Briere, J., Gisselbrecht, C., et al. (2003) Rituximab plus CHOP (R-CHOP) overcomes bcl-2-associated resistance to chemotherapy in elderly patients with diffuse large B-cell lymphoma (DLBCL). *Blood* 101: 4279–4284.

Nadler, L.M., Ritz, J., Hardy, R., et al. (1981) A unique cell surface antigen identifying lymphoid malignancies of B cell origin. *J Clin Invest* 67: 134–140.

O'Brien, S., Kantarjian, H., Thomas, D.A., et al. (2001) Rituximab dose-escalation trial in chronic lymphocytic leukemia. *J Clin Oncol* 19: 2165–2170.

Oertel, S.H., Verschuuren, E., Reinke, P., et al. (2005) Effect of anti-CD 20 antibody rituximab in patients with post-transplant lymphoproliferative disorder (PTLD). *Am J Transplant* 5: 2901–2906.

Olszewski, A.J., Grossbard, M.L. (2004) Empowering targeted therapy: lessons from rituximab. *Sci STKE* July 6 (241), pe30.

Onrust, S.V., Lamb, H.M., Barman Balfour, J.A. (1999) Rituximab. *Drugs* 58: 79–88.

Pfreundschuh, M., Trümper, L., Kloess, M., et al. (2004a) Two-weekly or 3-weekly CHOP chemotherapy with or without etoposide for the treatment of elderly patients with aggressive lymphomas: results of the NHL-B2 trial of the DSHNHL. *Blood* 104: 634–641.

Pfreundschuh, M., Trümper, L., Kloess, M., et al. (2004b) Two-weekly or 3-weekly CHOP chemotherapy with or without etoposide for the treatment of young patients with good-prognosis (normal LDH) aggressive lymphomas: results of the NHL-B1 trial of the DSHNHL. *Blood* 104: 626–633.

Pfreundschuh, M.G., Trümper, L., Ma, D., et al. (2004c) Randomised intergroup trial of first line treatment for patients ≤60 years with diffuse large B-cell non-Hodgkin's lymphoma (DLBCL) with a CHOP-like regimen with or without the anti-CD20 antibody rituximab – early stopping after the first interim analysis. Abstract 6500, updated in an oral presentation at the 40th Annual Meeting of the American Society of Clinical Oncology, 5–8 June 2004, New Orleans, USA.

Pfreundschuh, M., Trümper, L., Österborg, A., et al. (2006) CHOP-like chemotherapy plus rituximab versus CHOP-like chemotherapy alone in young patients with good-prognosis diffuse large B-cell lymphoma: a randomised controlled trial by the Mab Thera International Trial (MInT) Group. *Lancet Oncol* 7: 379–391.

Pfreundschuh, M.G., Ho, A., Wolf, M., et al. (2005a) Treatment results on CHOP-21, CHOEP-21, MACOP-B and PMitCEBO with and without rituximab in young good-prognosis patients with aggressive lymphomas: rituximab as "equalizer" in the MinT (MabThera International Trial Group) study. Abstract 6529, updated in an oral presentation at the 41st Annual Meeting of the American Society of Clinical Oncology (ASCO), 13–17 May, 2005, Orlando, Florida, USA.

Pfreundschuh, M., Kloess, M., Schmits, R., et al. (2005b) Six, Not Eight Cycles of Bi-Weekly CHOP with Rituximab (R-CHOP-14) Is the Preferred Treatment for Elderly Patients with Diffuse Large B-Cell Lymphoma (DLBCL): Results of the RICOVER-60 Trial of the German High-Grade Non-Hodgkin Lymphoma Study Group (DSHNHL). Abstract 13, updated in an oral presentation at the 47th Annual Meeting of the American Society of Hematology (ASH), 10–13 December 2005, Atlanta, USA.

Pijpe, J., van Imhoff, G.W., Spijkervet, F.K., et al. (2005) Rituximab treatment in patients with primary Sjögren's syndrome: an open-label phase II study. *Arthritis Rheum* 52: 2740–2750.

Pileri, S.A., Milani, M., Fraternali-Orcioni, G., et al. (1998) From the R.E.A.L. Classification to the upcoming WHO scheme: a step toward universal categorization of lymphoma entities? *Ann Oncol* 9: 607–612.

Piro, L.D., White, C.A., Grillo-López, A.J., et al. (1999) Extended rituximab (anti-CD20 monoclonal antibody) therapy for relapsed or refractory low-grade or follicular non-Hodgkin's lymphoma. *Ann Oncol* 10: 655–661.

Pranzatelli, M.R., Tate, E.D., Travelstead, A. L., Longee, D. (2005) Immunologic and clinical responses to rituximab in a child with opsoclonus-myoclonus syndrome. *Pediatrics* 115: e115–119.

Press, O.W., Appelbaum, F., Ledbetter, J.A., et al. (1987) Monoclonal antibody 1F5 (anti-CD20) serotherapy of human B cell lymphomas. *Blood* 69: 584–591.

Rambaldi, A., Lazzari, M., Manzoni, C., et al. (2002) Monitoring of minimal residual disease after CHOP and rituximab in previously untreated patients with follicular lymphoma. *Blood* 99: 856–862.

Rambaldi, A., Carlotti, E., Oldani, E., et al. (2005) Quantitative PCR of bone marrow BCL2/IgH+ cells at diagnosis predicts treatment response and long-term outcome in follicular non-Hodgkin lymphoma. *Blood* 105: 3428–3433.

Ratanatharathorn V., Ayash L., Reynolds C., et al. (2003) Treatment of chronic graft-versus-host disease with anti-CD20 chimeric monoclonal antibody. *Biol Blood Marrow Transplant* 9: 505–511.

Ravandi-Kashani, F., O'Brien, S., Keating, M., et al. (2005) Complete Eradication of Minimal Residual Disease (MRD) in Patients with Hairy Cell Leukemia (HCL) after Cladribine (2CDA) Followed by an Extended Course of Rituximab. Abstract 3258, updated in a poster presentation at the 47th Annual Meeting of the American Society of Hematology (ASH), December 10–13, 2005, Atlanta, USA.

Reff, M.E., Carner, K., Chambers, K.S., et al. (1994) Depletions of B cells in vivo by a chimeric mouse human monoclonal antibody to CD20. *Blood* 83: 435–445.

Rigacci, L., Nassi, L., Alterini, R., et al. (2006) Dose-dense CHOP plus rituximab (R-CHOP-14) for the treatment of elderly patients with high-risk diffuse large B cell lymphoma. *Acta Haematologica* 115: 22–27.

Rivas-Vera, S., Báez, E., Sobrevilla-Calvo, P., et al. (2005) Is first line single agent rituximab the best treatment for indolent non-Hodgkin lymphoma? Update of a multicentric study comparing rituximab versus CNOP versus rituximab plus CNOP. Abstract 2431, updated in a poster presentation at the 47th Annual Meeting of the American Society of Hematology (ASH), December 10–13, 2005, Atlanta, USA.

Robak, T. (2004) Monoclonal antibodies in the treatment of autoimmune cytopenias. *Eur J Haematol* 72: 79–88.

Rodriguez, M.A., Dang, N.H., Fayad, L., et al. (2005) Phase II Study of Sphingosomal Vincristine in CHOP+/− Rituximab for Patients with Aggressive Non-Hodgkin's Lymphoma (NHL): Promising 3 Year Follow-Up Results in Elderly Patients. Abstract 943, updated in a poster presentation at the 47th Annual Meeting of the American Society of Hematology (ASH), 10–13 December, Atlanta, USA.

Rohatiner, A.Z., Gregory, W.M., Peterson, B., et al. (2005) Meta-analysis to evaluate the role of interferon in follicular lymphoma. *J Clin Oncol* 23: 2215–2223.

Romaguera, J.E., Fayad, L., Rodriguez, M.A., et al. (2005a) High rate of durable remission after treatment of newly diagnosed aggressive mantle-cell lymphoma with rituximab plus hyper-CVAD alternating with rituximab plus high-dose methotrexate and cytarabine. *J Clin Oncol* 23: 7013–7023.

Romaguera, J.E., Fayad, L.E., Wang, M., et al. (2005b) High (95%) response rates in relapsed/refractory mantle cell lymphoma after R-HCAD alternating with R-methotrexate/cytarabine (R-M-A). Abstract 2446, updated in a poster presentation at the 47th Annual Meeting of the American Society of Hematology (ASH), 10–13 December, Atlanta, USA.

Rothe, A., Schulz, H., Elter, T., et al. (2004) Rituximab monotherapy is effective in patients with poor risk refractory aggressive non-Hodgkin's lymphoma. *Haematologica* 89: 875–876.

Rubenstein, J.L., Shen, A.,, Abrey, L., et al. (2004) Results from a phase I study of intraventricular administration of rituximab in patients with recurrent lymphomatous meningitis. Abstract 6593, updated in a poster presentation at the 40th Annual Meeting of the American Society of Clinical Oncology (ASCO), 5–8 June 2004, New Orleans, USA.

Rummel, M.J., Al-Batran, S.E., Kim, S.-Z., et al. (2005) Bendamustine plus rituximab is effective and has a favorable toxicity profile in the treatment of mantle cell and low-grade non-Hodgkin's lymphoma. *J Clin Oncol* 23: 3383–3389.

Sacchi, S., Tucci, A., Merli, F., et al. (2003) Efficacy controls and long-term follow-up of patients (pts) treated with FC plus rituximab for relapsed follicular NHL. Abstract 4901, 45th Annual Meeting of the American Society of Hematology (ASH), 6–9 December 2003, San Diego, USA.

Sacchi, S., Federico, M., Vitolo, U., et al. (2001) Clinical activity and safety of combination immunotherapy with interferon-α2a and rituximab in patients with relapsed low grade non-Hodgkin's lymphoma. *Haematologica* 86: 951–958.

Salles, G., Foussard, C., Mounier, N., et al. (2004) Rituximab added to αIFN-CHVP improves the outcome of follicular lymphoma patients with a high tumor burden: first analysis of the GELA-GOELAMS FL-2000 randomised trial in 359 patients. Abstract 160, updated in an oral presentation at the 46th Annual Meeting of the American Society of Hematology (ASH), 3–7 December 2004, San Diego, USA.

Savage, D.G., Cohen, N.S., Hesdorffer, C.S., et al. (2003) Combined fludarabine and rituximab for low grade lymphoma and chronic lymphocytic leukemia. *Leuk Lymphoma* 44: 477–481.

Savage K.J., Al-Rajhi N., Voss N., et al. (2006) Favorable outcome of primary mediastinal large B-cell lymphoma in a single institution: the British Columbia experience. *Ann Oncol* 17: 123–130.

Schulz, H., Klein, S.K., Rehwald, U., et al. (2002) Phase 2 study of a combined immunochemotherapy using rituximab and fludarabine in patients with chronic lymphocytic leukemia. *Blood* 100: 3115–3120.

Schulz, H., Pels, H., Schmidt-Wolf, I., Zeelen, U., Germing, U., Engert, A. (2004) Intraventricular treatment of relapsed central nervous system lymphoma with the anti-CD20 antibody rituximab. *Haematologica* 89: 753–754.

Schulz, H., Skoetz, N., Bohlius, J., et al. (2005a) Does Combined Immunochemotherapy with the Monoclonal Antibody Rituximab Improve Overall Survival in the Treatment of Patients with Indolent Non-Hodgkin Lymphoma? Preliminary Results of a Comprehensive Meta-Analysis. Abstract 351, updated in an oral presentation at the 47th Annual Meeting of the American Society of Haematology (ASH), 10–13 December 2005, Atlanta, USA.

Schulz, H., Trelle, S., Reiser, M., et al. (2005b) Rituximab in Relapsed Lymphocyte Predominant Hodgkin's Disease (LPHD). Long-Term Results of a phase-II Study from the German Hodgkin Lymphoma Study Group (GHSG). Abstract 1503, updated in a poster presentation at the 47th Annual Meeting of the American Society of Hematology (ASH), 10–13 December 2005, Atlanta, USA.

Sirohi, B., Cunningham, D., Norman, A., et al. (2005) Use of Gemcitabine, Cisplatin and Methylprednisolone (GEM-P) with or without rituximab in relapsed and refractory patients with Diffuse Large B Cell Lymphoma (DLBCL). Abstract 939, updated in a poster presentation at the 47th Annual Meeting of the American Society of Haematology (ASH), 10–13 December 2005, Atlanta, USA.

Sehn, L.H., Donaldson, J., Chhanabhai, M., et al. (2005) Introduction of combined CHOP plus rituximab therapy dramatically improved outcome of diffuse large B-cell lymphoma in British Columbia. *J Clin Oncol* 23: 5027–5033.

Solal-Celigny, P., Salles, G.A., Brousse, N., et al. (2004) Single 4-Dose Rituximab Treatment for Low-Tumor Burden Follicular Lymphoma (FL): Survival Analyses with a Follow-Up (F/Up) of at Least 5 Years. Session Type: Abstract 585 updated in an oral presentation at the 46th Annual meeting of the American Society of Hematology (ASH), 3–7 December 2004, San Diego, USA.

Solal-Celigny, P., Imrie, I., Belch, A., et al. (2005) Mabthera (Rituximab) Plus CVP Chemotherapy for First-Line Treatment of Stage III/IV Follicular Non-Hodgkin's Lymphoma (NHL): Confirmed Efficacy with Longer Follow-Up. Abstract 350 updated in an oral presentation at the 47th Annual meeting of the American Society of Hematology (ASH), 10–13 December 2005, Atlanta, USA.

Sonneveld, P., van Putten, W., Holte, H., et al. (2005) Intensified CHOP with Rituximab for Intermediate or High-Risk Non-Hodgkin's Lymphoma: Interim Analysis of a Randomized phase III Trial in Elderly Patients by the Dutch HOVON and Nordic Lymphoma Groups. Abstract 16, updated in an oral presentation at the 47th Annual Meeting of the American Society of Hematology (ASH), 10–13 December 2005, Atlanta, USA.

Spina, M., Jaeger, U., Sparano, J.A., et al. (2005) Rituximab plus infusional cyclophosphamide, doxorubicin, and etoposide in HIV-associated non-Hodgkin lymphoma: pooled results from 3 phase 2 trials. *Blood* 105: 1891–1897.

Stewart, W.P., Crowther, D., McWilliam, L.J., et al. (1988) Maintenance chlorambucil after CVP in the management of advanced stage, low-grade histologic type non-Hodgkin's lymphoma. A randomized prospective study with an assessment of prognostic factors. *Cancer* 61: 441–447.

The Non-Hodgkin's Lymphoma Classification Project. (1997a) A clinical evaluation of the International Lymphoma Study Group classification of non-Hodgkin's lymphoma. *Blood* 89: 3909–3918.

The Non-Hodgkin's Lymphoma Classification Project. (1997b) Effect of age on the characteristics and clinical behavior of non-Hodgkin's lymphoma patients. *Ann Oncol* 8: 973–978.

Thomas, D.A., O'Brien, S., Giles, F.J., et al. (2001) Single agent rituximab in early stage chronic lymphocytic leukaemia. Abstract 1533, updated in a poster presentation at the 43rd Annual meeting of the American Society of Hematology (ASH), 7–11 December 2001, Orlando, USA.

Thomas, D.A., O'Brien, S., Bueso-Ramos, C., et al. (2003) Rituximab in relapsed or refractory hairy cell leukemia. *Blood* 102: 3906–3911.

Thomas, D.A., Faderl, S., O'Brien, S., et al. (2006) Chemoimmunotherapy with hyper-CVAD plus rituximab for the treatment of adult Burkitt and Burkitt-type lymphoma or acute lymphoblastic leukemia. *Cancer* 106: 1569–1580.

Tobinai, K., Igarashi, T., Itoh, K., et al. (2004) Japanese multicenter phase II and pharmacokinetic study of rituximab in relapsed or refractory patients with aggressive B-cell lymphoma. *Ann Oncol* 15: 821–830.

Trappe, R., Oertel, S., Choquet, S., et al. (2005) Sequential Treatment with the Anti-CD 20 Antibody Rituximab and CHOP+GCSF Chemotherapy in Patients with Post-Transplant Lymphoproliferative Disorder (PTLD): First Interim Analysis of a Multicenter phase II Study. Abstract 932, updated in a poster presentation at the 47th Annual Meeting of the American Society of Hematology (ASH), 10–13 December 2005, Atlanta, USA.

Treon, S.P., Agus, D.B., Link, B., et al. (2001) CD20-directed antibody-mediated immunotherapy induces responses and facilitates hematologic recovery in patients with Waldenström's macroglobulinemia. *J Immunother* 24: 272–279.

Treon, S.P., Kelliher, A., Keele, B., et al. (2003) Expression of serotherapy target antigens in Waldenström's macroglobulinemia: therapeutic applications and considerations. *Semin Oncol* 30: 248–252.

Treon, S.P., Emmanouilides, C., Kimby, E., et al. (2005a) Extended rituximab therapy in Waldenström's macroglobulinemia. *Ann Oncol* 16: 132–138.

Treon, S.P., Hunter, Z., Barnagan, A.R. (2005b) CHOP plus rituximab therapy in Waldenström's macroglobulinemia. *Clin Lymphoma* 5: 273–277.

Treon, S.P., Gertz, M.A., Dimopoulos, M., et al. (2006) Update on treatment recommendations from the Third International Workshop on Waldenström's Macroglobulinemia. *Blood* 107: 3442–3446.

Trneny M., Belada D., Vasova I., et al. (2005) Rituximab combination with anthracycline based chemotherapy significantly improved the outcome of young patients with diffuse large B-cell lymphoma in low as well in high risk subgroups. Abstract 2444, updated in a poster presentation at the 47th Annual Meeting of the American Society of Hematology (ASH), 10–13 December 2005, Atlanta, USA.

Van der Kolk, L.E., Grillo-López, A.J., Baars, J.W., et al. (2003) Treatment of relapsed B-cell non-Hodgkin's lymphoma with a combination of chimeric anti-CD20 monoclonal antibodies (rituximab) and G-CSF: final report on safety and efficacy. *Leukemia* 17: 1658–1664.

van Oers, M.H., Klasa, R., Marcus, R.E., et al. (2006) Rituximab maintenance improves clinical outcome of relapsed/resistant follicular non-Hodgkin's lymphoma, both in patients with old without rituximab during induction: results of a prospective randomized phase III intergroup trial. *Blood* 108: 3295–3301.

Venugopal, O., Gretory, S.A., Showel, J., et al. (2004) Rituximab (Rituxan) Combined with ESHAP Chemotherapy Is Highly Active in Relapsed/ Refractory Aggressive Non-Hodgkin's Lymphoma. Abstract 4636, 46th Annual Meeting of the American Society of Hematology (ASH), 3–7 December 2004, San Diego, USA.

Vitolo, U., Boccomini, C., Ladetto, M., et al. (2004) Sequential brief chemo-immunotherapy FND + rituximab in elderly patients with advanced stage follicular lymphoma (FL): high clinical and molecular remission rate associated with a prolonged failure-free survival. Abstract 1320, updated in a poster presentation at the 46th Annual Meeting of the American Society of Hematology (ASH), 4–7 December 2004, San Diego, USA.

Vose, J.M., Link, B.K., Grossbard, M.L., et al. (2001) Phase II study of rituximab in combination with CHOP chemotherapy in patients with previously untreated, aggressive non-Hodgkin's lymphoma. *J Clin Oncol* 19: 389–397.

Vose, J.M., Link, B.K., Grossbard, M.L., et al. (2002) Long term follow-up of a phase II study of rituximab in combination with CHOP chemotherapy in patients with previously untreated aggressive non-Hodgkin's lymphoma (NHL). Abstract 1396, updated in a poster presentation at the 44th American Society of Hematology Annual Meeting, 6–10 December 2002, Philadelphia, USA.

Vose, J.M., Link, B.K., Grossbard, M.L., et al. (2005) Long-term update of a phase II study of rituximab in combination with CHOP chemotherapy in patients with previously untreated, aggressive non-Hodgkin's lymphoma. *Leuk Lymphoma* 46: 1569–1573.

Weber, D.M., Dimopoulos, M.A., Delasalle, K., et al. (2003) 2-Chlorodeoxyadenosine alone and in combination for previously untreated Waldenstrom's macroglobulinemia. *Semin Oncol* 30: 243–247.

Weide, R., Heymanns, J., Thomalla, J., et al. (2004) Bendamustine/mitoxantrone/rituximab (BMR): a very effective and well tolerated immuno-chemotherapy for relapsed and refractory indolent lymphomas – results of a multicentre phase-II-study of the German Low Grade Lymphoma Study Group (GLSG). Abstract 2482, updated in a poster presentation at the 46th Annual Meeting of the American Society of Hematology (ASH), 4–7 December 2004, San Diego, USA.

Wierda, W., O'Brien, S., Wen, S., et al. (2005a) Chemoimmunotherapy with fludarabine, cyclophosphamide, and rituximab for relapsed and refractory chronic lymphocytic leukemia. *J Clin Oncol* 23: 4070–4078.

Wierda, W.G., O'Brien, S., Faderl, S., et al. (2006) A retrospective comparison of three sequential groups of patients with recurrent/refractory chronic lymphocytic leukemia treated with fludarabine-based regimens. *Cancer* 106: 337–345.

Wierda, W., O'Brien, S., Ferrajoli, A., et al. (2005b) Salvage Therapy with Combined Cyclophosphamide (C), Fludarabine (F), Alemtuzumab (A), and Rituximab (R)

(CFAR) for Heavily Pre-Treated Patients with CLL. Abstract 719, updated in an oral presentation at the 47th Annual Meeting of the American Society of Hematology (ASH), 10–13 December 2005, Atlanta, USA.

Wierda, W.G., Kipps, T.J., Keating, M.J. (2005c) Novel immune-based treatment strategies for chronic lymphocytic leukemia. *J Clin Oncol* 23: 6325–6332.

Wilson, W.H., Grossbard, M.L., Pittaluga, S., et al. (2002). Dose-adjusted EPOCH chemotherapy for untreated large B-cell lymphomas: a pharmacodynamic approach with high efficacy. *Blood* 99: 2685–2693.

Wilson, W.H., Pittaluga, S., Gutierrez, G., et al. (2003) Dose-adjusted EPOCH-rituximab in untreated diffuse large B-cell lymphoma: Benefit of rituximab appears restricted to tumors harboring anti-apoptotic mechanisms. Abstract 356, updated in an oral presentation at the American Society of Hematology Annual Meeting, 6–9 December 2003, San Diego, USA.

Wilson, W.H., Porcu, P., Hurd, D., et al. (2005) Phase II study of dose-adjusted EPOCH-R in untreated de novo CD20+ diffuse large B-cell lymphoma – CALGB 50103. Abstract 6530, updated in an oral presentation at the 41st Annual Meeting of the American Society of Clinical Oncology, May 13–17, Orlando, Florida, USA.

Witzig, T.E. (2005) Current treatment approaches for mantle-cell lymphoma. *J Clin Oncol* 23: 6409–6414.

Witzig, T.E., Vukov, A.M., Habermann, T. M., et al. (2005) Rituximab therapy for patients with newly diagnosed, advanced-stage, follicular grade I non-Hodgkin's lymphoma: a phase II trial in the North Central Cancer Treatment Group. *J Clin Oncol* 23: 1103–1108.

Wong, E.T., Tishler, R., Barron, L., Wu, J.K. (2004) Immunochemotherapy with rituximab and temozolomide for central nervous system lymphomas. *Cancer* 101: 139–145.

Woods, A., Buckstein, R., Mangel, J., et al. (2005) Prolonged remissions in patients with recurrent follicular lymphoma treated with ASCT and rituximab as in vivo purging and consolidative immunotherapy. Abstract 404, 9th International Conference on Malignant lymphoma (ICML), 8–11 June 2005, Lugano, Switzerland.

Younes, A., McLaughlin, P., Fayad, L.E., et al. (2005a) Six Weekly Doses of Rituximab Plus ABVD for Newly Diagnosed Patients with Advanced Stage Classical Hodgkin Lymphoma: Depletion of Reactive B-Cells from the Microenvironment by Rituximab. Abstract 1499, updated in a poster presentation at the 47th Annual Meeting of the American Society of Hematology (ASH), 10–13 December 2005, Atlanta, USA.

Younes, A., Fayad, L., Pro, B., et al. (2005b) Gemcitabine Plus Rituximab Therapy of Patients with Relapsed and Refractory Classical Hodgkin Lymphoma. Abstract 1498, updated in an oral presentation at the 47th Annual Meeting of the American Society of Hematology (ASH), 10–13 December 2005, Atlanta, USA.

Yuen, A.R., Kamel, O.W., Halpern, J., et al. (1995) Long-term survival after histologic transformation of low-grade follicular lymphoma. *J Clin Oncol* 13: 1726–1733.

Yunus, F., George, S., Smith, J., et al. (2003) Phase II Multicenter Trial of Pentostatin and Rituximab in Patients with Previously Treated and Untreated Chronic Lymphocytic Leukemia (CLL). Abstract 5168, 45th Annual Meeting of the American Society of Hematology (ASH), 6–9 December 2003, San Diego, USA.

Zinzani, P.L., Storti, S., Zaccaria, A., et al. (1999) Elderly aggressive-histology non-Hodgkin's lymphoma: first-line VNCOP-B regimen experience on 350 patients. *Blood* 94: 33–38

Zinzani P.L., Pulsoni A., Perrotti A., et al. (2004) Fludarabine plus mitoxantrone with and without rituximab versus CHOP with and without rituximab as front-line treatment for patients with follicular lymphoma. *J Clin Oncol* 22: 2654–2661.

Zaja, F.F., Bacigalupo, A.A., Bosi, A.A., et al. (2005) Rituximab may allow to improve multi-organ involvement in patients with cGVHD. Abstract 3100, updated in a poster presentation at the 47th Annual Meeting of the American Society of Hematology (ASH), 10–13 December 2005, Atlanta, USA.

Websites

Surveillance Epidemiology and End Results (SEER), National Cancer Institute. Cancer stat fact sheets: non-Hodgkin's lymphoma. http://seer.cancer.gov/statfacts/html/nhl.html (accessed January 2006)

Surveillance Epidemiology and End Results (SEER), National Cancer Institute. Cancer stat fact sheets: chronic lymphocytic leukemia. http://seer.cancer.gov/statfacts/html/clyl.html (accessed January 2006)

European Medicines Agency (EMEA). MabThera Summary of Product Characteristics (SmPC). Available from URL: http://www.emea.eu.int/humandocs/Humans/EPAR/mabthera/mabthera.htm (accessed January 2006)

Genentech, Inc. Rituxan prescribing information. http://www.gene.com/gene/products/information/pdf/rituxan-prescribing.pdf (accessed January 2006)

National Comprehensive Cancer Network Physician Prescribing Guidelines. Non-Hodgkin's lymphoma, v1 2006 http://www.nccn.org/professionals/physician_gls/PDF/nhl.pdf (accessed February 2006)

Clinicaltrials.gov. A service of the US National Institutes of Health http://www.clinicaltrials.gov/ (accessed February 2006)

14
Trastuzumab (Herceptin)
A Treatment for HER2-Positive Breast Cancer

Paul Ellis

14.1
Introduction

Worldwide, it is estimated that more than 1 million women have breast cancer, of whom around two-fifths will die from the disease (Parkin et al. 2005). The progression of breast cancer can be separated into five stages depending on the size of tumor, the involvement of lymph nodes, and the spread of tumor cells to other tissues/organs. Stages 0 to 2 are classified as early, stage 3 as locally advanced, and stage 4 as advanced or metastatic breast cancer (MBC). Human epidermal growth factor receptor 2 (HER2) is overexpressed in approximately 20 to 30% of breast cancers (Owens et al. 2004; Penault-Llorca et al. 2005; Ross et al. 2004; Slamon et al. 1987, 1989) and is associated with aggressive tumor behavior and poor prognosis (Ménard et al. 2001; Press et al. 1997; Slamon et al. 1987). The HER2 status of a tumor can be assessed using immunohistochemistry (IHC), which provides a score of 0 to 3+ based on HER2 protein expression (IHC 3+ indicating HER2 overexpression), or by DNA-based methodologies such as fluorescence *in-situ* hybridization (FISH) or chromogenic *in-situ* hybridization, which measure HER2 gene amplification. The term "HER2-positive" is used to describe a tumor that has HER2 protein overexpression and/or HER2 gene amplification.

Trastuzumab (Herceptin) is a biologically engineered, humanized immunoglobulin-1, anti-HER2 monoclonal antibody (mAb) developed by Genentech/Roche that is directed against the extracellular domain of HER2 (Carter et al. 1992). Antibody production uses a genetically engineered Chinese hamster ovary (CHO) cell line grown on a commercial scale. The DNA coding sequence of trastuzumab is inserted into these cells using standardized recombinant techniques. The antibody is secreted into the culture medium by the cells, and is purified using standard chromatographic and filtration methods. Trastuzumab is 95% human and 5% murine, and is supplied as a lyophilized powder to be

Handbook of Therapeutic Antibodies. Edited by Stefan Dübel
Copyright © 2007 WILEY-VCH Verlag GmbH & Co. KGaA, Weinheim
ISBN 978-3-527-31453-9

reconstituted with bacteriostatic or sterile water for either multidose or single-dose vials.

Trastuzumab is the first commercially available mAb-based therapy for the treatment of breast cancer, and is currently available in more than 90 countries worldwide. It has been shown to markedly inhibit the growth of breast tumors (Baselga et al. 1998; Pegram et al. 1999; Pietras et al. 1998). Although the exact mechanisms of this effect are not yet clear, recent studies have suggested a number of possible mechanisms by which trastuzumab exerts its anti-tumor effects. A high infiltration of leukocytes and natural killer (NK) cells has been noted in HER2-positive patients following trastuzumab treatment (Arnould et al. 2006; Gennari et al. 2004), suggesting that immune cells are recruited by trastuzumab to attack the tumor cells by antibody-dependent cellular cytotoxicity. When combined with standard chemotherapy agents, trastuzumab has been found to inhibit the anti-apoptotic effect of HER2 overexpression, and thereby trigger apoptosis (Kim et al. 2005). Other possible mechanisms of action for trastuzumab are decreasing HER2 phosphorylation and downstream signaling (Ekerljung et al. 2006), interacting with other signaling pathways (Holbro et al. 2003; Le et al. 2006), HER2 receptor down-modulation and internalization from the cell membrane (Klapper et al. 2000), and the inhibition of constitutive HER2 cleavage/shedding mediated by metalloproteinase which prevents homodimerization of HER2 derivatives (Molina et al. 2001).

Approval was granted by the US Food and Drug Administration (FDA) in September 1998, and by the European Medicines Agency in September 2000, for the use of trastuzumab in women with HER2-positive MBC. The initial license was based on data from two pivotal trials, one using trastuzumab as second-/third-line monotherapy (see Section 14.2.1) (Cobleigh et al. 1999), and a second using first-line trastuzumab plus chemotherapy (the H0648g trial, including paclitaxel for patients relapsing after prior adjuvant anthracycline chemotherapy; see Section 14.2.1) (Slamon et al. 2001).

In 2006, the European Commission approved the use of trastuzumab for the adjuvant treatment of HER2-positive early breast cancer (EBC), based on results from the international HERA (HERceptin Adjuvant) study (see Section 14.3). Trastuzumab given after standard chemotherapy was found significantly to reduce the risk of cancer remission by 46%, and to be well tolerated compared with chemotherapy alone (Piccart-Gebhart et al. 2005). This was supported by the results of three other major global and US studies: the North Central Cancer Treatment Group (NCCTG) N9831 trial, the National Surgical Adjuvant Breast and Bowel Project (NSABP) B-31 trial and the Breast Cancer International Research Group (BCIRG) 006 trial (see Section 14.3).

14.2 Metastatic Breast Cancer

14.2.1 Trastuzumab Monotherapy

The efficacy and safety of trastuzumab (4 $mg\,kg^{-1}$ body weight loading dose followed by 2 $mg\,kg^{-1}$ weekly) was studied in a multicenter, open-label, single-arm trial comprising 222 women with HER2-positive (IHC 2+/3+) MBC that had progressed after one or two chemotherapy regimens (Cobleigh et al. 1999). Two or more lines of therapy had previously been given to 68% of patients; 94% had received anthracyclines and 67% taxanes. In the intent-to-treat population, there were eight complete and 26 partial responses, giving an objective response rate of 15%. The median duration of response was 9.1 months, which was substantially longer than the 5.2 months achieved with the chemotherapy regimens used previously by these women. Median survival was 13 months (Fig. 14.1), and median time to progression was 3.1 months. These results highlight the clinical benefits of trastuzumab as second- or third-line monotherapy in HER2-positive MBC. Patients with IHC 3+ and/or FISH-positive disease derived the most clinical benefit from trastuzumab in this study (Cobleigh et al. 1999).

After promising results with trastuzumab as second-/third-line monotherapy, the efficacy and safety of trastuzumab as first-line monotherapy (4 $mg\,kg^{-1}$ loading dose followed by 2 $mg\,kg^{-1}$ weekly, or 8 $mg\,kg^{-1}$ loading dose followed by 4 $mg\,kg^{-1}$

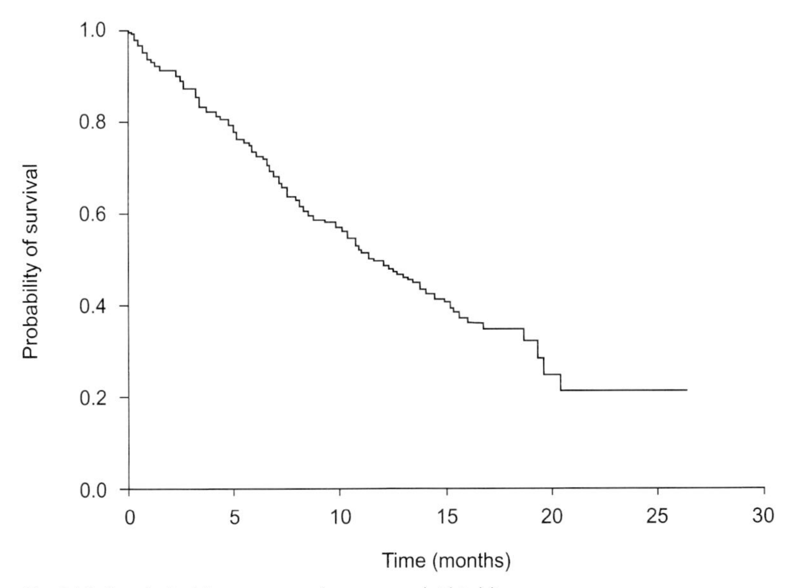

Fig. 14.1 Survival with trastuzumab as second-/third-line monotherapy given weekly (Cobleigh et al. 1999).

weekly) was investigated in a phase II multicenter trial involving 114 patients with previously untreated HER2-positive (IHC 2+/3+) MBC (Vogel et al. 2002). The objective response rate was 26%, with seven complete and 23 partial responses. For patients with IHC 3+ tumors (n = 111), the response rate was 35%; however, no response was seen in IHC 2+ patients. Clinical benefit (complete, partial or minor response, or stable disease for longer than 6 months) rates of 48% and 7% were observed in IHC 3+ and 2+ patients, respectively. Response rates in 108 assessable patients with and without HER2 gene amplification by FISH were 34% and 7%, and clinical benefit rates were 48% and 10%, respectively. At 12-month follow-up or later, 57% of patients with objective response and 51% with clinical benefit had not progressed. The median overall survival was 24.4 months. There was no clear evidence of any dose-dependent relationship for response or survival. As was also seen in the Cobleigh study (Cobleigh et al. 1999), IHC 3+ and/or FISH-positive patients had the greatest clinical benefit in this study (Vogel et al. 2002).

Ideally, treatment should be convenient for the patient, and this can be achieved with less-frequent dosing, which may also have a positive health-economic impact because of reduced hospital and clinic visits. Thus, in addition to the standard weekly regimen, a 3-weekly regimen of trastuzumab monotherapy was investigated. The efficacy, safety and pharmacokinetics of 3-weekly trastuzumab ($8\,mg\,kg^{-1}$ loading dose followed by $6\,mg\,kg^{-1}$) were assessed in a phase II single-arm trial comprising 105 patients with previously untreated HER2-positive MBC (Baselga et al. 2005). The median time to progression was 3.4 months, and median duration of response 10.1 months. The overall response rate was 19% and the clinical benefit rate 33%. In patients with measurable, centrally confirmed HER2-positive disease (per-protocol subset), these rates were even higher: 23% response rate and 36% clinical benefit rate. After a further 12 months of follow-up, the overall response rate was 24% and the clinical benefit rate remained at 36% in the per-protocol subset. These data are within the range observed in other trials of weekly trastuzumab plus paclitaxel, and indicate that the 3-weekly regimen is as potentially efficacious as the weekly regimen when used in combination with paclitaxel. Furthermore, the 3-weekly schedule did not compromise the safety of trastuzumab in women with HER2-positive MBC.

14.2.2 Trastuzumab in Combination with Taxanes

Pivotal randomized combination trials of trastuzumab have demonstrated that trastuzumab plus a taxane (paclitaxel or docetaxel) is associated with superior clinical benefit compared with a taxane alone. In a randomized, multicenter, phase III trial comprising 469 women with previously untreated HER2-positive (IHC 2+/3+) MBC (Slamon et al. 2001; Eiermann and International Herceptin Study Group, 2001), patients who had previously received anthracyclines in the adjuvant setting were randomized to receive paclitaxel ($175\,mg\,m^{-2}$, 3-weekly) either alone (n = 96) or with trastuzumab ($4\,mg\,kg^{-1}$ loading dose followed by

2 mg kg^{-1} weekly; n = 92) (Slamon et al. 2001). All other patients were randomized to receive anthracycline (doxorubicin 60 mg m^{-2} or epirubicin 75 mg m^{-2}) plus cyclophosphamide (600 mg m^{-2}) either alone (n = 138) or with trastuzumab (4 mg kg^{-1} loading dose followed by 2 mg kg^{-1} weekly; n = 143).

At 30-month median follow-up, the addition of trastuzumab to paclitaxel improved all clinical endpoints compared with paclitaxel alone. The improvement in median overall survival from 18.4 to 22.1 months (18 to 25 months in IHC 3+ patients) with addition of trastuzumab is particularly notable because of the aggressive nature of HER2-positive disease and the poor prognosis of these patients. Furthermore, the crossover design of the study, in which 73% of patients subsequently received trastuzumab after progressing on single-agent paclitaxel, biased against observing a survival advantage (Slamon et al. 2001). In a similar pattern to that seen in the weekly monotherapy trials (Vogel et al. 2002), the trastuzumab-paclitaxel combination provided the greatest clinical benefit in patients with IHC 3+ and/or FISH-positive disease (Baselga, 2001; Slamon et al. 2001).

The combination of trastuzumab and docetaxel was investigated in the pivotal randomized phase II trial (M77001) comprising 186 patients with previously untreated MBC (Extra et al. 2005; Marty et al. 2005). Patients received 3-weekly docetaxel (100 mg m^{-2}) with or without weekly trastuzumab (4 mg kg^{-1} loading dose followed by 2 mg kg^{-1}). Patients in the docetaxel-alone arm could cross over to receive trastuzumab on disease progression. At 24-month follow-up, the addition of trastuzumab to docetaxel significantly improved all clinical outcomes investigated (Table 14.1), including an increase in median overall survival from 22.7 to 31.2 months (treatment benefit of 8.5 months; p = 0.0325) (Fig. 14.2). Twice as many patients who received trastuzumab plus docetaxel lived 3 years or longer compared to those given docetaxel only. Furthermore, patients who crossed over from docetaxel monotherapy to receive trastuzumab on progression (57%) appeared to survive longer than those who did not receive trastuzumab (Marty et al. 2005).

Table 14.1 Clinical outcomes with trastuzumab plus docetaxel, and docetaxel alone (Marty et al. 2005).

	Trastuzumab + docetaxel (n = 92)	*Docetaxel alone (n = 94)*	*p-value*
ORR, %	61.0	34.0	0.0002
Median DoR [months]	11.7	5.7	0.0009
Median TTP [months]	11.7	6.1	0.0001
Median OS[a] [months]	31.2	22.7	0.0325

a Kaplan-Meier estimates.
DoR = duration of response; ORR = overall response rate; OS = overall survival; TTP = time to progression.

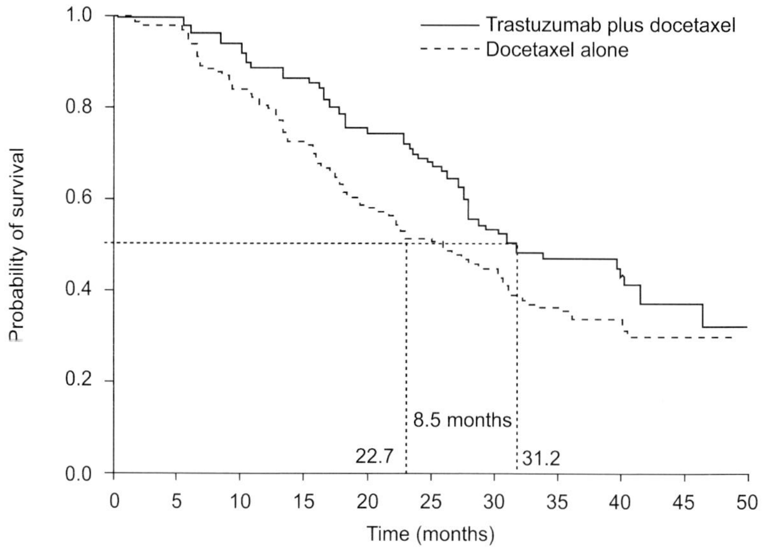

Fig. 14.2 Overall survival with trastuzumab plus docetaxel and docetaxel alone (Extra et al. 2005; Marty et al. 2005).

In addition to the pivotal trials, several phase II MBC trials have confirmed that the combination of trastuzumab plus a taxane improves clinical outcomes compared with taxane alone. In these trials, response rates for trastuzumab given weekly or 3-weekly in combination with paclitaxel or docetaxel ranged from 36 to 81% and from 44 to 75% for paclitaxel and docetaxel combinations, respectively.

14.2.3
Trastuzumab in Combination with other Standard Chemotherapy

Trials have been reported that investigated trastuzumab in combination with several chemotherapeutic agents commonly used in the treatment of MBC. Phase II trials of trastuzumab plus vinorelbine as first- or subsequent-line therapy showed objective response rates ranging from 43% to 85% (Bayo et al. 2004; Bernardo et al. 2004; Burstein et al. 2001, 2003, 2006; Chan et al. 2005; de Wit et al. 2004; Glogowska et al. 2004; Jahanzeb et al. 2002; Papaldo et al. 2006), indicating that this combination is highly active in the treatment of MBC. The combination of trastuzumab with gemcitabine in patients with HER2-positive MBC has also been reported as effective, with objective response rates of 36 to 40% in phase II trials (Christodoulou et al. 2003; O'Shaughnessy et al. 2004; Stemmler et al. 2005). A phase II study of the 5-fluorouracil prodrug capecitabine plus trastuzumab in patients with pretreated MBC achieved an objective response rate of 60% (Schaller et al. 2005), and this was mirrored in a phase II study of first-line trastuzumab-capecitabine therapy (objective response rate, 63%; n = 43) (Xu et al. 2006).

Combinations of trastuzumab with each of the aforementioned agents were generally well tolerated.

14.2.4 Trastuzumab in Triple Combination

Trastuzumab-containing triple combinations (e.g., trastuzumab/carboplatin/paclitaxel, trastuzumab/cisplatin/docetaxel, trastuzumab/epirubicin/docetaxel, trastuzumab/vinorelbine/docetaxel, or trastuzumab/gemcitabine/paclitaxel) can produce high response rates and are generally feasible if overlapping toxicity profiles are avoided. A phase II randomized trial of trastuzumab plus docetaxel with or without capecitabine in patients with HER2-positive metastatic or locally advanced breast cancer (CHAT) is currently in progress. An initial safety analysis of 110 patients reported an incidence of adverse events similar to that seen in trials of trastuzumab plus docetaxel or trastuzumab plus capecitabine (Wardley et al. 2005). The randomized phase III trial BCIRG 007 investigated the treatment of HER2-positive MBC with trastuzumab either in combination with docetaxel, or with both docetaxel and carboplatin (Forbes et al. 2006). Although addition of the platinum compound carboplatin did not improve clinical outcome, this combination remains a valid therapeutic option and is widely used in the United States.

Both regimens were well tolerated, but differed in their toxicity profiles. In patients treated with trastuzumab and docetaxel significant increases in sensory neuropathy, myalgia, skin and nail changes were reported. In triple combination therapy patients, the incidence of thrombocytopenia was significantly higher, and there was an increase in nausea and vomiting episodes.

14.2.5 Trastuzumab in Combination with Hormonal Therapies

Trastuzumab is still being assessed in combination with hormonal therapies, including tamoxifen, and the aromatase inhibitors letrozole, anastrozole (the TAnDEM trial) and exemestane (the eLEcTRA trial). Several pivotal trials using trastuzumab either as a single agent or combined with chemotherapy drugs have shown that hormone-receptor status does not affect the efficacy of trastuzumab (Marty et al. 2005).

14.3 Early Breast Cancer

Primary systemic therapy (PST) – also referred to as “neoadjuvant therapy” – is being used increasingly in women with EBC to reduce tumor size, thereby providing more opportunity for breast-conserving surgery, and to decrease the number of positive nodes. Several PST studies have evaluated trastuzumab treatment for 12 weeks before surgery and up to 40 weeks after surgery, provided that the patient is responding or has stable disease.

The overall response rate (70–95%) and the pathological complete response rate (7–42%) in these studies compare favorably with results from phase III and/or randomized trials of PST anthracyclines and/or taxane-containing regimens in patients with unselected (HER2-positive or negative) breast cancer (Bear et al. 2003; Evans et al. 2002; Perez and Rodeheffer 2004; Vinholes et al. 2001). Furthermore, a phase II study evaluated the efficacy of a triple regimen of trastuzumab, docetaxel and cisplatin PST for HER2-positive breast cancer patients with untreated locally advanced disease (Hurley et al. 2006). Progression-free and overall survival were reported to have significantly improved for these patients following treatment with this triple combination.

Based on excellent data in MBC and the promising results in the PST trial, there was a clear rationale to investigate trastuzumab in the adjuvant setting. Four major randomized, multicenter trials examined different adjuvant trastuzumab-based treatment options, both sequential to and concomitant with chemotherapy: HERA (Piccart-Gebhart et al. 2005), NSABP B-31 (Romond et al. 2005), NCCTG N9831 (Romond et al. 2005), and BCIRG 006 (Slamon et al. 2005) (Table 14.2). Results from a smaller trial of HER2-positive women in Finland (FinHer) have also been made available (Joensuu et al. 2006). More than 14 000 women with EBC were enrolled into these five trials. All patients were required to have HER2-positive invasive breast cancer and to have undergone either lumpectomy or mastectomy in order to be included. The regimens of chemotherapy and trastuzumab treatment differed between the trials and are outlined in Fig. 14.3.

In the HERA trial, patients completed a regimen of chemotherapy, with or without radiotherapy, and were randomized to 3-weekly trastuzumab or observation only over 1 to 2 years (Piccart-Gebhart 2005). A significant disease-free survival improvement of 6.3% ($p < 0.0001$) (Fig. 14.4) and an overall survival benefit of 2.7% ($p = 0.0115$) (Fig. 14.5) were reported at 3 years following treatment with trastuzumab. All patient subgroups showed a significant treatment effect.

A joint analysis of the NSABP B-31 and NCCTG N9831 trials in North America assessed the standard adjuvant chemotherapy regimen of doxorubicin plus cyclophosphamide (AC), followed initially with paclitaxel with or without 1 year of trastuzumab (Romond et al. 2005). Significantly longer disease-free survival was observed among those patients who were treated with trastuzumab compared with those who had not received trastuzumab. A significant improvement in overall survival was also noted at 2 years' median follow-up.

The BCIRG 006 trial compared the three treatment regimens of AC followed by docetaxel (as reference treatment), AC followed by docetaxel plus trastuzumab, and an anthracycline-free regimen of docetaxel in combination with carboplatin and trastuzumab (Slamon et al. 2005). The two trastuzumab-containing regimens were associated with significantly improved disease-free survival compared with that not including trastuzumab. Patients receiving trastuzumab with

Table 14.2 Summary of patient characteristics from the adjuvant trastuzumab early breast cancer (EBC) trials (Joensuu et al. 2006; Piccart-Gebhart et al. 2005; Romond et al. 2005; Slamon et al. 2005).

	HERA (n = 5090)	*NSABP B-31 and NCCTG N9831 combined analysis (n = 5535)*	*BCIRG 006 (n = 3222)*	*FinHer (n = 232)*[a]
Age <50 years [%]	51	51	52	51
Node-negative disease [%]	32[b]	5.7	29[c]	16[d]
Grade III tumors [%]	60	69	n.a.	65
Taxane-based chemotherapy [%]	26	100	100	50
Planned endocrine therapy [%]	46	52	54 (with ER+ and/or PgR+ tumors)	
Normal cardiac function [%]	At completion of locoregional therapy and chemotherapy	At completion of doxorubicin + cyclophosphamide	After surgery	After surgery

a HER2-positive subgroup.
b Only if tumor >1 cm.
c Only if other concomitant risk factors.
d Only if tumor >2 cm and PgR-negative.
ER = estrogen receptor; PgR = progesterone receptor; n.a. = not available.

docetaxel had a 51% reduced risk of relapse compared with those receiving only docetaxel. A 39% decrease in risk was seen in patients receiving docetaxel plus carboplatin and trastuzumab.

The small randomized FinHer trial aimed to compare docetaxel and vinorelbine as part of a chemotherapy regimen in 1010 patients with EBC (Joensuu et al. 2006). Patients were tested for HER2 status, and those shown to be positive by IHC (3+ or 2+; n = 232) were further randomized either to receive (n = 116) or not receive (n = 116) trastuzumab over 9 weeks. At 3 years' follow-up, patients who were given trastuzumab had better recurrence-free survival than those who did not.

Overall, these trials indicate that trastuzumab adjuvant therapy can decrease the risk of recurrence by approximately 50% for women with EBC. HERA and the joint analysis of NSABP B-31 and NCCTG N9831 revealed a significant improvement in overall survival after 1 year of adjuvant trastuzumab treatment (a 33% reduction in the risk of death). This is of significance because the ultimate aim of all adjuvant trials is to improve the overall survival of patients, potentially leading to more patients being cured. The results from analysis after 1 to 3 years' follow-up indicate a disease-free survival benefit for trastuzumab treatment in all

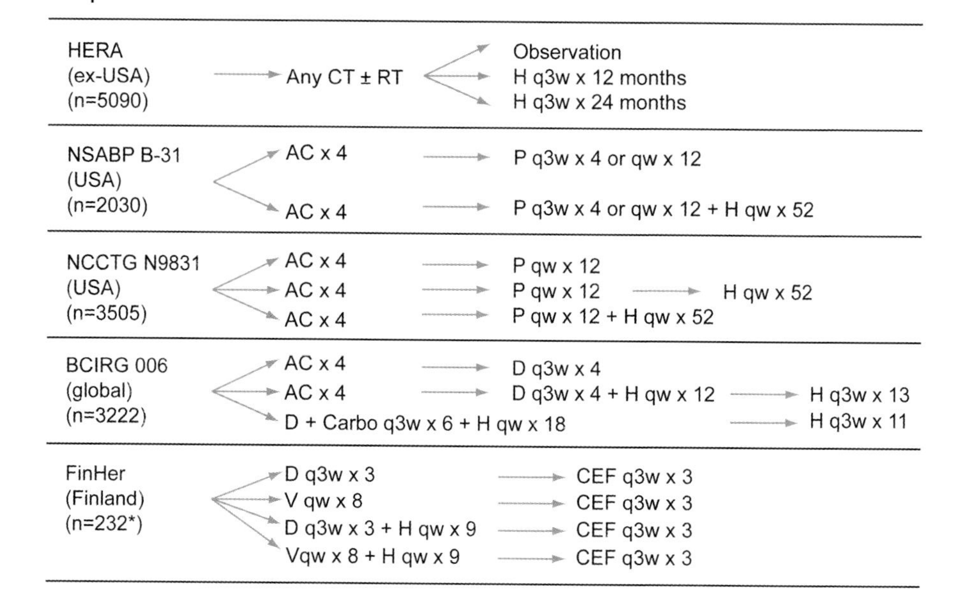

Fig. 14.3 Study designs of trastuzumab EBC trials: HERA, NSABP B-31, NCCTG N9831, BCIRG 006 and FinHer (Joensuu et al. 2006; Piccart-Gebhart et al. 2005; Romond et al. 2005; Slamon et al. 2005).

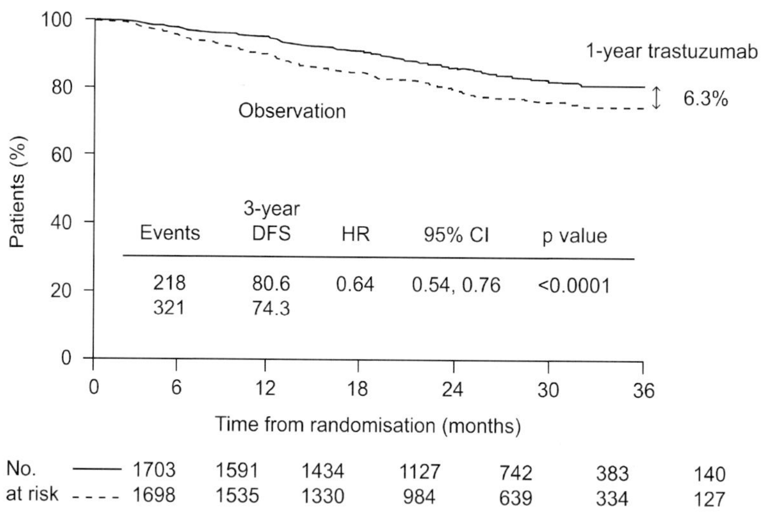

Fig. 14.4 Disease-free survival progression of patients in the HERA trial at 2-year median follow-up (Piccart-Gebhart 2005).

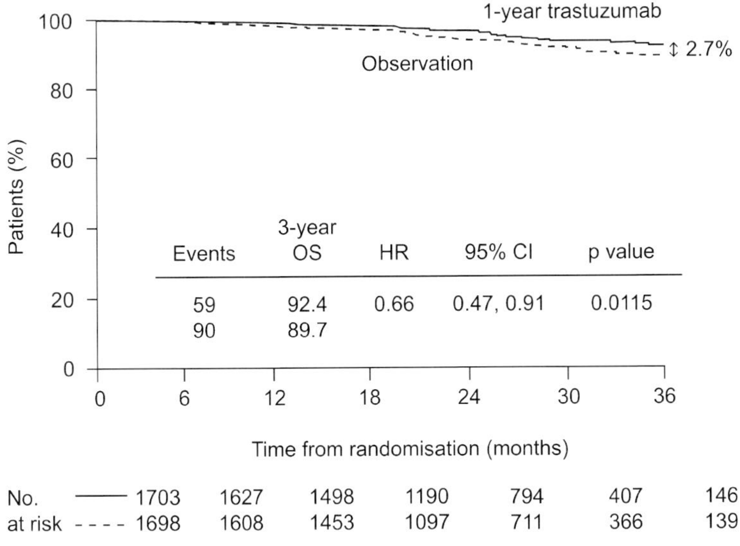

Fig. 14.5 Overall survival progression of patients in the HERA trial at 2-year median follow-up (Piccart-Gebhart 2005).

five EBC trials (Fig. 14.6). Following the success of these trials, trastuzumab has recently gained approval from the European Commission for use as an adjuvant treatment for HER2-positive EBC.

14.4
Trastuzumab Treatment in other Tumor Types

In addition to breast cancer, overexpression of HER2 has been seen in a number of other tumor types, including ovarian, bladder, salivary gland, endometrial, esophageal, gastric, pancreatic and non-small-cell lung (NSCLC).

A large randomized phase III trial was conducted in NSCLC (Gatzemeier et al. 2004). Unfortunately, the majority of patients recruited into this trial were not HER2-positive according to the standards in place today, and tumors were diagnosed as IHC 2+. Currently, knowledge regarding the definition of HER2 positivity has improved, and IHC 2+ cases are now classified as equivocal, requiring further analysis by ISH. Cases identified as IHC 2+ and ISH negative are regarded as HER2-negative. This helps to explain why the trial failed its endpoints.

ToGA, a phase III, open-label, randomized, multicenter study, is an ongoing clinical development program to evaluate the efficacy, safety, pharmacokinetics and effect on quality of life of first-line fluoropyrimidine and cisplatin with or without trastuzumab in 374 patients with HER2-positive advanced gastric cancer. Patient recruitment began in September 2005, with treatment given until progression. Recruitment for this study is currently ongoing.

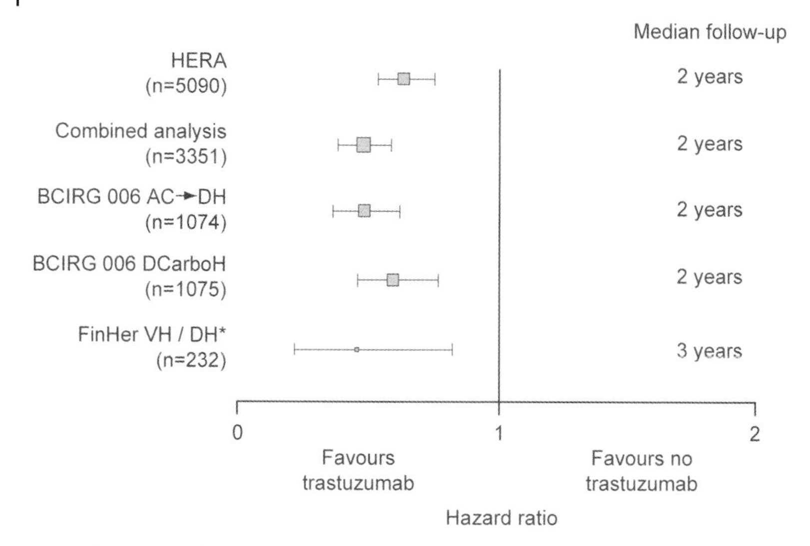

Fig. 14.6 Summary of disease-free survival in the five pivotal EBC trials (Joensuu et al. 2006; Piccart-Gebhart et al. 2005; Romond et al. 2005; Slamon et al. 2005).

14.5 Safety

Trastuzumab monotherapy or combination therapy is generally well tolerated, with only mild to moderate adverse side effects; the majority of patients are not expected to experience grade 3 or 4 adverse events. Cardiac events and serious infusion-related reactions are the most clinically important side effects; however, these occur relatively infrequently and are readily manageable with standard therapeutic interventions (see Sections 14.5.1 and 14.5.2). Trastuzumab monotherapy is not associated with cumulative toxicity (Cook-Bruns 2001), and does not induce the adverse events such as alopecia, myelosuppression or mucositis that are common with chemotherapy (Cobleigh et al. 1999; Vogel et al. 2002). Also, routine use of anti-emetics is not required with trastuzumab. The decision to continue trastuzumab therapy in any patients who experience adverse events will depend on the outcome of individual benefit: risk assessment.

In two monotherapy trials, the most commonly reported adverse events were mild to moderate infusion-related reactions, which occurred in around 40% of patients (Vogel et al. 2002). These reactions were mostly associated with the first infusion, did not lead to infusion interruption, and resolved with standard treatment. Hematological toxicities frequently seen with chemotherapy, such as neutropenia, anemia and thrombocytopenia, were uncommon following trastuzumab monotherapy (Cobleigh et al. 1999; Vogel et al. 2002), and severe adverse events were rare.

As with trastuzumab monotherapy, the most common adverse events in the pivotal phase III combination trial were mild infusion-related reactions (fever and chills) (Slamon et al. 2001). Trastuzumab does not generally appear to exacerbate the toxicity of chemotherapy (Slamon et al. 2001; Marty et al. 2005), apart from increases in anemia and leucopenia when used in combination with paclitaxel or anthracycline plus cyclophosphamide (Slamon et al. 2001). These events generally resolved with standard treatments.

A similar pattern of adverse events was observed more recently in a randomized phase II pivotal trial (M77001) in which 188 HER2-positive (IHC 3+) MBC patients were given 3-weekly docetaxel ($100\,mg\,m^{-2}$) with or without weekly trastuzumab ($2\,mg\,kg^{-1}$) until disease progression (Marty et al. 2005). There was little difference in safety profile between the two treatment groups, although some nonhematological toxicities usually seen with docetaxel were more frequently reported within the combination group. A higher incidence of adverse events was reported in combination therapy patients; however, fewer patients from this group were withdrawn from the trial due to serious adverse events.

Trastuzumab is an immunoglobulin and is, therefore, secreted into human milk. Hence, breastfeeding is not recommended for the duration of trastuzumab therapy and for 6 months after the last dose of trastuzumab.

14.5.1 Cardiac Adverse Events

Trastuzumab has been associated with an increased risk of cardiac events (Perez and Rodeheffer 2004; Seidman et al. 2002; Slamon et al. 2001; Suter et al. 2004), and careful cardiac monitoring is recommended for all patients undergoing treatment. Clinical trial data showed that these events were more common in patients who received concomitant anthracyclines or who had previously received anthracyclines (Seidman et al. 2002).

In the pivotal clinical trial conducted by Slamon and colleagues, cardiac events were reported for 27% of patients who received trastuzumab in combination with anthracyclines and cyclophosphamide compared with 8% who received only anthracyclines and cyclophosphamide (Seidman et al. 2002; Slamon et al. 2001). In women with prior anthracycline therapy, cardiac events were reported for 13% of patients who received trastuzumab plus paclitaxel compared with 1% who received only paclitaxel (Seidman et al. 2002). Most trastuzumab-treated patients who developed cardiac dysfunction were symptomatic (75%) and the majority improved with standard treatment for congestive heart failure (79%) (Seidman et al. 2002). A Cardiac Task Force analyzed the cardiac-event data from this trial and concluded that: ". . . in most cases the cardiac events observed may reflect an exacerbation of anthracycline-induced cardiotoxicity; symptomatic heart failure was associated with concomitant or previous anthracycline use; and the cardiac events are generally reversible and can usually be managed with standard medical treatment." (Suter et al. 2004).

It is recommended that all patients are assessed for cardiac disease risk factors before receiving trastuzumab in order to identify those susceptible to cardiac events. Evaluation of the left ventricular ejection fraction (LVEF), an indicator of cardiac function, is also recommended before and during trastuzumab therapy (Suter et al. 2004). In 2006, an expert group met to discuss the prevalence of cardiac events and to prepare some recommendations for clinicians to prevent and monitor such events. The three proposed algorithms (Figs. 14.7–14.9) are applicable only to those patients receiving trastuzumab for 12 months. It is possible that these recommendations may alter when 24-month data from the HERA trial become available. The first algorithm (Fig. 14.7) outlines the recommended cardiac criteria to be considered when determining the appropriate course of trastuzumab therapy. A number of recommendations were made concerning asymptomatic LVEF decline during adjuvant trastuzumab treatment (Fig. 14.8), and the occurrence of New York Heart Association (NYHA) class II or III/IV cardiac events during adjuvant trastuzumab (Fig. 14.9).

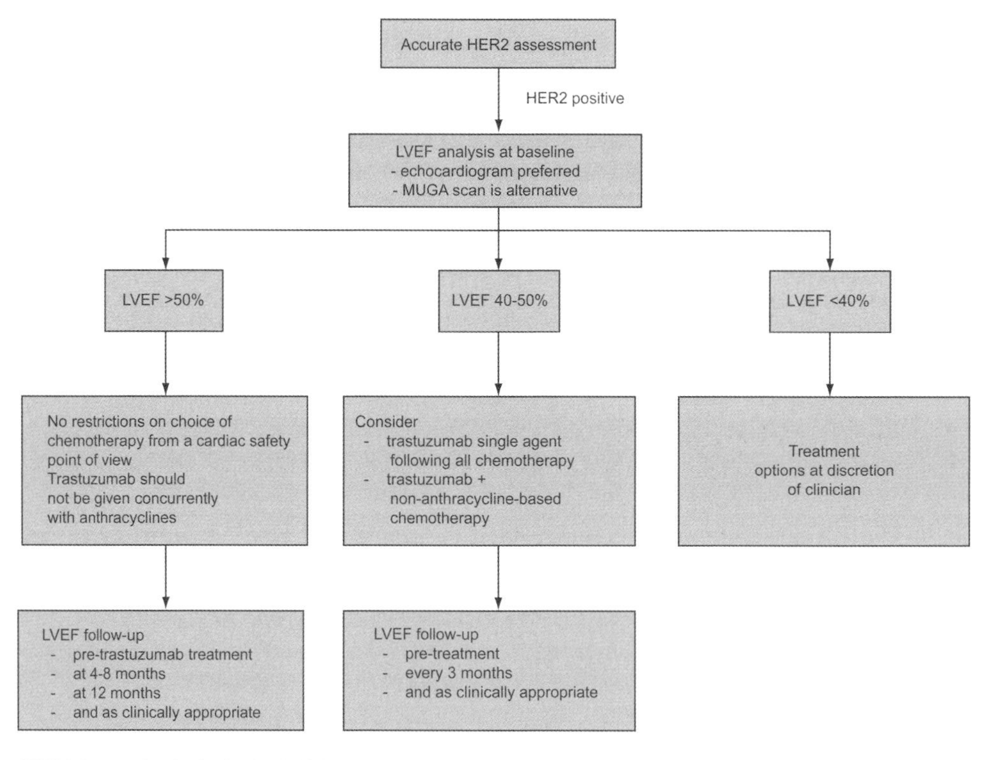

Fig. 14.7 Cardiac criteria for initiating adjuvant trastuzumab. LVEF = left ventricular ejection fraction; MUGA = multiple-gated acquisition.

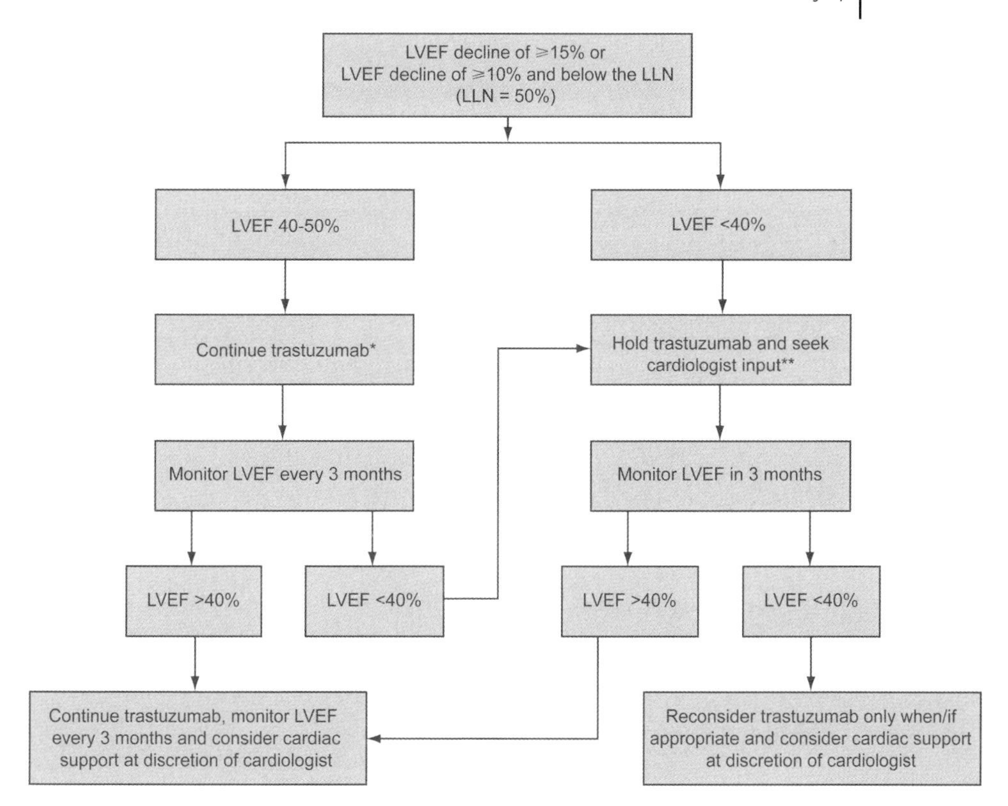

*Proceed with increased caution and surveillance
**Proceed with great caution
LNN, lower limit of normal

Fig. 14.8 Treatment choices for asymptomatic declines in LVEF occurring during adjuvant trastuzumab therapy.

Prospective cardiac monitoring was included in trastuzumab trials conducted after the pivotal combination trial. An analysis of data pooled from six phase II and III trastuzumab trials in HER2-positive MBC comprising a total of 629 patients (418 of whom received trastuzumab) showed that the incidence of clinically significant cardiac events (congestive heart failure) in patients who received trastuzumab was only 2.7% (Marty et al. 2003). This incidence is much lower than that seen in the pivotal trial of trastuzumab plus paclitaxel (12%) (Slamon et al. 2001), and is a consequence of stringent cardiac eligibility criteria and regular cardiac monitoring.

The increased risk of cardiac events observed with trastuzumab when given in combination with anthracyclines has prevented filing and approval of this treatment for use in HER2-positive MBC. However, new anthracyclines with improved cardiac safety (epirubicin and liposomal doxorubicin) have been investigated to address this situation.

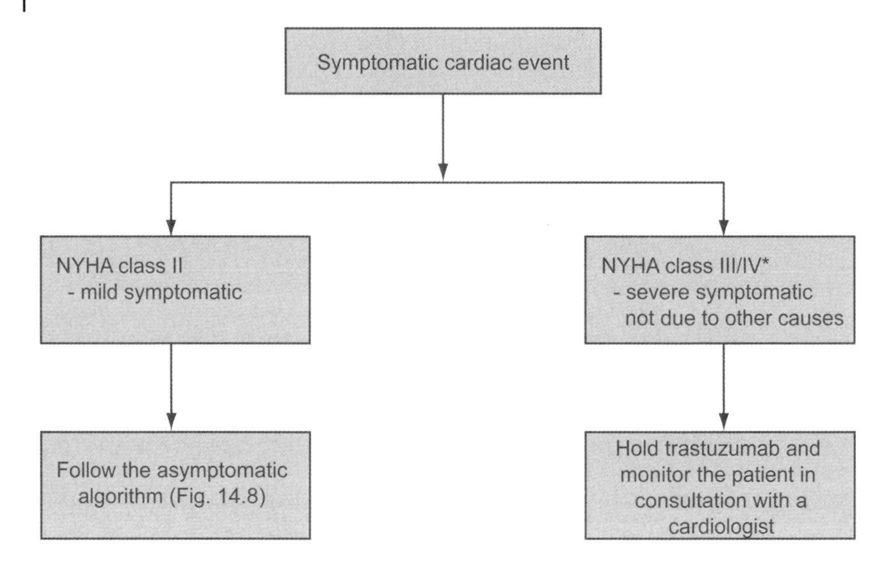

Fig. 14.9 Treatment choices based on symptomatic cardiac events occurring during adjuvant trastuzumab therapy. NYHA = New York Heart Association.

The HERCULES multicenter phase II trial examined the cardiac safety of trastuzumab plus epirubicin-cyclophosphamide (n = 25), compared with epirubicin-cyclophosphamide alone (n = 24), in patients with MBC. Only two patients in the triple-combination drug arm experienced a cardiac event, compared with one in the epirubicin-cyclophosphamide arm (Langer et al. 2003).

Patients in the four major adjuvant EBC trials (HERA, NSABP B-31, NCCTG N9831 and BCIRG 007) showed similar and acceptable levels of cardiac events. A small overall increase of 0.6 to 3.3% in the incidence of congestive heart failure was seen, the majority of which improved with treatment.

14.5.2 Infusion-related Reactions

Infusion-related reactions in the pivotal trastuzumab clinical trials were common but generally mild, and occurred mostly with the first infusion; severe cases were rare (Slamon et al. 2001; Vogel et al. 2002). Initial post-marketing surveillance, however, has revealed more serious trastuzumab-associated infusion-related reactions, which presented as fatal complications in some cases (Cook-Bruns 2001). The incidence of infusion-related deaths (death within 24 h of infusion) has been reported as 0.04% (9 of 25 000 patients; cut-off March 2000) and that of infusion-related events overall as 0.3% (74 of 25 000 patients). No new concerns have arisen since, and a further analysis is not indicated. Most of the nine patients with fatal infusion-related reactions had significant pre-existing pulmonary compromise secondary to advanced malignancy, and several were receiving supportive oxygen therapy at the time of their first trastuzumab infusion.

Serious infusion-related reactions are characterized by respiratory symptoms, such as dyspnea, bronchospasm and respiratory distress, and may be accompanied by an anaphylactoid reaction with hypotension and rash. These symptoms generally appear within 2 h following the start of the first trastuzumab infusion, and can be managed through cessation, treatment with antihistamines, corticosteroids and β-antagonists, and administration of oxygen. Further infusions of trastuzumab to patients who experienced serious reactions have not been associated with any recurrence. Patients who are at risk of developing severe infusion-related reactions can be identified before the initiation of treatment. Overall, therefore, serious infusion-related events associated with trastuzumab occur infrequently and do not impact on the favorable benefit:risk profile of this drug.

14.5.3 Age Considerations

In general, older age is associated with poorer clinical outcome in breast cancer; however, for patients with MBC, the addition of trastuzumab to chemotherapy (paclitaxel or anthracycline-based) has been shown to increase overall response rate and survival compared with chemotherapy alone, irrespective of patient age (Fyfe et al. 2001). The overall response rate increased from 33% to 52% in patients aged less than 60 years, and from 28% to 44% in patients aged more than 60 years. Corresponding figures for survival are from 23 to 26 months and from 14 to 19 months, respectively. These clinical benefits indicate that trastuzumab combination therapy should be considered for patients, whatever their age. The safety of trastuzumab in patients aged less than 18 years has not been established.

14.5.4 Patient Considerations

Trastuzumab is not recommended for use in patients with known hypersensitivity to murine proteins, trastuzumab, or any of its excipients. Patients with severe dyspnea at rest due to complications of advanced malignancy, or those requiring supplementary oxygen, are also not recommended to receive trastuzumab because of the possibility of serious infusion-related reactions. Patients with symptomatic heart failure, history of hypertension or coronary artery disease, or LVEF < 50% should undergo a careful benefit:risk assessment before they receive treatment.

14.6 Dosing/Scheduling

Two dosing schedules are labeled and used for trastuzumab. Based upon an initial assumption of a half-life of 5.8 days, a weekly schedule was developed and licensed for MBC: $4\,mg\,kg^{-1}$ loading dose followed by $2\,mg\,kg^{-1}$ weekly.

A retrospective analysis of pooled pharmacokinetic data using population pharmacokinetics indicated a mean half-life of 28.5 days. Subsequently, a 3-weekly schedule was developed with the same total doses as the weekly schedule: $8\,mg\,kg^{-1}$ loading dose followed by $6\,mg\,kg^{-1}$ 3-weekly from Week 3 onwards.

Since the 3-weekly dosage had been used in the adjuvant trial program, this schedule was licensed by the European Agency for the Evaluation of Medicinal Products (EMEA) for the treatment of patients with HER2-positive EBC following surgery, chemotherapy (neoadjuvant or adjuvant) and radiotherapy (if applicable).

14.7 Conclusions

Trastuzumab has become a backbone therapy for patients with HER2-positive breast cancer. It has been shown to provide benefits with respect to all clinical outcomes, but the survival benefits achieved with trastuzumab in MBC and EBC are the most impressive. In both settings, clinical trials provide level 1 evidence for improvement of overall survival.

At the same time, trastuzumab in general adds little to the toxicities of respective combination partners. The adverse events which have caused the most concern are cardiac events. Specific inclusion/exclusion criteria and cardiac monitoring have helped to control the incidence of cardiac events to an acceptable level. Overall, trastuzumab is characterized by an excellent benefit : risk ratio.

References

Arnould, L., Gelly, M., Penault-Llorca, F., Benoit, L., Bonnetain, F., Migeon, C., et al. (2006) Trastuzumab-based treatment of HER2-positive breast cancer: an antibody-dependent cellular cytotoxicity mechanism? *Br J Cancer* 94: 259–267.

Baselga, J. (2001) Herceptin alone or in combination with chemotherapy in the treatment of HER2-positive metastatic breast cancer: pivotal trials. *Oncology* 61(Suppl.2): 14–21.

Baselga, J., Norton, L., Albanell, J., Kim, Y.-M., and Mendelsohn, J. (1998) Recombinant humanized anti-HER2 antibody (Herceptin) enhances the antitumor activity of paclitaxel and doxorubicin against HER2/*neu* overexpressing human breast cancer xenografts. *Cancer Res* 58: 2825–2831.

Baselga, J., Carbonell, X., Castañeda-Soto, N.-J., Clemens, M., Green, M., Harvey, V., et al. (2005) Phase II study of efficacy, safety, and pharmacokinetics of trastuzumab monotherapy administered on a 3-weekly schedule. *J Clin Oncol* 23: 2162–2171.

Bayo, J., Mayordomo, J.I., Sanchez-Rovira, P., Perez-Carrion, R., Illaramendi, J., Gonzalez Flores, E., et al. (2004) Trastuzumab and vinorelbine combination in the treatment of Her2 positive metastatic breast cancer. *Proc Am Soc Clin Oncol* 23: 67, abstract 763.

Bear, H.D., Anderson, S., Brown, A., Smith, R., Mamounas, E.P., Fisher, B., et al. (2003) The effect on tumor response of adding sequential preoperative docetaxel to preoperative doxorubicin and cyclophosphamide: preliminary results from National Surgical Adjuvant Breast and Bowel Project Protocol B-27. *J Clin Oncol* 21: 4165–4174.

Bernardo, G., Palumbo, R., Bernardo, A., Villani, G., Melazzini, M., Poggi, G., et al. (2004) Final results of phase II study of weekly trastuzumab and vinorelbine in chemonaive patients with HER2-overexpressing metastatic breast cancer. *Proc Am Soc Clin Oncol* 23: 59, abstract 731.

Burstein, H.J., Kuter, I., Campos, S.M., Gelman, R.S., Tribou, L., Parker, L.M., et al. (2001) Clinical activity of trastuzumab and vinorelbine in women with HER2-overexpressing metastatic breast cancer. *J Clin Oncol* 19: 2722–2730.

Burstein, H.J., Harris, L.N., Marcom, P.K., Lambert-Falls, R., Havlin, K., Overmoyer, B., et al. (2003) Trastuzumab and vinorelbine as first-line therapy for HER2-overexpressing metastatic breast cancer: multicenter phase II trial with clinical outcomes, analysis of serum tumor markers as predictive factors, and cardiac surveillance algorithm. *J Clin Oncol* 21: 2889–2895.

Burstein, H.J., Keshaviah, A., Baron, A., Hart, R., Lambert-Falls, R., Marcom, P.K., et al. (2006) Trastuzumab and vinorelbine or taxane chemotherapy for HER2+ metastatic breast cancer: The TRAVIOTA study. *J Clin Oncol* (Meeting Abstracts) 24: 40s, abstract 650.

Carter, P., Presta, L., Gorman, C.M., Ridgway, J.B.B., Henner, D., Wong, W.L.T., et al. (1992) Humanization of an anti-p185^{HER2} antibody for human cancer therapy. *Proc Natl Acad Sci USA* 89: 4285–4289.

Chan, A., Petruzelka, L., Untch, M., Martin, M., Gil, M., Guillem Porta, V., et al. (2005) Long term survival of vinorelbine (N) and trastuzumab (H) as first line therapy for HER2-positive metastatic breast cancer patients (HER2 + MBC) (pts). *J Clin Oncol* (Meeting Abstracts) 23: 25s, abstract 587.

Christodoulou, C., Fountzilas, G., Razi, E., Tsavdaridis, D., Karina, M., Lambropoulos, S., et al. (2003) Gemcitabine and trastuzumab combination as salvage treatment in patients with HER 2-positive metastatic breast cancer. *Proc Am Soc Clin Oncol* 22: 42, abstract 166.

Cobleigh, M.A., Vogel, C.L., Tripathy, D., Robert, N.J., Scholl, S., Fehrenbacher, L., et al. (1999) Multinational study of the efficacy and safety of humanized anti-HER2 monoclonal antibody in women who have HER2-overexpressing metastatic breast cancer that has progressed after chemotherapy for metastatic disease. *J Clin Oncol* 17: 2639–2648.

Cook-Bruns, N. (2001) Retrospective analysis of the safety of Herceptin immunotherapy in metastatic breast cancer. *Oncology* 61(Suppl.2): 58–66.

de Wit, M., Becker, K., Thomssen, C., Harbeck, N., Hoffmann, R., Villena, C., et al. (2004) Vinorelbine and trastuzumab as first line therapy in patients with HER2-positive metastatic breast cancer – interim analysis of a prospective, open-label, multicentre phase II trial. *Ann Oncol* 15(Suppl.3): 37, abstract 138P.

Eiermann, W., International Herceptin Study Group (2001) Trastuzumab combined with chemotherapy for the treatment of HER2-positive metastatic breast cancer: pivotal trial data. *Ann Oncol* 12(Suppl.1): S57–S62.

Ekerljung, L., Steffen, A.C., Carlsson, J., Lennartsson, J. (2006) Effects of HER2-binding affibody molecules on intracellular signaling pathways. *Tumour Biol* 27: 201–210.

Evans, T., Gould, A., Foster, E., Crown, J.P., Leonard, R., Mansi, J.L. (2002) Phase III randomised trial of adriamycin (A) and docetaxel (D) versus A and cyclophosphamide (C) as primary medical therapy (PMT) in women with breast cancer: an ACCOG study. *Proc Am Soc Clin Oncol* 21: 35a, abstract 136.

Extra, J.-M., Cognetti, F., Maraninchi, D., Snyder, R., Mauriac, L., Tubiana-Hulin, M., et al. (2005) Long-term survival demonstrated with trastuzumab plus docetaxel: 24-month data from a randomised trial (M77001) in HER2-positive metastatic breast cancer. *J Clin Oncol* 23: 17s, abstract 555.

Forbes, J.F., Kennedy, J., Pienkowski, T., Valero, V., Eiermann, W., von Minckwitz, G., et al. (2006) BCIRG 007: randomized phase III trial of trastuzumab plus docetaxel with or without carboplatin first line in HER2 positive metastatic breast cancer (MBC): main time to progression (TTP) analysis. *J Clin Oncol* 24: 7s, abstract LBA516.

Fyfe, G.A., Mass, R., Murphy, M., Slamon, D. (2001) Survival benefit of trastuzumab (Herceptin) and chemotherapy in older (age >60) patients. *Proc Am Soc Clin Oncol* 20: 48a, abstract 189.

Gatzemeier, U., Groth, G., Butts, C., Van Zandwijk, N., Shepherd, F., Ardizzoni, A., et al. (2004) Randomized phase II trial of gemcitabine–cisplatin with or without trastuzumab in HER2-positive non-small-cell lung cancer. *Ann Oncol* 15: 19–27.

Gennari, R., Menard, S., Fagnoni, F., Ponchio, L., Scelsi, M., Tagliabue, E., et al. (2004) Pilot study of the mechanism of action of preoperative trastuzumab in patients with primary operable breast tumors overexpressing HER2. *Clin Cancer Res* 10: 5650–5655.

Glogowska, I., Sienkiewicz-Kozlowska, R., Bauer-Kosinska, B., Jaczewska, S., Pienkowski, T. (2004) Trastuzumab (T) plus vinorelbine (VNR) as first combination in Her-2 overexpressing patients with metastatic breast cancer. *Proc Am Soc Clin Oncol* 23: 235, abstract 3165.

Holbro, T., Beerli, R.R., Maurer, F., Koziczak, M., Barbas III, C.F., Hynes, N.E. (2003) The ErbB2/ErbB3 heterodimer functions as an oncogenic unit: ErbB2 requires ErbB3 to drive breast tumor cell proliferation. *Proc Natl Acad Sci USA* 100: 8933–8938.

Hurley, J., Doliny, P., Reis, I., Silva, O., Gomez-Fernandez, C., Velez, P., et al. (2006) Docetaxel, cisplatin, and trastuzumab as primary systemic therapy for human epidermal growth factor receptor 2–positive locally advanced breast cancer. *J Clin Oncol* 24: 1831–1838.

Jahanzeb, M., Mortimer, J.E., Yunus, F., Irwin, D.H., Speyer, J., Koletsky, A.J., et al. (2002) Phase II trial of weekly vinorelbine and trastuzumab as first-line therapy in patients with $HER2^+$ metastatic breast cancer. *Oncologist* 7: 410–417.

Joensuu, H., Kellokumpu-Lehtinen, P.-L., Bono, P., Alanko, T., Kataja, V., Asola, R., et al. (2006) Adjuvant docetaxel or vinorelbine with or without trastuzumab for breast cancer. *N Engl J Med* 354: 809–820.

Kim, C., Bryant, J., Horne, Z., Geyer, C.E., Wickerham, D.L., Wolmark, N., et al. (2005) Trastuzumab sensitivity of breast cancer with co-amplification of HER2 and cMYC suggest pro-apoptotic function of dysregulated cMYC in vivo. Presented at San Antonio Breast Cancer Symposium, abstract 46.

Klapper, L.N., Waterman, H., Sela, M., Yarden, Y. (2000) Tumor-inhibitory antibodies to HER-2/ErbB-2 may act by recruiting c-Cbl and enhancing ubiquitination of HER-2. *Cancer Res* 60: 3384–3388.

Langer, B., Muscholl, M., Pauschinger, M., Thomssen, Ch., Eidtmann, H., Untch, M., et al. (2003) A prospective study of NT-pro brain natriuretic peptide (NT-proBNP) and troponin T (TnT) in the HERCULES (M77003) study (epirubicin + cyclophosphamide ± Herceptin) in patients with metastatic breast cancer (MBC). Poster 223 presented at the 26th Annual San Antonio Breast Cancer Symposium, San Antonio, TX, USA, December 3–6, 2003.

Le, X.F., Bedrosian, I., Mao, W., Murray, M., Lu, Z., Keyomarsi, K., et al. (2006) Anti-HER2 antibody trastuzumab inhibits CDK2-mediated NPAT and histone H4 expression via the PI3K pathway. *Cell Cycle* 5: 1654–1661.

Marty, M., Baselga, J., Gatzemeier, U., Leyland-Jones, B., Suter, T., Klingelschmitt, G., et al. (2003) Pooled analysis of six trials of trastuzumab (Herceptin): exploratory analysis of changes in left ventricular ejection fraction (LVEF) as a surrogate for clinical cardiac events. *Breast Cancer Res Treat* 82(Suppl.1): S48, abstract 218.

Marty, M., Cognetti, F., Maraninchi, D., Snyder, R., Mauriac, L., Tubiana-Hulin, M., et al. (2005) Randomized phase II trial of the efficacy and safety of trastuzumab combined with docetaxel in patients with human epidermal growth factor receptor 2-positive metastatic breast cancer administered as first-line treatment: the M77001 study group. *J Clin Oncol* 23: 4265–4274.

Ménard, S., Fortis, S., Castiglioni, F., Agresti, R., Balsari, A. (2001) HER2 as a prognostic factor in breast cancer. *Oncology* 61(Suppl.2): 67–72.

Molina, M.A., Codony-Servat, J., Albanell, J., Rojo, F., Arribas, J., Baselga, J. (2001) Trastuzumab (Herceptin), a humanized anti-HER2 receptor monoclonal antibody, inhibits basal and activated HER2 ectodomain cleavage in breast cancer cells. *Cancer Res* 61: 4744–4749.

O'Shaughnessy, J.A., Vukelja, S., Marsland, T., Kimmel, G., Ratnam, S., Pippen, J.E. (2004) Phase II study of trastuzumab plus gemcitabine in chemotherapy-pretreated patients with metastatic breast cancer. *Clin Breast Cancer* 5: 142–147.

Owens, M.A., Horten, B.C., Da Silva, M.M. (2004) HER2 amplification ratios by fluorescence in situ hybridization and correlation with immunohistochemistry in a cohort of 6556 breast cancer tissues. *Clin Breast Cancer* 5: 63–69.

Papaldo, P., Fabi, A., Ferretti, G., Mottolese, M., Cianciulli, A.M., Di Cocco, B., et al. (2006) A phase II study on metastatic breast cancer patients treated with weekly vinorelbine with or without trastuzumab according to HER2 expression: changing the natural history of HER2-positive disease. *Ann Oncol* 17: 630–636.

Parkin, D.M., Bray, F., Ferlay, J., Pisani, P. (2005) Global cancer statistics, 2002. *CA Cancer J Clin* 55: 74–108.

Pegram, M., Hsu, S., Lewis, G., Pietras, R., Beryt, M., Sliwkowski, M., et al. (1999) Inhibitory effects of combinations of HER-2/*neu* antibody and chemotherapeutic agents used for treatment of human breast cancers. *Oncogene* 18: 2241–2251.

Penault-Llorca, F., Vincent-Salomon, A., Mathieu, M.C., Trillet-Lenoir, V., Khayat, D., Marty, M., et al. (2005) Incidence and implications of HER2 and hormonal receptor overexpression in newly diagnosed metastatic breast cancer (MBC). *J Clin Oncol* (Meeting Abstracts) 23: 69s, abstract 764.

Perez, E.A., Rodeheffer, R. (2004) Clinical cardiac tolerability of trastuzumab. *J Clin Oncol* 22: 322–329.

Piccart-Gebhart, M.J. (2005) First results of the HERA trial. A randomized three-arm multi-centre comparison of: 1 year Herceptin, 2 years Herceptin or no Herceptin in women with HER-2 positive primary breast cancer who have completed adjuvant chemotherapy. Slide presentation at the 41st ASCO Annual Meeting, Orlando, FL, USA, May 13–17, 2005. http://www.asco.org/ac/1,1003,_12-002511-00_18-0034-00_19-005816-00_21-001,00.asp. 2005. 7-7-2005.

Piccart-Gebhart, M.J., Procter, M., Leyland-Jones, B., Goldhirsch, A., Untch, M., Smith, I., et al. (2005) Trastuzumab after adjuvant chemotherapy in HER2-positive breast cancer. *N Engl J Med* 353: 1659–1672.

Pietras, R.J., Pegram, M.D., Finn, R.S., Maneval, D.A., Slamon, D.J. (1998) Remission of human breast cancer xenografts on therapy with humanized monoclonal antibody to HER-2 receptor and DNA-reactive drugs. *Oncogene* 17: 2235–2249.

Press, M.F., Bernstein, L., Thomas, P.A., Meisner, L.F., Zhou, J.Y., Ma, Y., et al. (1997) HER-2/neu gene amplification characterized by fluorescence in situ hybridization: poor prognosis in node-negative breast carcinomas. *J Clin Oncol* 15: 2894–2904.

Romond, E.H., Perez, E.A., Bryant, J., Suman, V.J., Geyer Jr., C.E., Davidson, N. E., et al. (2005) Trastuzumab plus adjuvant chemotherapy for operable HER2-positive breast cancer. *N Engl J Med* 353: 1673–1684.

Ross, J.S., Fletcher, J.A., Bloom, K.J., Linette, G.P., Stec, J., Symmans, W.F., et al. (2004) Targeted therapy in breast cancer. The HER-2/*neu* gene and protein. *Mol Cell Proteomics* 3: 379–398.

Schaller, G., Bangemann, N., Weber, J., Kleine-Tebbe, A., Beisler, G.K., Conrad, B., et al. (2005) Efficacy and safety of trastuzumab plus capecitabine in a German multicentre phase II study of pre-treated metastatic breast cancer. *J Clin Oncol* (Meeting Abstracts) 23: 57s, abstract 717.

Seidman, A., Hudis, C., Pierri, M.K., Shak, S., Paton, V., Ashby, M., et al. (2002) Cardiac dysfunction in the trastuzumab clinical trials experience. *J Clin Oncol* 20: 1215–1221.

Slamon, D.J., Clark, G.M., Wong, S.G., Levin, W.J., Ullrich, A., McGuire, W.L. (1987) Human breast cancer: correlation of relapse and survival with amplification of the HER-2/*neu* oncogene. *Science* 235: 177–182.

Slamon, D.J., Godolphin, W., Jones, L.A., Holt, J.A., Wong, S.G., Keith, D.E., et al. (1989) Studies of the HER-2/neu proto-oncogene in human breast and ovarian cancer. *Science* 244: 707–712.

Slamon, D.J., Leyland-Jones, B., Shak, S., Fuchs, H., Paton, V., Bajamonde, A., et al. (2001) Use of chemotherapy plus a monoclonal antibody against HER2 for metastatic breast cancer that overexpresses HER2. *N Engl J Med* 344: 783–792.

Slamon, D., Eiermann, W., Robert, N., Pienkowski, T., Martin, M., Pawlicki, M., et al. (2005) Phase III randomized trial comparing doxorubicin and cyclophosphamide followed by docetaxel (AC→T) with doxorubicin and cyclophosphamide followed by docetaxel and trastuzumab (AC →TH) with docetaxel, carboplatin and trastuzumab (TCH) in HER2 positive early breast cancer patients: BCIRG 006 study. *Breast Cancer Res Treat* 94(Suppl.1): S5, abstract 1.

Stemmler, H.J., Kahlert, S., Brudler, O., Beha, M., Müller, S., Stauch, B., et al. (2005) High efficacy of gemcitabine and cisplatin plus trastuzumab in patients with HER2-overexpressing metastatic breast cancer: a phase II study. *Clin Oncol (R Coll Radiol)* 17: 630–635.

Suter, T.M., Cook-Bruns, N., Barton, C. (2004) Cardiotoxicity associated with trastuzumab (Herceptin) therapy in the treatment of metastatic breast cancer. *Breast* 13: 173–183.

Vinholes, J., Bouzid, K., Salas, F., Mickiewicz, E., Valdivia, S., Ostapenko, V., et al. (2001) Preliminary results of a multicentre phase III trial of taxotere and doxorubicin (AT) versus 5-fluorouracil, doxorubicin and cyclophosphamide (FAC) in patients (pts) with unresectable locally advanced breast cancer (ULABC). *Proc Am Soc Clin Oncol* 20: 26a, abstract 101.

Vogel, C.L., Cobleigh, M.A., Tripathy, D., Gutheil, J.C., Harris, L.N., Fehrenbacher, L., et al. (2002) Efficacy and safety of trastuzumab as a single agent in first-line treatment of *HER2*-overexpressing metastatic breast cancer. *J Clin Oncol* 20: 719–726.

Wardley, A., Antón-Torres, A., Otero Reyes, D., Jassem, J., Toache, L.M.Z., Alcedo, J.C., et al. (2005) CHAT – an open-label, randomised, Phase II study of trastuzumab plus docetaxel with or without capecitabine in patients with advanced and/or metastatic HER2-positive breast cancer: second interim safety analysis. Poster 6094 presented at the 28th Annual San Antonio Breast Cancer Symposium, San Antonio, TX, USA, 8-11 December, 2005.

Xu, L., Song, S., Zhu, J., Luo, R., Li, L., Jiao, S., et al. (2006) A phase II trial of trastuzumab (H) + capecitabine (X) as first-line treatment in patients (pts) with HER2-positive metastatic breast cancer (MBC). *J Clin Oncol* (Meeting Abstracts) 24: 577s (abstract 10615).

15
Abciximab, Arcitumomab, Basiliximab, Capromab, Cotara, Daclizumab, Edrecolomab, Ibritumomab, Igovomab, Nofetumomab, Satumomab, Sulesomab, Tositumomab, and Votumumab

Christian Menzel and Stefan Dübel

Summary

This chapter offers an overview of all of the approved antibodies which are not described in separate chapters in this handbook. This includes antibodies for immunotherapy and *in-vivo* imaging which were approved but later withdrawn, or the marketing of which has been stopped. Interestingly, among the latter is a majority of the approved *in-vivo* imaging agents using radionuclides. Here, a development is seen from the pure diagnostic tool towards the combined imaging and treatment regime of Bexxar or Zevalin. Further information on immunotargeted radionuclides can be found in Chapter 1 of Volume II (Immunoscintigraphy and Radioimmunotherapy).

Note: The data presented in this chapter represent opinions compiled from public domain sources. Despite some parts of this chapter citing the drug leaflets, they are not to be construed as recommendations for treatments. Full prescribing information must be consulted for any of the drugs or procedures discussed herein.

15.1
Abciximab (Reopro)

Abciximab is the Fab fragment of a chimeric mouse/human antibody, 7E3. Reopro is produced by continuous perfusion culture of the antibody 7E3 by mammalian cells and subsequent papain digestion to obtain the Fab fragment. Reopro binds to the platelet surface protein GPIIb/IIIa receptor, a member of the integrin family and the major factor involved in platelet aggregation. By inhibiting the binding of GPIIb/IIIa receptor to a number of serum proteins – such as von

Handbook of Therapeutic Antibodies. Edited by Stefan Dübel
Copyright © 2007 WILEY-VCH Verlag GmbH & Co. KGaA, Weinheim
ISBN 978-3-527-31453-9

Willebrand factor, fibrinogen, and other adhesion proteins – Abciximab inhibits the platelet aggregation.

Abciximab is FDA approved as an adjunct to percutaneous coronary intervention for the prevention of cardiac ischemic complications in patients undergoing percutaneous coronary intervention, or in patients with unstable angina not responding to conventional medical therapy when percutaneous coronary intervention is planned within 24 h.

Information Sources/References

http://www.fda.gov.
http://www.reopro.com.
Drug leaflet.
Oster, Z.H., Srivastava, S.C., Som, P., Meinken, G.E., Scudder, L.E., Yamamoto, K., Atkins, H.L., Brill, A.B., Coller, B.S. (1985) Thrombus radioimmunoscintigraphy: an approach using monoclonal antiplatelet antibody. *Proc Natl Acad Sci USA* 82: 3465–3468.
Faulds, D., Sorkin, E.M. (1994) Abciximab (c7E3 Fab). A review of its pharmacology and therapeutic potential in ischaemic heart disease. *Drugs* 48: 583–598.
Schneider, D.J., Aggarwal, A. (2004) Development of glycoprotein IIb-IIIa antagonists: translation of pharmacodynamic effects into clinical benefit. *Expert Rev Cardiovasc Ther* 2: 903–913.

15.2
Arcitumomab (CEA-Scan)

The targeting component of CEA-Scan (Arcitumomab) consists of an antibody Fab fragment generated from a murine IgG_1, called IMMU-4. Arcitumomab is further covalently labeled with the radioactive metastable technetium isotope ^{99m}Tc (half-life 6.01 h) via exposed sulfhydryl (SH)-groups. These are chemically generated by reduction of the disulfide bonds linking the two Fab fragments in the Fab_2 fragments generated by pepsin digest from the IMMU-4 IgG preparation produced from murine ascites.

Arcitumomab binds specifically to carcinoembryonic antigen (CEA, CD66e, *CEACAM5*), a protein expressed by colorectal and many other tumors, but which is also present in certain inflammatory conditions such as Crohn's disease or inflammatory bowel disease. CEA can be shed, and its serum level is usually determined for the prognosis of patients with recurrent disease after having undergone resection of colorectal cancer. Arcitumomab exhibits no cross-reactivity to any genetically distinct CEA variants or surface granulocyte nonspecific cross-reacting antigen (NCA). Even though shedding occurs, the antibody still detects cell membrane-located CEA. Elevated CEA levels of up to 2000 $ng\,mL^{-1}$ led to 50% antibody complexation with serum CEA, whereas no complexation was observed with plasma levels below 250 $ng\,mL^{-1}$.

CEA-Scan was reconstituted with ^{99m}Tc shortly before application, with imaging normally taking place between 2 and 5 h after injection, using a standard nuclear camera. Delayed imaging might interfere with kidney, gallbladder and intestinal activity. Pharmacokinetic studies revealed a terminal half-life time after intravenous injection of CEA-Scan of approximately 14 h.

Four clinical trials were conducted to determine the imaging efficacy and safety of CEA-Scan. Joint application of CEA-Scan and standard detection methods such as computed tomography (CT) scanning significantly increased the surgical confirmation of scan-identified lesions as cancer. The positive prediction value (PPV) rose to from 83% for CEA-Scan and 86% from CT scanning alone, to 97% by complementary analysis.

Due to heterologous protein administration, the development of human anti-mouse antibodies (HAMA) may evolve, but has been reported in less than 1% of the patients. Observed adverse side effects included transient eosinophilia, fever, urticaria, and headache. No long-term studies of possible mutagenic or cancerogenic effects from the ionizing radiation from ^{99m}Tc have yet been carried out.

CEA-Scan was approved in 1996 by FDA for radioimmunoscintigraphy. The indication for CEA-Scan is solely in patients with histologically demonstrated carcinomas of the colon or rectum to evaluate the presence, location and extent of recurrence and/or metastases. The technique also provided additional information to standard noninvasive imaging techniques such as CT scanning or ultrasonography. CEA-Scan is also indicated in cases of suspected recurrence or metastasis of colon or rectum carcinomas due to rising serum levels of CEA if standard detection methods fail. However, CEA-Scan is indicated neither for differential diagnosis of colorectal carcinomas nor as a screening tool.

Previously, CEA-Scan was marketed by Immunomedics Inc. in Europe and USA, until 2005 when the product was withdrawn from EU market for commercial reasons. It is also no longer marketed in the USA.

Information Sources/References

http://www.fda.gov.
http://www.emea.eu.int.
http://www.dpgonline2.com/pkginserts/ceascan.pdf (CEA-Scan label).
Goldenberg, D.M., Nabi, H.A. (1999) Breast cancer imaging with radiolabeled antibodies. *Semin Nucl Med* 29: 41–48.
Swayne, L.C., Goldenberg, D.M., Diehl, W. L., Macaulay, R.D., Derby, L.A., Trivino, J.Z. (1991) SPECT anti-CEA monoclonal antibody detection of occult colorectal carcinoma metastases. *Clin Nucl Med* 16: 849–852.

15.3 Basiliximab (Simulect)

Basiliximab, a recombinant mouse/human chimeric IgG$_1$k, binds specifically to the interleukin (IL)-2 receptor alpha chain (IL-2Ra, or Tac or CD25) on the surface of activated T lymphocytes. The production of basiliximab is achieved in a mouse myeloma cell line transformed with two plasmids encoding the mouse heavy and light chain variable region genes genetically fused to human heavy and light chain constant region genes. Interestingly, the V-region genes of the chimeric antibody are derived from Hybridoma RFT5, a cell line originally identified by its capability to secrete an IgG_1 which, *in vitro*, can kill a Hodgkin's lymphoma cell line (L450) when chemically coupled to deglycosylated ricin A-chain to form an immunotoxin (Engert et al. 1991). Immunotoxins based on this antibody underwent further evaluation for cancer treatment, including single chain antibody fusion proteins with a *Pseudomonas* exotoxin A derivative (see Chapter 3, Volume II).

Basiliximab acts as an inhibitor of IL-2 receptor function, by binding to its alpha subunit with a sub-nanomolar affinity ($10^{10}\,M^{-1}$). The expression of this subunit is found exclusively on activated T cells. Binding of the antibody to this receptor prevents the immunostimulatory signaling induced by IL-2 molecules. In a double-blind, randomized, placebo-controlled study with 340 patients for the prophylaxis of acute renal transplant rejection in adults, when used in combination with a triple immunosuppressive regimen (cyclosporine, corticosteroids and azathioprine), basiliximab reduced acute rejection (at 0–6 months) from 35% to 21%.

Basiliximab is approved for the prophylaxis of acute organ rejection in patients undergoing renal transplantation when used as part of an immunosuppressive regimen that includes cyclosporine and corticosteroids. An extension of indications has been attempted in a number of studies, with some controversial results (for a review, see Buhaescu et al. 2005). Basiliximab is the active ingredient of Simulect, marketed by Novartis.

References

Buhaescu, I., Segall, L., Goldsmith, D., Covic, A. (2005) New immunosuppressive therapies in renal transplantation: monoclonal antibodies. *J Nephrol* 18(5): 529–536.

Engert, A., Martin, G., Amlot, P., Wijdenes, J., Diehl, V., Thorpe, P. (1991) Immunotoxins constructed with anti-CD25 monoclonal antibodies and deglycosylated ricin A-chain have potent anti-tumour effects against human Hodgkin cells *in vitro* and solid Hodgkin tumours in mice. *Int J Cancer* 49(3): 450–456.

Sources

http://www.fda.gov.

Nashan, B., Moore, R., Amlot, P., Schmidt, A.-G., Abeywickrama, K., Soulillou, J.-P. (1997) Randomised trial of basiliximab versus placebo for control of acute cellular rejection in renal allograft recipients. CHIB 201 International Study Group. *Lancet* 350: 1193–1198.

Ponticelli, C., Yussim, A., Cambi, V., Legendre, C., Rizzo, G., Salvadori, M., Kahn, D., Kashi, H., Salmela, K., Fricke, L., Heemann, U., Garcia-Martinez, J., Lechler, R., Prestele, H., Girault, D. (2001) Simulect Phase IV Study Group. A randomized, double-blind trial of basiliximab immunoprophylaxis plus triple therapy in kidney transplant recipients. *Transplantation* 72(7): 1261–1267.

15.4 Capromab Pendetide (ProstaScint)

The key ingredient of ProstaScint is a radiolabeled murine IgG_1 kappa monoclonal antibody (mAb) called CYT-351 (designation by Cytogen) or 7E11-C5 (designation by Horoszewicz et al., 1987). The antibody conjugated with a tripeptide linker-chelator (GYK-DTPA) is called Capromab Pendetide, which binds a radioactive isotope of indium (^{111}In, half-life 2.8 days) via the chelator. The murine IgG is produced by a hybridoma cell line derived from a fusion of spleen cells of mice, immunized with whole cells and membrane extracts from human prostate adenocarcinoma cells, with myeloma cells. 7E11-C5 binds specifically to the prostate-specific membrane antigen (PSMA). PSMA, which is present on normal and neoplastic prostate epithelial cells, is a glycosylated membrane protein (MW ~ 100 kDa). The antibody was found to be most reactive on malignant primary and metastatic prostate cells, whereas benign prostatic hypertrophy and normal prostate cells reacted to a lesser extent. No cross-reactivity was found for other carcinoma, lymphoma, sarcoma or melanoma cell lines.

ProstaScint is injected intravenously as a single dose of 0.5 mg. Single photon emission computed tomography (SPECT) imaging follows 30 min after injection, not for determination of prostate cancer cells but for background staining of the vascular and pelvic anatomical structure. The actual imaging takes place 72 to 120 h later. Pharmacokinetic studies revealed an elimination half-life for ProstaScint of approximately 67 h.

Clinical studies revealed an accuracy of ProstaScint from 55% to 70%. However, the PPV for patients with low risk for lymph node metastases or recurrence of disease is below 20%. The PPV rises to 75% when ^{111}In-Capromab Pendetide was administered in scans with patients at very high risk of recurrence or nodal metastases.

Mild adverse side effects were reported in 4% of the patients, ranging from liver toxicity (hyperbilirubinemia), hypo- and hypertension, and hypersensitivity. However, HAMA development was reported after single injection in 1% of the patients leading to a high level of anti-mouse antibodies ($100\,ng\,mL^{-1}$) and 4% of patients with levels of $\sim 8\,ng\,mL^{-1}$. Therefore, repeated administration is not recommended.

ProstaScint was approved by the FDA in 1996 for radioimmunoscintigraphy. ^{111}In-labeled Capromab Pendetide is indicated for radiodiagnostic imaging in patients, with biopsy-proven prostate cancer, who are at high risk of pelvic lymph node metastasis, recurrent and/or metastatic prostate cancer, and also in patients after prostatectomy with a high risk of occult metastatic disease. ProstaScint allows the detection, staging and follow-up of prostate adenocarcinomas. It is also found to improve the diagnosis of nodal metastasis and prostate carcinomas when used in conjunction with other diagnostic methods for nodal metastasis. The radioimmunoconjugate is manufactured and distributed by Cytogen Corp. in the USA.

Information Sources/References

http://www.fda.gov (Label; SBA).

Horoszewicz, J.S., Kawinski, E., Murphy, G.P. (1987) Monoclonal antibodies to a new antigenic marker in epithelial prostatic cells and serum of prostatic cancer patients. *Anticancer Res* 7: 927–935.

Seo, Y., Franc, B.L., Hawkins, R.A., Wong, K.H., Hasegawa, B.H. (2006) Progress in SPECT/CT imaging of prostate cancer. *Technol Cancer Res Treat* 5: 329–336.

15.5
^{131}I-chTNT-1/B (Cotara)

The antibody is a chimeric mAb labeled with the radioactive iodine isotope ^{131}I (half-life 8.02 days). Synonyms are Tumor Necrosis Therapy-1 (TNT-1) and ^{131}I-chTNT-1/B. This agent targets double-stranded (ds) DNA and DNA/histone H1 complexes in the necrotic inner core of solid tumors. Cotara is based on the Tumor Necrosis Therapy concept, where the antibody is injected into the bloodstream or directly into the brain in order to bind to the inner core of necrotic tumors. Rapidly growing cancer cells start degenerating in the inner core of a tumor, with their cell membrane becoming porous. This facilitates the access of antibodies to the intracellular space and, in the case of Cotara, to bind ds-DNA. Upon radiation, neighboring viable cells are destroyed, thereby increasing the necrotic area of the malignant tumor.

The clinical trial jointly conducted with the brain cancer consortium "New Approaches to Brain Tumor Therapy" (NABTT) is evaluating the safety and efficacy of a single infusion of Cotara for treatment of patients with first or second recurrence of glioblastoma multiforme (GBM). The designated indication for Cotara is for two types of brain tumor in patients either with GBM (an extremely deadly and the most frequent form of brain cancer with a 1-year survival rate below 30%) or anaplastic astrocytoma [1,2]. The drug was granted orphan drug status in the USA in 1999 for the treatment of GBM and anaplastic astrocytoma, and two years later in the EU. Currently, Cotara is manufactured and marketed by Peregrine Pharmaceuticals Inc., and received approval from the FDA in 2003 for a clinical phase III study. In 2003, the Chinese company MediPharm Biotech Co Ltd. received marketing approval for licensed ^{131}I-chTNT-1/B in China from the State Food and Drug Administration (SFDA) of China. The approved indication is for treatment of refractory advanced lung cancer (stages II and IV), with the regimen occurring in two doses given between 2 and 4 weeks apart.

Currently, Cotara is produced in NS0 murine myeloma cells. The radiolabeled antibody is administered either intravenously or intratumorally with additional potassium iodine to avoid thyroid uptake of any radioactive iodine. According to MediPharm's published clinical data of 107 enrolled patients, 3.7% achieved complete remission and 30.8% partial remission [3]. Side effects included an adverse effect on the formation of mature hematopoietic cells in the bone marrow.

Neutrophil, platelet and hemoglobin toxicity was also observed to a lesser extent when Cotara was administered intratumorally. To date, however, no development of HAMA has been detected.

The approval of Cotara in China made it the third approved radiolabeled antibody for human radioimmunotherapy. The other two antibodies for radiotherapy are Bexxar and Zevalin for the treatment of human malignant B-cell non-Hodgkin lymphoma (see below).

Information Sources/References

http://www.fda.gov.

1 Central Brain Tumor Registry of the US (CBTRUS) data, 1998–2002.

2 http://www.peregrineinc.com.

3 Chen, S., Yu, L., Jiang, C., Zhao, Y., Sun, D., Li, S., Liao, G., Chen, Y., Fu, Q., Tao, Q., Ye, D., Hu, P., Khawli, L.A., Taylor, C.R., Epstein, A.L., Ju, D.W. (2005) Pivotal study of iodine-131-labeled chimeric tumor necrosis treatment radioimmunotherapy in patients with advanced lung cancer. *J Clin Oncol* 3: 1538–1547.

Street, H.H., Goris, M.L., Fisher, G.A., Wessels, B.W., Cho, C., Hernandez, C., Zhu, H.J., Zhang, Y., Nangiana, J.S., Shan, J.S., Roberts, K., Knox, S.J. (2006) Phase I study of 131I-chimeric(ch) TNT-1/B monoclonal antibody for the treatment of advanced colon cancer. *Cancer Biother Radiopharm* J21: 243–256.

Shapiro, W.R., Carpenter, S.P., Roberts, K., Shan, J.S. (2006) (131)I-chTNT-1/B mAb: tumour necrosis therapy for malignant astrocytic glioma. *Expert Opin Biol Ther* 6: 539–545.

15.6 Daclizumab (Zenapax)

Daclizumab (Zenapax) is a humanized IgG_1 mAb that binds specifically to the alpha subunit (~55 alpha, CD25, or Tat subunit) of the human high-affinity interleukin-2 (IL-2) receptor that is expressed on the surface of activated lymphocytes. Daclizumab is a composite of human and murine antibody sequences. The human sequences were derived from the constant domains of human IgG_1 and the variable framework regions of the Eu myeloma antibody. The murine sequences were derived from the complementarity-determining regions of a murine anti-Tat antibody. Daclizumab immunosuppressive function results from its binding to an IL-2 receptor. In particular, it binds with high affinity to the Tat subunit of the high-affinity IL-2 receptor complex expressed on activated (but not resting) lymphocytes, and inhibits IL-2 binding. The administration of Zenapax inhibits IL-2-mediated activation of lymphocytes, a critical pathway in the cellular immune response involved in allograft rejection.

Zenapax is indicated for the prophylaxis of acute organ rejection in patients receiving renal transplants. It is typically used as part of an immunosuppressive regimen that includes cyclosporine and corticosteroids. It is further under evaluation for the treatment of multiple sclerosis (study NCT00071838).

Information Sources/References

Drug leaflet.
http://www.clinicaltrials.gov.
http://www.rochetransplant.com.
Swiatecka-Urban, A. (2003) Anti-interleukin-2 receptor antibodies for the prevention of rejection in pediatric renal transplant patients: current status. *Paediatr Drugs* 5: 699–671.
Church, A.C. (2003) Clinical advances in therapies targeting the interleukin-2 receptor. *Q J Med* 96: 91–102
Carswell, C.I., Plosker, G.L., Wagstaff, A.J. (2001) Daclizumab: a review of its use in the management of organ transplantation. *BioDrugs* 15: 745–773.

15.7
Edrecolomab (Panorex 17-1A)

Edrecolomab is a murine monoclonal IgG_{2a} antibody to tumor-associated epithelial cell adhesion molecule (EpCAM, or CO17-1A) antigen. Edrecolomab (marketing name Panorex) was the first mAb to be approved for cancer therapy (1994 in Germany, for the treatment of Dukes' C colorectal cancer). It attaches to EpCAM, a human cell-surface glycoprotein that is found on normal epithelial cells and on a number of tumor cells found in colon and breast carcinomas, and a vide variety of other tumors. The loss of the Ep-CAM CO17-1A epitope expression predicts survival in patients with gastric cancer. The antibody's mode of action is believed to rely both on antibody-dependent cell-mediated cytotoxicity (ADCC) and the induction of a host anti-idiotypic response, as a survival advantage was reported for edrecolomab-treated patients who developed anti-anti-idiotypic antibodies compared with those who did not.

Following the approval of edrecolomab, a significant number of studies were conducted with the agent; in particular, several smaller studies were performed where edrecolomab did not demonstrate consistent benefit either in monotherapy or in combination with other anticancer agents. In particular, the results of a large randomized multi-center study showed that edrecolomab monotherapy was associated with significantly shorter overall survival and disease-free survival compared to that with 5-fluorouracil (5-FU) and folinic acid for the treatment of stage III colon cancer. Moreover, when it was added to 5-FU and folinic acid it did not result in a superior outcome to treatment with 5-FU and folinic acid alone. At present, the role of edrecolomab in the management of colorectal cancer remains uncertain, as the clinical studies were halted and marketing suspended by the distributor during the summer of 2000.

Information Sources/References

NCI Thesaurus, 2004_11_17.
Glaxo Wellcome Press Release, July 2000.
http://clinicaltrials.gov.
Wils, J.A. (2001) Therapy strategies for colorectal cancer: state of the art and beyond 5-FU/leucovorin. In: *Oncology*

Biotherapeutics. Cancer Communications Limited. ISSN 1470-7217. pp. 3–22.
Adkins, J.C., Spencer, C.M. (1998) Edrecolomab (monoclonal antibody 17-1A). *Drugs* 56: 619–626.
Mellstedt, H., Frodin, J.E., Biberfeld, P., Fagerberg, J., Giscombe, R., Hernandez, A., Masucci, G., Li, S.L., Steinitz, M. (1991) Patients treated with a monoclonal antibody (ab1) to the colorectal carcinoma antigen 17-1A develop a cellular response (DTH) to the 'internal image of the antigen' (ab2). *Int J Cancer* 30: 344–349.
Lobuglio, A.F., Saleh, M., Peterson, L., Wheeler, R., Carrano, R., Huster, W., Khazaeli, M.B. (1986) Phase I clinical trial of CO17-1A monoclonal antibody. *Hybridoma* 5(Suppl.1): S1171–S1123.

15.8 Gemtuzumab Ozogamicin (Mylotarg)

Gemtuzumab is a recombinant humanized IgG_4, kappa antibody (hP67) that binds specifically to the CD33 antigen, a sialic acid-dependent adhesion protein found on the surface of leukemic blasts and immature normal cells of myelomonocytic lineage, but not on normal hematopoietic stem cells. CD33 is expressed on the surface of leukemic blasts in more than 80% of patients with acute myeloid leukemia (AML). CD33 is also expressed on normal and leukemic myeloid colony-forming cells, including leukemic clonogenic precursors, but it is not expressed on pluripotent hematopoietic stem cells or on nonhematopoietic cells. For clinical use, Gemtuzumab is conjugated to the cytotoxic antitumor antibiotic, *N*-acetyl-gamma calicheamicin (ozogamicin). This antibody is conjugated to *N*-acetyl-gamma calicheamicin via a bifunctional linker, to form the drug available as Mylotarg. *N*-Acetyl-gamma calicheamicin is a low molecular-weight chemical compound produced by the bacterium, *Micromonospora echinospora* ssp. *calichensis*. The antibody portion of Mylotarg, the anti-CD33 hP67.6 antibody, is produced by mammalian cell suspension culture using a myeloma NS0 cell line. Mylotarg contains amino acid sequences of which approximately 98.3% are of human origin. The constant region and framework regions contain human sequences, while the complementarity-determining regions are derived from a murine antibody (p67.6) that binds to CD33. Gemtuzumab ozogamicin has approximately 50% of the antibody loaded with 4–6 moles calicheamicin per mole antibody, which results in a molecular mass of 151 to 153 kDa. The remaining 50% of the antibody is not linked to the calicheamicin derivative. Binding of Mylotarg to CD33 results in the formation of a complex that is internalized. Upon internalization, the calicheamicin derivative is released in the lysosomes of the tumor cell. The released calicheamicin derivative binds to DNA in the minor groove, resulting in DNA double strand breaks and cell death.

Mylotarg is indicated for the treatment of patients with CD33-positive AML in first relapse who are aged ≥60 years and are not considered candidates for other cytotoxic chemotherapy.

Information Sources/References

Drug leaflet.
http://www.wyeth.com.
Giles, F., Estey, E., O'Brien, S. (2003) Gemtuzumab ozogamicin in the treatment of acute myeloid leukemia. *Cancer* 98: 2095–2104.
Tsimberidou, A.M., Giles, F.J., Estey, E., O'Brien, S., Keating, M.J., Kantarjian, H. M. (2006) The role of gemtuzumab ozogamicin in acute leukaemia therapy. *Br J Haematol* 132: 398–409.
Fenton, C., Perry, C.M. (2006) Spotlight on gemtuzumab ozogamicin in acute myeloid leukaemia. *BioDrugs* 20: 137–139.
Press, O.W., Shan, D., Howell-Clark, J., Eary, J., Appelbaum, F.R., Matthews, D., King, D.J., Haines, A.M., Hamann, P., Hinman, L., Shochat, D., Bernstein, I.D. (1996) Comparative metabolism and retention of iodine-125, yttrium-90, and indium-111 radioimmunoconjugates by cancer cells. *Cancer Res* 56: 2123–2129.

15.9
Ibritumomab (Ibritumomab Tiuxetan, Zevalin)

Zevalin (ibritumomab tiuxetan) is an immunoconjugate resulting from a thiourea covalent bond between the mouse mAb ibritumomab (clone: IDEC-2B8) and the linker-chelator tiuxetan [*N*-[2-bis(carboxymethyl)amino]-3-(*p*-isothiocyanatophenyl)-propyl]-[*N*-[2-bis(carboxymethyl)amino]-2-(methyl)-ethyl]glycine. This linker-chelator provides a high-affinity, conformationally restricted binding site for a radioactive isotope of indium (^{111}In, half-life 2.8 days) or a radioactive isotope of yttrium, ^{90}Y (half-life 64.1 h). The indium- and yttrium-labeled antibodies are also named IDEC-In2B8 and IDEC-Y2B8, respectively, in some publications. The chimerized version of the antibody used in the Zevalin preparation is antibody IDEC-C2B8 [1] (rituximab; see Chapter 13), which is also part of the treatment regime with Zevalin.

Ibritumomab is a murine IgG_1 kappa mAb directed against the CD20 antigen, a nonglycosylated phosphoprotein on the cell surface of normal and malignant B lymphocytes. The antibody is produced in Chinese hamster ovary (CHO) cells. It is thought to induce apoptosis, as this activity was observed in CD20-positive B-cell lines *in vitro*, in addition to the effect of the radionuclides in the therapeutic regimen, which affect the neighboring cells (bystander effect) by inducing highly reactive radicals that damage a variety of biomolecules.

Zevalin is FDA approved for the treatment of relapsed or refractory low-grade, indolent non-Hodgkin's lymphoma (NHL), including patients with rituximab-refractory follicular NHL. The treatment regimen is complex and requires the administration of two different doses in two steps: An infusion of rituximab (see Chapter 13) preceding ^{111}In- Zevalin (emitting gamma for imaging) by not more than 4 h is followed 7 to 9 days later by a second infusion of rituximab followed by ^{90}Y Zevalin within 4 h of completion of the rituximab infusion. Whole-body gamma camera images are required at 48 to 72 h following infusion of the ^{111}In-labeled ibritumomab to confirm correct biodistribution before the therapeutic dose labeled with ^{90}Y can be given. This is necessary to detect those patients who show an excessive uptake of the antibody by the reticuloendothelial system (resulting in significant accumulation of radioactivity in the liver, spleen, and bone marrow) or other normal organs. Patients with these altered biodistributions cannot be treated with the therapeutic yttrium conjugate.

Clinical studies are either ongoing or have been conducted for the use of ^{90}Y-ibritumomab tiuxetan to treat mantle cell lymphoma (MCL), relapsed or refractory diffuse large B-cell lymphoma (DLBCL), and as conditioning regimen for stem cell transplantation and novel combination regimens.

Information Sources/References

Drug leaflet.
http://www.zevalin.com.
http://www.fda.gov.

1 Reff, M.E., Carner, K., Chambers, K.S., Chinn, P.C., Leonard, J.E., Raab, R., Newman, R.A., Hanna, N., Anderson, D. R. (1994) Depletion of B cells in vivo by a chimeric mouse human monoclonal antibody to CD20. *Blood* 83 435–445.

Cheung, M.C., Haynes, A.E., Stevens, A., Meyer, R.M., Imrie, K. and Members of the Hematology Disease Site Group of the Cancer Care Ontario Program in Evidence-Based Care. (2006) Yttrium 90 ibritumomab tiuxetan in lymphoma. *Leuk Lymphoma* 47: 967–977.

15.10 Igovomab (Indimacis-125)

Indimacis 125 consists of an indium-radiolabeled $(Fab)_2$ fragment from a murine monoclonal IgG_1 specific for the cancer antigen 125 (CA125). The antibody (Igovomab; clone name OC125) is produced in a murine hybridoma cell line. The $(Fab)_2$ fragment is generated by proteolytic cleavage with pepsin from the complete IgG, and becomes covalently linked to the chelating agent diethylenetriamine penta-acetic acid (DTPA). Prior to application, labeling must be carried out with a radioactive isotope of indium (^{111}In, half-life 2.8 days) via the chelator DTPA.

The mucin CA125 (*MUC16*; 2353 kDa) as an oncofetal protein is present on more than 90% of all ovarian serous adenocarcinomas. Due to low level of CA125 in early stages of cancer, Indimacis is unsuitable for routine screening.

Indimacis was used for radioimmunoscintigraphic imaging of ovarian cancer in cases of relapse with increased serum levels of CA125 detected, but a lack of confirmation by ultrasound or CT scanning. In these cases, Indimacis scintigraphy revealed 50% of cases as being positive for relapse. The radiolabeled antibody was given intravenously and did not produce any severe side effects such as allergic reactions. Indimacis 125 was approved in the EU in 1996. Indimacis-125 was manufactured and marketed by CIS Bio International/Schering until 1999, when it was withdrawn by the company.

Information Sources/References

Spada, S., Walsh, G. (2005) *Approved Biopharmaceutical Products*. CRC Press, Boca Raton, USA.

Uttenreuther-Fischer, M.M., Feistel, H., Wolf, F., Jäger, W. (1997) Distribution of radiolabelled anti-CA125 monoclonal antibody OC125-F(ab)2-fragment following resection guided by antibodies (REGAJ) in ovarian cancer patients. *J Clin Lab Anal* 11: 94–103.

15.11 Nofetumomab (Verluma)

Verluma was marketed as a kit to label the CD20-specific antibody nofetumomab with the radioactive metastable technetium isotope ^{99m}Tc (half-life 6.01 h). It was approved by the FDA in 1996. The radiolabeled antibody was used for radioimmunoscintigraphy of tumors, especially small cell lung cancer (SCLC) metastases. Nofetumomab was a murine Fab derived from a monoclonal IgG_{2b} antibody (NR-LU-10) produced from a hybridoma cell line. The Fab fragment is generated by proteolytic cleavage of whole IgG by papain. The labeling is performed by complexing the radioisotope with a phenothioate ligand, which is subsequently linked to the Fab fragment.

The serum half-life of Verluma was 1.5 h, and the elimination half-life 10.5 h. Imaging was carried out at between 14 and 17 h after injection. In the clinical trial involving 89 patients with confirmed SCLC, Tc-labeled Verluma accurately determined whether the disease was extensive or limited on 82% of occasions. If the test indicated extensive disease, the result was true in 94% of the patients. However, if the test indicated limited disease, it was less valuable as a diagnostic aid, failing to image tumors in some body organs in approximately 23% of patients. Because of these false negative readings, additional standard diagnostic tests, such as a bone or CT scan or a bone marrow biopsy should be performed when limited disease is found.

Unspecific accumulation of ^{99m}Tc-Nofetumomab at nontumor sites such as the organs of excretion (e.g., gallbladder, intestine, kidneys, urinary bladder), regions of inflammation and areas of recent surgery were observed. Further tests (e.g., CT examinations, bone scan) were necessary to exclude extensive-stage disease in cases of imaging interpretation for limited-stage disease. Mild adverse reactions such as fever and skin rash were reported.

The clinical indication of Verluma was radiodiagnostic detection of extensive-stage disease in patients with biopsy-confirmed and untreated SCLC.

Verluma was developed and manufactured by Dr. Karl Thomae GmbH (affiliated to Boehringer Ingelheim Pharma KG). Marketing was carried out by NeoRx Corp. and DuPont Merck in the USA. Verluma is no longer on the market after DuPont Merck terminated its licensed distribution in 1999.

Information Sources/References

http://www.fda.gov.
Breitz, H.B., Tyler, A., Bjorn, M.J., Lesley, T., Weiden, P.L. (1997) Clinical experience with Tc-99m nofetumomab merpentan (Verluma) radioimmunoscintigraphy. *Clin Nucl Med* 22: 615–620.
Spada, S., Walsh, G. (2005) *Approved Biopharmaceutical Products*. CRC Press, Boca Raton, USA.

15.12 Satumomab (OncoScint/Oncorad: B72.3n)

OncoScint was the first FDA-approved mAb for tumor imaging [1], passing regulatory processes in 1992. OncoScint consists of a murine monoclonal IgG_1 kappa antibody (satumomab, MAb B72.3) conjugated with a tripeptide linker chelator module (GYK-DTPA). Antibody production was achieved by cultivation of a murine hybridoma cell line in an airlift bioreactor. The antibody linker conjugate is loaded prior to application with the radioisotope indium (^{111}In, half-life 2.8 days). Satumomab binds specifically to the tumor-associated antigen TAG-72, which is highly glycosylated. TAG-72 is expressed on 94% of colorectal and almost all common epithelial ovarian carcinomas, but it is also frequently associated with non-small cell lung, pancreatic, gastric and other carcinomas [2].

The elimination half-life of OncoScint is 56 h. Clinical studies revealed a PPV of 68% for patients with ovarian adenocarcinoma, and of 70% in patients with colorectal cancer. Adverse effects observed were, to a low extent, fever and in less-frequent cases allergic reaction including skin rash, hypo-, and hypertension. Due to the mouse origin of the antibody, development of human anti-mouse antibodies (HAMA) was detected in more than 55% of the patients after a single dose injection.

The approved indication for OncoScint is to determine the location, extent and follow-up of confirmed colorectal and ovarian carcinomas, especially in cases for presurgical imaging, providing complementary information when used with CT scan.

OncoScint was distributed by Cytogen in the USA, but is no longer commercially available.

Information Sources/References

1 Bohdiewicz, P.J. (1998) Indium-111 satumomab pendetide: the first FDA-approved monoclonal antibody for tumor imaging. J Nucl Med Technol 26: 155–163.
2 Thor, A., Ohuchi, N., Szpak, C.A., Johnston, W.W., Schlom, J. (1986) Distribution of oncofetal antigen tumor-associated glycoprotein-72 defined by monoclonal antibody B72.3. *Cancer Res* 46: 3118–3124.

15.13
Sulesomab (LeukoScan, MN-3)

The targeting component of LeukoScan is the Fab fragment of the murine IgG Sulesomab (anti-NCA-90, IMMU-MN3). Sulesomab binds the carcinoembryonic antigen (CEA) and the surface granulocyte nonspecific cross-reacting antigen (NCA90). The mAb is produced in murine ascites fluid. Antibodies were produced by ascites technology from a hybridoma cell line derived from a fusion of lymphocytes isolated from CEA-immunized mice with the SP2/0 mouse myeloma cell line. The full IgG is proteolytically cleaved by pepsin, generating a Fab_2 which is further chemically reduced resulting in a mixture of Fab-SH, Fab_2, heavy and light chains. The antibody is labeled with the radioactive metastable technetium isotope ^{99m}Tc (half-life 6.01 h) shortly before administering the radioimmunoconjugate.

The antigen NCA90 is present on almost all neutrophils. Being highly motile, neutrophils are rapidly found at points of infection and inflammation, whereby they become a suitable target to visualize infection/inflammation.

Radioimmunoscintigraphy is performed between 1 and 8 h after the injection of LeukoScan, using a standard nuclear camera for planar imaging or SPECT.

Clinical studies of efficacy and safety revealed a significantly increased sensitivity and accuracy of LeukoScan when compared to conventional imaging with *in-vitro*-labeled autologous white blood cells (WBC). The sensitivity of LeukoScan was 88% compared to 72% in WBC, and accuracy 76.6% compared to 70.9%, whereas specificity decreased.

In clinical safety evaluation involving over 350 patients, no incidents of HAMA development were detected. However, HAMA titers must be determined before repeated administration of LeukoScan. No animal studies were performed to determine the carcinogenic and mutagenic potential of LeukoScan due to the low-energy gamma radiation of ^{99m}Tc.

LeukoScan's approved indication is diagnostic imaging to determine the location and extent of infection/inflammation in the bone of patients with probable osteomyelitis, as well as in patients with diabetic foot ulcers. When a bone scan is positive and imaging with LeukoScan negative, then infection is unlikely. When a bone scan is negative, imaging with LeukoScan may rarely show a positive response, and this may indicate early osteomyelitis.

The potential for LeukoScan to identify inflammatory bowel disease, pelvic inflammatory disease, fever of unknown origin, subacute endocarditis and acute, atypical appendicitis was demonstrated in recent years, suggesting possible future indications.

LeukoScan is manufactured by Eli Lilly Pharma and marketed by Immunomedics GmbH in Europe and Australia. It was approved in the EU in 1997, but is not approved by the FDA.

Information Sources/References

http://www.emea.eu.int.
Gratz, S., Schipper, M.L., Dorner, J., Hoffken, H., Becker, W., Kaiser, J.W., Behe, M., Behr, T.M. (2003) LeukoScan for imaging infection in different clinical settings: a retrospective evaluation and extended review of the literature. *Clin Nucl Med* 28: 267–276.

15.14 Tositumomab; Iodine ^{131}I Tositumomab (Bexxar)

Bexxar is one of two FDA-approved radioimmunotherapy agents; approval was received in 2003. Bexxar is composed of a mixture of a radiolabeled and an unlabeled version of the mouse mAb tositumomab. The radioactive iodine isotope ^{131}I (half-life 8.02 days) is covalently bound to the monoclonal murine IgG_2 lambda antibody (initially inconsistently both referred to as B1 or anti-B1 in the literature). Tositumomab is produced from a murine hybridoma cell line by Boehringer Ingelheim Pharma KG. The antibody recognizes an epitope located on the extracellular domain of CD20 (also known as human B-lymphocyte-restricted differentiation antigen; synonyms Bp35 or B1) present on B cells, including non-Hodgkin B-cell lymphoma (NHL). According to the National Cancer Institute, NHL was the sixth leading cause for cancer-related deaths in the USA in 2003, with an estimated 18 840 deaths and 59 000 new cases in 2006 [1]. CD20 is a transmembrane phosphoprotein which is present on pre-B lymphocytes as well as mature B lymphocytes; it is also expressed on more than 90% of B-cell NHLs. Neither shedding nor internalization of CD20 occurs. The mechanism of action of Bexxar, which is responsible for a sustained depletion of transformed and nontransformed CD20-positive lymphocytes, is assumed to be complement-dependent cytotoxicity (CDC), antibody-dependent cellular cytotoxicity (ADCC), and ionizing radiation from the radioisotope. All of these mechanisms trigger cell death, including induction of apoptosis leading to the killing of malignant B-cell lymphocytes [2].

The therapeutic regimen is applied in a two-step approach. In the first step, nonlabeled tositumomab for improved biodistribution of radiolabeled antibody is administered, followed by an ^{131}I-tositumomab infusion to determine optimal dosing. The second (therapeutic) step follows 7 to 14 days after the dosimetric step.

Two multicenter studies were conducted with 40 patients. In study 1, patients did not respond to rituximab or progressive disease after rituximab treatment. Study 2 comprised 60 patients who were refractory to chemotherapy. The overall response rate ranged from 68% (Study 1) to 47% (Study 2), with a median duration of response of between 12 and 18 months, and a pathologically and clinically complete responses of 33% and 20%, respectively. The most common adverse side reactions observed were cytopenias (neutropenia, thrombocytopenia and anemia) as well as secondary leukemia were the most frequently observed events in clinical trials (a total of 230 patients). The most common nonhematologic side effects were fever, weakness infections and allergic reactions such as bronchospasm and angioedema. Further associated risks of Bexxar include hypothyroidism, HAMA formation and (due to the ionizing radiation) not only secondary leukemia but also solid tumors and myelodysplasia.

The indication of Bexxar is not for the initial treatment of CD20-positive NHL but only for the treatment of patients with CD20-positive, follicular NHL, with and without transformation, whose disease is refractory to rituximab and has relapsed following chemotherapy. Bexxar is manufactured and marketed by GlaxoSmithKline in the USA.

Information Sources/References

http://www.bexxar.com.
http://www.fda.gov.
1 http://www.cancer.gov.
2 Cardarelli, P.M., Quinn, M., Buckman, D., Fang, Y., Colcher, D., King, D.J., Bebbington, C., Yarranton, G. (2002) Binding to CD20 by Anti-B1 Antibody or F(ab')2 is sufficient for induction of apoptosis in B-cell lines. *Cancer Immunol Immunother* 51: 15–24.

Nowakowski, G.S., Witzig, T.E. (2006) Radioimmunotherapy for B-cell non-Hodgkin lymphoma. *Clin Adv Hematol Oncol* 4: 225–231.
Press, O.W., Eary, J.F., Appelbaum, F.R., Martin, P.J., Badger, C.C., Nelp, W.B., Glenn, S., Butchko, G., Fisher, D., Porter, B., Matthews, D.C., Fisher, L.D., Bernstein. I.D. (1993) Radiolabeled-antibody therapy of B-cell lymphoma with autologous bone marrow support. *N Engl J Med* 329: 1219–1224.

15.15
Votumumab (Humaspect)

Votumumab is a human IgG_{3k} (clone 88BV59H21-2V67-66, in the literature frequently referred to as 88BV59) derived from a monoclonal lymphoblastoid cell line, which has been immortalized by infection with Epstein–Barr virus (EBV). The antibody binds to a cytokeratin tumor-associated antigen, CTA 16.88, which is found in non-necrotic areas of most epithelial-derived tumors including carcinomas of the colon, pancreas, breast, ovary, and lung. Conjugated to the radioactive metastable technetium isotope ^{99m}Tc (half-life 6.01 h), the agent was approved in 1998 for the detection of carcinoma of the colon or rectum.

Marketing authorization in Europe was granted in September 1998, for use in the treatment of patients with histologically proven carcinoma of the colon or rectum for imaging of recurrence or metastases. This authorization expired in September 2003 when the supplier opted not to renew it.

Information Sources/References

Wolff, B.G., Bolton, J., Baum, R., Chetanneau, A., Pecking, A., Serafini, A. N., Fischman, A.J., Hoover, H.C., Jr., Klein, J.L., Wynant, G.E., Subramanian, R., Goroff, D.K., Hanna, M.G. (1998) Radioimmunoscintigraphy of recurrent, metastatic, or occult colorectal cancer with technetium Tc 99m 88BV59H21-2V67-66 (HumaSPECT-Tc), a totally human monoclonal antibody. Patient management benefit from a phase III multicenter study. *Dis Colon Rectum* 41: 953–962.

Statement EMEA 3885/04 released by the European Agency for the Evaluation of Medical Products.

Index

Handbook of Therapeutic Antibodies. Edited by Stefan Dübel
Copyright © 2007 WILEY-VCH Verlag GmbH & Co. KGaA, Weinheim
ISBN 978-3-527-31453-9

C

D

N

O

Q

R

S

Related Titles

Dembowsky, K., Stadler, P. (eds.)

Novel Therapeutic Proteins

Selected Case Studies

2001

ISBN 978-3-527-30270-3

Breitling, F., Dübel, S.

Recombinant Antibodies

1999

ISBN 978-0-471-17847-7

Chamow, S. M., Ashkenazi, A. (eds.)

Antibody Fusion Proteins

1999

ISBN 978-0-471-18358-7

Mountain, A., Ney, U. M., Schomburg, D. (eds.)

Biotechnology, Second, Completely Revised Edition, Volume 5a

Recombinant Proteins, Monoclonal Antibodies, and Therapeutic Genes

1999

ISBN 978-3-527-28315-6